# PILOT'S
# HANDBOOK OF NAVIGATION

# PILOT'S
# HANDBOOK OF NAVIGATION

by
*Lt. Col. James C. Elliott*
*and*
*Lt. Col. Gene Guerny*

## 1977

**AERO PUBLISHERS, INC.**
329 West Aviation Road, Fallbrook, CA 92028

ISBN 0-8168-7330-5

**Library of Congress Cataloging in Publication Data**

Elliott, James C
   Pilot's handbook of navigation.

   Edition for 1967 published under title: Private pilot's handbook of navigation.
   1. Navigation (Aeronautics) 2. Private flying.
     I. Guerny, Gene, joint author.   II. Title.
     TL586.E4 1977     629.132'51     76-62739
     ISBN 0-8168-7330-5

*Printed and Published in the United States by Aero Publishers, Inc.*

# TABLE OF CONTENTS

Chapter

# ACKNOWLEDGEMENTS

Greatest appreciation is acknowledged for the navigation materials supplied from the Department of the Air Force's Manual 51-40, Air Navigation and the Air Force Manual 5112, Dead Reckoning Computer; from the Department of the Navy's Manual NAVAIR 00-80T-60, All-Weather Flight Manual; from the Department of the Army's Technical Manual 1-225, Navigation for Army Aviation; from the Federal Aviation Administration's Airman's Information Manual and Instrument Flying Handbook; and Departments of the Air Force and the Navy AF Manual 51-40, NAVAIR 00-80V-49.

Lt. Col. James C. Elliott
Lt. Col. Gene Guerny

# INTRODUCTION

By definition, air navigation is the art of safely and efficiently directing an aircraft from one place to another, and determining its position at any time. Actually it's more of a science that involves a wide variety of influential factors which must be taken into account by the pilot.

By itself, the word "navigation" is derived from the Latin words, "navis" meaning ship and the verb "egere" meaning to move or direct. Today's aviators, of course, owe their seamen forebears a great deal of credit. It's estimated that our seafaring friends took about 500 years to perfect their system of navigation. And the contributions they have made to air navigation have served as tremendously helpful springboards toward progress.

To the pilot, navigation becomes largely the gathering of data and the use of this data to solve the navigation problem in a more or less mechanical manner. Judgment, however, plays a significant part of successful navigation, and the more hours of flying one has the more judgment will enter into both his planning and his in-flight decisions. In time, one learns that the cornerstone upon which accuracy and reliability are built lies in this judgment based upon experience.

Navigation, of course, has become an indispensable part of most of today's flying missions. The aviator must be able to plan a mission covering any eventuality. During flight, he must be able to assess his progress and determine his future course of action. It is important that one learn that changing flight conditions will require changes in planning. Thus, one must always try to think ahead of the aircraft and to make decisions immediately based on the changes encountered.

For that reason, knowledge of the environment in which the aviator flies is essential. Only by understanding the behavior of the great masses of air in which the aircraft operates can an aviator learn to cope with the complexity of navigation. A basic understanding of meteorology will increase the effectiveness of coping with navigational problems tremendously.

In general, approximately one-third of the earth's surface is composed of dry land on which vastly different types of terrain prevail. The remaining two-thirds of the earth's surface is covered with water. Both land and water surfaces have definite effects on the weather. This is so because of the major complications in the general air circulation brought about by the irregular distribution of oceans and continents, the relative effectiveness of different surfaces in transferring heat to the atmosphere, the daily variations in temperature, seasonal changes and other factors.

These factors lead to the establishment of quasi-permanent regions of low pressure called "lows" and regions of high pressure called "highs." From experience, one knows that one can normally associate good weather with high pressure areas and poor weather with low pressure areas.

The majority of the mountain ranges on the earth are oriented in a north-south direction, always across the path of the prevailing circulation in a given area either causing the air to be lifted aloft or to veer or shift in one direction or another.

Water on the earth's surface is continually evaporating into the atmosphere. The air is cooled them by various processes. This cooling produces condensation which causes precipitation in the form of rain, sleet, snow, etc. This moisture is then evaporated into the air again. The addition of the moisture into the air by evaporation and the removal of moisture by condensation and precipitation is a continuing process. Officially, it's known as the Hydrostatic Cycle, and it's this cycle, accompanied by temperature and pressure changes and other phenomena, that has become generally referred to the world over as "weather."

The great mass of air which completely envelops the earth is actually part of the earth, and we call it atmosphere. The earth's atmosphere does not end abruptly but becomes increasingly less dense (fewer molecules per unit volume) with the increasing distance away from the earth's surface.

The earth does not rotate within the atmosphere. The atmosphere rotates with the earth in space and can be considered a gaseous outer cover of the earth. The atmosphere, by the way, is in continuous motion which is completely separate from its motion of rotation with the earth. This atmospheric motion is referred to as circulation and is due primarily to differential heating of the earth's surface.

The atmosphere, for our purposes, can be divided into two parts — the troposphere and the stratosphere. The boundary separating these two divisions is known as the tropopause.

Practically all weather phenomena occur within the troposphere. Although there are factors which sometimes cause variations, temperatures in the troposphere normally decrease at an average rate of 2 degrees Centigrade for every 1,000 feet of altitude. This is known as a standard lapse rate.

At the point where the temperature ceases to decrease with altitude, we find the top of the troposphere or the tropopause. The height of the tropopause will vary with the season, being higher in the summer than in winter. It also varies with latitude, being approximately 60,000 feet at the equator and only 24,000 over the poles.

Above that, the stratosphere is characterized by the absence of vertical air currents. The air motion is mainly horizontal, and the temperatures vary only slightly with altitude.

The atmosphere is a mixture of several gases. At sea level, dry, pure air will contain approximately 78 per cent nitrogen, 21 per cent oxygen and smaller concentrations of other gases such as carbon dioxide, hydrogen ozone, helium, neon, krypton and argon. Water vapor in the atmosphere will vary from very near zero to four per cent by volume.

As a private pilot, of course, you can recognize that the troposphere is your ballpark, and the more you know and understand it, the better off you'll be.

For navigational purposes, the earth is assumed to be a perfect sphere. A perfect sphere is a body whose surface is at all points equidistant from a point within called center. Any straight line which passes from one side, through the center to the opposite side is called the diameter of the sphere. In actuality, the earth really is a spheroid—slightly flattened at the poles—but the equatorial diameter exceeds the polar diameter to such a small degree (23.35 nautical miles) that we can still consider it a perfect sphere for navigational purposes.

The earth makes one complete rotation around an imaginary line called its axis every 24 hours. The ends of this imaginary line or axis are known as the North and South Poles. In addition, the earth revolves around the sun in an elliptical path. The earth's axis is inclined approximately 23½ degrees from a perpendicular to the plane of its orbit around the sun. This inclination is such that the North Pole points almost directly to the North Star (Polaris).

graphic features is a necessity. This is known as a coordinate or grid system.

This system designates location or position in terms of angular magnitude with respect to two reference lines which intersect at right angles. This is done through the use of imaginary reference lines known as parallels of latitude and meridians of longitude. Latitude, of course, is measured north and south with reference to the Equator. Longitude is measured east and west of the Greenwich Meridian.

Latitude' is expressed in degrees up to 90, and longitude is expressed in degrees up to 180. The total number of degrees in any one circle cannot exceed 360. A degree of arc, too, can be subdivided into smaller units by dividing each degree into 60 minutes of arc. Each minute may be further subdivided into 60 seconds of arc.

Naming the parallel and meridian which passes through a point is essentially the same

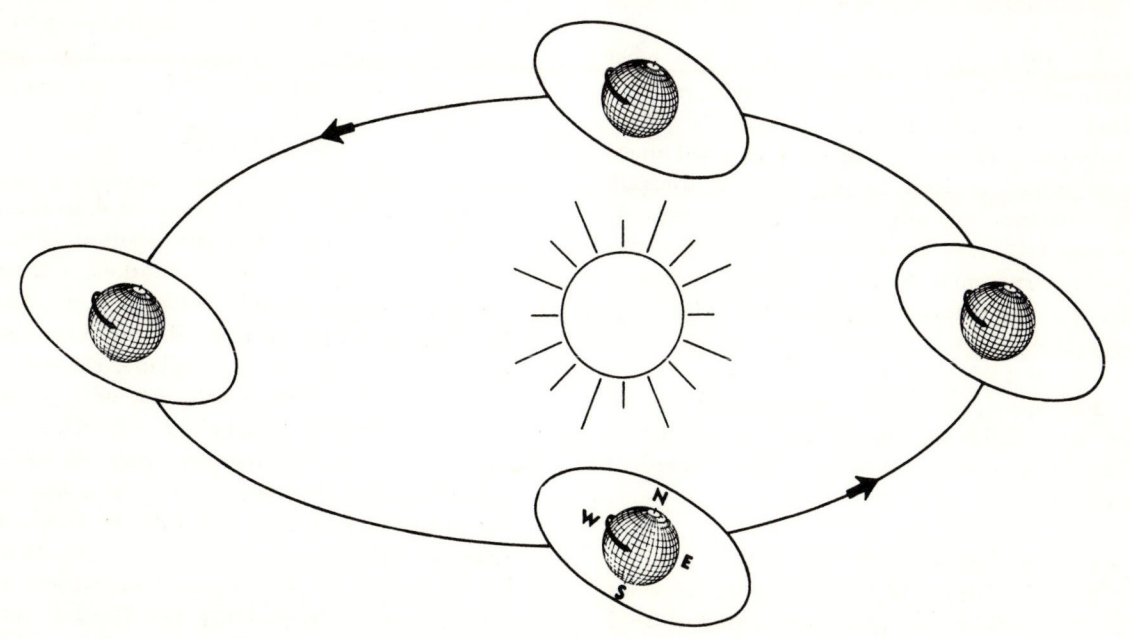

*Fig. 1-1 West to east rotation of the earth and its revolution around the sun.*

To facilitate the location of a point on the earth's surface, a universal system of expressing position without reference to geo-

as giving its coordinates. Each is named according to its angular measurement from the Equator or prime meridian. A meridian

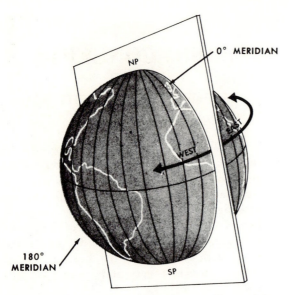

*Fig. 1-2A Longitude is measured east and west of the Greenwich Meridian.*

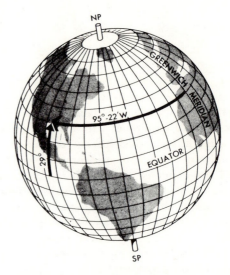

*Fig. 1-2B Latitude is measured with reference to the Equator; Longitude with reference to the Prime Meridian.*

of longitude is a line, but longitude is an angle: a parallel of latitude is a line, but latitude is an angle. In giving the coordinates of a place, latitude is given first, followed by the longitude, such as 29°45'N 95°22'W.

## DISTANCE

In navigation, the most common unit used for measuring distances is the nautical mile.

In the United States, one mile has been defined by statute to be 1,760 yards, or 5,280 feet. That's called the U.S. Statute Mile. And unless you do your navigating by road map, you might as well forget it. Most aircraft instruments and other navigational measurements are given in nautical miles. Recently, the International Nautical Mile was adopted by the Department of Commerce. This mile equals 6,076.10 U.S. feet.

For most practical navigation purposes all of the following units are used interchangeably as the equivalent of one nautical mile:

1.  1,852 meters (6,076.10 feet).
2.  One minute of arc of a great circle on a sphere having an area equal to that of the earth.
3.  One minute of arc on the earth's equator.
4.  One minute of arc on a meridian.
5.  Two thousand yards (only for short distances).

Sometimes it is necessary to convert statute and nautical miles. This conversion is easily made with the following ratio: In a given distance:

$$\frac{\text{Number of statute miles - 76}}{\text{Number of nautical miles - 66}}$$

Closely related to the concept of distance is speed. It is usually expressed in miles per hour, be they statute or nautical. You'll be using nautical miles, for all practical purposes. And remember, it's incorrect to say knots per hour. It's 110 knots, not 110 knots per hour!

## DIRECTION

One of the most widely accepted definitions of direction is that it is the position of one point in space relative to another without reference to the distance between them. And that system might be okay if you're on foot or horseback. But in an airplane, the use of directions like north, south, east or west are about as good as a nickel in a quarter slot machine. It's too cumbersome and too easily misunderstood. For that reason, aviators use the numerical system and talk in terms of degrees. The degrees, of course, are those measurements of a compass rose, which, in effect, is a circle dividing the horizon into 360 degrees. Starting with north as 000 degrees and continuing clockwise through east, south and west back to north. East is 090 degrees. South is 180 degrees, and west is 270 degrees.

Because direction is one of the most important parts of the aviator's work, the various

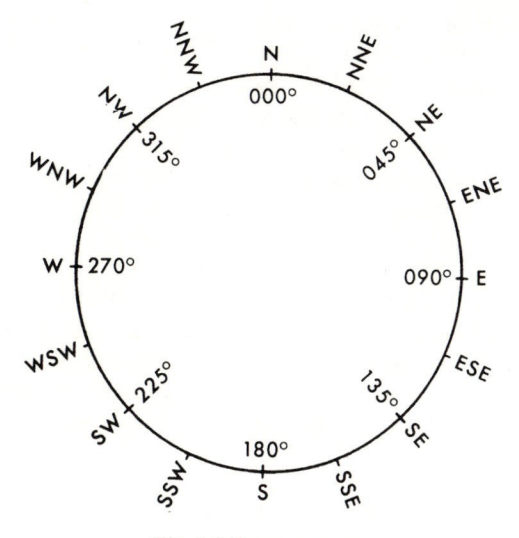

Fig. 1-3 Compass rose.

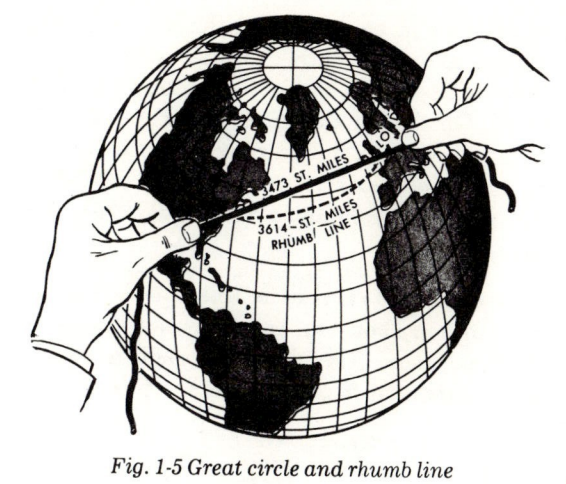

Fig. 1-5 Great circle and rhumb line

*Great Circle and Rhumb Line Direction*

terms involved should be well understood:

Course — the intended horizontal direction of travel.

Heading — the horizontal direction in which an aircraft is pointed.

Track — the actual horizontal direction made by the aircraft over the earth.

Bearing — the horizontal direction of one terrestrial point from another.

Bearings, by the way, usually are expressed in terms of one of two reference directions—true north or the direction in which the aircraft is pointed. If true north is the reference direction, the bearing is called a true bearing. If the reference direction is the heading of the aircraft, the bearing is called a relative bearing. These are shown in the illustration:

The direction of the great circle route changes constantly as progress is made along a given flight. Flying such a route requires constant change of direction and would be difficult to fly under ordinary conditions without either special equipment or special knowledge. Great circle routes, however, are the most desirable because they follow the shortest distance between any two points.

A line which makes the same angle with each meridian is called a rhumb line. An aircraft holding a constant true heading would be flying a rhumb line. Flying this sort of path results in a greater distance traveled. It is, however, easier to steer.

As shown on the illustration, the great circle and rhumb line courses will take an

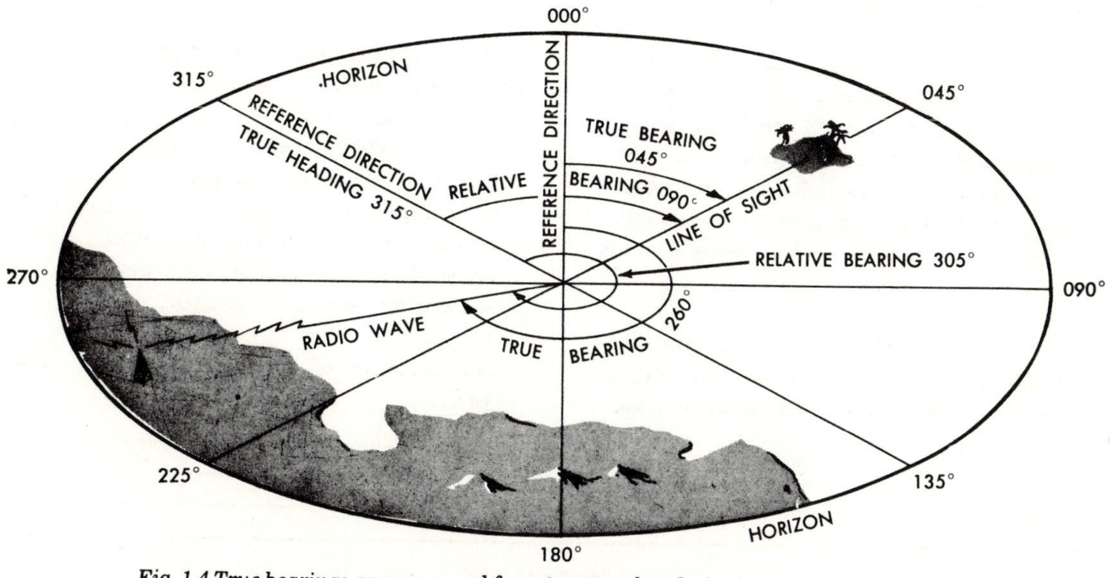

Fig. 1-4 *True bearings are measured from true north; relative bearings from heading.*

aircraft on a completely different route. As shown, the great circle route covers 3,473 miles while the rhumb line route is 3,614 miles. On shorter distances, the difference in mileage over the two different routes would be negligible, except in high latitudes or if the route approximates a meridian or the equator.

If continued, a rhumb line spirals toward the poles in a constant true direction but never reaches them. The spiral formed is called a loxodrome or loxodromic curve.

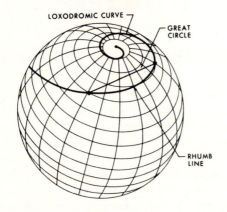

Fig. 1-6 Rhumb line, loxodromic curve, and great circle.

## Maps and Charts

Map making is one of the most important branches of geography, and most other areas of geography are dependent upon it. From time to time, nearly every person makes use of maps of one kind or another, and in general, ideas with regard to the relative areas of the various portions of the earth's surface are derived from this source. A good map which displays the various land forms is necessary to study the structure of any region. For navigation and flight operations, charts, plans and maps are indispensable. Therefore, some knowledge of the essential qualities inherent in various map structures is highly desirable. A thorough understanding of maps and charts is an absolute prerequisite for today's aviators.

An air navigation chart is a diagrammatic representation of the earth's surface, or a part of it, on a flat surface. An air navigation chart and map, by the way, will be used synonymously in this book. For all practical purpose, they're the same, and you'll hear your fellow pilots ask or order a "map" a lot more frequently, we imagine, than you'll hear him ask for a "chart". Nevertheless, some hold that a "map is to look at, and a chart is to work on," so you can take your choice.

At any rate, in greater or lesser detail, the map shows elevation, cities and towns, principal highways and railroads, oceans, lakes, rivers, radio aids to navigation, danger and restricted areas and other features useful to the pilot.

The ratio between the distance on a chart and the distance it represents on the earth is the scale of the chart. Scales will vary with the various maps. For instance, a map showing the entire surface of the earth would be drawn to small scale for convenient size. A chart covering a small area would be drawn to a larger scale.

The scale of a chart may be expressed rather simply, such as "1 inch equals 30 miles," which means that a ground distance of 30 miles is one inch in length on the map. On aeronautical charts, the scale is shown in representative fractions or in graphic scales. With the representative fraction, the scale may be given such as 1 : 500,000 or 1/500,000. This means that one unit on the chart represents 500,000 units of the same dimension on the earth. For example, 1 inch on the chart would represent 500,000 inches on the earth or approiximately 6.9 nautical or 8 statute miles.

The graphic scale, on the other hand, shows the distance on a chart labeled in terms of the actual distance it represents on the earth. The distance between parallels of latitude is a convenient graphic scale since one degree of latitude always equals 60 nautical miles. Meridians are often divided into minutes of latitude with each division representing one nautical mile.

An example of a graphic scale is shown below:

| KILOMETERS | | | 10 | | | | 0 | 10 | | 20 | 30 | | 40 | |
|---|---|---|---|---|---|---|---|---|---|---|---|---|---|---|
| NAUTICAL MILES | 10 | | | | | | 0 | | 10 | | 20 | | | |
| STATUTE MILES | | 10 | | | | | 0 | | 10 | | 20 | | | 30 |

Fig. 1-7 Graphic scale.

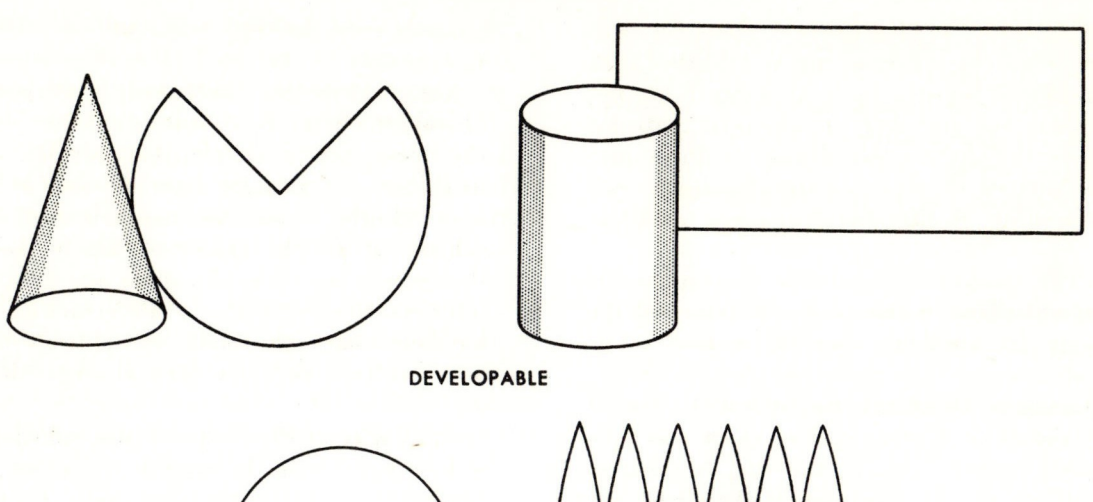

DEVELOPABLE

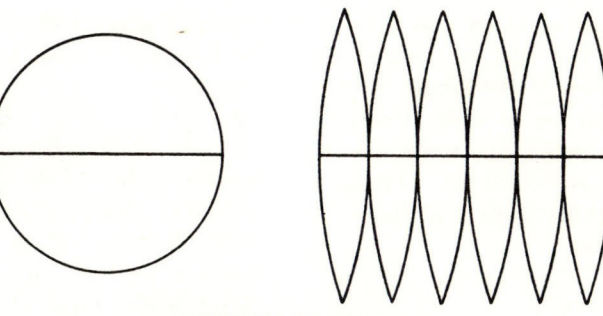

NONDEVELOPABLE

*Fig. 1-8 Developable and nondevelopable surfaces.*

### Distortion

What people refer to as a developable surface is a curved surface that can be flattened without tearing, stretching or wrinkling. Surfaces like a plane, cylinder or cone are developable surfaces. The surface of a sphere or spheroid is said to be nondevelopable because no part of it can be laid out on a flat surface without distortion or a misrepresentation of direction, shape and relative size of the features on the earth's surface. It's more easily understood, perhaps, by attempting to flatten half of an orange peel. A small piece of the orange peel because it is nearly flat can be flattened with little stretching or tearing. Likewise, a small area of the earth's surface which is nearly flat, can be represented on a flat surface with little distortion. Distortion becomes a serious problem in charting large areas and can never be completely eliminated. It can, however, be controlled. A chart for a particular purpose can be drawn so as to minimize the type of distortion which is most detrimental. The globe is the only means of representing the entire surface of the earth without distortion.

Each type of chart has distinctive features which make it preferable for certain uses; no one chart is best for all. If it were possible to construct the perfect chart, it would have:

1. True shape of all physical features.
2. Correct angular relationships, or conformality.
3. Representation of areas in their correct relative proportions.
4. True scale value for measuring distances.
5. Great circles and rhumb lines represented as straight lines.

Of course, it's possible to preserve one and sometimes more of the above properties in any one projection, but it's impossible to preserve all of them. A chart, for instance, can't have both conformality and equal area. It is desirable, however, to have charts which offer some ease in finding and plotting coordinates of points, ease in joining two or more charts, cardinal directions parallel throughout the chart and simplicity and ease in construction.

### GRATICULE

The coordinates of any point on earth may be found by astronomical means. With reference to control points established in this manner, the exact location of nearby features may be found by geographic survey or by aerial photography. A chart can then be made by drawing the established geographical features on a framework of meridians and parallels known as a graticule. Once the graticule is drawn, features may be plotted in their correct positions with references to the meridians and parallels.

The form of the graticule determines the general characteristics and appearance of the chart. Its size determines the scale.

Because meridians and parallels cannot be shown on a plane surface exactly as they would appear on a sphere, no perfect graticule can be constructed. Meridians, for example, may be shown as straight lines, as variously curved lines, or some as straight and some as curved lines. They may be spaced in various ways and may intersect at various angles.

## PROJECTION

The method used to represent all or part of the surface of a sphere or spheroid on a plane surface is known as projection. The actual projection of a graticule is accomplished by application of mathematical formulas. Projections are classified primarily as to the type of developable surface to which the spherical or spheroidal surface is transferred. They are sometimes further classified as to whether the projection is centered on the Equator (equatorial), a pole (polar), some point or line between the Equator and the poles (oblique) or tangent at a meridian (transverse). Charts most commonly used in air navigation are Lambert, Mercator and Polar Stereographic.

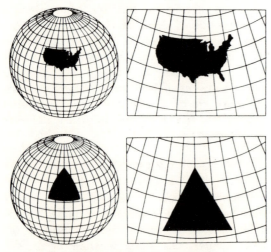

*Fig. 1-9 Areas and angles (Lambert).*

### Lambert Projection

The Lambert conformal projection (Lambert chart) is a conic projection using the cone for a developable surface. All meridians are straight lines meeting at the apex of the cone. All parallels are concentric circles, the center of which also is the apex of the cone. Meridians and parallels intersect at right angles. Scale errors on the Lambert chart are small, so the scale may be considered constant over a single sheet.

For a map of the United States, maximum scale error would be slightly in excess of one percent. A straight line approximately represents a great circle on this chart, while a rhumb line will be a curve concave to the nearer pole. Because it is conformal, the Lambert chart cannot be equal area. However, because it is conformal, the scale remains uniform in all directions about any point, and because of the uniformity of scale, areas do retain true shape.

As cited earlier, any straight line on a Lambert chart is nearly a great circle. In the distance of 2,572 statute miles between San Francisco and New York, a great circle and a straight line connecting them on a Lambert chart are only 9½ miles apart at midlongitude. For shorter distances, the difference is negligible. Thus, for all practical purposes, if a flight is only a few hundred miles long, a straight line drawn on a Lambert chart may be considered a great circle.

A rhumb line is a curved line on a Lambert chart. The closer its direction is to east-west, the more a rhumb line departs from a straight line. Over distances of 100 to 200 miles, in the latitude of the United States, a rhumb line departs little from a

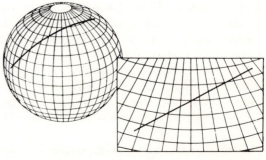

*Fig. 1-10 Great circle vs. straight line on a Lambert chart.*

straight line, but over long distances, the difference is large. Between San Francisco and New York, for example, the length of a rhumb line differs from a straight line by approximately 170 miles.

The constant scale and conformality of the Lambert charts place them among the best for air navigation. They are suitable for use with long-distance radio bearings and, for problems involving distances and true directions, they are superior to the Mercator charts. For plotting positions and measuring rhumb line directions, they are inferior to the Mercator.

## MERCATOR PROJECTION

The Mercator chart is a cylindrical chart upon which meridians appear as straight lines which are equidistant and parallel. Parallels of latitude are parallel to each other and perpendicular to the meridians. The distance between parallels increases with an increase in latitude. Because the meridians are parallel to each other, the east-west scale is increased with increase in latitude. Consequently, parallels must be placed in such a manner that the north-south scale increase proportionately. As a result, the scale at any point is constant in all directions. Since meridians and parallels intersect at right angles as on the earth, all angles are shown correctly. Every rhumb line appears as a straight line, and every straight line is constant in direction. The Equator and the meridians are the only

great circles which appear as straight lines. All other great circles appear as curved lines.

The Mercator is used in air navigation only for long-range over-water flying. Its greatest advantage is that a rhumb line on the chart is a straight line. Plotting is easier because of the rectangular graticule. By the same token, however, long range radio bearings cannot be plotted without special corrections.

Because of the expanding scale on the Mercator chart, distances are difficult to measure.

## POLAR STEREOGRAPHIC PROJECTION

The Polar Stereographic projection is based on a plane, tangent at the pole, with the point of projection at the opposite pole. Meridians are straight lines converging at the pole. Parallels are concentric circles with the pole as their common center. And the polar stereographic chart, as its name would imply, is the best chart for navigation in the polar regions.

The Polar Stereographic chart may include an entire hemisphere, although for air navigation, a chart will not extend more than 20 or 30 degrees from the pole.

*Fig. 1-12 The polar sterographic projection.*

Because the interval between parallels increases with distance from the pole the north-south scale also increases away from the pole. The east-west scale increases in the same proportion, so that at any point the scale is constant in all directions. On the Polar Stereographic chart, all angles are correctly shown since meridians appear as radii of circles representing parallels, and the meridians and parallels intersect at 90

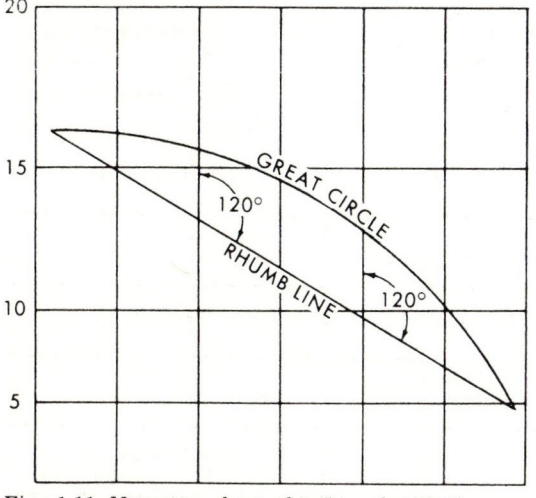

*Fig. 1-11 Mercator chart showing rhumb line vs. great circle.*

degrees on the chart.

Meridians, which are great circles, appear

as straight lines on this projection, thus, any great circle passing through the center of the chart appears as a straight line.

## S U M M A R Y   O F   P R O J E C T I O N S

| CHARACTERISTIC | POLAR GNOMONIC | EQUATORIAL GNOMONIC | OBLIQUE GNOMONIC |
|---|---|---|---|
| PARALLELS | CONCENTRIC CIRCLES UNEQUALLY SPACED | CURVED LINES UNEQUALLY SPACED | CURVED LINES UNEQUALLY SPACED |
| MERIDIANS | STRAIGHT LINES RADIATING FROM THE POLE | PARALLEL STRAIGHT LINES UNEQUALLY SPACED | STRAIGHT LINES CONVERGING AT THE POLE |
| APPEARANCE OF GRID |  | 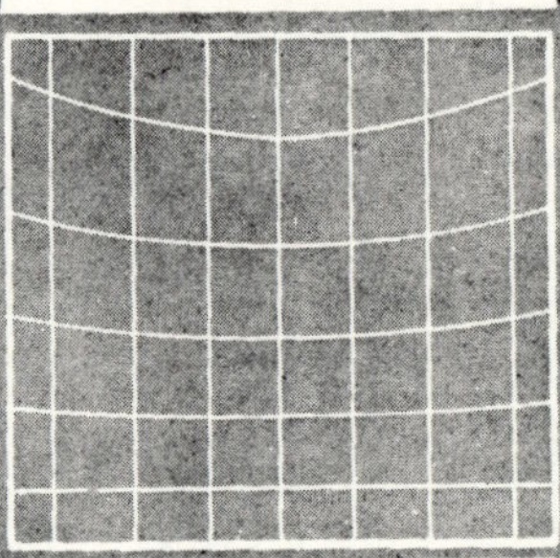 | 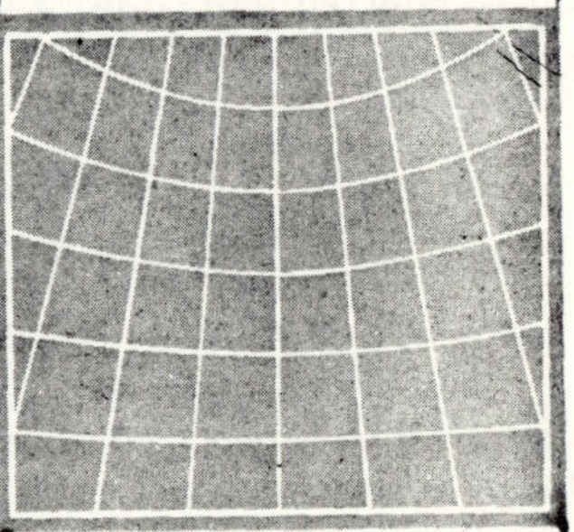 |
| ANGLE BETWEEN PARALLELS & MERIDIANS | 90 | VARIABLE | VARIABLE |
| STRAIGHT LINE CROSSES MERIDIANS | VARIABLE ANGLE (Great Circle) | CONSTANT ANGLE (Great Circle) | VARIABLE ANGLE (Great Circle) |
| GREAT CIRCLE | STRAIGHT LINE | STRAIGHT LINE | STRAIGHT LINE |
| RHUMB LINE | CURVED LINE | CURVED LINE | CURVED LINE |
| DISTANCE SCALE | VARIABLE | VARIABLE | VARIABLE |
| GRAPHIC ILLUSTRATION | 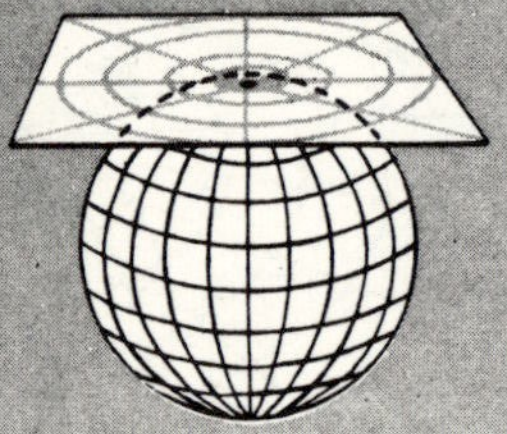 |  |  |
| ORIGIN OF PROJECTORS | CENTER OF SPHERE | CENTER OF SPHERE | CENTER OF SPHERE |
| DISTORTION OF SHAPES & AREAS | INCREASES AWAY FROM POLE | INCREASES AWAY FROM POINT OF TANGENCY | INCREASES AWAY FROM POINT OF TANGENCY |
| METHOD OF PRODUCTION | GRAPHIC OR MATHEMATICAL | GRAPHIC OR MATHEMATICAL | GRAPHIC OR MATHEMATICAL |
| NAVIGATIONAL USES | GREAT CIRCLE ROUTE DETERMINATION | GREAT CIRCLE ROUTE DETERMINATION | GREAT CIRCLE ROUTE DETERMINATION |
| CONFORMALITY | NOT CONFORMAL | NOT CONFORMAL | NOT CONFORMAL |

*Fig. 1-13*

Other great circles appear as slightly curved, however; the closer they are to center, the straighter they appear. Rhumb lines appear as curved lines.

There are, of course, a number of other types of projections used in map and chart making. They include the azimuthal or zenithal projection on which points on the earth are transferred directly to a plane tangent to the earth, gnomonic which are direct geometric projections with the graticule of

| POLAR STEREOGRAPHIC | MERCATOR | LAMBERT CONFORMAL | POLYCONIC |
|---|---|---|---|
| CONCENTRIC CIRCLES UNEQUALLY SPACED | PARALLEL STRAIGHT LINES UNEQUALLY SPACED | ARCS OF CONCENTRIC CIRCLES NEARLY EQUALLY SPACED | ARCS OF NON-CONCENTRIC CIRCLES EQUALLY SPACED ON MID-MERIDIAN |
| STRAIGHT LINES RADIATING FROM THE POLE | PARALLEL STRAIGHT LINES EQUALLY SPACED | STRAIGHT LINES CONVERGING AT THE POLE | MID-MERIDIAN STRAIGHT OTHER CURVED |
| | | | |
| 90° | 90° | 90° | VARIABLE |
| VARIABLE ANGLE (Approximates Great Circle) | CONSTANT ANGLE (Rhumb Line) | VARIABLE ANGLE (Approximates Great Circle) | VARIABLE ANGLE (Approximates Great Circle Near Mid-Meridian) |
| APPROXIMATED BY STRAIGHT LINE | CURVED LINE (Except Equator and Meridians) | APPROXIMATED BY STRAIGHT LINE | APPROXIMATED BY STRAIGHT LINE NEAR MID-MERIDIAN |
| CURVED LINE | STRAIGHT LINE | CURVED LINE | CURVED LINE |
| NEARLY CONSTANT EXCEPT ON SMALL SCALE CHARTS | MID-LATITUDE | NEARLY CONSTANT | CONSTANT FOR SMALL AREAS VARIABLE FOR LARGER AREAS |
| | | | |
| OPPOSITE POLE | CENTER OF SPHERE (For Illustration Only) | CENTER OF SPHERE | CENTER OF SPHERE |
| INCREASES AWAY FROM POLE | INCREASES AWAY FROM EQUATOR | VERY LITTLE | INCREASES AWAY FROM MID-MERIDIAN |
| GRAPHIC OR MATHEMATICAL | MATHEMATICAL | GRAPHIC OR MATHEMATICAL | MATHEMATICAL |
| POLAR NAVIGATION, ALL TYPES | DEAD RECKONING AND CELESTIAL (Suitable for All Types) | PILOTAGE AND RADIO (Suitable for All Types) | GROUND FORCES MAPS |
| CONFORMAL | CONFORMAL | CONFORMAL | NOT CONFORMAL BUT IS USED AS SUCH ON VERY LARGE SCALE MAPS |

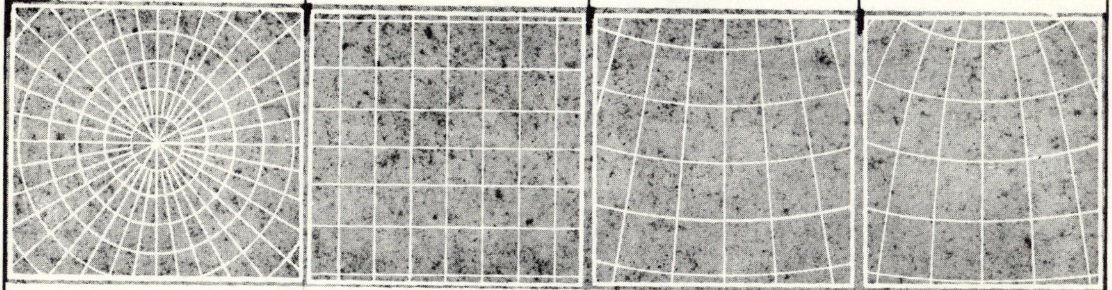

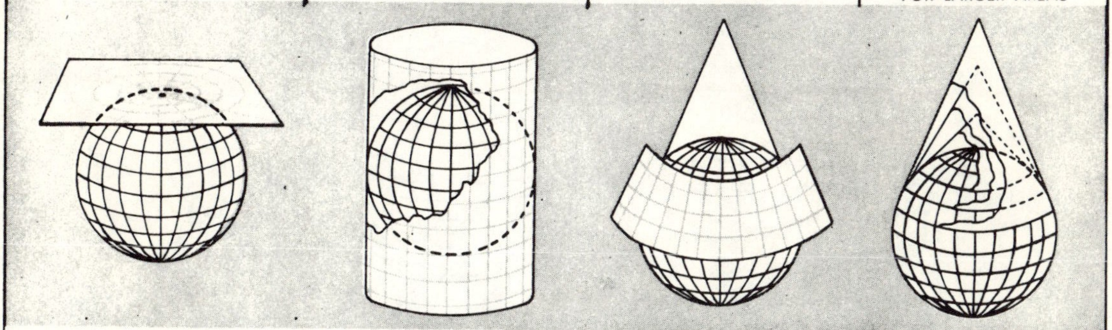

*Fig. 1-14*

the reduced earth projected from a light source at the center of the earth onto a flat surface held tangent to it at a given point, and orthographic projections whereupon a light source at infinity projects the graticule

of the reduced earth onto a plane tangent to the earth at some point perpendicular to the projecting rays. A summary of the various projections is shown (figs. 1 - 13, 1 - 14, and 1 - 15)

| CHARACTERISTICS | TRANSVERSE MERCATOR | MODIFIED LAMBERT | AZIMUTHAL EQUIDISTANT |
|---|---|---|---|
| PARALLELS | ARCS OF CONCENTRIC CIRCLES NEARLY EQUALLY SPACED | ARCS OF CONCENTRIC CIRCLES NEARLY EQUALLY SPACED | CURVED LINES UNEQUALLY SPACED |
| MERIDIANS | STRAIGHT LINES CONVERGING AT THE POLE | STRAIGHT LINES CONVERGING AT THE POLE | CURVED LINES CONVERGING AT THE POLE |
| APPEARANCE OF GRID |  | 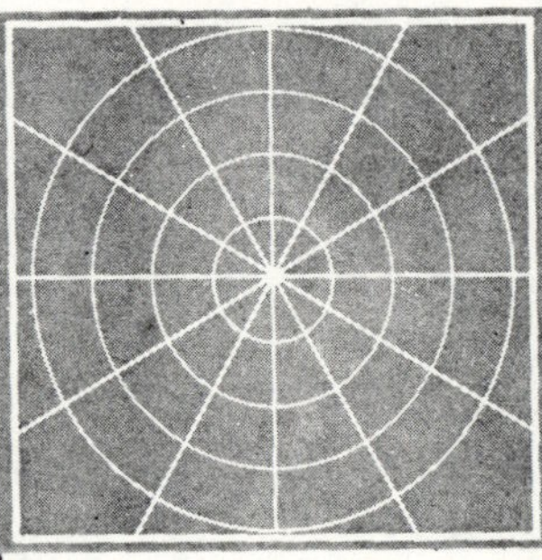 | 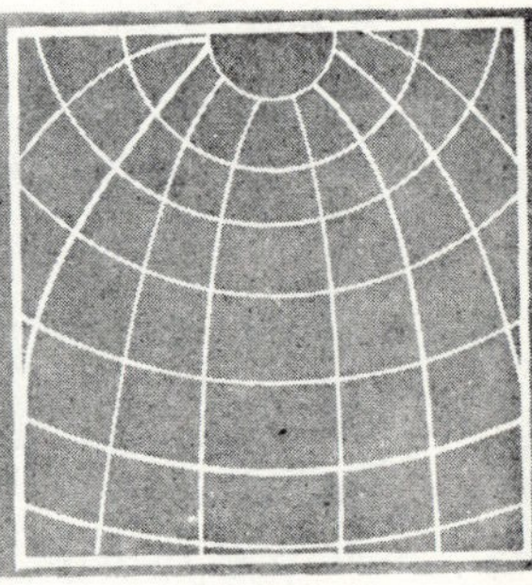 |
| ANGLE BETWEEN PARALLELS & MERIDIANS | 90° | 90° | VARIABLE |
| STRAIGHT LINE CROSSES MERIDIANS | VARIABLE ANGLE (Approximates Great Circle) | VARIABLE ANGLE (Approximates Great Circle) | VARIABLE ANGLE |
| GREAT CIRCLE | APPROXIMATED BY STRAIGHT LINE | APPROXIMATED BY STRAIGHT LINE | ANY STRAIGHT LINE FROM CENTER OF PROJECTION |
| RHUMB LINE | CURVED LINE | CURVED LINE | CURVED LINE |
| DISTANCE SCALE | NEARLY CONSTANT | NEARLY CONSTANT | CORRECT SCALE ON ALL STRAIGHT LINES THROUGH CENTER |
| GRAPHIC ILLUSTRATION |  |  |  |
| ORIGIN OF PROJECTORS | CENTER OF SPHERE | CENTER OF SPHERE | NOT PROJECTED |
| DISTORTION OF SHAPES & AREAS | INCREASE OUTWARDLY FROM CIRCUMFERENCE OF TANGENCY | INCREASES AWAY FROM STANDARD PARALLELS | INCREASES FROM CENTER OUTWARD TO EDGES |
| METHOD OF PRODUCTION | MATHEMATICAL | MATHEMATICAL | MATHEMATICAL |
| NAVIGATIONAL USES | GROUND FORCE MAPS AND AIR SUPPORT CHARTS GREAT CIRCLE NAVIGATION | POLAR NAVIGATION | STRATEGIC PLANNING |
| CONFORMALITY | CONFORMAL | CONFORMAL | NOT CONFORMAL |

*Fig. 1-15*

An aeronautical chart is a pictorial representation of the earth and its culture, and it can provide a picture of any region of the earth. Properly used, it is a vital adjunct to navigation. Improperly used, however, it might even prove a hazard. Because of the great importance of charts, aviators must become thoroughly familiar with the aeronautical charts they employ and understand their uses.

Because of the many different purposes of aeronautical charts, various scales will be used. If a map is to show the whole world, for instance, it must be drawn to small scale or it would simply be too large. If a map is to show much detail, it must be drawn to large scale; then it shows a smaller area than does a map of the same size drawn to a small scale. Remember: large area, small scale; small area, large scale.

As it already has been mentioned, the scale on aeronautical charts is indicated in one of two ways, representative fraction or graphic scale. The representative fraction, such as 1 : 500,000 or 1/500,000 means that any unit of one on the chart represents 500,000 of the same unit on the earth. A representative fraction can be converted into a statement of miles to the inch. Thus, if the scale is 1 : 1,000,000, 1 inch on the chart stands for 1,000,000 inches or 1,000,000 divided by (6,080 x 12) equals about 13.7 nautical miles. Similarly, if the scale is 1 : 500,000, 1 inch on the chart represents about 6.85 nautical miles.

The graphic scale may be shown by a graduated line, usually found printed along the border of a chart. Take a measurement on the chart and compare it with the graphic scale of miles. The number of miles that the measurement represents on the earth may be read directly from the graphic scale.

Another convenient scale for distance measurement can be found by the distance between parallels of latitude. One degree of latitude always equals sixty nautical miles, and one minute of latitude equals one nautical mile. Here's an example: (fig. 1 - 16)

Aeronautical charts are differentiated on a functional basis by the type of information they contain. The name of the chart is a reasonable indication of its intended purpose. A Minimal Flight planning chart is primarily

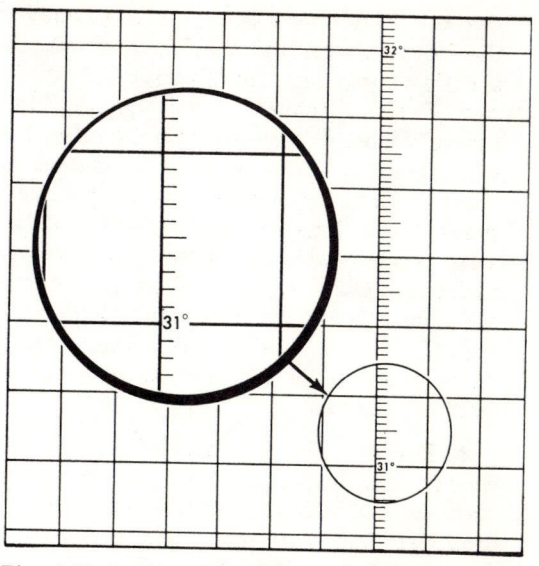

*Fig. 1-16 Latitude provides a convenient graphic scale.*

used in minimal flight planning. A jet navigation chart has properties that are adaptable to the speed, altitude and instrumentation of jets.

Charts will vary also according to the amount of information displayed. Charts designed to facilitate long-range planning, for example, carry less detail than required for the navigation en route. Local charts present great detail and approach charts contain a lot of specific information relating to the area surrounding the airport.

Most of the land and adjacent areas of the earth are represented on the World Aeronautical Charts (WAC) because the information contained on them is basic for understanding and interpreting most of the other charts.

## AERONAUTICAL CHARTS
### United States Sectional Charts

*a. General.* United States sectional charts are Lambert conformal charts published by U.S. Coast and Geodetic Survey. The scale is 1:500,000 (1 inch equals approximately 6.9 nautical miles or about 8 statute miles). The purpose of the charts is to provide coverage for the United States and the Hawaiin Islands at a scale appropriate for flights of short duration. They are intended primarily for pilotage (visual flight) but are suitable for other forms of navigation. The large scale of the sectional chart permits information to be included in great detail. This series of charts will be revised semiannually and will be available from authorized chart agents or from the chart distribution office of the Coast and Geodetic Survey (b through f below list some of the features incorporated in this series of visual aeronautical charts.)

b. *Terrain.* The portrayal of terrain on the charts will emphasize land forms by relief shading and will also include contours, elevation tints and a generous depiction of a number of spot elevations. The highest terrain elevation located within each 30' longitude and 30' latitude area will be dramatically shown in large size type.

c. *Cultural.* Cultural features, such as railroads and major roads and, in sparsely settled areas, even dirt roads or paths, may be shown. Cities and towns, mines, lookout towers, and many other good landmarks are indicated by symbols.

d. *Aeronautical.* Aeronautical features such as airports, airways, radio ranges, etc., are shown in these charts. In a rectangle near each airport, information relative to the airport is listed and includes such data as lighting, type and length of longest runway, frequencies available for contacting the tower, etc.

e. *Coverage.* There are 37 charts printed back to back on a sheet 20⅝ x 55 inches. These charts cover the United States and are designated by name and series (e.g., New Orleans—Sectional). Each chart covers an area containing approximately 4° of latitude and 6° to 8° of longitude. Overlapping coverage on adjoining charts will be provided on the north and east sides to facilitate transition between charts. Charts are printed to the edge of the sheet

eliminating the need for cutting and folding of border areas when matching data on adjacent charts is required.

f. *Fold.* Each chart will be folded to convenient dimensions of 5 x 10 $\frac{5}{16}$ inches providing for easy handling and stowage in small cockpit. Unique fold is designed to provide rapid switch over from front to back, insuring availability of navigation information for continuous and uninterrupted flight.

## World Aeronautical Chart

The world aeronautical chart (WAC) is published by the U.S. Coast and Geodetic Survey. Scale is 1:1,000,000 (1 inch equals approximately 14 nautical or 16 statute miles). From 0° to 80° latitude, WAC charts are based on the Lambert conformal projection; from 80° to the poles, they are based on the modified polar stereographic projection. Their purpose is to provide a standard series of aeronautical charts covering the world at a size and scale convenient for navigation over land surfaces. The smaller scale of the WAC chart does not permit as much detail as the sectional chart. All types of topographical and cultural features, including railroads and major roads, are shown. All important navigational aids and air facilities are included on the overprint. Scheduled revisions supersede previously printed charts. Time for the next scheduled edition is shown below the date in the lower right-hand corner of the margin.

## Photomaps

Photomaps prepared by the Army Map Service, Corps of Engineers, are used for air navigation over small areas. These maps may be constructed by using a single photomap or a mosaic of several photomaps. They may be printed on the reverse side of tactical maps. The scale varies from 1:5,000 to 1:60,000. Meridians and parallels are indicated on the margin of the map. Positions are located by reference to a system of horizontal and vertical grid lines.

## SUMMARY OF CHARTS

| Name | Code | Projection | Scale | Coverage | Description | Purpose |
|---|---|---|---|---|---|---|
| USAF Jet Navigation Charts | JN | Charts 4, 5, 6, and 7 transverse Mercator. All others Lambert conformal conic | 1 to 2,000,000 | Northern Hemisphere | Charts show all pertinent hydrographic and cultural high altitude features. Complete transportation network in the areas surrounding cities. A maximum of radar significant detail as required for long range navigation and a suitable aeronautical information overprint including runway pattern. | For long range, high speed, high altitude aircraft; for radar navigation, celestial navigation, DR and visual pilotage. |
| World Aeronautical Charts | WAC | 0°–80° Lambert conformal conic 80°–90° polar stereographic | 1 to 1,000,000 | World-wide | Charts show all types hydrographic and cultural features. All important navigation aids and air facilities are included. | Standard series of aeronautical charts for world coverage at a size and scale convenient for overland navigation. |
| USAF Pilotage Charts | PC | 0°–80° Lambert conformal conic 80°–90° polar stereographic | 1 to 500,000 | World-wide | Charts show all types hydrographic and cultural features. All important navigation aids and air facilities are included. | To provide charts for selective areas of the world at a size and scale appropriate for flights of short duration. |
| USAF Pilotage Charts Low Level | PCL | 0°–80° Lambert conformal conic 80°–90° polar stereographic | 1 to 500,000 | World-wide | Similar to PC charts except for critical changes to satisfy low level, high speed navigation requirement. Uses pictorial symbolization of cultural and natural ground features. | To provide major check points for preflight and inflight use on low level, high speed missions. |
| Minimal Flight Planning Charts | MPC MFC | Lambert conformal conic | 1 to 10,000,000 | Mostly Northern Hemisphere covering water areas. | Coastline, primary levels and land contours, weather navigation stations, geostrophic distance scales, and international boundaries are shown. | For use in pressure pattern navigation for preflight planning, enroute replanning, and postflight briefing. |

Fig. 1-17

## SUMMARY OF CHARTS (Cont'd)

| Name | Code | Projection | Scale | Coverage | Description | Purpose |
|---|---|---|---|---|---|---|
| Route Charts | RT | Oblique Mercator and Lambert conformal conic | 1 to 2,000,000 | Parts of U.S., Canada and Alaska | Charts show major cities, hydrographic features, roads, etc., as well as all important navigational aids and air facilities. | For high altitude and long range flights on established local and commercial air routes. |
| Navigation Flight Charts | NF | Mercator | 1 to 3,000,000 or smaller | Western Europe, England, Ireland, central U.S., the Aleutians, central Pacific Islands, North Atlantic, West Indies, and northeast coast of South America. | Charts only wide enough to allow a reasonable safety margin on either side of course. Show as many cultural and hydrographic features as possible in this scale. Air facilities are included. | Strip sheets suitable for navigational plotting inflight. Not suitable for pilotage navigation. In most cases are designed for use on large water areas. |
| Aeronautical Planning Charts | AP | 0°–75° Lambert conformal conic 75°–90° polar stereographic | 1 to 5,000,000 | World-wide | Show only essential hydrographic and cultural features. Includes isogonic lines and aircraft facilities. | Charts covering the world for long range planning purposes. |
| Consol-Loran Navigation Charts | CJC LJC | Transverse Mercator and Lambert conformal conic | 1 to 2,000,000 | Northern Europe and Arctic | Show land areas in subdued tone. Consul, loran, and radio aids are emphasized. | Long range flight planning and electronic navigation. |
| Outline Planning Charts | PO | 0°–75° Lambert conformal conic 75°–90° polar stereographic | 1 to 5,000,000 | World-wide with exception of some large water areas. | Charts show projection, cultural, and marginal information in black; open water and drainage in blue. Covers same areas and same scale as planning charts. | Charts serve as worksheets for operational planning. |
| Navy Air Navigation Charts | V 30 | Mercator | 1 to 2,188,800 | World-wide with relating Arctic coverage. | Charts contain essential topographical, hydrographical, and aeronautical information. Roads and railroads not shown. Land areas are tinted gray. | Basic long range air navigation plotting series designed primarily for use in larger aircraft. |

*Fig. 1-18*

**SUMMARY OF CHARTS (Cont'd)**

| Name | Code | Projection | Scale | Coverage | Description | Purpose |
|---|---|---|---|---|---|---|
| Navy Loran Air Navigation Charts | VL 30 | Mercator | 1 to 2,188,800 | World-wide with graphical, relating Arctic coverage | Contain essential topographical, hydrographical navigation and aeronautical information. Roads and railroads not shown. Land areas tinted gray. Loran curve interval 20, 50, and 100 microseconds. | Basic long range Loran air navigation plotting series designed primarily for use in larger aircraft. |
| Special Navigation Charts | NS | Mercator and Lambert conformal conic | 1 to 5,000,000, 1 to 1,000,000, 1 to 3,000,000 | Special areas. | Charts show essential hydrographic and cultural features. Roads are omitted but major railroads are shown. Important air facilities included. | Large sheet, small scale charts suitable for long range navigation over areas of frequent operations. |
| Navy Airways Plotting Charts | VR | Mercator | Varies from 1 to 3,000,000 to 1 to 6,000,000 | Pacific area | Show land areas tinted in gray. Include primary aircraft facilities, radio navigational aids, magnetic variation, and compass rose. | Provide plotting charts for flights of various air routes in the Pacific area. |
| Loran Chart (Atlantic) | LS LN | Mercator except LS 210 (LC) | 1 to 2,000,000 to 1 to 5,000,000 | Atlantic Ocean | Base chart printed in gray with limited cultural features. Loran line of position are overprinted in colors. | To provide charts suitable for utilization of Loran or Consol. |
| Gnomonic Tracking Charts | GT | Gnomonic | 1 to 4,000,000 and 1 to 10,500,000 | From North Pole to approximately 40° south latitude. | Majority of 2-color outline charts with blue for graticule and buff for all land areas. Majority of charts are overprinted with special airways and air communication services. | Series of plotting charts suitable for accurate tracking of aircraft by electronic devices and small scale plotting charts for accurate great circle courses. |
| Universal Plotting Charts | UP | Mercator | 1 to 3,000,000, 1 to 5,000,000 | World-wide | For use within certain bands of latitude. Longitude must be inserted. | Charts suitable for celestial in selected bands of latitude. |

Fig. 1-19

# 2 CHART READING, PILOTAGE, AND LOW-LEVEL NAVIGATION

## Section I. CHART READING AND PILOTAGE

### General

Chart reading is the identification of landmarks with their representation on a chart. The degree of success in navigating by observation of landmarks (pilotage) depends upon the aviator's proficiency in chart interpretation.

### Accuracy of Charts

The latest revised aeronautical charts of the United States are accurate and complete. Charts of other parts of the world may not be as accurate or complete due to lack of information or utility. Since aeronautical charts undergo repeated revision, the chart used by the aviator should be the latest revision. Current aeronautical charts are listed in current navigation publications.

### Chart Content

Although charts do not picture all details, those particular features useful to the aviator are emphasized, including those features which have the most distinctive appearance from the air. For emphasis, many features are shown out of proportion to their true size, though centered in their correct positions. *For example,* the line representing a road on a WAC chart may appear to be a quarter of a mile wide according to the scale of the chart. Radio stations are prominently shown even though they are inconspicuous from the air. Many lines, such as meridians, parallels, isogonics, airways, and contours take up space on the chart even though they are invisible on the ground.

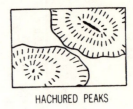

HACHURED PEAKS

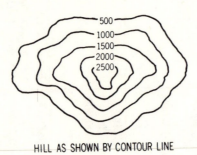

HILL AS SHOWN BY CONTOUR LINE

*Figure* 2-1 *Relief.*

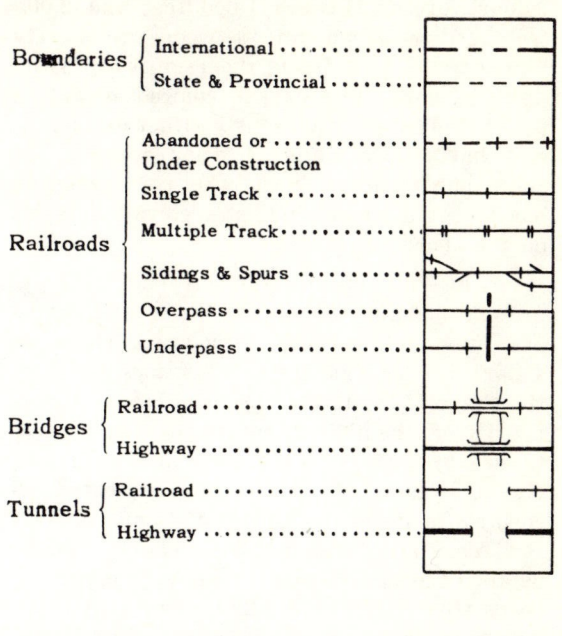

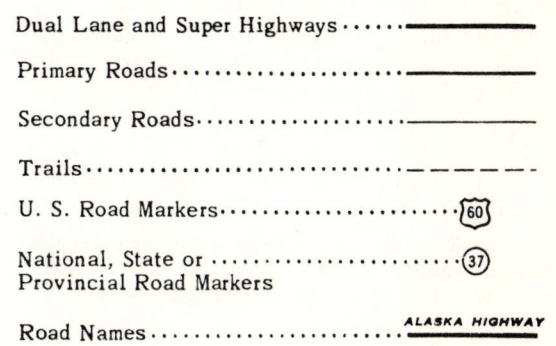

Figure 2-2  Cultural symbols.

## Symbols

Since a chart is a diagram, it consists of symbols which do not necessarily resemble the shape or appearance of the objects they represent. Skill in chart reading depends on a complete understanding and interpretation of these symbols.

*a. Relief.*

(1) Aeronautical charts show elevation and inequalities of the earth's surface, which are known collectively as *relief*. Mountains are good landmarks but are also a hazard to flying. Elevations of the highest peaks and other significant spot elevations are shown in italics.

(2) Most charts represent relief by means of contours (lines connecting points of equal elevation (fig. *2-1* )). The shoreline of the ocean might be thought of as the 0-foot contour, since every point on it is at sea level. The 1,000-foot contour is a line connecting all points

that are 1,000 feet above the average (mean) sea level. On a steep slope, contours are close together; on a gentle slope, they are farther apart.

(3) On sectional and WAC charts, contours are brown lines, each line labeled with the elevation it represents (fig. *2-2* ). Contour intervals vary from chart to chart. On charts where only low elevations exist, the contours are at 500-foot intervals; where high elevations exist, the contours are 1,000 feet apart. On charts where unexplored areas are shown, mountains may be indicated by hachures or shading, with elevations of peaks shown as accurately as they are known.

(4) Hachures may be used on contour charts to show prominent hills or buttes too small to depict by contours. The relief shown by contours is further emphasized on sectional and WAC charts by a gradient system of coloring. The area between sea level and 1,000 feet

is dark green; between 1,000 feet and 2,000 feet, medium green; and between successively higher contours, different shades of brown from light to dark. The darker colored mountain peaks stand out conspicuously. Other aeronautical charts have different color schemes, or show only contour lines. The color shading block of the legend indicates elevation levels on the chart.

*b. Cultural.* Cultural symbols (fig. *2-2* ) represent manmade features. Cities and towns are shown by several methods. A circle or square denotes a small town, but does not show the shape of the town. The town can be recognized from the air only by its position relative to nearby features such as roads, railroads, rivers, streams, etc. A city is represented according to its shape and size. Comparatively few roads are shown on WAC charts, and on detailed sectional charts many conspicuous roads are omitted, especially in congested areas. All railroads are shown on aeronautical charts; normally they are more permanent than automobile roads and are more likely to be accurately depicted. A chart may or may not shown a bridge where a road or railroad crosses a body of water. Many cultural features, such as racetracks, oil fields, tank farms, and ranger stations, are shown by special symbols.

*Note.* There are no standard symbols for many conspicuous features, such as smokestacks, water towers, monuments, and prominent buildings. These are often indicated by brief descriptive notes, and each feature is indicated with an arrow and perhaps a dot showing the location.

*c. Aeronautical Information.* Aeronautical information is printed in magenta or blue color on sectional and WAC charts. Classes of airports are distinguished by different symbols, and the elevation of each airport is given (fig. *2-3* ). Light beacons (with their code signals) and radio stations (with their call letters and frequencies) are also shown (figs. *2-4* and *2-5* ). Airways, danger areas, and isogonic lines are clearly marked (fig. *2-6* ).

*d. Water and Forest.* Bodies of water are valuable to the navigator because they are relatively permanent and easily seen from the air. Conventionally, water is shown in blue with coastlines accurately drawn. (The color does not mean that the water looks blue from the air.) In arid regions, the most important rivers, often having little water showing above the ground, may have their courses outlined by comparatively dense vegetation. *European charts* show forest areas in green; on charts of the United States, forests are not shown.

*e. Reverse Side of Charts.* The following items are shown on the reverse side of sectional and WAC charts:

(1) Aeronautical symbols.

(2) A list of all airports within the limits

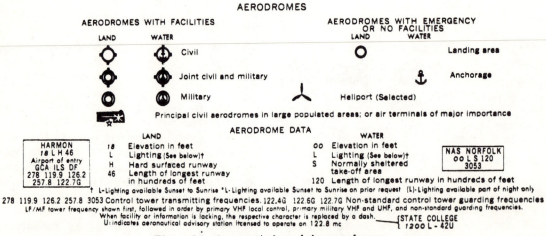

*Figure 2-3 Aeronautical symbols—aerodromes.*

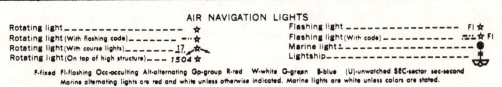

*Figure 2-4 Aeronautical symbols—air navigation lights.*

of the chart, with information concerning latitude and longitude, elevation, number and length of runways, etc.

(3) A list of prohibited, restricted, caution, and warning areas in the chart area.

(4) A chart of the United States, showing the sectional chart numbers required for each area.

(5) Other information of value to the aviator.

## Checkpoints (Daylight Hours)

*a.* A landmark used to fix the aircraft's position is called a *checkpoint*. A checkpoint must be a unique feature, or group of features, in a given area. The value of the type of checkpoint will vary with the geographical and cultural features of an area. In open areas or farm country, any town, railroad, or highway can be used for a checkpoint. In more densely populated areas, minor features such as roads and small towns are difficult to distinguish. Important highways, large towns with distinctive shapes, or towns near lakes or rivers are easily identified. In forested areas, swaths cut for pipelines and powerlines are easily traced. In mountainous areas, mines, ranger stations, prominent peaks, passes, and gorges can be used. In a desert, where checkpoints are few, minor features may be satisfactory checkpoints.

*b.* If there is any uncertainty about the position, every possible detail should be checked before identifying a checkpoint. The aviator may have to check back and forth *from chart to ground* to compare the angles at which roads or railroads leave a town, the position of bridges and intersections, or bends in streams and roads. Because the chart shows only significant details, it is essential that the aviator select reliable features on the chart to compare with features on the ground. Figure *2-7* shows the characteristics of good and poor checkpoints.

## Estimating Distance

Ground distance can be estimated by comparison with the known distance between two other points measured on the chart.

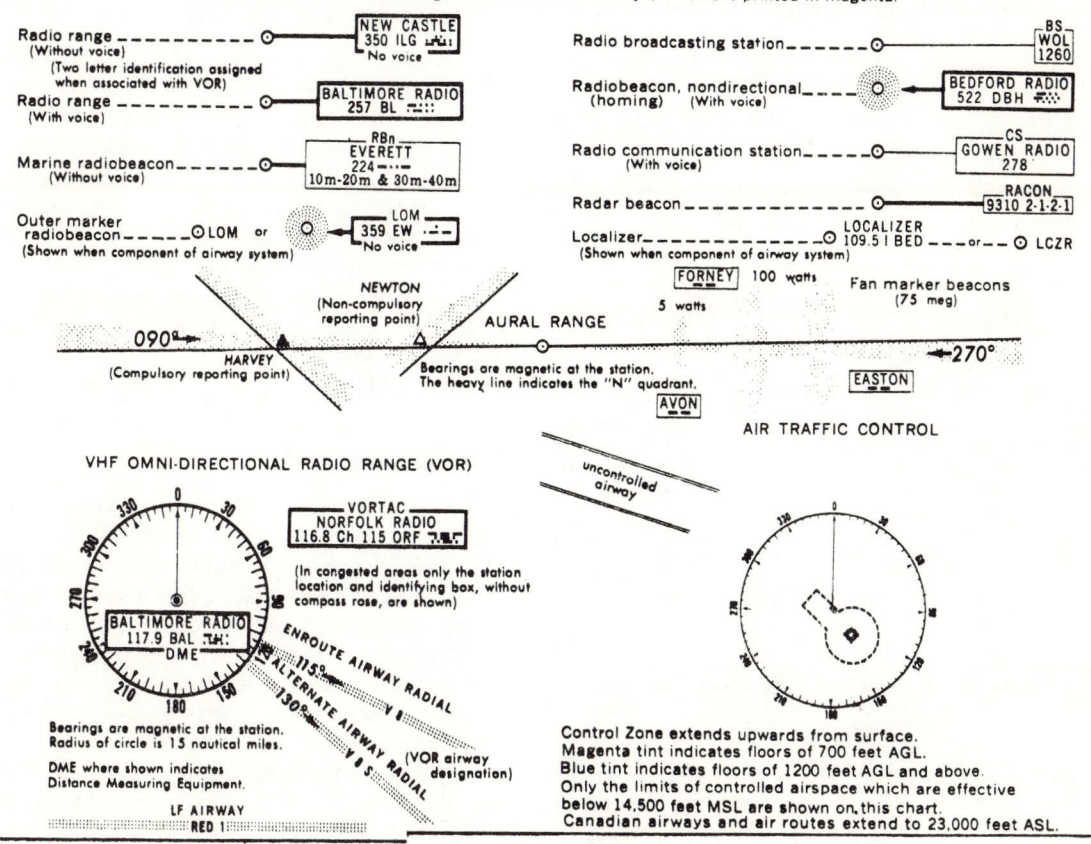

*Figure* 2-5 *Aeronautical symbols—radio facilities.*

## Appearance of the Terrain

*a. Effects of the Sun.* When the sun is low, long shadows cause strong terrain contrasts and emphasize relief. At noon, or when the sun is obscured, the absence of shadows causes the terrain to appear flat.

*b. Ostructions to Visibility.* Smoke, haze, and dust reduce visibility and often restrict observation of the terrain to the area directly beneath the aircraft. Clouds below the aircraft may block the ground view completely.

*c. Seasonal Changes.* Snow on the ground may conceal a landmark. The shape and size of lakes, rivers, and ponds often change with the seasons, especially in low, flat country.

*d. Low-Level Flight.* When flying at low levels, only small areas of the terrain can be seen. Because of the oblique angle of sight, apparent object depth is increased and relief detail is pronounced. The ground appears to move rapidly, and only brief glimpses of checkpoints are possible. An aircraft flying at 100 feet above the terrain with a groundspeed of 70 knots has the same apparent speed over the ground as an aircraft flying at 2,000 feet with a groundspeed of 1,400 knots.

*e. High Altitude Flight.* From high altitudes

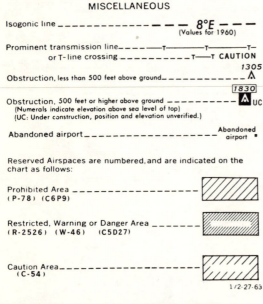

*Figure 2-6 Aeronautical symbols—miscellaneous.*

the ground appears to move slowly, and the aviator has difficulty determining the exact time of checkpoint passage. When visibility is good, a large area can be seen, distances appear

to contract, and the terrain looks flat with little detail.

## Night Pilotage

*a. General.* During hours of darkness, an unlighted landmark may be difficult or even impossible to see, and lights can be very confusing because they appear to be closer than they really are. Stars near the horizon may be confused with lighted landmarks; however, locating the North Star (Polaris) will assist in orientation. Polaris is always due north and its altitude above the horizon is approximately equal to the latitude. Objects can be seen more easily at night by looking at them from the side or rods of the eye. Starting directly at objects during night flight impairs night vision and can cause vertigo and disorientation.

*b. Unlighted Landmarks.* In moonlight, and occasionally on moonless nights, some of the more prominent unlighted landmarks such as coastlines, lakes, and rivers are visible from the air. Reflected moonlight causes a stream or lake to stand out brightly for a moment; however, this view may be too brief to permit recognition. By close observation, roads and railroads may be seen after the eyes have become accustomed to darkness.

*c. Lighted Landmarks.* Cities and large towns are usually well-lighted and are more visible at night than in the daytime. They can often be identified by their distinctive shapes and frequently can be seen at great distances, often appearing closer than they actually are. Smaller towns that are darkened early in the evening are hard to see and difficult to recognize. Busy highways are discernible because of automobile headlights, especially in the early hours of darkness.

## Chart Reading in Flight

*a. Preparation on the Ground.* Proper ground preparation for navigation will save much time and map searching in flight. The course line should be drawn so that a quick glance at the chart will give an indication of the aircraft's location with respect to the desired course. If both departure and destination are on the same chart, the course line is drawn along a straight edge between them. If they do not appear on the same chart, the aeronautical planning chart may be used to determine the principal points over which the flight will proceed. The course line is then drawn on each of the charts being

| CONDITIONS | GOOD CHECKPOINTS | POOR CHECKPOINTS |
|---|---|---|
| **MOUNTAINOUS AREAS** | Prominent peaks, cuts and passes. General profile of ranges, transmission lines, railroads, large bridges over gorges, highways, lookout stations. Tunnel openings and mines. Clearings and grass valleys. Radio Aids. | Smaller peaks and ridges, similar in size and shape. |
| **COASTAL AREAS** | Coastline with unusual features. Lighthouses, marker buoys, towns and cities, structures. Radio Aids. | General rolling coastline with no distinguishing points. |
| **SEASONAL CHANGES** | Unusually shaped wooded areas in winter. Dry river beds if they contrast with surrounding terrain. Dry lakes. | Open country and frozen lakes in winter unless in forested areas. Small lakes and rivers in arid sections of country - in summer - when they may dry up. Lakes (small) in wet seasons in lake areas, where ponds may form by surface waters. |
| **HEAVILY POPULATED AREAS** | Large cities with definite shape. Small cities with some outstanding check point; river, lake, structure, easy to identify from others. Radio aids, prominent structures, speedways, railroad yards, underpasses, rivers and lakes. Race tracks and stadia, grain elevators, etc. | Small cities and towns, close together with no definite shape on chart. Small cities or towns with no outstanding check points to identify them from others. Regular highways and roads, single railroads, transmission lines. |
| **OPEN AREAS FARM COUNTRY** | Any city, town, or village with identifying structures or prominent terrain features adjacent. Prominent paved highways, large railroads, prominent structures, race tracks, fairgrounds, factories, bridges, and underpasses. Lakes, rivers, general contour of terrain; coastlines, mountains, and ridges where they are distinctive. Radio Aids. | Farms, small villages rather close together, and with no distinguishing characteristics. Single railroads, transmission lines and roads through farming country. Small lakes and streams in sections of country where such are prevalent, ordinary hills in rolling terrain. |
| **FORESTED AREAS** | Transmission lines and railroad right-of-ways. Roads and highways, cities, towns and villages, forest lookout towers, farms. Rivers, lakes, marked terrain features, ridges, mountains, clearings, open valleys. Radio Aids. | Trails and small roads without cleared right-of-ways. Extended forest areas with few breaks or outstanding characteristics of terrain. |

*Figure 2-1  Good and poor checkpoints.*

used. The total distance should be measured. If the course line is marked off in increments (for example) of 20 miles, the trouble of unfolding the chart and measuring distance in flight is avoided.

*Note.* Since aircraft cockpit lights are red, a red pencil mark on maps is invisible at night.

*b. Orienting the Chart.* When in flight, orient the chart so that north on the chart is toward true or magnetic north. The course line on the chart will then parallel the intended course of the aircraft, and all surface objects will be directionally oriented with the chart.

## Section II.  LOW-LEVEL NAVIGATION

### Definition

*Low-level navigation* is the technique of directing an aircraft along a desired course at low altitudes and determining position along this course at any time. Low-level altitudes generally are considered to be below 500 feet.

### General

*a.* Navigation at low levels differs considerably from that at higher altitudes. Low-level navigation requires comprehensive flight planning, accurate dead reckoning and/or pilotage, and extensive use of available checkpoints. The aviator and observer must work very rapidly in order to make and interpret in-flight observations with accuracy.

*b.* Reduced visibility, turbulence, and inconsistency of winds close to the terrain compound the problem of low-level navigation. In addition, the fundamentals of navigation, such as writing, measuring, computing, and plotting, are more difficult since the aviator must devote most of his attention to flying the aircraft.

*c.* The aviator-observer team must function in harmony and prearrange the exact duties to be performed by each. Normally, the aviator concentrates on flying the aircraft while the observer performs navigational duties. However, when operating at extremely low levels, as in nap-of-the-earth flight, the aviator must concentrate on clearing obstacles while maintaining the proper heading and airspeed; maximum viewing outside the cockpit is a necessity. The observer, in addition to navigating, should monitor the engine, transmission, and other instruments. He must remain oriented at all times and announce to the aviator the headings to be flown and time to the next checkpoint. When time permits, he describes the next checkpoint so that the aviator may assist in identifying the checkpoint. The aviator-observer team should be sufficiently familiar with the route to avoid study of the navigational chart for long periods of time during flight.

### Pilotage and Dead Reckoning

*a. Maps and Charts.* There are no published charts designed specifically for the type of low-level navigation defined above. As the flight level lowers and visual distances lessen, surface features become more prominent. The terrain perspective changes from that of a plane surface to that of a silhouette, and the interpretation of relief becomes more important. In most cases aeronautical charts are too small in scale for this purpose, lacking detailed features of terrain. Tactical maps (available in scales of 1:25,000, 1:50,000, and 1:100,000) are the most satisfactory maps available for low-level navigation. These are large-scale maps, however, and even the slowest aircraft traverse the distance represented on an entire map sheet in a relatively short time. Therefore, the scale selected for a given operation will depend upon the distance to be traveled during the mission, the degree of detail required, and the speed of the aircraft. Whenever possible, scaled aerial photos should be used when adequate maps and charts are not available for a specific area, or at any other time for supplementary reference. In some instances, a combination of scales may be required. *For example,* a map of 1:100,000 scale or an aeronautical chart may be adequate for the flight from the departure airfield to an intermediate checkpoint; but beyond that point to the landing zone, as in an airmobile operation, more precise navigation may require using a map of 1:50,000 or 1:25,000 scale. To obtain greater detail, recent aerial photos of the target area or landing zone should be studied to detect terrain changes since the time of mapping. A strip map prepared during preflight planning will enhance in-flight pilotage.

*b. Routes.* During the conduct of tactical operations, low-level flight routes are selected which afford cover and concealment from visual and electronic detection to exploit surprise to the fullest.

(1) *Route selection.* Flight planning

should take advantage of directional characteristics of natural or cultural features which will simplify navigation. Rather than planning long straight legs, slight deviations in the route may offer a better selection of navigation checkpoints. Roads, railroads, power transmission lines, canals, stream beds, and natural corridors are excellent for navigation; however, in steep, narrow valleys, enemy cables or other obstacles may be encountered. Turning points should be close to terrain features which can be identified at maximum range. Air control points (ACP), which are points of positive control and coordination between air and ground elements, are excellent choices for turning points. In operations employing formations of appreciable size, turns must be started prior to reaching checkpoints to insure departure on desired course after completion of the turn. The radius of the turn will vary with the type aircraft and size of formation involved. To insure adequate en route positioning, turns should be plotted in the flight planning. When formations are employed, turns may extend ETE's to the next checkpoint.

(2) *Route detours*. Deviations from a selected route may be necessary to avoid air-defended areas, cities and towns, or short sections of a route overly exposed to enemy observation. Route deviation can be effected by using an off-course checkpoint as a link between the usable portions of the route. If a **checkpoint** is not available, a geometrical

pattern should be flown to insure return to the original course at the proper point (fig. 2-8 ).

*c. Altitude.* When selecting a flight altitude, particular attention must be given to terrain elevation, both along the intended flightpath and adjacent to it. The highest terrain feature, as well as abrupt or irregular changes in terrain elevation along the route, must be noted to insure clearance, particularly in the event of unforeseen weather. The selected altitude should also provide concealment from visual and electronic detection.

*d. Checkpoints.* Checkpoints should be selected not more than 5 minutes apart along the route. A 2-minute interval is recommended; however, this will vary with the specific mission, the speed of the aircraft, and the number of available checkpoints. Due to the perspective presented at low levels, checkpoints are divided into two general categories, which may overlap according to the terrain flown.

(1) *Distant*. Distant checkpoints keep the aviator oriented and offer a general course to follow. Both natural and manmade terrain features which stand out above the horizon and have distinctive profiles can be readily identified at long distances and will remain in view a relatively long period of time. *Examples are—*
Prominent mountains and hilltops.
Passes and cuts through high terrain.
Lakes.
Water and communication towers.

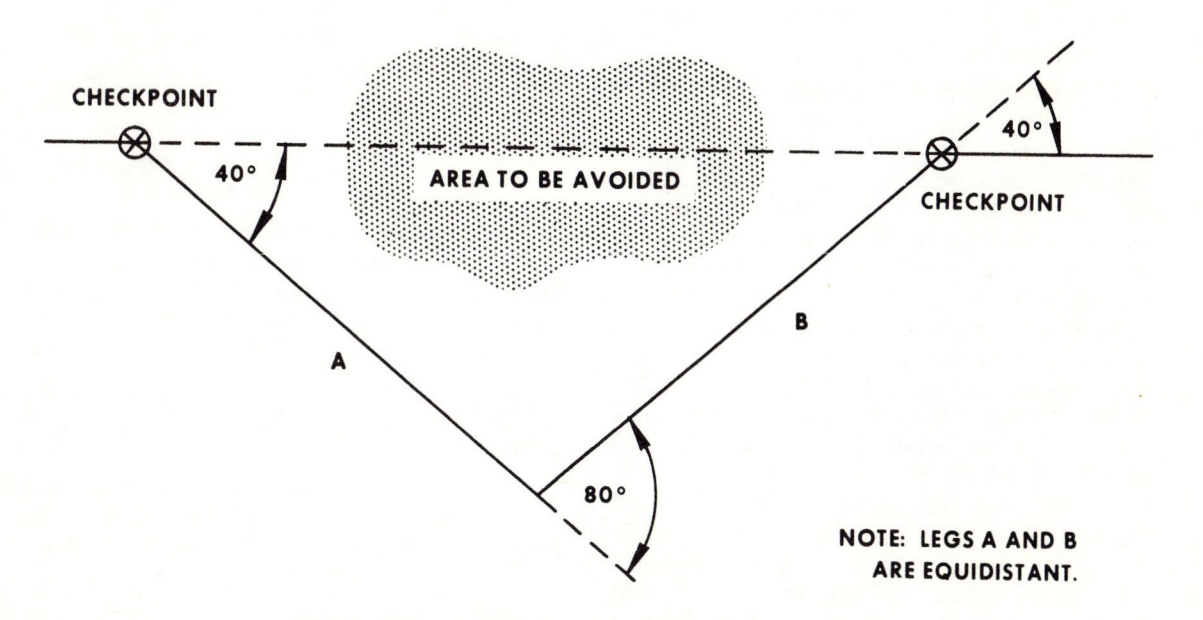

*Figure* 2-8 *Geometrical pattern to return to original course.*

Gaps in the tree line in forested areas.

Powerline and pipeline rights-of-way.

(2) *Near*. Some checkpoints must be selected on or very near the intended flightpath in order to obtain time/distance factors along a given course. These checkpoints are essential in obtaining accurate data on groundspeed and exact location at a given time. Such points include—

Railroad and highway bridges.

Junctions, crossings, and prominent curves and turns in railroads and highways.

Stream junctions and other prominent configurations.

Lakes and ponds.

Churches and schools.

Various patterns formed by the combination of timbered and adjacent cleared lands.

Roads, railroads, and streams usually can be detected in vegetation when approached at a shallow angle; however, approaches from the perpendicular may make them difficult to see due to the masking effect of vegetation.

## Flight Plan Graph

Adherence to preplanned routes, and accuracy in estimated times of arrival for rendezvous, turning points, and initial points are of paramount importance to successful low-level navigation. Since one of the principal difficulties of low-level navigation is the physical manipulation of the ordinary navigational tools, a flight plan graph (FPG) will reduce the demand upon time and effort without sacrificing accuracy and reliability.

*a. Description.* The flight plan graph (fig. 2-9) is a device for monitoring flight progress of a mission. It precludes the necessity for carrying an excessive number of maps, or drawing unnecessary lines and notes on necessary maps. It is prepared during the mission planning phase and requres little attention during the flight. Basically, the flight plan graph consists of a line representing the flight plan time from departure to destination or turning point, and roughly parallels the true course of the flight. Predicted times to various points along the true course, together with departure and destination times, are plotted on this time scale. Thus, the flight plan graph represents a visual time line comparable to the predicted track. Using the time line, the predicted estimated time of arrival to any point on the predicted track can be determined at any time without computation. Comparison with a fix, checkpoint,

or obstacle gives the aviator or observer an indication of the time he is ahead or behind his flight plan. In addition, this time difference can be applied to destination and/or intermediate checkpoints to maintain accurate running estimated times and arrival for these positions.

*b. Preparation.* The flight plan graph can be prepared as follows:

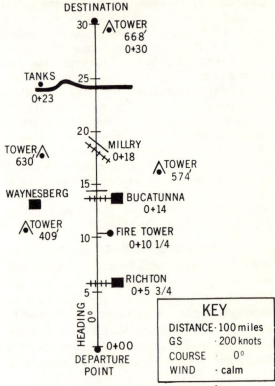

*Figure 2-9 Flight plan graph.*

(1) Draw a line to represent the true course. This line should be drawn parallel to the long axis of the paper.

(2) On the true course line, using a suitable scale, indicate departure and destination points. All time intervals are evenly spaced since groundspeed is held constant.

(3) Using conventional signs and symbols, indicate, at properly scaled times, the checkpoints, fixes, and terrain obstacles along and adjacent to the route of flight.

(4) With lines perpendicular to the true course and flight plan graph lines, connect checkpoints, fixed, and obstacles to the flight plan graph line.

(5) Label obstructions, checkpoints, and fixes, as deemed necessary, along with their elevations if known.

(6) Using a computer, and based on pre-calculated groundspeed, compute the time to each checkpoint, fix, or obstacle, recording

this time outside the flight plan graph line where it intersects with the line drawn from the checkpoint, fix, or obstacle.

(7) If the route to be flown has several legs, this same procedure can be carried out for each leg.

## Radio Navigation

The same principles of radio navigation found elsewhere in this manual apply to low-level navigation. The capabilities of navigational aids such as VOR, LF/MF radio beacons, and FM homers are greatly reduced at low altitudes, particularly in mountainous terrain; but these aids should be used when possible. Use of radio navigational aids for some portion of a low-level flight should be considered during preflight planning. As an example, ADF or FM homers might be used from a departure airfield or a loading zone to an intermediate checkpoint on outbound flights, and again for return flight to the base of operations. During airmobile operations, radio aids may also be employed by pathfinders as navigation aids from release points or other predetermined points to the objective area or landing zone; however, enemy capabilities of duplicating radio signals and establishing false stations must be considered.

## Weather

During a tactical situation in which the security and success of the mission demand low-level flight, weather considerations increase in importance.

a. *Visibility*. Visibility is the primary weather factor in low-level navigation. During periods of restricted visibility, distant checkpoints essential to orientation and course selection become vague or obliterated, often limiting observation to the ground directly beneath the aircraft. During periods of reduced visibility, the aviator must rely heavily on dead reckoning.

b. *Winds*. In the conduct of low-level flight, the same attention must be given to the effects of wind as in navigation at higher altitudes. Winds at low levels, as well as those at higher levels, are subject to unexpected changes. Should in-flight observations indicate changed wind conditions, corrections should be calculated as quickly as possible.

(1) One advantage of low-level flight is the visual indication of surface wind from smoke, dust, etc. Because the aircraft is operating close to the surface, it will usually be in the same wind conditions. This allow direct reading of the wind direction rather than by

computation as at higher altitudes.

(2) The combination of wind and certain types of terrain may produce turbulence intense enough to be a hazard to light aircraft. In rugged terrain, average or spot wind measurements frequently are nonrepresentative and should be used with caution. Under certain wind conditions, routes through such areas as deep valleys, gorges, and mountain passes may have to be avoided because of severe turbulence.

c. *Temperature*. Temperature (as well as humidity or dewpoint) and its influence on density altitude and icing conditions must alway be considered in low-level navigation. The margin of safety and room for maneuvering to overcome the hazards of high density altitude or icing conditions decrease as the flight level lowers. Density altitude is more important when flying at low altitudes, particularly in areas where turbulence and downdrafts may be encountered. Low-level flights on hot days may also increase pilot fatigue due to increased cockpit temperatures. Areas with predicted or suspected icing conditions should be avoided.

## Low-Level Navigation at Night

a. The feasibility of attempting low-level flight at night depends on the geographical area, available natural light, and the weather. The most important factor of low-level navigation at night is the increased danger of collision with terrain or manmade objects. A chart should be maintained listing the heights of all known and suspected obstructions. In areas where low-level navigation at night may be necessary, as in a combat zone, there would be a serious lack of obstruction lights. Many cultural features do not appear the same at night as during the day; however, river, lakes, coast lines, and most manmade objects with distinctive outlines are good checkpoints at night.

b. When flying in mountainous terrain, the aviator must realize that the actual horizon is near the *base* of the mountains. The summit of peaks used a a horizon would place the aircraft in an attitude of constant climb. To prevent vertigo, distraction of attention, and loss of night vision, cockpit lights should be used only when necessary. When possible, a copilot or observer should make any in-flight computation necessary and assist in monitoring flight and engine instruments. Aviators should be proficient in daytime low-level operations before attempting extensive low-level navigation at night. If a reasonable degree of safety is afforded, low-level flight can be conducted at night using the same navigational procedures as in day light operations.

# 3 DEAD RECKONING PROCEDURES

The pilot builds his knowledge upon certain fundamentals as is true in learning any profession. In navigation these basic ideas or methods are called dead reckoning (DR) procedures. Correct use of the plotter, dividers, computer, and charts to solve the three basic problems of navigation (position of the aircraft, direction to destination, and time of arrival) all make up dead reckoning procedures.

It is possible, using only basic instruments such as the pressure altimeter, compass, airspeed indicator, and driftmeter, to navigate directly to any place in the world. But this would take conditions approaching the ideal such as good weather and perfect instruments. Therefore aids to dead reckoning or DR such as celestial, radar, loran, map reading, etc., have been developed to correct inadequacies of basic navigation. These aids are primarily known as fixing aids, and provide information from which the fundamental skills of DR can be used to compute winds, ground speeds, and alterations to the path of the aircraft in order to complete an assigned mission.

Thus, proficiency in dead reckoning procedures is indispensable if full use is to be made of the aids to navigation.

## PLOTTING

Proper chart work must be learned before a pilot can fly an aircraft from one point to another. The structure and properties of maps and charts have been covered. Now, the next step of plotting positions and directions or lines on a chart will be taken. Chart work should be an accurate and graphic picture of the progress of the aircraft from departure to destination and with the log, should serve as a complete record of the flight. Thus it also follows that the pilot must be familiar with and use accepted standard symbols and labels on his chart.

### Definition of Terms

Several terms have been mentioned in earlier portions of this handbook. Precise definitions of these terms must now be understood before the mechanics of chart work are learned.

True Course (TC) is the intended horizontal direction of travel over the surface of the earth, expressed as an angle measured clockwise from true north (000°) through 360 degrees.

Course line is the horizontal component of the intended path of the aircraft comprising both direction and magnitude or distance.

Track (Tr) is the horizontal component of the actual path of the aircraft over the surface of the earth. Track may, but very seldom does, coincide with the true course or intended path of the aircraft. The difference between the two is caused by an inability to predict perfectly all inflight conditions.

True Heading (TH) is the horizontal direction in which an aircraft is pointed.

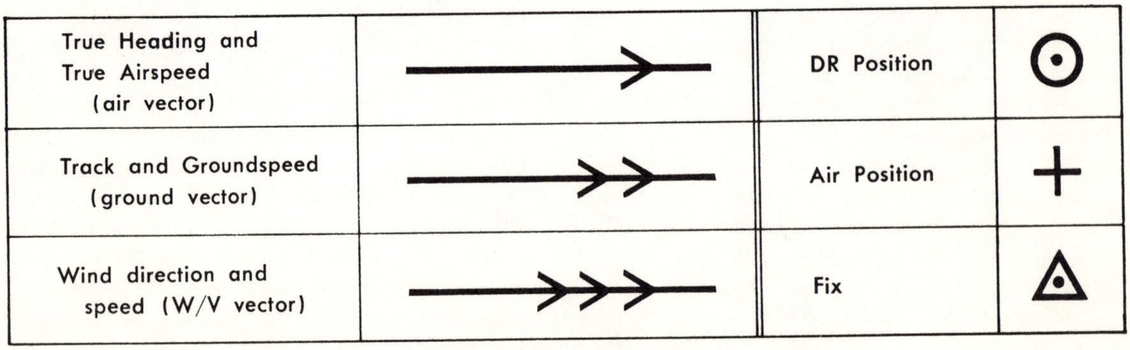

| | | | |
|---|---|---|---|
| True Heading and True Airspeed (air vector) | ————————→ | DR Position | ⊙ |
| Track and Groundspeed (ground vector) | ———————≫ | Air Position | + |
| Wind direction and speed (W/V vector) | ——————⋙ | Fix | △ |

*Fig. 3-1 Standard plotting symbols.*

More precisely, it is the angle measured clockwise from true north through 360 degrees to the longitudinal axis of the aircraft. The difference between track and true heading is caused by wind and will be explained fully under wind triangles.

Groundspeed (GS) is the rate of motion of the aircraft relative to the earth's surface or, in simpler terms, it is the speed of the aircraft over the ground. It may be expressed in nautical miles, statute miles, or kilometers per hour, but, as a pilot, you will be expected to use nautical miles per hour (knots).

True Airspeed (TAS) is the rate of motion of an aircraft relative to the air mass surrounding it. Since the air mass is usually in motion in relation to the ground, airspeed and groundspeed seldom are the same.

Dead Reckoning Position (DR position) is a point in relation to the earth established by keeping an accurate account of time, groundspeed, and track since the last known position. It may also be defined as the position obtained by applying wind effect to the true heading and true airspeed of the aircraft.

A Fix is an accurate position determined by one of the aids to DR. One method of fixing (map reading) is discussed in this chapter.

Air Position (AP) is the location of the aircraft in relation to the air mass surrounding it. True heading and true airspeed are the components of the vector used to establish an air position.

Most Probable Position (MPP) is a position determined with partial reference to a DR position and partial reference to a fixing aid.

## PLOTTING EQUIPMENT

PENCIL AND ERASER. Probably the most elementary but indispensable articles of plotting equipment are pencils and erasers. The pencil should be fairly soft and well sharpened. A hard pencil is undesirable, not only because it makes lines which are light and difficult to see, but also because it makes lines which are difficult to erase. An experienced pilot keeps his chart clean by erasing all unnecessary lines. It is good practice to carry several well-sharpened wooden pencils rather than a mechanical pencil. Mechanical pencils generally make broader, less precise lines.

Use a soft eraser which will not smudge or damage the chart. Hard, gritty erasers tend to wear away printed information as well as the paper itself. Not all erasers will remove all pencil lines, so find a pencil-eraser combination that will work satisfactorily. Nothing reflects a pilot's ability more than keeping neat, clean, accurate charts and logs. An untidy chart, smudged and worn through in places, often causes one to conclude that the navigation performed was careless.

DIVIDERS. A pair of dividers is an instrument used for measuring distances on a chart by separating the points of the dividers to the desired distance on the proper scale (usually the latitude scale) and transferring this distance to the working area of the chart. In this way, lines of a desired length can be marked off. By reversing the process, unknown distances on the chart can be spanned and compared with the scale.

It is desirable to manipulate the dividers with one hand, leaving the other free to use the plotter, pencil, or chart as necessary. Most navigation dividers have a tension screw which can be adjusted to prevent the dividers from becoming either too stiff or too loose for convenient use. The points of the dividers should be adjusted to approximately equal length. A small screw driver, required for

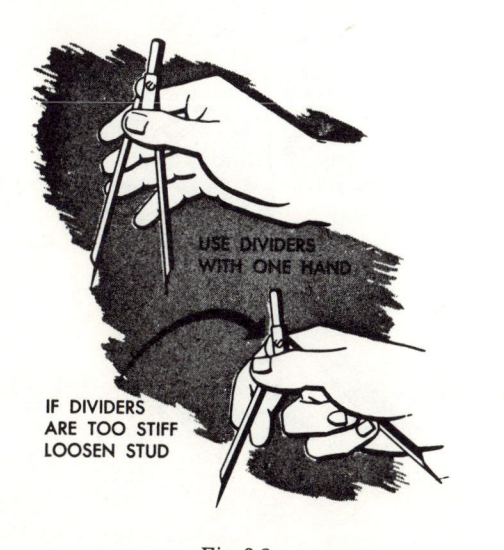

USE DIVIDERS WITH ONE HAND

IF DIVIDERS ARE TOO STIFF LOOSEN STUD

*Fig. 3-2*

these adjustments, should be a part of the pilot's equipment.

While measurement is being made, the chart should be flat and smooth between the points of the dividers, as a wrinkle may cause an error of several miles. Points on the chart may be marked by applying slight pressure on the dividers so that they prick the chart. Too much pressure results in large holes in the chart and tends to spread the points of the dividers farther apart, reducing measurement accuracy. Precision navigation requires precision measurement.

PLOTTERS. A plotter is an instrument designed primarily to aid in drawing and measuring lines in desired directions. Plotters vary from complicated drafting machines and complete navigator's drafting sets to a simple plotter combining a kind of protractor and straight edge. The plotter in most common use in private flying is the Weems plotter.

The Weems is a semicircular protractor with a straight edge attached to it. A small hole at the base of the protractor portion indicates the center of the arc of the angular

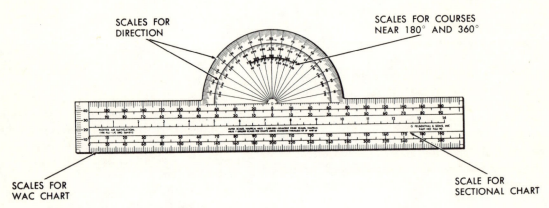

*Fig. 3-3*

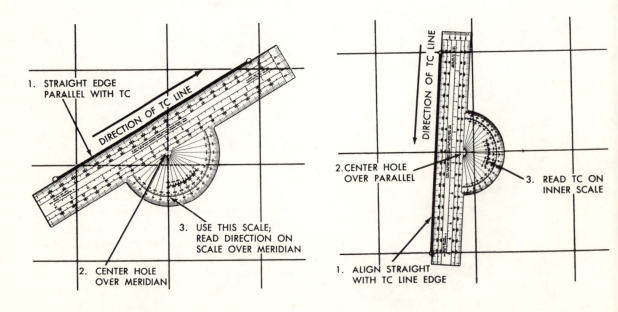

*To measure true course*    – *Fig. 3-4* –    *Measure true course near 180° or 360°*

scale. Two complete scales cover the outer edge of the protractor and are graduated in degrees. A third, inner scale measures the angle from the vertical.

The outer scale increases from 000 degrees to 180 degrees counter-clockwise, and the inner scape increases from 180 to 360 degrees counter-clockwise. An angle is measured by placing the vertical line (that line containing the small hole) on a meridian, and aligning the base of the plotter with the line to be measured as shown in the illustrations measuring true course.

The angle measured is the angle between the meridian and the straight line. The outer scale is used to read all angles between north through east to south, and the inner scale is used to read all angles between south through west to north as in the related illustration.

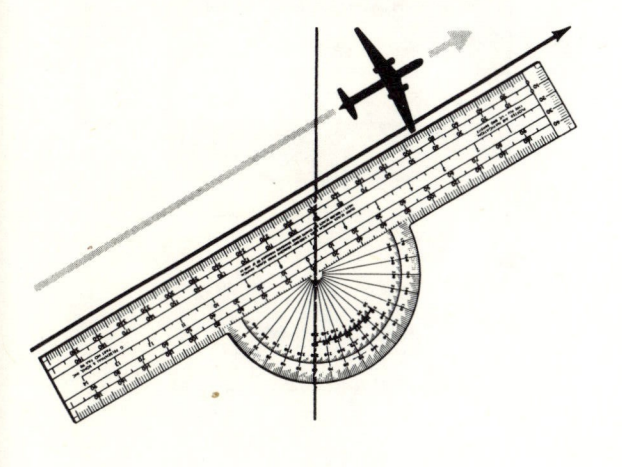 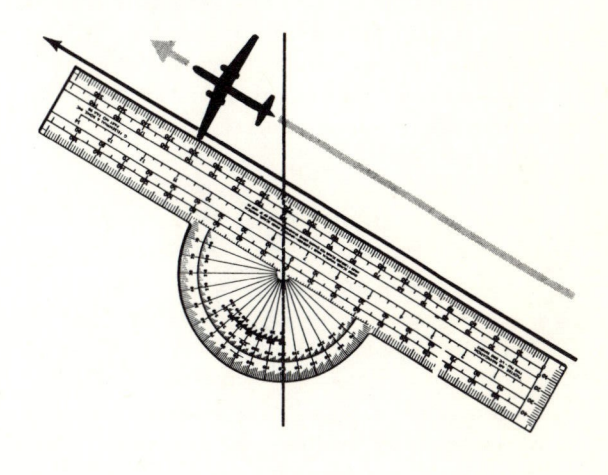

*The inner scale reads from 180° to 360°* — Fig. 3-5 — *The outer scale reads from 0° to 180°.*

# 4 INSTRUMENTS USED FOR DEAD RECKONING NAVIGATION

Instruments used by the aviator for dead reckoning navigation are the magnetic compass, heading indicator, outside air temperature gage, airspeed indicator, altimeter, and clock. From these, information can be read concerning direction, airspeed, altitude, and time, each of which must be correctly interpreted for successful navigation. Information on the instruments discussed in this handbook is general in nature and this knowledge is presented as an aid to navigation.

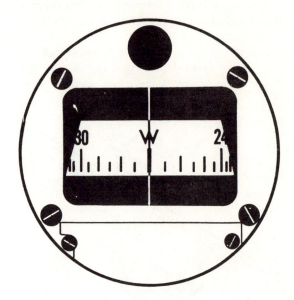

Fig. 4-1 Magnetic compass.

## MAGNETIC COMPASS

The magnetic compass is the only direction seeking instrument in an aircraft. It has a compass card marked at 5° increments and numbered 30° apart. A fixed line known as the lubberline, is located on a glass aperture of the compass case and is a reference line for reading the compass. The reading on the compass card, under the lubberline, indicates the compass heading of the aircraft.

Navigation charts are printed in geographical true directions. Since the magnetic compass derives its directional qualities by the needle aligning itself with the direction of the earth's magnetic field, it points to magnetic north, not true north. The angular difference between the true and magnetic meridian is called magnetic variation or simply variation. It is necessary to correct for this error in order to maintain true direction as plotted on a chart. Magnetic variation affects the compass on any heading and varies with latitude and longitude by the amount of existing magnetic variation at any particular point. Magnetic variation is indicated on navigation charts by means of agonic and isogonic lines:

*Agonic Line.* An agonic line is a line on a chart connecting all points where no magnetic variation exists and is labeled 0°.
*Isogonic Lines.* An isogonic line is a line on a chart connecting all points of equal magnetic variation. They are drawn one or more degrees of variation apart according to the size of the chart. Each line is labeled

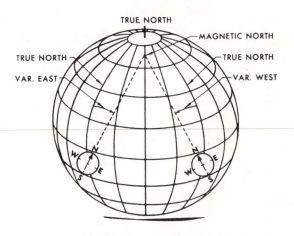

Fig. 4-2 Magnetic variation.

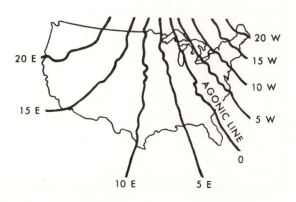

Fig. 4-3 Isogonic chart of the United States.

according to the number of degrees of variation and shows the east or west deflection of the compass needle. If the magnetic pole is to the east of the true north pole, variation is east; if to the west, it is west. A true course corrected for variation becomes a magnetic course.

A magnetic compass is affected by magnetic fields other than those of the earth. Any piece of ferrous material or electrical equipment close to the compass tends to deflect the needle away from magnetic north.

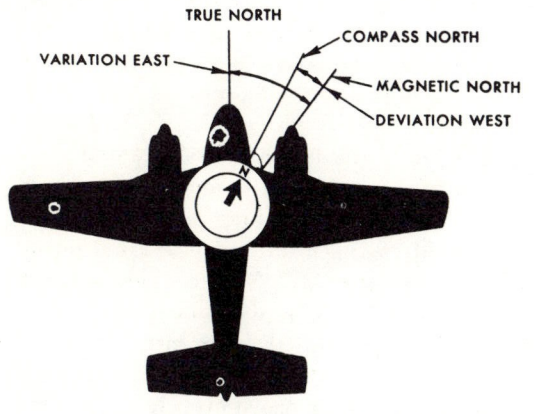

*Figure 4-4A   Variation and Deviation*

| AIRCRAFT _15088_ | |
|---|---|
| SWUNG: _5 JUNE 58_ | |
| **TO FLY** | **STEER** |
| N | *001* |
| 045 | *047* |
| 090 | *093* |
| 135 | *136* |
| 180 | *181* |
| 225 | *224* |
| 270 | *268* |
| 315 | *315* |

*Figure 4-4B   Deviation Card*

This angular deflection of the compass needle from magnetic north is called magnetic deviation. If the needle is deflected to the east, deviation is east; if to the west, deviation is west. A magnetic course corrected for deviation becomes a compass course:

*Compass North.* The direction in which the compass needle points is called compass north. Compass directions may be expressed relative to compass north just as true or magnetic directions are expressed relative to true or magnetic north.
*Deviation Card.* A deviation card records the errors in the compass indications. The card is mounted next to the compass and includes the aircraft number, date on which the compass was swung, and a compass heading to be flown for each magnetic heading in increments of 45°.

> NOTE. Deviation may change with each change of heading of the aircraft, whereas, variation changes with change of locality.

APPLYING COMPASS CORRECTIONS

To find what the compass should read to follow a given course, corrections for drift, variation, and deviation are applied. When drift correction is applied to a true course (TC ± DC = TH), it becomes a true heading. A good method for recording application of variation and deviation is as follows:

(1) Write the equation

TH ± V = MH ± D = CH

in which TH is the true heading; V is the variation; MH the magnetic heading; D the deviation; and CH the compass heading.

(2) Below each factor, place the known information.

TH ± V = MH ± D = CH
168°   12°E         5°W

(3) When making calculations from a true heading to a compass heading, easterly error is substracted and westerly error is added. Completing the above problem, substracting the 12°E from the TH gives a magnetic heading (MH) of 156° adding the 5°W to the MH gives a compass heading (CH) of 161°. Place these headings in their proper places in the equation.

To find a true heading when the compass heading is known, the same equation is written as in the above problem. Placing the known information in the proper places, it would appear as follows:

$$TH \pm V = MH \pm D = CH$$
$$12°E \qquad 5°W \qquad 161°$$

(1) When changing from a compass heading to a true heading, easterly error is added and westerly error is substracted. It is the reverse of changing from TH to CH.
(2) Substract the 5°W from the CH (161°). Place this figure (156°) below the MH.
(3) Add the 12°E to the MH (156°) to obtain the TH (160°). Place this figure below the TH.

Errors in Flight. Unfortunately, variation and deviation are not the only errors of a magnetic compass. Additional errors are introduced by the motion of the aircraft itself. These errors may be classified as:
1. Northerly turning error.
2. Speed error.
3. Heeling error.
4. Swirl error.
5. Yaw error.

Northerly turning error is maximum during a turn from a north or a south heading, and is caused by the action of the vertical component of the earth's field upon the compass magnets. When the aircraft is in straight and level flight, the compass card is balanced on its pivot in approximately a horizontal position. In this position, the vertical component of the earth's field is not effective as a directive force. However, if the aircraft is on a north heading (in the Northern Hemisphere) and turns to the right in a properly banked attitude, the compass card also banks as a result of the centrifugal force acting upon it. The vertical component of the earth's field now causes the compass card to rotate toward the right (east). This temporary deflection is called northerly turning error. Had the turn been to the left, the error would have been westerly. If the turn had been made from a heading of south, the direction of the above errors would have been reversed. It is apparent that precise turns are difficult by reference to such a compass. This error can be virtually eliminated by reference to a gyro stabilized system.

If an aircraft flying an east or west heading

changes speed, the compass card tends to tilt forward or backward. The vertical component then has an effect as in a turn, causing an error which is like northerly turning error, but is known as speed error. If an aircraft is flying straight but in a banked attitude, a heeling error may be introduced due to the changed position, with respect to the compass magnets, of the small compensating magnets in the case and other magnetic fields within the aircraft. The heading on which heeling error is maximum depends upon the particular situation. Swirl error may occur when the liquid within the compass case is set in motion during a turn. Finally, yaw error may occur if a multi-engined aircraft is yawing about its vertical axis due to difference in power settings of the engines.

Although a basic magnetic compass or standby compass, has numerous shortcomings, it is simple and reliable. The compass is very useful to the pilot and is carried on all aircraft as an auxiliary. Because most modern compass systems are dependent upon the electrical system of the aircraft, a loss of power means a loss of the compass system. For this reason, a constant check on the standby compass will provide a good check on the electrial compass systems of the aircraft.

## THE GYROCOMPASS

The gyrocompass system was developed to offset some of the errors in the magnetic compass system and to attempt to provide a more reliable and more universal directional device. Nearly all of today's aircraft are equipped with some sort of gyro stabilized steering device. Many different compass systems have been developed using combined gyro and magnetic references. This section deals with only the most elementary directional gyro.

*The gyro.* Any spinning body exhibits gyroscopic properties. A wheel designed and mounted to utilize these properties is called a gyroscope or gyro. Basically, a gyro is a rapidly rotating mass which is free to move about one or both axes perpendicular to the axis of rotation and to each other. The three axes of a gyro, namely, spin axis, drift axis, and topple axis are defined as follows:
1. In a directional gyro, the spin axis or axis of rotation is mounted horizontally as shown.

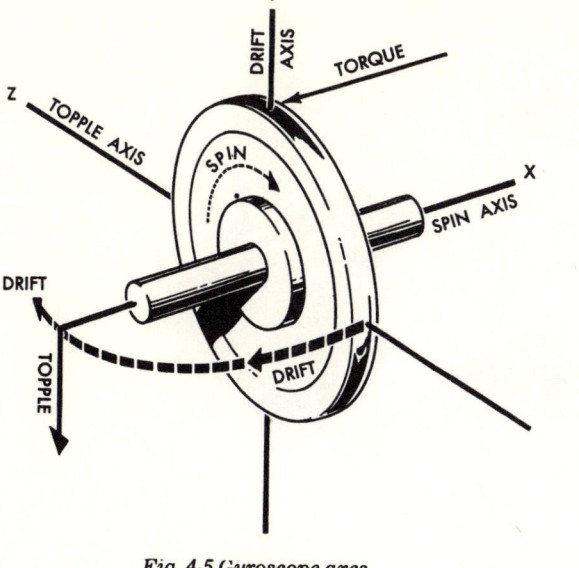

*Fig. 4-5 Gyroscope axes.*

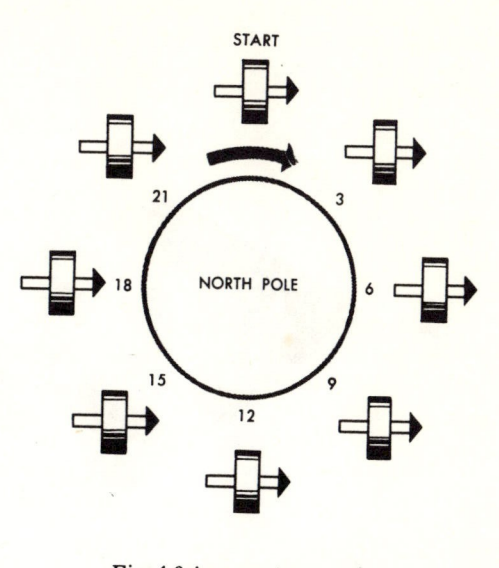

*Fig. 4-6 Apparent precession.*

2.  The topple axis is that axis in the horizontal plane that is 90 degrees from the spin axis.
3.  The drift axis is that axis 90 degrees vertically from the spin axis.

Gyroscopic drift is the horizontal rotation of the spin axis about the drift axis. Topple is the vertical rotation of the spin axis about the topple axis. These two component drifts result in motion of the gyro called precession.

A freely spinning gyro tends to maintain its spin axis in a constant direction in space, a property known as rigidity in space of gyroscopic inertia. Thus, if the spin axis of a gyro were pointed toward a star, it would rotate at the same rate as the star and keep pointing at the star. Actually, the gyro does not move, but the earth moving beneath it gives it an apparent motion. This apparent motion shown in the captioned illustration is called apparent precession. The magnitude of apparent precession is dependent upon latitude. The horizontal component, drift, is equal to 15 degrees per hour times the sine of the latitude, and the vertical component, topple, is equal to 15 degrees per hour times the cosine of the latitude.

These computations assume the gyro is stationary with respect to the earth. However, if the gyro is to be used in a high speed aircraft, it is readily apparent that its speed with respect to a point in space may be more or less than the speed of rotation of the earth. If the aircraft in which the gyro is mounted is moving in the same direction as the earth, the speed of the gyro with respect to space will be greater than the earth's speed. The opposite is true if the aircraft is flying in a direction opposite to that of the earth's rotation. This difference in the magnitude of apparent precession caused by transporting the gyro over the earth is called transport precession.

A gyro may precess because of factors other than the earth's rotation. When this occurs, the precession is labeled real precession. When a force is applied to the plane of rotation of a gyro, the plane tends to rotate, not in the direction of the applied force, but 90 degrees around the spin axis from it. This torquing action, shown in the illustration, may be used to control the gyro by bringing about a desired reorientation of the spin axis, and most directional gyros are equipped with some sort of device to introduce this force.

However, friction within the bearings of a gyro may have the same effect and cause a certain amount of unwanted precession. Great care is taken in the manufacture and maintenace of gyroscopes to eliminate this factor as much as possible, but, as yet, it has not been possible to eliminate it entirely. Precession caused by the mechanical limitations of the gyro is called real or induced precession. The combined effect of apparent precession, transport precession, and real precession produce the total precession of the gyro.

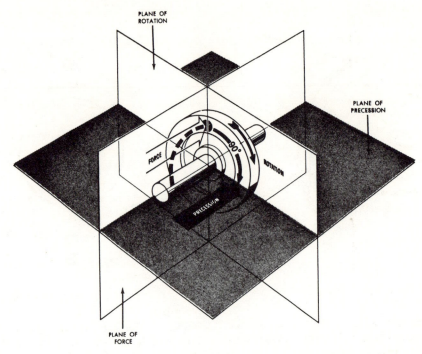

*Fig. 4-7A Precession of a gyroscope resulting from an applied deflective force.*

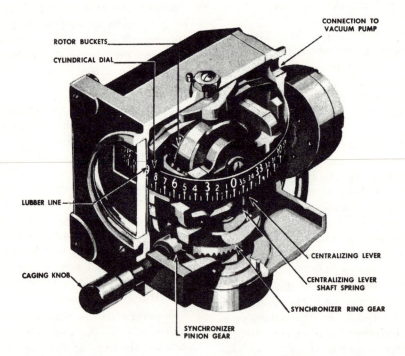

*Fig. 4-7B Cutaway view of a directional gyro.*

The properties of the gyro that most concern the pilot are rigidity and precession. By understanding these two properties, the navigator is well equipped to use the gyro as a reliable steering guide.

The directional gyro. The discussion thus far has been of a universally mounted gyro, free to turn in the horizontal or vertical, or any component of these two. This type of gyro is seldom, if ever, used as a directional gyro. When the gyro is used as a steering instrument, it is restricted so that the spin axis remains parallel to the surface of the earth. Thus, the spin axis is free to turn only in the horizontal plane (assuming the aircraft normally flies in a near level attitude), and only the horizontal component (drift) will affect a steering gyro. In the terminology of gyro steering, precession always means the horizontal component of precession.

The operating of the instrument depends upon the principle of rigidity in space of the gyroscope. Fixed to the plane of the spin axis is a circular compass card, similar to that of the magnetic compass as shown in the cutaway view. Since the spin axis remains rigid in space, the points on the card hold the same position in space relative to the horizontal plane. The case, to which the lubber line is attached, simply revolves about the card.

It is important at this point to understand that the numbers on the compass card have no meaning within themselves, as on the magnetic compass. The fact that the gyro may indicate 100 degrees under the lubber line is not an indication that the instrument is actually oriented to magnetic north, or any other known point. In order to steer by the gyro, it must first be set to a known direction or point. Usually, this is magnetic north or geographic north, although it can be at any known point. If, for example, magnetic north is set as the reference, all headings on the gyro read relative to the position of the magnetic poles.

The actual setting of the initial reference heading is done by utilizing the principle discussed earlier of torque application to the spinning gyro. By artificially introducing precession, the gyro can be set to whatever heading is desired and can be reset at any time using the same technique.

Gyrocompass errors. The major errors affecting the gyro and its use as a steering instrument is precession. Apparent precession will cause an apparent change of heading equal to 15 degrees per hour times the sine of the latitude. Real precession, due to defects in the gyro, may occur at any rate. This type of precession has been greatly reduced by the high precision of modern manufacturing methods. Apparent precession is a known value depending upon location and can be compensated. In some of the more complex gyro systems, apparent precession is compensated by setting in a constant correction equal to and in the opposite direction to the precession caused by the earth's rotation.

### ALTITUDE AND THE ALTIMETER

Altitude is the height of an aircraft in the air. Knowledge of the aircraft's altitude is important for several reasons. In order to remain a safe distance above dangerous mountain peaks, the altitude of the aircraft and the elevation of the surrounding terrain must be known at all times. This is especially important when visibility is poor and the terrain cannot be seen. Also, it is often desirable to fly a certain altitude to take advantage of favorable winds and weather conditions.

Altitude may be defined as vertical distance above some point or plane used as a reference. It follows, then, that there may be as many kinds of altitude as there are reference planes from which to measure. The navigator is concerned, generally, with the three kinds of altitude illustrated in Types of Altitude.

True altitude (TA) is height above mean sea level; absolute altitude (AA) is height above the terrain directly below the aircraft; and altitude above the standard datum plane is called pressure altitude (PA).

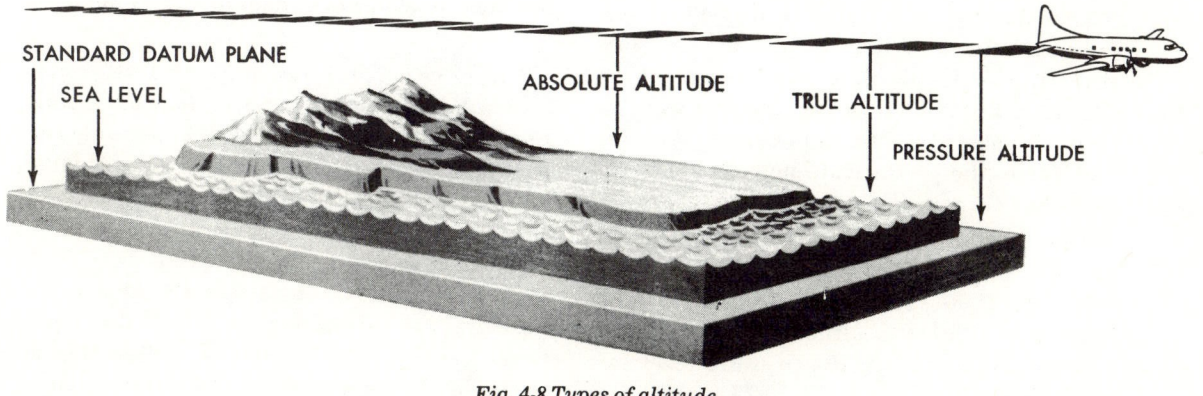

STANDARD DATUM PLANE
SEA LEVEL
ABSOLUTE ALTITUDE
TRUE ALTITUDE
PRESSURE ALTITUDE

*Fig. 4-8 Types of altitude.*

The standard datum plane is a theoretical plane where the atmosphere pressure is 29.92 inches of mercury (Hg) and the temperature is + 15°C. The standard datum plane is the zero elevation level of an imaginary atmosphere known as the standard atmosphere. In the standard atmosphere, pressure is 29.92' Hg at 0 feet and decreases upward at .the standard pressure lapse rate. The temperature is + 15°C at 0 feet and decreases upward at the standard temperature lapse rate. Both the pressure and temperature lapse rates are given in the table.

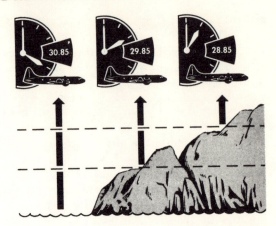

Fig. 4-10 Altimeter pressure setting affects altitude measured.

| Feet | Pressure | Temperature |
|---|---|---|
| 16,000 | 16.21″ Hg | −17° C |
| 15,000 | 16.88 | −15 |
| 14,000 | 17.57 | −13 |
| 13,000 | 18.29 | −11 |
| 12,000 | 19.03 | − 9 |
| 11,000 | 19.79 | − 7 |
| 10,000 | 20.58 | − 5 |
| 9,000 | 21.38 | − 3 |
| 8,000 | 22.22 | − 1 |
| 7,000 | 23.09 | 1 |
| 6,000 | 23.98 | 3 |
| 5,000 | 24.89 | 5 |
| 4,000 | 25.84 | 7 |
| 3,000 | 26.81 | 9 |
| 2,000 | 27.82 | 11 |
| 1,000 | 28.86 | 13 |
| Sea Level | 29.92 | 15 |

Fig. 4-9 Lapse rate table.

The standard atmosphere is theoretical. It was derived by averaging the readings taken over a period of many years. The list of altitudes and their corresponding values of temperature and pressure given in the table were determined by these averages.

There are two main types of altimeters: the pressure altimeter which is installed in every aircraft, and the absolute altimeter or radio altimeter. Few, if any private planes use the radio altimeter, so the explanation will be confined to the pressure altimeter.

The pressure altimeter is designed to measure altitude above any pressure level from 28.00″ to 31.00″ Hg. Notice in the illustration that the altimeter is measuring altitude above the specific pressure level for which it has been set.

True altitude is comparatively easy to find using an altimeter, the free air temperature gage, and the DR computer. It is important to understand some facts about the atmosphere to understand how and why the computations are made.

### ATMOSPHERE WEIGHT AND DENSITY

The atmosphere is an ocean of air surrounding the earth. The air has weight and density, just as water has, and this weight produces pressure. The pressure is heaviest at the bottom of the "ocean" of air; that is, at the surface of the earth, just as the pressure of water is greatest at the bottom of the ocean.

The pressure altimeter measures atmospheric pressure at flight level. Consequently, to understand the altimeter, you must know something of the relationship of density and pressure to altitude.

Density is weight per unit volume. One cubic foot of water weighs about 62.4 pounds. Thus, the density of water is about 62.4 pounds per cubic foot. The density of air at sea level is about 0.08 pounds per cubic foot.

Pressure is force per unit area. If water stands 1 foot deep in a tank, there is 1 cubic foot of 62.4 pounds of water standing on each square foot of the bottom of the tank. The pressure on the bottom is 62.4 pounds per square foot. If the water is twice as deep, twice the weight of water rests on each square foot. In fact, whatever the depth, the pressure on the bottom is equal to depth times density. This is true regardless of the shape of the container or the shape or area of the bottom. It is also true of any submerged surface other than the bottom. In a tank of water, pressure decreases evenly from the bottom of the tank upward.

The atmosphere can be compared to a tank of water. Because of its weight, air exerts pressure on the surface of the earth which is the bottom of the atmosphere. Atmospheric pressure at sea level is about 14.7 pounds per square inch.

Atmospheric pressure decreases upward from the earth as pressure decreases upward from the bottom of a tank of water.

Air differs from water in one important respect. Water is practically incompressible but air can be compressed into a smaller volume. When air is compressed, a given volume contains more air and hence weighs more. As air is compressed its density increases.

A cubic foot of air at the surface of the earth is compressed by the weight of a column of air·extending directly above it to the upper limit of the atmosphere. A cubic foot of air above the surface of the earth is compressed by the lesser weight of a shorter column of air. Air at the surface is under the greatest pressure because it has the greatest weight on it. Therefore, the density of the air is

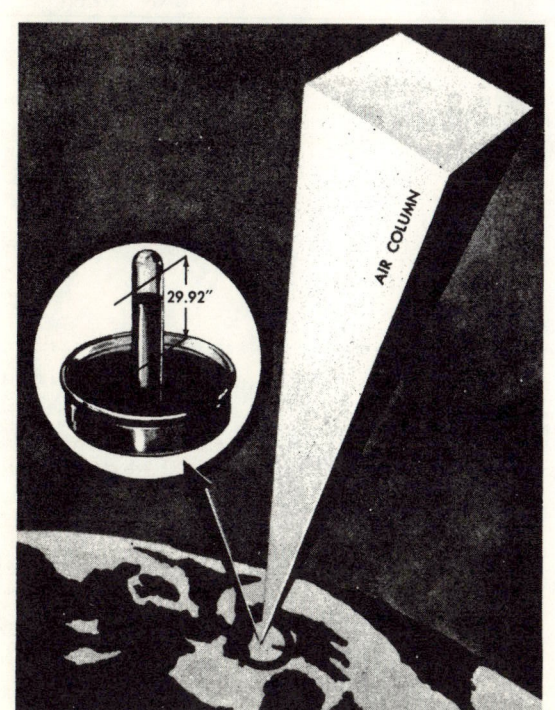

*Fig. 4-11 Atmospheric pressure supports the column of mercury.*

greatest at the surface of the earth and decreases upward. Both pressure and density decrease with altitude.

In a tank of water, a cubic foot of water at the bottom weighs essentially the same as a cubic foot at the top. Though pressure decreases with distance from the bottom, density is uniform throughout. In the atmosphere, dense air contributes more pressure than thin air. Thus, a layer of air near the surface of the earth contributes more pressure than an equally thick layer higher in the atmosphere. Therefore, pressure decreases most rapidly just above the surface of the earth. Actually, half the weight of the atmosphere is below an altitude of approximately 18,000 feet. This means that the pressure at 18,000 feet is approximately half that at sea level.

Pressure decreases with altitude. If pressure were constant at each level, an instrument which measures pressure could be calibrated in feet of altitude to serve as an accurate altimeter. Actually, the pressure at any level varies continuously.

A pressure altimeter measures pressure and can be read in feet of altitude. Because of the variation of pressure at different altitudes, the pressure altimeter may be considerably in error; but by correcting its reading for atmospheric conditions, a fairly accurate true altitude can be obtained.

MEASUREMENT OF ATMOSPHERIC PRESSURE

Atmospheric pressure is measured by means of a barometer. There are two types of barometers: the mercurial barometer and the aneroid barometer. Only the aneroid barometer is used as an altimeter; but an understanding of the mercurial barometer will aid in understanding one of the scales of the altimeter.

A mercurial barometer is a tube of mercury sealed at the top and inverted in an open dish of mercury. Atmospheric pressure on the surface of the mercury in the dish supports a column of mercury in the tube. The weight of the column of mercury equals the weight of a column of air of equal diameter, extending from the surface of the mercury in the dish to the upper limits of the atmosphere (see related illustration). Thus, the column of mercury always exerts a pressure equal to atmospheric pressure. An increase in atmospheric pressure forces the mercury higher in the tube. Because the higher column of mercury exerts a greater pressure, a perfect balance is thus maintained between the column of mercury and the column of air.

The height of the column of mercury is proportional to atmospheric pressure: the greater the pressure, the higher the column. The pressure exerted by a column of liquid is equal to the density of the liquid times the height of the column. Therefore, atmospheric pressure can be found by multiplying the density of mercury by the height of the column in the barometer. However, it is simpler to express atmospheric pressure in terms of heights of the column. For example, the pressure at sea level at a certain time may be 29.92 inches Hg. This means that the mercury in a barometer stands at a height of 29.92 inches above the surface of the mercury in the dish. Although pressure at sea level varies, 29.92 inches is taken as a standard value for sea-level pressure.

Aneroid barometer. A toy balloon filled with gas expands until the pressure of the gas inside equals the pressure of the air outside, plus the pressure exerted inward by the rubber. If the air pressure decreases, the gas pressure expands the balloon until a new equilibrium is reached.

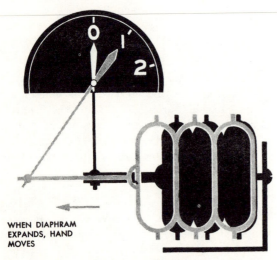

WHEN DIAPHRAM EXPANDS, HAND MOVES

*Fig. 4-12 Aneroid cell.*

The heart of an aneroid barometer is an aneroid cell, which works much like a toy balloon. It is an airtight metal box with thin, flexible sides. The air is partially evacuated, and the sides are held apart by a spring or the elasticity of the metal. When air pressure decreases, the sides of the cell move outward; when it increases, they move inward. In the illustration of the Aneroid Cell, you can see that the motion of the sides of the cell is amplified mechanically and translated into rotary motion of pointers on a dial. The dial may be calibrated in inches of mercury so that it reads units of atmospheric pressure; or, if the instrument serves as an altimeter, the dial is calibrated in feet of altitude. Regardless of how the dial is calibrated, remember that the aneroid barometer actually measures pressure.

Pressure altimeter. The pressure altimeter is an aneroid barometer. Though calibrated in feet, it really measures atmospheric pressure at flight level and interprets this value in terms of feet above a certain pressure level. If the pressure and temperature decrease in accord with the standard lapse rate, an altimeter could be designed to indicate the true altitude. However, the pressure and temperature changes seldom follow the standard lapse rate. Therefore, the altimeter cannot indicate true altitude directly. What then is the meaning of the altitude indicated by an altimeter?

The altimeter is designed to indicate changes of altitude as the pressure varies according to an arbitrary rate called the standard pressure lapse rate. It would be difficult to arrive at an accurate average pressure lapse rate for all conditons of the atmosphere, but the standard pressure lapse rate may be thought of as an attempt at such an average. Thus, for any change of pressure, the altimeter indicates the corresponding change of altitude according to the standard pressure lapse rate.

The altimeter is designed to measure altitude above any pressure level from 28.00" Hg to 31.00" Hg. The pressure set in the window on the dial face shown in Pressure Altimeter is the pressure level above which the altimeter measures pressure altitude.

The face of the altimeter is a clocklike dial calibrated in tens and hundreds of feet. There are ten numbers, 0 to 9, for hundreds of feet, with the intervening spaces marked into 5 or 10 subdivisons. There are three pointers. The long narrow one, corresponding to the sweep second hand on a watch, indicates tens

*Fig. 4-13 Pressure altimeter.*

and hundreds of feet. One revolution of this pointer indicates a change of 1,000 feet. A short, broad pointer, corresponding to the minute hand of a watch, indicates thousands of feet, thus counting the revolutions of the long pointer. One revolution of this pointer indicates a change of 10,000 feet. A shorter narrow pointer indicates tens of thousands of feet. It never makes more than one-half revolution because most altimeters are built for a maximum altitude of 35,000 to 50,000 feet.

Altimeter errors. Altimeters, like other navigational instruments, are subject to certain instrument errors. These errors can

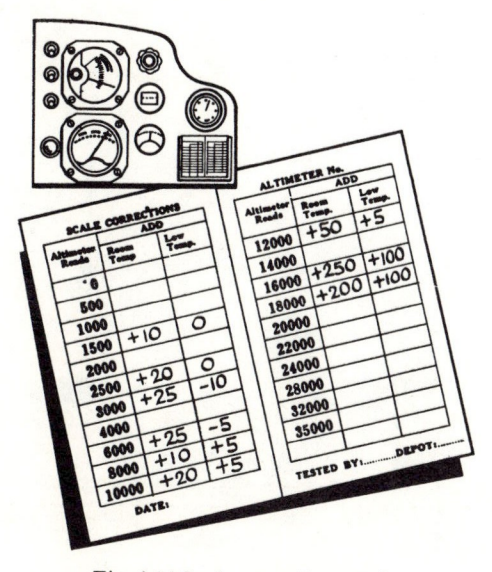

*Fig. 4-14 Scale correction card.*

be divided into four general classes: (1) mechanical errors, (2) elastic errors, (3) temperature error, and (4) installation error.

*Mechanical Errors.* Mechanical errors consist of three kinds:

1. Scale error is the difference between the indicated altitude and the basic altitude corresponding to the pressure level at which the instrument makes its measurement. Scale error is recorded on a card like that illustrated and placed near the altimeter.

2. Friction error is caused by the friction of moving parts within the instrument, resulting in irregular or jerky movements of the indicator needle and lost motion or hesitation upon reversal of direction of vertical motion. This can be corrected by gently tapping the altimeter. If the aircraft is in flight, the natural vibration is usually sufficient for this purpose.

3. Position error is that error caused by an unusual attitude of the instrument. Altimeters are designed to be in static balance with the dial vertical. If the aircraft assumes an abnormal attitude, a slight error is introduced into the altimeter. Such an error seldom exceeds 20 feet.

*Elastic Errors.* Elastic errors consist of three kinds:

1. Hysteresis causes a lag in the response of the instrument and is measured by the difference in two readings at a given altitude, the first reading obtained when altitude is increasing, and the second when it is decreasing. Hysteresis errors should be less than plus or minus 50 feet at zero altitude, and less than plus or minus 300 feet at 30,000 feet altitude.

2. Drift is a slow increase in the reading of an altimeter without increase in altitude after leveling off following a climb. The reading returns to the correct value after descent. In flights of more than one hour duration, drift should not be more than approximately 0.2 percent of the altitude for every 15,000 feet change of altitude.

3. Secular error is the slow change with time of the entire scale error curve. It is caused largely by internal stresses in the metal of the instrument. Secular error is corrected by

resetting the altimeter and need not be allowed for in flight.

*Temperature Error*. Temperature error is a change in reading caused by a change in the temperature of the instrument. In most instruments this error has been reduced to a negligible amount of temperature compensation features of the design. This should not be confused with temperature effects of the atmosphere.

*Installation Error*. Installation error arises when the altimeter is not exposed to the true static pressure of the atmosphere. In modern installations this error is usually avoided by connecting the cases of the instrument to a suitable static tube so placed on the surface of the aircraft that compression or rarefaction due to its forward motion is negligible. During climb or descent some change in pressure may result in the static tube opening being subjected to positive or negative pressure, resulting in a slight inaccuracy of reading. Such error should seldom exceed 10 feet.

Of the errors listed above, only secular error and friction error are likely to be objectionable in flight. Secular error is eliminated by resetting the zero position, and friction is overcome by tapping the instrument. Modern altimeters should indicate true altitude within 20 feet at ground level.

## ALTITUDE CORRECTIONS

Basic altitude is accurate only under one particular set of conditions; that is, when the atmosphere assumes a certain pressure and temperature at sea level, and a specific lapse rate. Since such standard conditions rarely exist, the altimeter reading usually requires correction. Remember, the altimeter is a pressure measuring device. It will indicate 10,000 feet when the pressure is 20.58″ Hg even though the actual altitude may be more or less than 10,000 feet.

Atmospheric temperature and pressure vary continuously. Rarely is the pressure at sea level exactly 29.92″ Hg or the temperature exactly + 15°C. Furthermore, the temperature lapse rate and the pressure lapse rate both deviate from the standard. As shown in the illustration, on a warm day the expanded air is lighter in weight per unit volume than on a standard day or on a cold day, and the pressure levels are raised. Therefore, the pressure level where the altimeter will indicate 4,000 feet will be higher than it would under standard conditions. On a cold day the reverse would be true, and the 4,000-foot level would be lower. This is the condition that must be closely watched with regard to deviation from the standard temperature lapse rate.

The basic altitude must be corrected for the temperature aloft by means of the conversion scale on the DR computer.

Changes in surface pressure also affect the pressure levels at altitude even under standard

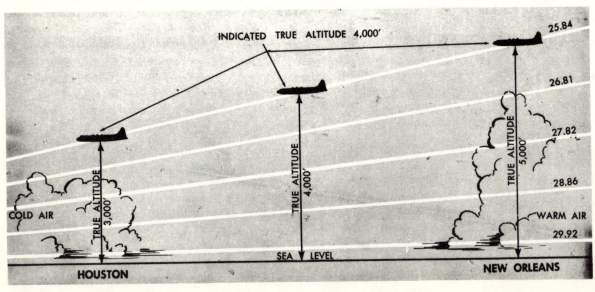

*Fig. 4-15 True altitude decreases when going into cold air.*

conditions. The illustration If Surface Changes, So Does the Pressure Altitude shows that if the surface pressure changes to a lower reading, the pressure level where the altimeter will indicate 4,000 feet pressure altitude will be lower than under standard conditions.

altitude), as shown in The Altimeter Indicates Altitude above the Value Set in the Window. The actual pressure at sea level, however, may be more or less than 29.92" Hg. This means that the standard datum plane may be above or below sea level. For the altimeter to indicate

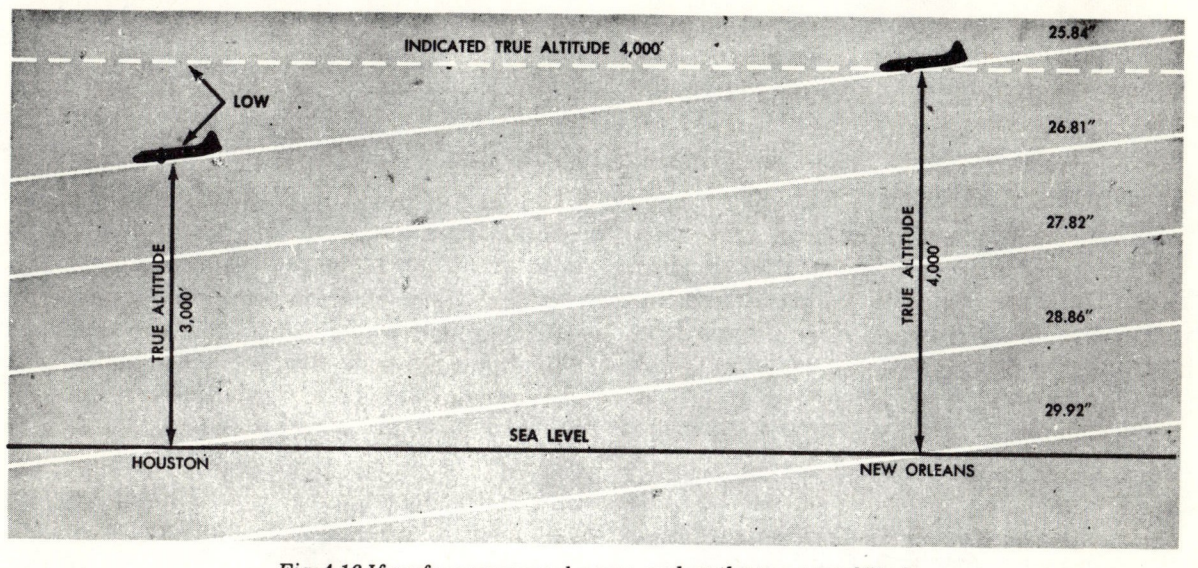

Fig. 4-16 If surface pressure changes, so does the pressure altitude.

To compute true altitude, the altimeter reading must be corrected for the difference between the existing atmospheric pressure and standard atmospheric pressure for the same level, and also for the difference between the actual temperature at flight level and the corresponding standard-atmosphere temperature for that altitude.

**Altimeter setting.** Basic altitude must be corrected for the difference between existing atmospheric pressure and standard atmospheric pressure for the same level. This correction can be made mechanically by adjusting the altimeter.

A barometric pressure scale, graduated in inches of mercury rather than in feet, has been incorporated in all pressure altimeters. This pressure scale is graduated from 28.0 to 31.0 inches of mercury and is visible through a small window on the right side of the dial face called the "Kollsman" window. The scale is set by means of a pressure setting knob. The pressure set on the barometric pressure scale is the pressure level above which the altimeter will measure standard altitude.

With 29.92" Hg set on the barometric scale, the altimeter will indicate altitude above the standard datum plane (indicated pressure

true altitude or altitude above sea level the actual pressure at mean sea level must be set on the barometric scale.

The value which is set in the window on the barometric scale to enable the altimeter to indicate true altitude (no instrument error considered) is the altimeter setting.

For stations at sea level, the altimeter setting is simply the atmospheric pressure. If the barometer reads 28.92" Hg, set this figure in the window. As you turn the counters to 28.92", you also turn the pointers to 0 feet altitude.

At any elevation above sea level, the altimeter setting is not the atmospheric pressure at the ground level. Instead, it is the sum of the atmospheric pressure at the ground level and the pressure equivalent in the standard atmosphere of the elevation of the place. It is ground level pressure reduced to sea level pressure according to the standard atmosphere. At a place 1,000 feet above sea level, the altimeter setting is the pressure that would exist at the bottom of a 1,000-foot well. In the illustration, Altimeter Indicates Altitude Above the Value Set in the Window, 27.92 is the field barometric pressure, or

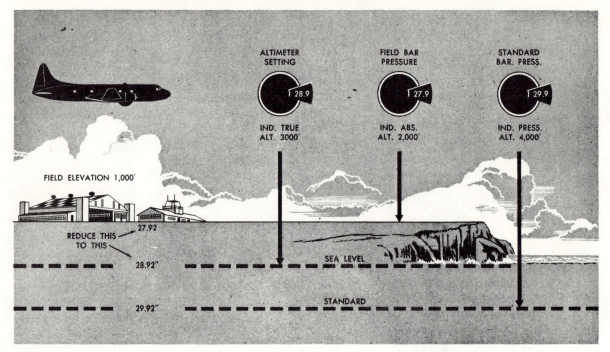

*Fig. 4-17 Altimeter indicates altitude above the value set in window.*

pressure at ground level. To get the altimeter setting this field barometric pressure must be reduced to sea level which, in this case, makes the altimeter setting 28.92.

The altimeter setting for any station is the pressure that must be set in the window of the altimeter in order for the altimeter to read the true elevation of the station. With altimeter setting set in the window, the altimeter will indicate true altitude at least at one pressure level, namely, at the ground level. Of course, the altimeter setting for any station changes as the atmospheric pressure changes, and altimeter setting varies from one station to another. Therefore, each station must make its own calculations at frequent intervals.

On the ground, by turning the counters to the altimeter setting, you also turn the pointers to the elevation of the station. Conversely, by turning the pointers to the elevation of the station, you turn the counters to the altimeter setting. In this way the altimeter setting can easily be found when the aircraft is on the ground. Since the altimeter is a little above the level of the runway, add the height of the aircraft above the runway to the elevation of the field as shown.

This method cannot be used in flight because the exact altitude of the aircraft is not

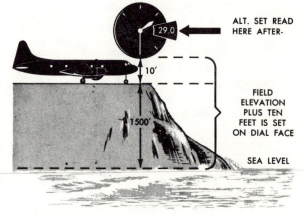

*Fig. 4-18 To determine altitude setting.*

known. However, altimeter settings are included in weather broadcasts so they can be obtained en route. The pilot always obtains the altimeter setting before landing so his altimeter will read the true elevation of the field as he lands.

With the altimeter setting in the "Kollsman" window, the altimeter indicates true altitude at least at one pressure level. If the pressure lapse rate is normal, the altimeter indicates true altitude at every level. However, since the pressure lapse rate varies, there usually is some error at other levels.

Pressure Altitude. When 29.92 is

set in the window, the reading of the altimeter is called indicated pressure altitude. Corrected for instrument error, it becomes basic pressure altitude. Basic pressure altitude corrected for temperature is known as density altitude.

Basic pressure altitude does not agree with true altitude unless the pressure at flight level happens to be the pressure in the standard atmosphere corresponding to that altitude. This is a rare occurrence. Basic pressure altitude usually is just an approximate expression of altitude. Actually, it is an exact expression of flight-level pressure.

The pilot is primarily interested in absolute altitude and hence in true altitude. Therefore, he keeps altimeter setting on his altimeter so that the direct reading is as close as possible to true altitude.

The navigator is also interested in true altitude, but in order to calculate accurate true altitude and also true airspeed, he must know the flight-level pressure expressed as pressure altitude.

Indicated true altitude may be found in either of two ways. The altimeter setting can be set in the "Kollsman" window and the indicated true altitude can be read directly, or a correction known as pressure altitude variation (PAV), can be applied to the indicated pressure altitude.

Pressure altitude variation (PAV) is the difference in feet between the standard datum plane and the datum plane above which indicated true altitude is measured. It is the standard-atmosphere distance equivalent to the pressure difference between 29.92 and the altimeter setting.

The table of Standard Pressure and Temperature shows that in the lower part of the atmosphere an altitude change of 1,000 feet corresponds to a pressure change of about 1 inch of mercury. Thus, 0.01 inch of mercury is the equivalent of about 10 feet of altitude. This is a convenient figure for converting pressure difference into pressure altitude variation. The pressure difference is multiplied by 1,000 and called "feet."

If the altimeter setting is 30.03, this is 0.11 more than 29.92; therefore, pressure altitude variation is 110 feet. But what is its sign? Is it added to or substracted from pressure altitude to get true altitude?

Remember that pressure decreases upward. Since 29.92 is less than 30.03, the standard datum plane is higher in the atmosphere than the datum plane above which indicated true altitude is measured. Therefore, as can be seen in the illustration, To Find True Altitude from Pressure Altitude, pressure altitude is less than true altitude and pressure altitude variation must be added to pressure altitude to get true altitude.

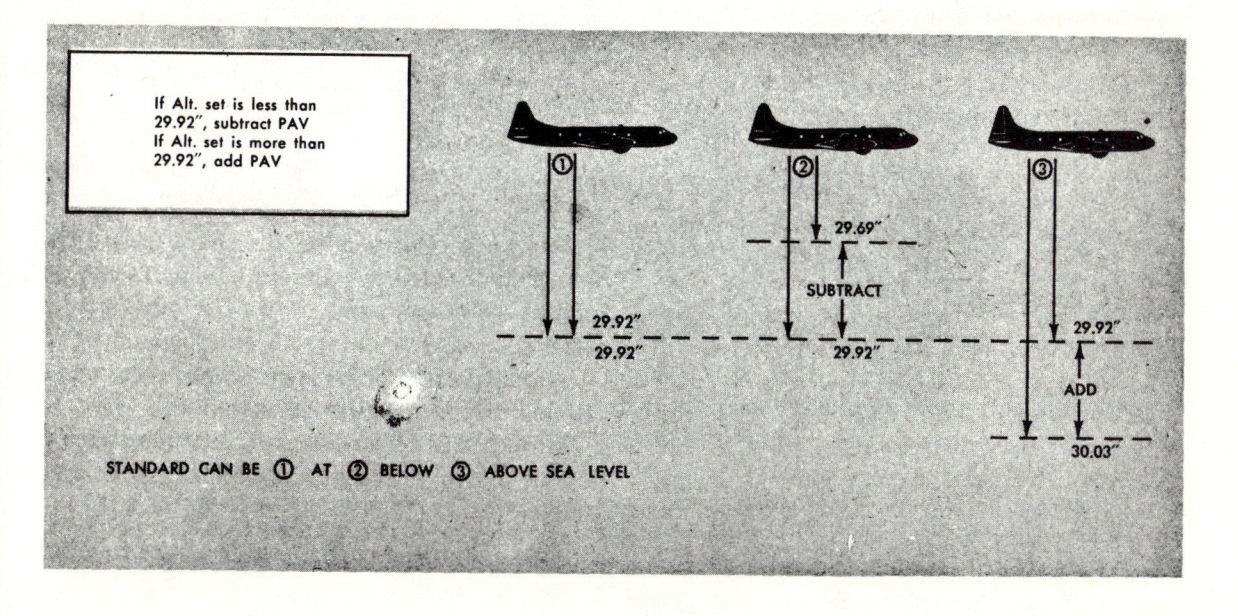

*Fig. 4-19 To find true altitude from pressure altitude.*

If the altimeter setting is 29.69, the pressure altitude variation is 230 feet. Since 29.92 is greater than 29.69, the standard datum plane is below the plane above which the true altitude is measured. Therefore, pressure altitude is greater than true altitude and pressure altitude variation must be substracted from pressure altitude to find true altitude.

If altimeter setting is greater than 29.92, pressure altitude variation must be added to pressure altitude. If the altimeter setting is less than 29.92, the pressure altitude variation must be substracted from pressure altitude.

Of course, to find pressure altitude from indicated true altitude, the pressure altitude variation must be applied with reverse signs. These rules are harder to remember than they are to figure out. Just remember that pressure decreases upward and think of the relative positions of the datum planes.

Density Altitude. Ordinarily, the pressure lapse rate above the ground is not standard; it varies with temperature. Therefore, basic pressure altitude must be corrected for temperature variation to find density altitude.

Air expands with increase in temperature. Consequently, warm air is less dense than cold air. As a result, the pressure lapse rate is less in a column of warm air than in cold air. That is, pressure decreases more rapidly with height in cold air than in warm air. Consequently, an altimeter tends to read too high in cold air and too low in warm air.

Density altitude is of interest primarily in the control of engine performance and is of direct concern to the navigator in connection with cruise control.

### COMPUTER ALTITUDE SOLUTIONS

The two altitudes most commonly accomplished on the computer are true altitude and density altitude. Nearly all DR computers have a window by which density altitude can be determined; however, be certain that the window is labeled *Density Altitude.*

True Altitude Determination. In the space marked For Altitude Computations are two scales: (1) a centigrade scale in the window and (2) a pressure altitude

scale on the upper disk. When a pressure altitude is placed opposite the temperature at that height, all values on the outer (miles) scale are equal to the corresponding values on the inner (minutes) scale increased or decreased by two percent for each 5.5.°C that the actual temperature differs from the standard temperature at that pressure altitude, as set in the window.

Although the pressure altitude is set in the window, the indicated true altitude is used on the inner (minutes) scale for finding the true altitude, corrected for difference in temperature lapse rate.

Example:

Given: Pressure altitude 8,500 feet.

Indicated true altitude 8,000 feet.

Air Temperature (°C) -16.

To find: True altitude

Procedure: Place PA (8,500 feet) opposite the temperature (-16) on the For Altitude Computations scale. Opposite the indicated true altitude (8,000 feet) on the inner scale, read the true altitude (7,500 feet) on the outer scale. The solution is illustrated.

TEMP. —16     READ 7600 ON MILES SCALE
ALT.  8500     OVER 8000 ON MIN. SCALE

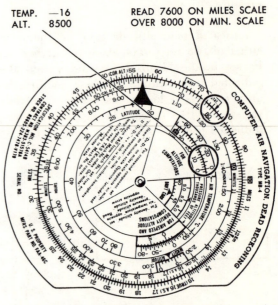

*Fig. 4-20 Finding true altitude.*

Density Altitude Determination. Density altitude determination on the computer is accomplished by using the window just above "For Airspeed and Density Altitude Computations" and the small window just above that marked "Density Altitude."

Example:

Given: Pressure altitude 9,000 feet.

Air temperature (°C) + 10

To Find: Density altitude

Procedure: Place pressure altitude (9,000 feet) opposite air temperature (+ 10) in window marked *For Airspeed and Density Altitude Computations*. Opposite index in *Density Altitude* window, read density altitude (10,400 feet). The solution is illustated.

TEMP. +10        READ 10400 IN
ALT. 9000        DENSITY ALTITUDE
                 WINDOW

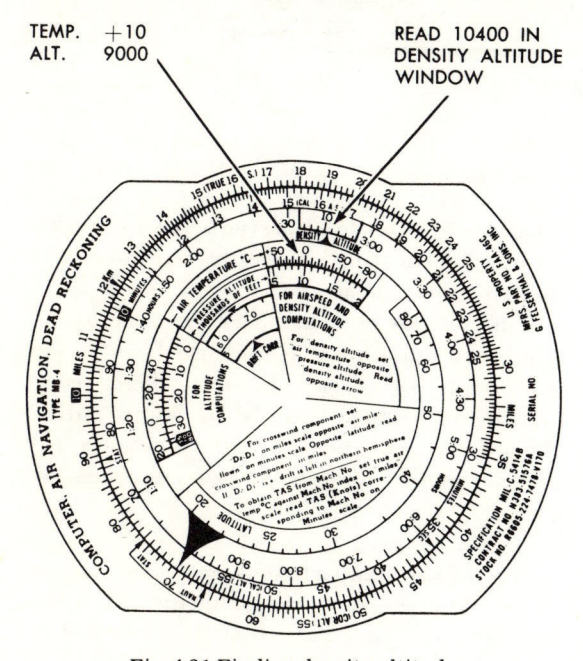

*Fig. 4-21 Finding density altitude.*

### TEMPERATURE

Determination of correct temperature is necessary for accurate computation of airspeed and altitude. Temperature, airspeed, and altitude are all closely interrelated, and the practicing navigator must be familiar with each in order to work effectively and accurately.

#### Temperature Gages

The temperature gage most commonly used employs a bimetallic element. The instrument, illustrated, is a single unit consisting of a stainless steel stem which projects into the air stream and a head which contains the pointer and scale. The sensitive element in the end of the stem—projected outside the aircraft—is covered by a radiation shield of brightly polished metal to cut down the amount of heat that the element might absorb by direct radiation from the sun.

The bimetallic element (called the sensitive element) is so named because it consists of two strips of different metal alloys welded together. When the element is heated, one alloy expands more rapidly than the other

causing the element, which is shaped like a coil spring, to turn. This, in turn, causes the indicator needle to move on the pointer dial. Temperature between -60°C and +50°C can be measured on this type of thermometer.

*Fig. 4-22 Free air temperature gage.*

### AIRSPEED, PITOT-STATIC SYSTEM, AND THE AIRSPEED INDICATOR

Speed is rate of motion or distance traveled per unit of time. A knowledge of speed is indispensable to navigation. The rate of travel over the surface of the earth is known as groundspeed (GS). The speedometer of an automobile indicates groundspeed directly. The wheels of the vehicle are touching the earth and, as they turn, give a direct indication of speed over the earth. Unfortunately, there is no simple aircraft instrument that gives a direct indication of groundspeed. However, there are many ways to determine the groundspeed of the aircraft. Each method requires the use of several instruments and careful computations.

The airspeed indicator is one of the most important of these instruments. As the name implies, this instrument indicates the speed of the aircraft through the air mass, and not the speed over the ground. Many corrections must be made to the indicated airspeed before the true speed through the air mass can be obtained. This section discusses the airspeed meter, causes of errors, and the corrections to be applied to the indicated air speed.

#### Pitot-Static System and Its Function

In order to understand the source of the forces which activate the airspeed indicator and other instruments such as the altimeter, it is necessary to understand the pitot-static system. This system consists of the pitot-static tube and the connecting lines to the instruments it serves.

The pitot-static tube is the beginning of the

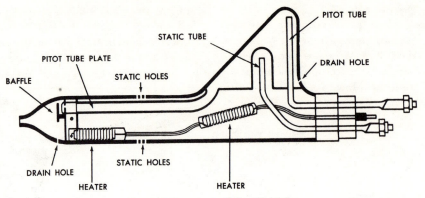

*Fig. 4-23 Structure of the pitot tube.*

system and serves as the source of the pressures for operation of the airspeed indicator. It is mounted on the aircraft in a location where it will receive the full impact of the air through which it is traveling. This location will vary with different types of aircraft. For most accurate operation, the pitot-static tube should be parallel to the line of the relative wind or the axis of motion, and be located in an area of minimum air stream turbulence. This condition can best be approximated by mounting it parallel to the longitudinal axis of the aircraft.

Mounted this way, the intake end of the tube points in the direction of flight. The forward section of the tube contains a baffle plate, shown in Structure of the Pitot Tube, to reduce turbulence and to keep rain, ice, and dirt particles from entering the tube. At the bottom there may be one or more drain holes to dispose of condensed moisture. The inside of the tube is designed to prevent any moisture or dirt from entering the instruments that are connected to it.

The inside of the pitot-static tube is divided into two compartments: the impact pressure chamber and lines, and the static pressure chamber and lines. An electrical heating element within the tube prevents the formation of ice. This heating element is controlled by a switch inside the aircraft.

Each of the two chambers has openings to allow air to enter. However, there is no continual air flow through these compartments because there is no return outlet. Therefore, as the aircraft moves, these compartments receive pressure against the already present air.

One of the compartments mentioned above is the total pressure compartment. The opening to this compartment is at the front end of the tube. Since the intake faces directly into the air through which the aircraft flies, the air in this compartment, as shown in Total Pressure Compartment, receives the full force of the entering air. That is, the pitot tube receives the total, or impact, pressure of the air through which the aircraft is moving.

The other compartment is the static pressure compartment. The air enters this compartment through small holes on the top and bottom of the tube. Usually there are three holes on the top and three on the bottom as shown. Since these holes do not face in the direction of flight, this compartment receives static (still air) pressure.

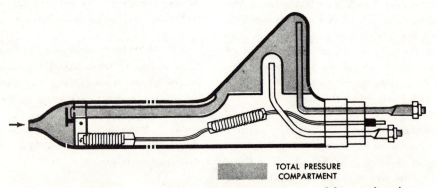

**TOTAL PRESSURE COMPARTMENT**

*Fig. 4-24 Total pressure compartment receives the full force of the entering air.*

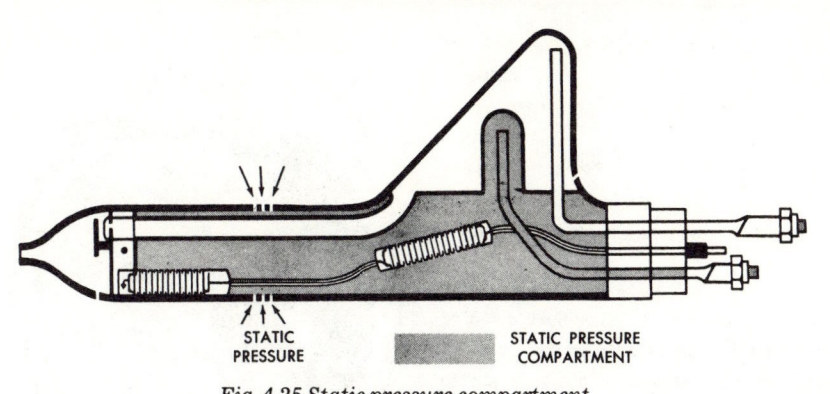

Fig. 4-25 Static pressure compartment.

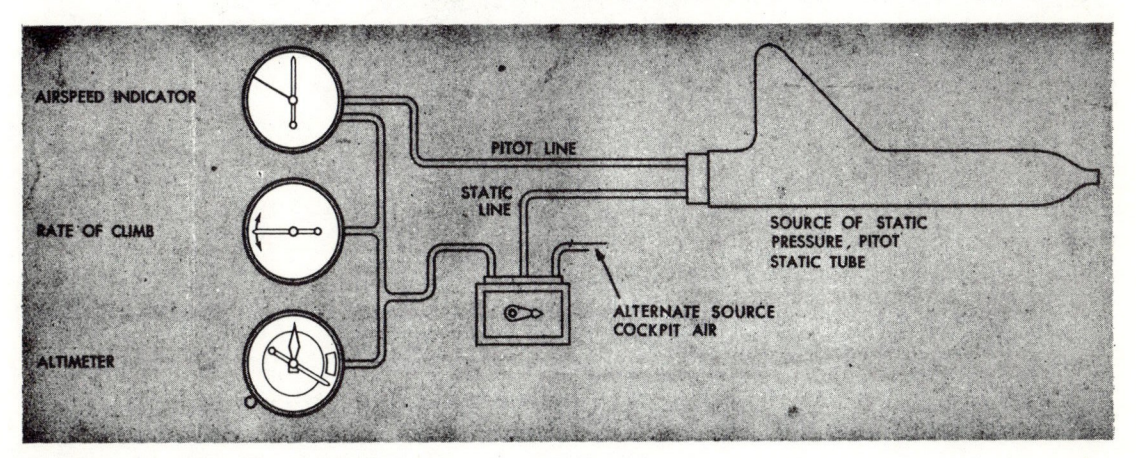

Fig. 4-26 Pitot static system.

The most common pitot-static installation in operational aircraft separates the impact and static sources. The impact pressure is taken from the pitot tube which is mounted in one of the usual positions. Static pressure is taken from an inlet that may be installed even, or flush, on the side of the aircraft, vertical stabilizer, etc.

Two lines supply the airspeed indicator; one supplies static pressure and the other supplies impact pressure. A typical complete pitot system is shown in Pitot Static System.

An alternate source for static pressure is provided for emergency use in most aircraft. In unpressurized aircraft the alternate source is usually located within the aircraft. This alternate static pressure is usually lower than the pressure provided by the pitot-static tube. Therefore, when the static source switch is placed in the alternate position, the altimeter usually reads higher than normal and the indicated airspeed usually reads greater because of the lower static pressure.

Reasonable care should be taken with the pitot-static tube. At the end of every flight, a cover should be placed on the intake end

of the tube to prevent dirt and moisture from collecting in the tube. Also, the drain holes should be checked regularly to see that they are not clogged.

*Construction and Operation of Airspeed Indicator*

Static pressure is undisturbed atmospheric pressure. Impact pressure is the pressure exerted on the nose of the aircraft (or pitot tube) while the aircraft is in motion. The result or difference between these two is called dynamic pressure. It is this dynamic pressure that the airspeed indicator actually measures.

The airspeed indicator has a cylindrical airtight case, which is connected to the static line from the pitot-static system as shown. The static pressure received from the static source fills the case, or housing of the airspeed indicator. The pressure within this case theoretically equals the barometric pressure of the air through which the aircraft is flying.

Inside the case is a small diaphragm made of phosphor-bronze or beryllium copper. This diaphragm is very sensitive to changes in pressure and is connected to the impact pressure line from the pitot-static tube.

The side of the diaphragm which is free to expand is connected to a series of levers and gears which operates the needle on the face of the instrument.

The airspeed indicator measures the difference between the pressure in the impact pressure line and the pressure in the static pressure line. The two pressures are equal when the aircraft is stationary on the ground, but movement through the air causes the pressure in the impact line to become greater than the pressure in the static line. The

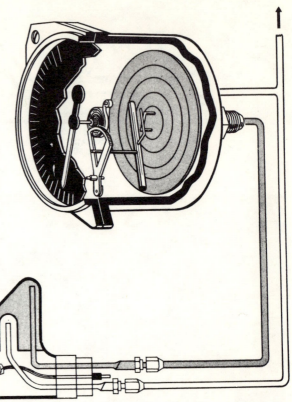

TO ALTIMETER

*Fig. 4-27 Expansion and contraction of the diaphragm is transmitted to the needle.*

diaphragm, having a direct connection to the impact pressure line, will expand with the increase in impact pressure. The expansion or contraction of the diaphragm is transmitted by mechanical linkage to the needle on the face of the instrument. This needle indicates dynamic pressure in reference to a sealed dial. The dial is scaled in units of speed rather than units of pressure. Older-type airspeed indicators are graduated in miles per hour. Newer-type indicators are graduated in knots.

*Fib. 4-28 Airspeed indicator.*

## AIRSPEED DEFINITIONS

There are many reasons for the difference between indicated air speed and true airspeed. Some of these reasons are: the error in the mechanical makeup of the instrument, the error caused by incorrect installation, and the fact that density and pressure of the atmosphere vary from standard conditions.

INDICATED AIRSPEED (IAS). Indicated airspeed is the uncorrected reading taken from the face of the indicator. It is the airspeed that the instrument shows on the dial. It can be read in miles per hour or in knots, depending upon the scale of the dial.

BASIC AIRSPEED (BAS). Basic airspeed is the indicated airspeed corrected for instrument error.

Each airspeed indicator has its own characteristics which cause it to differ from any other airspeed indicator. These differences may be caused by slightly different hair-spring tensions, flexibility of the diaphragm, accuracy of the scale markings, or even the effect of temperature on the different metals in the indicator mechanism. The effect of temperature introduces an instrument error due to the variance in the coefficient of expansion of the different metals comprising the working mechanism. This error can be removed by the installation of a bimetallic compensator within the mechanical linkage. The bimetallic compensator is installed and properly set at the factory, thereby eliminating the temperature error within the instrument.

The accuracy of the airspeed indicator is also affected by the length and curvature of the pressure line from the pitot tube. These installation errors must be corrected mathematically.

Installation, scale, and instrument errors are all combined under one title called instrument error. Instrument error can be determined by various calibration procedures, a correction card should be made up and used to find basic airspeed.

CALIBRATED AIRSPEED (CAS). Calibrated airspeed is basic airspeed corrected for pitot-static error and/or attitude of the aircraft. The pitot-static system of a moving aircraft will have some error. Minor errors will be found in the pitot section of the system. The major difficulty is encountered in the static pressure section.

As the flight attitude of the aircraft changes, the pressure at the static inlets will change. This is caused by the airstream striking the inlet at an angle.

Different types and locations of installations will cause different errors. It is immaterial whether the static source is located in the pitot-static head or at some flush mounting on the aircraft. This error will be essentially the same for all aircraft of the same model, and a correction table can be constructed. The table is entered with basic airspeed and the gross weight of the aircraft.

EQUIVALENT AIRSPEED (EAS): Equivalent airspeed is calibrated airspeed corrected for compressibility. Compressibility error has come into prominence with the advent of high speed aircraft. It becomes noticeable when the airspeed is great enough to create an impact pressure which will cause the air molecules to be compressed within the impact chamber of the pitot tube. The amount of compression is directly proportionate to the impact pressure. As the air is compressed, it causes the dynamic pressure to be greater than it should be. Therefore, the correction is a negative value.

DENSITY AIRSPEED (DAS): Density airspeed is calibrated airspeed corrected for pressure altitude and true air temperature.

Pitot pressure varies not only with airspeed but also with air density. As the density of the atmosphere decreases with height, pitot pressure for a given airspeed must also decrease with height. Thus, an airspeed indicator operating in a less dense medium than that for which it was calibrated will indicate an airspeed lower than the true speed. The higher the aircraft flies, the greater the discrepancy. The necessary correction can be found on the DR computer.

Using the window on the computer above the area marked for *Airspeed and Density Altitude Computations*, set the pressure altitude against the True Air Temperature (TAT). Opposite the calibrated airspeed on the minutes scale, read the density airspeed on the miles scale.

At lower airspeeds and altitudes, density airspeed may be taken as true airspeed with negligible error. However, at high speeds and altitudes, this is no longer true and compressibility error must be considered. The correction for compressibility is obtained by multiplying density airspeed by a ratio of the compressibility factors at flight altitude to the factors at sea level. This ratio of compressibility factors for various conditions is listed on the wind side of the computer. Extract the compressibility factor and multiply it by the density airspeed using the slide rule side of the computer. The result will be true airspeed.

TRUE AIRSPEED (TAS): True airspeed is equivalent airspeed corrected for density altitude (pressure and temperature). It is the actual speed of the aircraft through the air mass. The airspeed indicator does not compensate for the air density decrease with altitude, or temperature error.

To find true airspeed it is necessary to use the computer. The prerequisites are equivalent airspeed, true air temperature, and pressure altitude.

MACH NUMBER. Airspeed can be measured by a machmeter rather than an airspeed indicator. Mach number is a ratio between the speed of the aircraft and the speed of sound and is used in jet aircraft operations.

### SUMMARY OF TYPES OF AIRSPEEDS

Indicated airspeed (IAS) is the uncorrected reading taken directly from the indicator.

Basic airspeed (BAS) is the IAS corrected for instrument error.

Calibrated airspeed (CAS) is the BAS corrected for pitot position error.

Equivalent airspeed (EAS) is CAS corrected for compressibility.

True airspeed (TAS) is EAS corrected for pressure and temperature.

Density airspeed (DAS) is the CAS corrected for pressure and temperature.

True airspeed (TAS) is the DAS corrected for compressibility.

# WIND AND ITS EFFECT 5

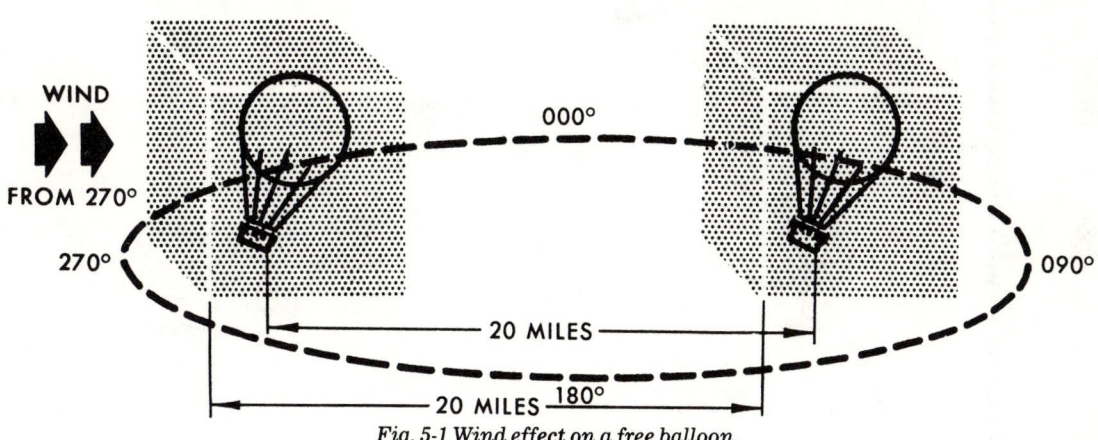

Fig. 5-1 Wind effect on a free balloon.

## WIND DIRECTION AND SPEED

Wind direction is named by the direction from which the wind blows. Wind speed is rate of wind motion without regard to direction. In the United States, wind speed is usually expressed in knots. Wind velocity (W/V) includes both direction and speed of the wind. For example, a west wind of 25 knots is recorded as W/V 270°/25 knots. "Downwind" is movement with the wind; "upwind" is movement against the wind.

## EFFECT OF WIND

*a. General.* Moving air exerts a force in the direction of its motion on any object within it. Objects that are free to move in air will move in a downwind direction at the speed of the wind. An aircraft will move as does the balloon in the example. In addition to its forward movement through the air, if an aircraft is flying in a 20 knot wind, it will move 20 nautical miles downwind in one hour. The path of the aircraft over the earth is determined by the motion of the aircraft through the air and the motion of the air over the earth's surface. The direction and movement of the aircraft through the air is determined by the direction in which the nose of the aircraft is pointed and the speed of the aircraft through the air.

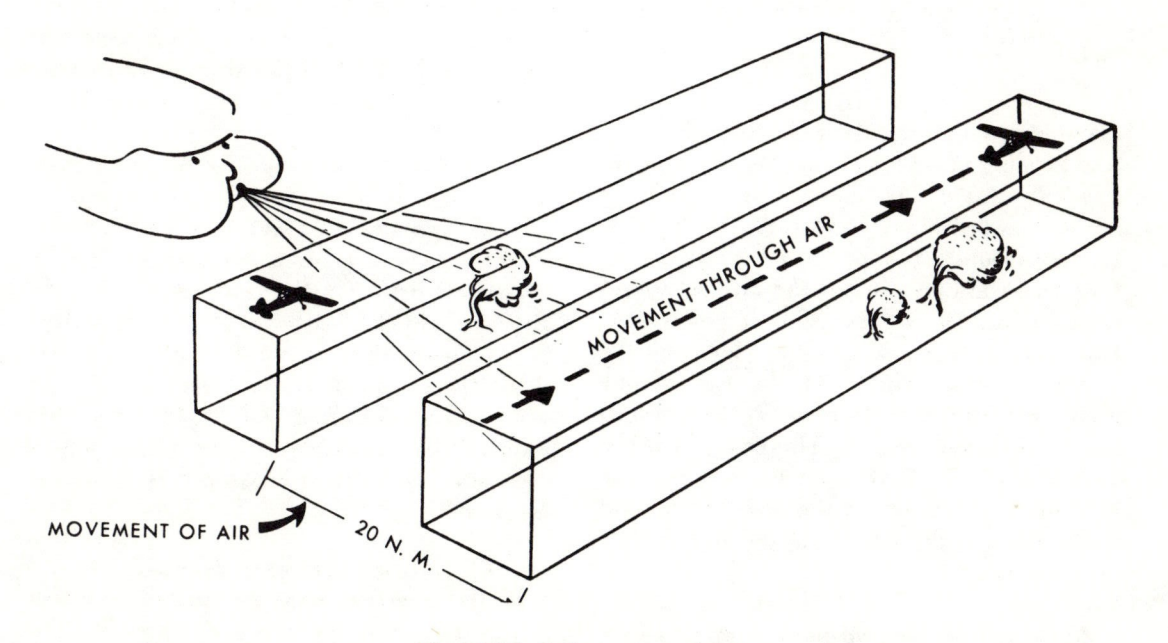

Fig. 5-2 Wind effect on an aircraft.

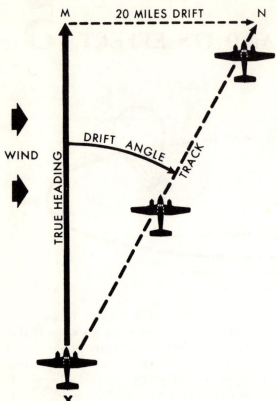

*Fig. 5-3 Drift.*

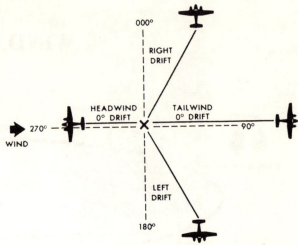

*Fig. 5-4 Effects of different headings on track and ground speed.*

*b. Drift.* The sideward displacement of the aircraft caused by the wind is called drift. Drift is measured by the angle between the true heading (true direction in which the nose is pointed) and the track (actual path of the aircraft over the earth).

> NOTE: Track must not be confused with true course which is the plotted course of intended track.

*c. Example of Drift.* As shown in the example, an aircraft departs point X on a true heading of 360° and flies for one hour in a wind of 270°/20 knots. The aircraft is headed toward point M directly north of X. Its true heading is represented by line XM. Under no-wind conditions, the aircraft would be at point M at the end of one hour. However, there is a wind of 20 knots and the aircraft moves with it. At the end of one hour, the aircraft is at point N, 20 nautical miles downwind from M. The line XM is the path of the aircraft through the air; the line MN shows the motion of the body of air; and the line XN is the actual path of the aircraft over the earth.

*d. Drift and Groundspeed Change With Heading Change.* A given wind causes a different drift on each heading and effects the distance traveled over the ground in a given time. With a given wind, the groundspeed (GS), varies on different headings.

*e. Effect of Wind on Different Headings With Respect to Track and Groundspeed.* Effect of wind on different headings in relation to track and groundspeed is shown in the illustration. A wind of 270°/20 knots is affecting the groundspeed and track of an airplane flying on headings of 000°, 090°, 180°, and 270°. On each heading the airplane flies from point X for one hour at a constant true airspeed. Length of each dashed line represents the distance the aircraft has traveled through the body of air or the distance it would have traveled over the ground in one hour had there been no wind. Each solid line represents the track of the aircraft. The length of each solid line represents groundspeed.

*f. Headwind, Tailwind, Crosswind Effect.* As shown in the example, the wind of 270°/20 knots causes right drift, on a heading of 000°, whereas on a heading of 180° it causes left drift. On the headings of 090°, the airplane, aided by a tailwind, travels farther in one hour than it would with no wind; thus, its groundspeed is increased by the wind. On the heading of 270°, the headwind reduces the groundspeed. On a heading of 000° and 180°, the groundspeed is somewhat increased.

### DRIFT CORRECTION

Drift correction must be applied to a true course to determine the true heading. The amount of drift correction must be just

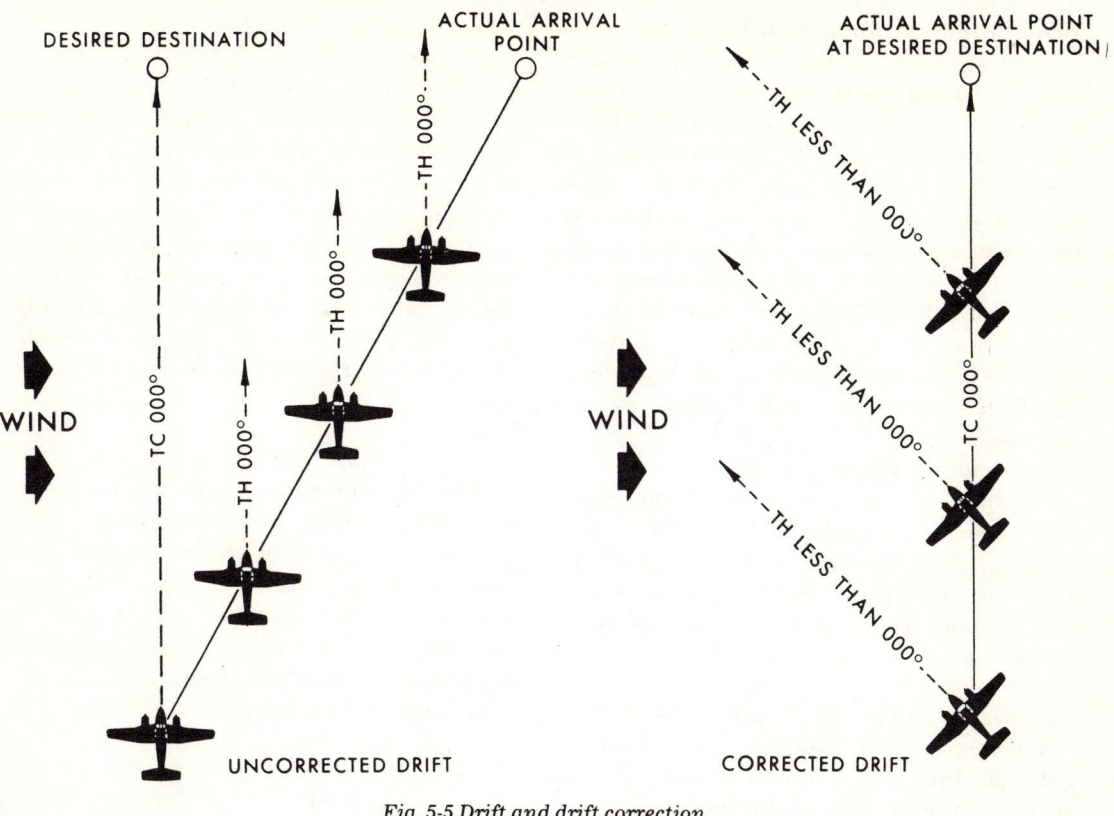

*Fig. 5-5 Drift and drift correction.*

enough to compensate for the amount of drift on a given heading. The drift correction angle (DCA) (sometimes called crab angle) is equal to but in the opposite direction from the drift angle (DA). As shown in the example, if a pilot attempts to fly to a destination due north of his point of departure (TC 000°) on a true heading of 000°, and a west wind were blowing, he would arrive somewhere east of his destination because of right drift. To correct for right drift so that the aircraft would remain on course and arrive at the desired destination, the nose would have to be pointed to the left of the true course or upwind.

## SUMMARY OF DRIFT AND DRIFT CORRECTION

*a.* Wind from the right causes drift to the left.

*b.* Wind from the left causes drift to the right.

*c.* If TH is greater than TR or TC, drift is to the left.

*d.* If TH is less than TR or TC, drift is to the right.

*e.* If drift is to the right, DC is to the left.

*f.* If drift is to the left, DC is to the right.

*g.* Drift is always downwind.

*h.* Drift correction is always upwind.

## APPLIED PROBLEMS OF DRIFT AND DRIFT CORRECTIONS

*a. Problem.* TH 160°, TR 170°. Is drift right or left? Is drift correction to be made to right or left?

*b. Solution.* Since TH is less than TR, DR is right; DC is left.

*c. Problem.* TH 350°, DR 4° left. What is the track? What is the drift correction?

*d. Solution.* Since DR is left, TH must be greater than TR. TR equals 346° (350° - 4°). DC. 4° right.

*e. Problem.* TR 005°, DR 10° right. What is the true heading? What drift correction is required?

*f. Solution.* Since DR is right, TH is less than TR.

TH equals 355° (005° - 10°)

DC equals 10° left.

## GROUNDSPEED (GS)

Groundspeed is the resultant of wind and the forward motion of the aircraft through the air. In calm air, the speed of the aircraft over the ground (GS) is equal to its true airspeed (TAS). If the aircraft is moving against the wind (headwind), the groundspeed is equal to the difference between the true airspeed and the windspeed. If the aircraft is moving with the wind (tailwind), the groundspeed is equal to the sum of the true airspeed and the windspeed. If the aircraft is moving at an angle to the wind, the groundspeed may be any speed between the extremes of the groundspeeds determined by headwind and tailwinds. Those groundspeeds that are less than the true airspeed are the result of hindering winds; those greater than the true airspeed are the result of helping winds. Wind directions that are approximately 90° to the longitudinal axis of the aircraft (beam winds) have a minimum effect on groundspeed. Winds may be classified as headwinds (hindering winds), tailwinds (helping winds) and cross winds.

In dead reckoning, problems involving speed and direction are primarily concerned with course, groundspeed, heading, true airspeed, wind direction, and wind speed. In order to solve these problems, it is necessary to understand the relationship of these six values.

### REPRESENTING VECTOR QUANTITY

A vector may be represented on paper by a straight line. The direction of the vector is shown by the bearing of the line with reference to north. It is usually drawn like an arrow, so that there can be no doubt as to direction. The magnitude of the vector is shown by the length of the line in comparison with an arbitrary scale. For example, if 1 inch equals 20 knots, then a velocity of 50 knots would be shown by a line 2½ inches long. Although the line is only a diagram of the vector, the term "vector" is loosely applied to the line itself.

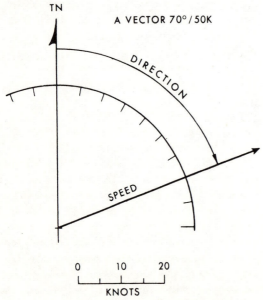

*Fig. 5-6 Representing vector quantities.*

### VECTOR

As used in air navigation, a vector is a velocity (speed in a given direction). A vector may be represented by a line segment which has magnitude, direction, and a point of origin. A series of line segments, representing vectors, are used in solving problems concerning winds, courses, headings, and speeds. The velocity of an aircraft relative to the air (air vector) includes heading and true airspeed; relative to the surface beneath it (ground vector), track or course and groundspeed.

### VECTOR DIAGRAMS

When two or more vectors are components of a third vector, this relationship may be shown by means of a vector diagram. If the components are drawn tail to head, in any order, a line from the tail of the first component to the head of the last component represents the resultant. A diagram of a vector sum forms a closed figure.

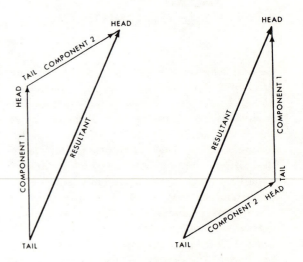

*Fig. 5-7 Vector diagrams.*

### TRIANGLE OF VELOCITIES

A triangle of velocities is a vector diagram drawn for the purpose of analyzing the effect of wind on the flight of an aircraft. The complete diagram includes an air vector, ground vector, and wind vector. The air vector is composed of the heading and true airspeed; the ground vector, track or course and groundspeed; the wind vector, wind direction and wind speed. Each vector must include the factors and only such factors as are required to form the component of the triangle of velocities. When any four of the above mentioned factors are known, the remaining two can be determined.

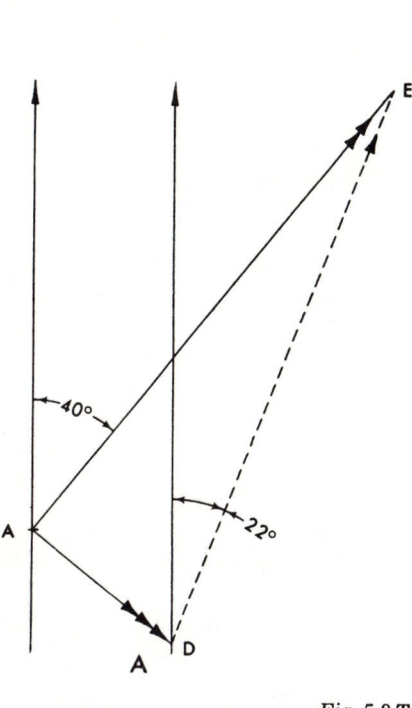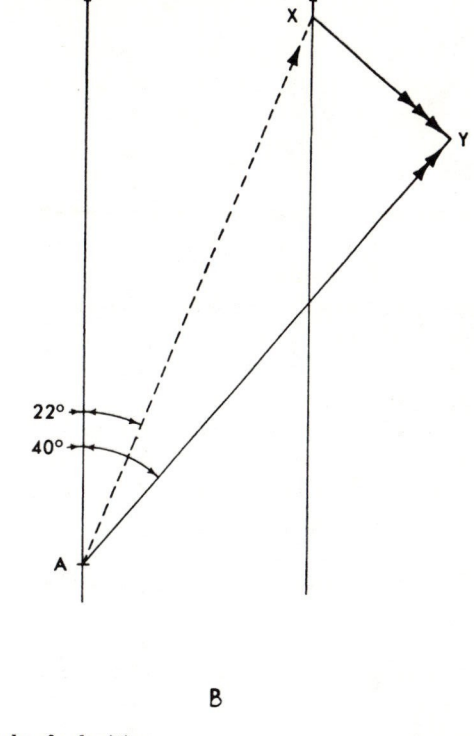

*Fig. 5-8 Triangle of velocities.*

The necessary steps for drawing the triangle of velocities are as follows:

(1) Draw a vertical datum or reference line with an arrow at the top indicating true north to facilitate angular and linear measurement. This is the theoretical true meridian passing through the point of origin to which all vectors are referred.

(2) Draw a very short line intercepting the reference line at a convenient point to indicate the point of origin in the diagram.

(3) Draw in the known vectors.

(4) Close the triangle to determine two unknown factors. Known and unknown factors will vary but each factor can be determined provided each vector includes its own factors, namely direction and length.

There are two basic methods of drawing a triangle of velocities. One method is used when the wind velocity is known; the other when solving the wind velocity. For quick recognition of the various vectors, some navigators use arrowheads along each vector. One arrowhead indicates the air vector; two arrowheads, the ground vector; and three arrowheads, the wind vector. A system of lettering various points is also essential.

Regardless of which four factors are known they are drawn first and the closing of the triangle determines the remaning two. Additional vertical reference lines may be drawn to facilitate angular measurements provided they are parallel to the first. The method used when wind velocity is known is shown in A; B, when determining the wind velocity.

Throughout this discussion, except when solving for wind, the following lettering is used:

(1) A is point of origin.

(2) B is point of intended landing.

(3) D is the downwind end of the wind vector.

(4) E is the end of the first hour of flight.

(5) When solving for wind, the letter X will be used for the upwind end of the wind vector; the letter Y for the downwind end.

(6) Letters C, F, G, and E are used only in radius of action problems and will be explained later.

With the exception of the problems solving for the wind, the vectors are lettered as follows:

(1) Line AD is the wind vector.

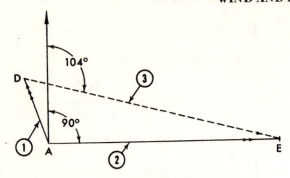

*Fig. 5-9 Finding true heading and groundspeed.*

(2) Line AE is the ground vector.
(3) Line DE is the air vector.
(4) In problems involving the solution for the wind vector, line AX is the air vector; line AY, the ground vector; line XY the wind vector.

## FINDING TRUE HEADING AND GROUND SPEED

*a. Problem.* True course 090°, wind velocity 160°/30 knots, true airspeed 120 knots. What is the true heading and groundspeed?
*b. Solution:*
(1) Draw vertical reference line and label point A.
(2) Using the plotter to determine angular measurement, draw in the wind vector at the reciprocal of 160° (160 + 180 = 340°).
(3) Determine the scale to be used and measure along the wind vector to find point D. Mark this point with a sharp cross line and Label D.
(4) Draw in the true course line ② at 090° for an indefinite distance.
(5) Using the same scale used in (3) above, draw a line from D intersecting the true course line at 120 nautical miles from D. Place a sharp mark at this point and label it E.
(6) Since the air vector crosses the vertical reference line, the true heading can be determined by measuring the angle between the vertical reference line and the air vector. (If the air vector did not cross the reference line, the air vector could be extended.) In this problem, the true heading is 104°.
(7) Measure distance AE to find the groundspeed (106 knots).

## FINDING TRUE HEADING AND TRUE AIRSPEED

*a. Problem.* True course is 120°, wind velocity 090°/20 knots, groundspeed 90 knots. What is the true heading and true airspeed?

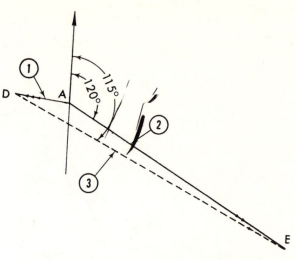

*Fig. 5-10 Finding true heading and true airspeed.*

*b. Solution:*
(1) Draw vertical reference line and label point A.
(2) Draw in wind vector ① at 270° (090° + 180°), 20 knots, and label point D.
(3) Draw in ground vector ② at 120° and measure 90 nautical miles to find and label point E.
(4) Draw a line ③ from D to E.
(5) Determine true heading by measuring the angle formed by the interception of the reference line and the true heading line (115°)
(6) Determine true airspeed by measuring line DE (108 knots).

## FINDING WIND VELOCITY

*a. Problem.* True heading 130°, true

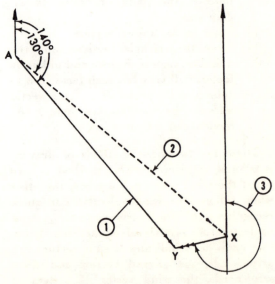

*Fig. 5-11 Finding wind velocity.*

airspeed 100 knots, track 140°, groundspeed 90 knots. What is the wind velocity?

*b. Solution:*

(1) Draw vertical reference line and label point A.

(2) Draw ground vector AY ① at 140° at 90 knots and label point Y.

(3) Draw air vector at AX ② at 130°, 100 knots and label point X.

(4) Draw a line from X to Y.

(5) Draw another vertical reference line through point X, parallel with the original to facilitate angular measurement.

(6) Measure the angle formed ③ between the vertical reference line and the wind vector and subtract 180° so as to name the wind by the direction from which it blows (075°). (Wind always blows from heading to track.)

(7) Measure line XY (20 knots).

RADIUS OF ACTION (FIXED BASE)

Radius of action refers to the maximum distance an aircraft can fly on a given course and still be able to return to the original point of departure within a given time. The radius of action is the distance out only.

*a. Problem.* A pilot is ordered to scout as far as possible on a true course of 090° returning to his point of departure in 2 hours. He maintains a true airspeed of 100 knots. The wind is 030°/20 knots. What is the true heading and groundspeed on each leg? What is the radius of action?

tor ① 030°/20 knots, and label point D.

(3) Through point A, draw in the true course line ②, 090° and 270°, for both the outbound and inbound flight. This line is indefinite in length in both directions.

(4) Draw a line from D intercepting the outbond course line at an airspeed distance (100 knots) from D and label E.

(5) Draw a line from D intercepting the inbound course line at an airspeed distance from D and label E.

(6) Draw a parallel reference line through D and measure the angle (080°) between the reference line and the outbound air vector. This is the true heading outbound.

(7) Measure line AE (89 knots). This is the groundspeed on the outbound leg.

(8) Measure the angle (280°) between the reference line and the inbound air vector. This is the true heading on the inbound leg.

(9) Measure line AE′ (108 knots). This is the groundspeed on the inbound leg.

(10) The distance out (radius of action) is the product of the total time in hours multiplied by the product of groundspeed out and the groundspeed back divided by the sum of these two speeds. The mathematical formula is -

$$\text{TX} \ \frac{\text{GS}_1 \times \text{GS}_2}{\text{GS}_1 + \text{GS}_2} = \text{R/A}$$

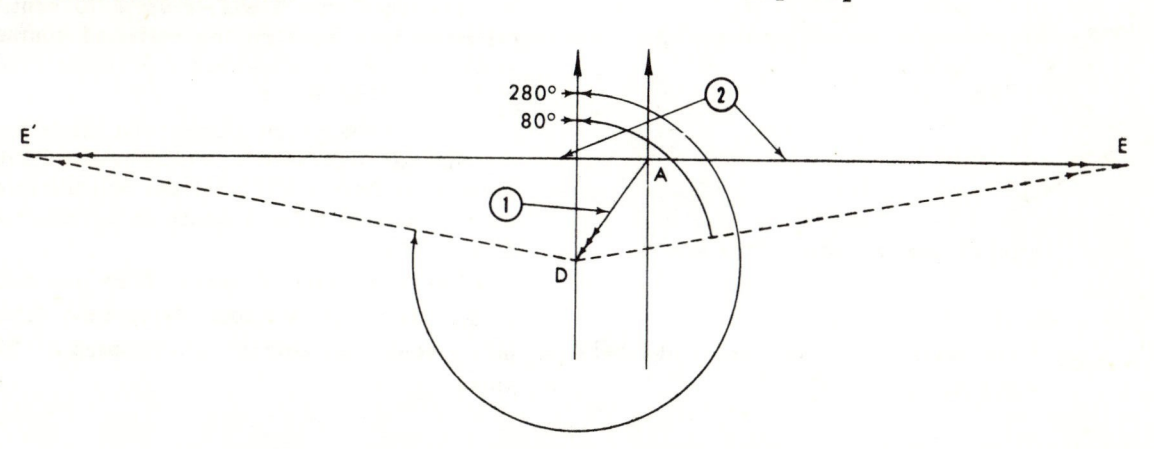

*Fig. 5-12 Radius of action (fixed base).*

*b. Solution:*

(1) Draw vertical reference line and label point A.

(2) From point A, draw in the wind vec-

in which T is the total time in hours; GS$_1$ is the groundspeed on the outbound leg; GS$_2$ is the groundspeed on the inbound leg; and R/A is the radius of action or, in the above case.

$$2 \times \frac{89 \times 108}{89 + 108} = 97.6 \text{ nautical miles (R/A)}.$$

*c. Checking Accuracy of Computations.* To check the accuracy of computations, the time on each leg is calculated and added. The sum must agree with the total time allowed. Calculate as follows: dividing the radius of action 97.6 nautical miles by the groundspeed out (89 knots) equals 1.1 hours or 66 minutes (time on outbound leg). Dividing 97.6 nautical miles by groundspeed back (108 knots) equals 0.9 hours or 54 minutes. Time utilized on both headings is equal to the total time (2 hours) allowed.

### FINDING TRACK AND GROUNDSPEED
*a. Problem.* Wind 300°/20 knots, true heading 045°, true airspeed 100 knots. What is the track and groundspeed?

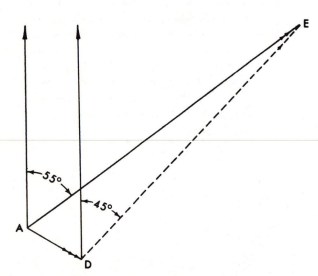

*Fig. 5-13 Finding track and groundspeed.*

*b. Solution:*
(1) Draw vertical reference line and label point A.

(2) Draw wind vector at 120° (the reciprocal of 300°), 20 nautical miles from A, label point D.
(3) Draw another vertical line through D that is parallel with the original reference line.
(4) From D, draw in the air vector at 045°, 100 nautical miles and label E.
(5) Close the triangle by drawing a line from A to E.
(6) Measure the angle between the reference line and the track line AE (055°).
(7) Measure the length of AE (107 knots) to determine the groundspeed.

## Average Groundspeed

Average groundspeed is calculated by dividing the total distance flown by the total time (in hours) required for the flight. Airspeed factors to be considered in computing average groundspeed include—

*a.* Climbing airspeed is usually less than cruising airspeed.

*b.* Descending airspeed may be greater than cruising airspeed.

*c.* Flying a constant true airspeed on the same outbound and return course with a constant wind velocity does not produce an average groundspeed equal to the average TAS. *For example,* figure 5-14 illustrates an aircraft flying a constant TAS (100 kt) for 1 hour against a 30-knot headwind and returning to the starting point.

(1) The aircraft will traverse 70 nautical miles in 1 hour on the outbound course ((A), fig.5-14); groundspeed is 70 knots (100 kt TAS − 30 kt W/V).

(2) On the return course ((B), fig.5-14), the aircraft will have a groundspeed of 130 knots (100 TAS + 30 W/V) and will traverse the 70 nautical miles distance in 32 minutes (0.53 hour).

(3) The total distance (140 nautical miles) divided by the total flying time (1.53 hour) equals an average groundspeed of 91 knots.

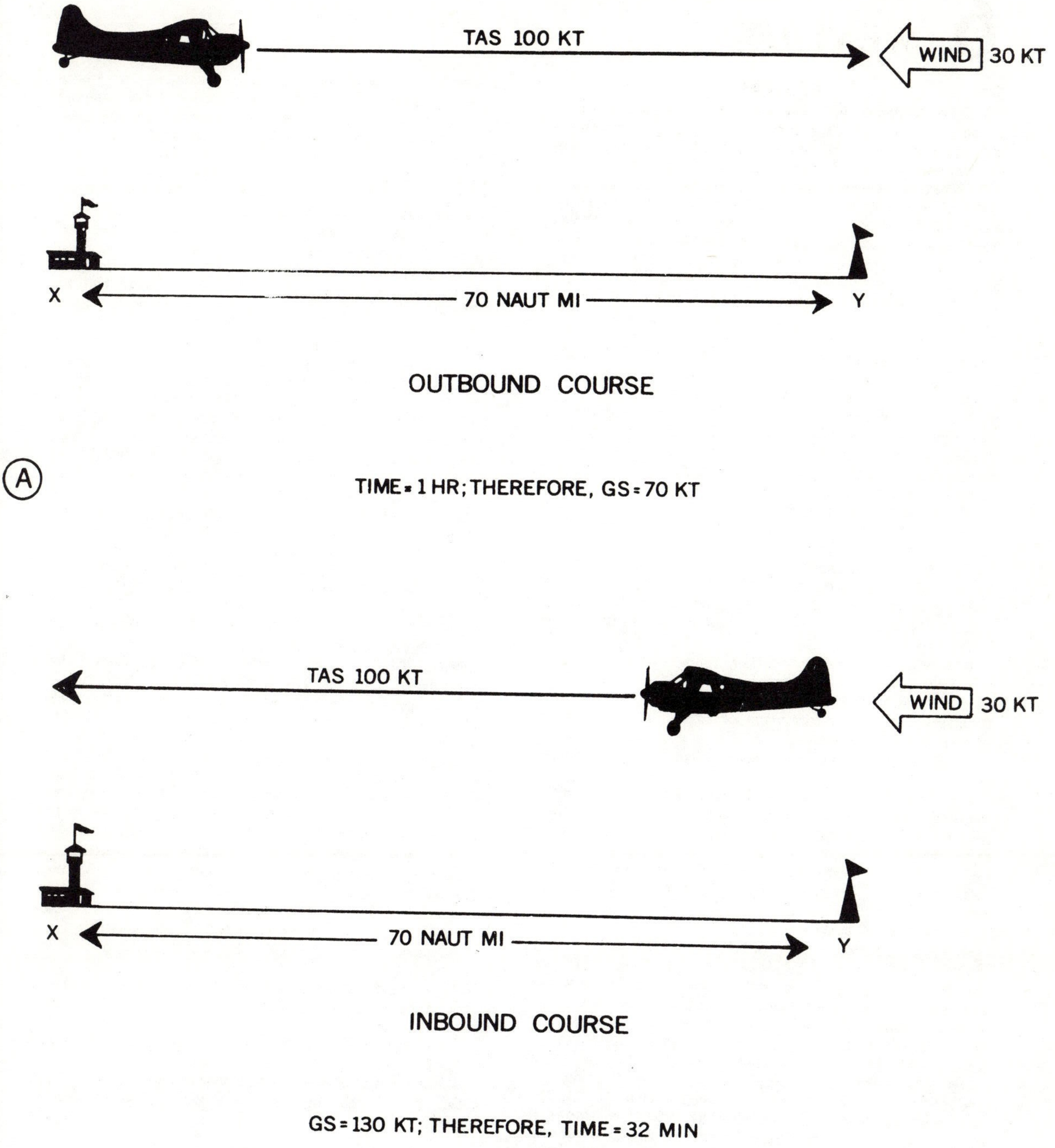

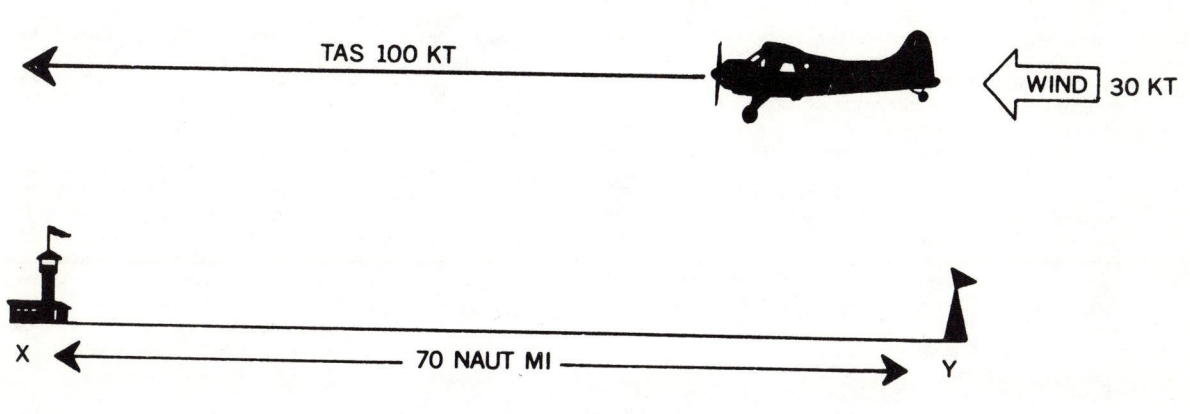

TAS 100 KT          WIND 30 KT

X ← 70 NAUT MI → Y

## OUTBOUND COURSE

Ⓐ      TIME = 1 HR; THEREFORE, GS = 70 KT

TAS 100 KT          WIND 30 KT

X ← 70 NAUT MI → Y

## INBOUND COURSE

GS = 130 KT; THEREFORE, TIME = 32 MIN

*Figure* 5-14 *Average groundspeed.*

# 6 WEATHER

## GENERAL

1. The National Weather Service (NWS) maintains a comprehensive surface and upper air weather observing program and a nation-wide aviation weather forecasting and pilot briefing service.

2. Weather observations are made each hour or more often at over 600 locations in the United States. These observations may be used to determine the present weather conditions for flight planning purposes.

3. Aviation forecasts are prepared by 51 Weather Service Forecast Offices (WSFO's). These WSFO's prepare and distribute a total of 452 terminal forecasts 3 times daily for specific airports in the 50 states and the Caribbean. They are valid to 24 hours with the last 6 hours as categorical outlooks of LIFR (limited), IFR, MVFR (marginal) VFR, and the cause of these conditions. They also prepare a total of 306 route forecasts and 39 synposes 3 times daily for the conterminuos U.S. used for PATWAS, TWEB, and briefing purposes. Forecasts issued morning and mid-day are valid for 12 hours, the forecast issued in the evening is valid for 18 hours. Twelve WSFO's also prepare area forecasts (FA) twice per day, each valid for 18 hours with an additional 12 hour categorical outlook. Winds aloft forecasts are also provided for 120 locations in the United States and Alaska for flight operational purposes. All of the above flying weather forecasts are given wide distribution via teletypewriter circuits and request reply procedures from the Weather Message Switching Center in Kansas City.

4. Available aviation weather reports and forecasts are displayed at each Weather Service Office and FAA Flight Service Station. Pilots should feel free to help themselves to this information or to ask the assistance of the duty employee.

5. When telephoning for information, use the following procedures :

   a. Identify yourself as a pilot and give aircraft identification if known. (Many persons calling Weather Service Offices want information for purposes other than flying.)

   ● b. State your intended route, destination, proposed departure time, estimated time en route and type of aircraft.

   c. Advise if prepared to fly IFR.

6. Flight Service Specialists are qualified and certificated by the *NOAA/NWS as Pilot Weather Briefers. They are not authorized to make original forecasts but are authorized to translate and interpret available forecasts and reports directly into terms of the weather conditions which you can expect along your flight route and at destination. They also will assist you in selecting an alternate course of action in the event adverse weather is encountered. It is not necessary to be thoroughly familiar with the standard phraseologies and procedures for air/ground communications. A brief call stating your message in your own words will receive immediate attention. If a complete weather briefing is desired, advise the FSS accordingly. The FSS only provides that weather information which is specifically requested.

7. Combined station/tower (CS/T) personnel are not certificated pilot weather briefers; however, they can assist you by providing factual data from weather reports and forecasts.

## EN ROUTE FLIGHT ADVISORY SERVICE

En Route Flight Advisory Service is a service specifically designed to provide the pilot with timely weather information pertinent to his type of flight intended, route of flight and altitude. It will be available throughout the conterminous U.S. along prominent and heavily traveled flyways at a service criterion of 5000' above ground level at 80 miles from an En Route Flight Advisory Service communications outlet. En Route Flight Advisory Service will be provided from selected FAA FSSs controlling one or more remote communications outlets covering a large geographical area. All communications will be conducted on the designated frequency, 122.0 MHz using the radio call (name of station) FLIGHT WATCH. Routine weather information plus current reports on the location of thunderstorms and other hazardous weather as observed and reported by pilots or observed on weather radar may be obtained from the nearest flight watch facility. Altimeter settings will not be routinely furnished by the Flight Watch specialist but will be provided upon request.

1. En Route Flight Advisory Service is not intended to be used for flight plan filing, routine position reporting or to obtain a complete preflight briefing in lieu of contacting an FSS or National Weather Service Office (WSO) prior to flight. In such instances the flight watch specialist will provide information of pertinent weather in his geographical area of responsibility and furnish the name and radio frequency of the FSS to contact for the other requested services.

2. To contact a flight watch facility on 122.0 MHz use the name of the controlling FSS and the words FLIGHT WATCH or, if the controlling FSS is unknown, simply call "FLIGHT WATCH" and give the aircraft position in relation to the nearest VOR.

   Example:
   OAKLAND FLIGHT WATCH, PIPER TWO SEVEN SIX THREE PAPA etc.
   FLIGHT WATCH, PIPER TWO SEVEN SIX THREE PAPA, OVER RED BLUFF VOR, OVER

3. Pilot participation is essential to the success of En Route Flight Advisory Service through a continuous exchange of weather information between pilots in flight and flight watch specialists on the ground. Pilots are encouraged to report weather encountered in flight between surface weather reporting points to the nearest flight watch facility or in the absence of a flight watch facility, to the nearest FSS.

4. En Route Flight Advisory Service was implemented on the West Coast of the U.S. at four FSSs: Los Angeles and Oakland, Californina; Portland, Oregon and Seattle, Washington. The remaining areas of the conterminous U.S. will be covered in incremental in the future.

   a. East Coast north of Charleston, S.C. and New England.

   b. Remaining area east of the Mississippi River.

   c. Area east of the Rocky Mountains to the Mississippi River.

   d. Rocky Mountain area and the remainder of western U.S.

   Commissioning dates and hours of service of individual Flight Watch locations will be announced by Notice to Airmen.

*National Oceanic & Atmospheric Administration*

## TRANSCRIBED WEATHER BROADCASTS

1. Equipment is provided at selected FAA FSSs by which meteorological and Notice to Airmen data is recorded on tapes and broadcast continuously over the low-frequency (200–415 kHz) navigational aid (L/MF range or H facility) and VOR.

2. Broadcasts are made from a series of individual tape recordings. The first three tapes identify the station, give general weather forecast conditions in the area, pilot reports (PIREP), radar reports when available, and winds aloft data. The remaining tapes contain weather at selected locations within a 400-mile radius of the central point. Changes, as they occur are transcribed onto the tapes.

## SCHEDULED WEATHER BROADCASTS

1. All flight service stations having voice facilities on radio ranges (VORs) or radio beacons (NDBs) broadcast weather reports and Notice to Airmen information at 15 minutes past each hour from reporting points within approximately 150 miles from the broadcast station.

### 2. UNSCHEDULED BROADCASTS

These broadcasts will be made at random times and will begin with the announcement "Aviation broadcast" followed by identification of the data.

**Example:**

Aviation Broadcast, Special Weather Report, (Notice to Airmen, Pilot Report, etc.) (location name twice) three seven (past the hour) observation . . . etc.

## IN-FLIGHT WEATHER ADVISORIES

1. The NWS issues in-flight safety advisories designated as SIGMETs (WS) AIRMET's (WA/WAC). These advisories are issued individually and their information may be included in relevant portions of Aviation Area Forecasts (FA). Normally, WSs and WAs/WACs issued separately automatically amend the relevant portion of an FA for the period of the advisory, but separate FA amendments can also be issued whenever a significant change occurs.

2. The purpose of this service is to notify en route pilots of the possibility of encountering hazardous flying conditions which probably were not provided in the preflight briefings. Whether or not the condition described is potentially hazardous to a particular flight is for the pilot himself to evalute on the basis of his own experience and the operational limits of his aircraft.

3. SIGMETs (WS) will be issued concerning weather phenomena of such severity as to be potentially hazardous to all categories of aircraft, specifically:

a. Tornadoes.
b. Line of thunderstorms (squall lines).
c. Embedded thunderstorms.
d. Hail ¾" or greater diameter.
e. Severe and extreme turbulence.
f. Severe icing.
g. Widespread duststorms/sandstorms, lowering visibilities to less than three miles.

NOTE.—In a, b, c, the fact that thunderstorms are the reason for issuance of the SIGMET implies severe or greater turbulence, and severe icing, and therefore, those items need not be specified in the advisory.

4. AIRMETs (WA) will be issued concerning weather phenomena that may be potentially hazardous to single engine and light aircraft and in some cases to all aircraft as well, specifically:

a. Moderate icing.
b. Moderate turbulence.
c. Sustained winds of 30 knots or more at, or within, 2000 feet of the surface.
d. Extensive areas of visibilities less than three (3) miles and/or ceilings less than 1,000 feet, including mountain ridges and passes.

NOTE.—Though neighboring ridges may be obscured, it is not necessary to issue an WA when conditions in mountain passes along established routes are expected to remain above ceiling 1,000 feet and visibility 3 miles.

NOTE.—SIGMETs apply to all categories of aircraft. When SIGMET and AIRMET weather categories apply simultaneously for approximately the same area, the advisories are combined and identified with the SIGMET designation. The words "Adnly" or "Also" are used to connect a combined advisory.

5. Identification of SIGMETs and AIRMETs—Advisories are identified by a letter and number beginning 0000 GMT by the WSFO where issued. The first SIGMET or AIRMET is identified as "Alfa 1"; each succeeding related advisory retains the same letter designator until cancelled, but is given the next number, i.e., "Alfa 2," etc. A SIGMET or AIRMET automatically cancels a preceding advisory of the same category and lettering. For example, SIGMET Bravo 2 supersedes

SIGMET Bravo 1 and AIRMET Alfa 4 cancels AIRMET Alfa 3. If a SIGMET or AIRMET condition develops in a second distinctly separate sector of the WSFO area, the advisory is identified as "Bravo 1," "Bravo 2," etc. Similarly, a third area is identified as "Charlie 1," "Charlie 2," etc.

6. AIRMETs (WAC) will be issued concerning weather phenomena that may be potentially hazardous to all aircraft, specifically:

a. Continuous low ceilings and/or visibilities.
b. Moderate turbulence over mountainous terrain.

7. FAA flight service stations (FSSs) broadcast SIGMETs and AIRMETs during their valid period when they pertain to the area within 150 NM of the FSS as follows:

a. SIGMETs—At 15 minute intervals (H+00, H+15*, H+30 and H+45); and AIRMETs—at 30 minute intervals (H+15* and H+45) during the first hour after issuance.

b. Thereafter, an alert notice will be broadcast at H+15* and H+45 during the valid period of the advisories.

**Example:**

"Washington SIGMET (or AIRMET) Bravo 3 is current."

*Included in the scheduled weather broadcast.

8. Pilots, upon hearing the alert notice, if they have not received the advisory or are in doubt, should contact the nearest FSS and ascertain whether the advisory is pertinent to their flights.

9. Air Route Traffic Control Centers and Terminal Control facilities, when announcing a SIGMET, do not include the WSFO name, SIGMET number or nature of the hazardous weather. Instead, the hazardous weather area (or general area affected) will be stated.

**Example:**

"Attention all aircraft, SIGMET, hazardous weather vicinity Los Angeles (or southern California). Monitor the VOR each quarter hour."

NOTE.—Control facilities do not announce AIRMETs.

## PILOT WEATHER REPORTS (PIREP's)

1. Whenever ceilings are at or below 5,000 feet, visibilities at or below five miles, or thunderstorms are reported or forecast, FAA Stations are required to solicit and collect PIREP's which describe conditions aloft. Pilots are urged to cooperate and volunteer reports of cloud tops, upper cloud layers, thunderstorms, ice, turbulence, strong winds, and other significant flight condition information. Such conditions observed between weather reporting stations are vitally needed. The PIREP's should be given to the FAA ground facility with which communication is established, i.e., FSS or Air Route Traffic Control Center. In addition to complete PIREP's, pilots can materially help round out the in-flight weather picture by adding to routine position reports, both VFR and IFR, the type of aircraft and the following phrases as appropriate:

ON TOP
BELOW OVERCAST
WEATHER CLEAR
MODERATE (or HEAVY) ICING
LIGHT, MODERATE, SEVERE, EXTREME TURBULENCE
FREEZING RAIN (or DRIZZLE)
THUNDERSTORM (location)
BETWEEN LAYERS
ON INSTRUMENTS
ON AND OFF INSTRUMENTS

2. If pilots are not able to make PIREP's by radio, reporting upon landing of the in-flight conditions encountered to the nearest Flight Service Station or Weather Service Office will be helpful. Some of the uses made of the reports are:

**a.** The airport traffic control tower uses the reports to expedite the flow of air traffic in the vicinity of the field and also forwards reports to other interested offices.

**b.** The Flight Service Station uses the reports to brief other pilots.

**c.** The Flight Weather Service Office uses the reports in briefing other pilots and in forecasting.

**d.** The Air Route Traffic Control Center uses the reports to expedite the flow of en route traffic and determine most favorable altitudes.

**e.** The Weather Service Forecast Office finds pilot reports very helpful in issuing advisories of hazardous weather conditions. This office also uses the reports to brief other pilots, and in forecasting.

## CLEAR AIR TURBULENCE (CAT)

Clear air turbulence (CAT) has become a very serious operational factor to flight operations at all levels and especially to jet traffic flying in excess of 15,000 feet. The best available information on this phenomena must come from pilots via the PIREP's procedures. All pilots encountering CAT conditions are urgently requested to report *time*, *location* and *intensity* (light, moderate, severe or extreme) of the element to the FAA facility with which they are maintaining radio contact. If time and conditions permit, elements should be reported according to the standards for other PIREP's and position reports. See Turbulence Reporting Criteria Table on preceding page.

## REPORTING OF CLOUD HEIGHTS

**1.** Ceiling, by definition in Part I Federal Aviation Regulations, and as used in Aviation Weather Reports and Forecasts, is the height *above ground* (or water) *level* of the lowest layer of clouds or obscuring phenomenon that is reported as "broken", "overcast", or "obscuration" and not classified as "thin" or "partial". For example, a forecast which reads "CIGS WILL BE GENLY 1 TO 2 THSD FEET" refers to heights *above ground level* (AGL). On the other hand, a forecast which reads "BRKN TO OVC LYRS AT 8 TO 12 THSD MSL" states that the height is *above mean sea level* (MSL).

**2.** Pilots usually report height values above mean sea level, since they determine heights by the altimeter. This is taken in account when disseminating and otherwise applying information received from pilots. ("Ceilings" heights are always above ground level.) In reports disseminated as PIREP's, height references are given the same as received from pilots, that is above mean sea level (MSL or ASL). In the following example, however, a pilot report of the heights of the bases and tops of an overcast layer in the terminal area is used in two ways in a surface aviation weather report:

A12 OVC 2FK 132/49/47/0000/002/ OVC 23

In this example the weather station has converted the pilot's report of the height of base of the overcast from the height (MSL) indicated on the pilot's altimeter to height above ground and has shown by the prefix "A" that the ceiling height was determined by an aircraft. The height of cloud tops shown in remarks (OVC 23) is *above mean sea level* (ASL or MSL) as initially reported by the pilot.

**3.** In aviation forecasts (Terminal, Area, or In-flight Advisories), ceiling are denoted by the prefix "C" when used with sky cover symbols as in "LWRG TO C5 OVC-1TRW", or by the contraction "CIG" before, or the contraction "AGL" after, the forecast cloud height value. When the cloud base is given in height above mean sea level, it is so indicated by the contraction "MSL" or "ASL" following the height value. The heights of clouds tops, freezing level, icing and turbulence are always given in heights above means sea level (ASL or MSL).

## WEATHER RADAR

**1.** The National Weather Service operates a 90-station network of weather radars. These stations are generally spaced in such a manner as to enable them to detect and identify the type and characteristics of most of the precipitation east of the Continental Divide. In addition, 32 radars of the Department of Defense and 18 FAA radars augment the network in the conterminous U.S. In Alaska data from 15 DOD radars are collected and summarized by the Anchorage Forecast Center.

**2.** When precipitation is being detected, a scheduled radar observation is taken at 40 minutes past each hour, and more often under certain conditions. These observations are transmitted to many Weather Service Stations and FAA stations and are available for use in pre-flight and in-flight planning. In addition, an hourly plain language radar summary and sixteen radar summary facsimile charts per day are prepared by the Radar Analysis and Development Unit in Kansas City, Missouri. The hourly radar summary is transmitted on Service A teletypewriter to all NWS and Flight Service Stations, and the radar chart is available to all subscribers to the facsimile circuit.

**3.** Some Flight Service Stations located near radar equipped Weather Service Stations have weather radar repeaterscopes. The FSS specialist at these locations are certified to make interpretations of the weather displayed on the radar scope. They can brief a pilot on the displayed weather pattern as to the area covered and the weather movement. However technical analysis of the intensity of precipitation and cells is made by the NWS, who in turn forwards this information to the FSS pilot briefer. The weather radar presents a very dependable display of the weather within 100 miles, and storms of considerable heights and intensities can be seen at ranges of more than 100 miles. Through the combined efforts of the NWS and the FSS, a pilot can receive a comprehensive picture of the weather to assist him in planning a safe flight.

There are 24 Flight Service Stations which provide this service. They are (alphabetically):

| | |
|---|---|
| Amarillo, Texas | Houston, Texas |
| Austin, Texas | Indianapolis, Indiana |
| Buffalo, New York | Jackson, Mississippi |
| Birmingham, Alabama | Minneapolis, Minnesota |
| Charleston, South Carolina | Mobile, Alabama |
| Chicago, Illinois | North Platte, Nebraska |
| Columbia, Missouri | Rochester, Minnesota |
| Columbus, Ohio | San Antonio, Texas |
| Dallas, Texas | St. Louis, Missouri |
| Des Moines, Iowa | Tulsa, Oklahoma |
| Garden City, Kansas | Wichita, Kansas |
| Goodland, Kansas | Wichita Falls, Texas |

## ARTCC RADAR WEATHER DISPLAY

Areas of weather clutter are radar echoes from rain or moisture. Radars cannot detect turbulence. The determination of the intensity of the weather displayed is based on its precipitation density. Generally the turbulence associated with a very heavy rate of rainfall will normally be significantly more severe than any associated with a very light rainfall rate.

ARTCCs are phasing in computer-generated digitized radar displays to replace the heretofore standard broadband radar display. This new system known as Narrowband Radar provides the controller with two distinct levels of weather intensity by assigning radar display symbols for specific precipitation densities measured by the narrowband system.

Radials (lines) show the lateral configuration of the lower density precipitation areas. "Hs" show areas of high density within the area outlined by the radials.

Refer to figure 1–6 in Chapter 1 for an illustration of the radar display symbology.

## THUNDERSTORMS

1. Turbulence, hail, rain, snow, lightning, sustained updrafts and downdrafts, icing conditions—all are present in thunderstorms. While there is some evidence that maximum turbulence exists at the middle level of a thunderstorm, recent studies show little variation of turbulence intensity with altitude.

2. There is no useful correlation between the external visual appearance of thunderstorms and the severity or amount of turbulence or hail within them. Too, the visible thunderstorm cloud is only a portion of a turbulent system whose updrafts and downdrafts often extend far beyond the visible storm cloud. Severe turbulence can be expected up to 20 miles from severe thunderstorms. This distance decreases to about 10 miles in less severe storms. These turbulent areas may appear as a well defined echo on weather radar.

3. Weather radar, airborne or ground based, will normally reflect the areas of most severe turbulence. The frequency and severity of turbulence generally increases with the radar reflectivity which is closely associated with the areas of highest liquid water content of the storm. NO FLIGHT PATH THROUGH AN AREA OF STRONG OR VERY STRONG RADAR ECHOS SEPARATED BY 20-30 MILES OR LESS MAY BE CONSIDERED FREE OF SEVERE TURBULENCE.

4. Turbulence beneath a thunderstorm should not be minimized. This is especially true when the relative humidity is low in any layer between the surface and 15,000 feet. Then the lower altitudes may be characterized by strong out-flowing winds and severe turbulence.

5. The probability of lightning strikes occuring to aircraft is greatest when operating at altitudes where temperatures are between −5°C and +5°C. Lightning can strike aircraft flying in the clear in the vicinity of a thunderstorm.

6. It is impossible to state rules for safe flight through a thunderstorm. A "soft" altitude in one thunderstorm may be a most severe altitude in another. THE ONLY SAFE RULE FOR THUNDERSTORM FLYING IS TO STAY OUT OF THEM AND GIVE THEM A WIDE BERTH.

## ATC INFLIGHT WEATHER-AVOIDANCE ASSISTANCE

For obvious reasons of safety, an IFR pilot must not deviate from the course or altitude/flight-level without a proper ATC clearance. When weather conditions encountered are so severe that an immediate deviation is determined to be necessary and time will not permit approval by ATC, the pilot's emergency authority may be exercised.

When the pilot requests clearance for a route deviation or for an ATC radar vector, the controller must evaluate the air traffic picture in the affected area, and coordinate with other controllers (if ATC jurisdictional boundaries may be crossed) before replying to the request.

It should be remembered that the controller's primary function is to provide safe separation between aircraft. Any additional service, such as weather avoidance assistance, can only be provided to the extent that it does not derogate the primary function. It's also worth noting that the separation workload is generally greater than normal when weather disrupts the usual flow of traffic. ATC radar limitations (see Chapter 1) and frequency congestion may also be a factor in limiting the controller's capability to provide additional service.

It is very important therefore, that the request for deviation or radar vector be forwarded to ATC as far in advance as possible. Delay in submitting it may delay or even preclude ATC approval or require that furnished to ATC when requesting clearance to detour around weather activity:

 a. Proposed point where detour will commence.

 b. Proposed route and extent of detour (direction and distance).

 c. Point where original route will be resumed.

 d. Flight conditions (IFR or VFR).

 e. Any further deviation that may become necessary as the flight progresses.

 f. Advise if the aircraft is equipped wth functioning airborne radar.

To a large degree, the assistance that might be rendered by ATC will depend upon the weather information available to controllers. Due to the extremely transitory nature of severe weather situations, the controller's weather information may be of only limited value if based on weather observed on radar only. Frequent updates by pilots giving specific information as to the area affected, altitudes intensity and nature of the severe weather can be of considerable value. Such reports are relayed by radio or phone to other pilots and controllers and also receive widespread teletype-writer dissemination.

Obtaining IFR clearance or an ATC radar vector to circumnavigate severe weather can often be accommodated more readily in the enroute areas away from terminals because there is usually less congestion and, therefore, greater freedom of action. In terminal areas, the problem is more acute because of traffic density, ATC coordination requirements, complex departure and arrival routes, adjacent airports, etc. As a consequence, controllers are less likely to be able to accommodate all requests for weather detours in a terminal area or be in a position to volunteer such route to the pilot. Nevertheless, pilots should not hesitate to advise controllers of any observed severe weather and should specifically advise controllers if they desire circumnavigation of observed weather.

## AIRFRAME ICING

The effects of ice accretion on aircraft are cumulative—Thrust is reduced, Drag increases, Lift lessens, Weight increases. The results are an increase in stall speed and a deterioration of aircraft performance. In extreme cases, 2 to 3 inches of ice can form on the leading edge of the airfoil in less than 5 minutes. It takes but ½ inch of ice to reduce the lifting power of some aircraft by 50% and increases the frictional drag by an equal percentage.

A pilot can expect icing when flying in visible precipitation such as rain or cloud droplets, and the temperature is 0 degrees Celsius or colder. When icing is detected, a pilot should do one of two things (particularly if the aircraft is not equipped with deicing equipment) he should get out of the area of precipitation or go to an altitude where the temperature is above freezing. This "warmer" altitude may not always be a lower altitude. Proper pre-flight action includes obtaining information on the freezing level and the above-freezing levels in precipitation areas. Report icing to ATC/FSS, and if operating IFR, request new routing or altitude if icing will be a hazard. Be sure to give type of aircraft to ATC when reporting icing. Following is a table that describes how to report icing conditions.

| INTENSITY | ICE ACCUMULATION |
|---|---|
| Trace | Ice becomes perceptible. Rate of accumulation slightly greater than rate of sublimation. It is not hazardous even though deicing/anti-icing equipment is not utilized, unless encountered for an extended period of time (over 1 hour). |
| Light | The rate of accumulation may create a problem if flight is prolonged in this environment (over 1 hour). Occasional use of deicing/anti-icing equipment removes/prevents accumulation. It does not present a problem if the deicing/anti-icing equipment is used. |
| Moderate | The rate of accumulation is such that even short encounters become potentially hazardous and use of deicing/anti-icing equipment or diversion is necessary. |
| Severe | The rate of accumulation is such that deicing/anti-icing equipment fails to reduce or control the hazard. Immediate diversion is necessary. |

Pilot Report: Aircraft Identification, Location, Time (GMT), Intensity of Type,* Altitude/FL, Aircraft Type, IAS.

* Rime Ice: Rough, milky, opaque ice formed by the instantaneous freezing of small supercooled water droplets.

Clear Ice: A glossy, clear, or translucent ice formed by the relatively slow freezing of large supercooled water droplets.

## ALTIMETRY

1. The accuracy of aircraft altimeters is subject to the following factors: (a) nonstandard temperature of the atmosphere; (b) aircraft static pressure systems (position error); and (c) instrument error. Pilots should disregard the effect of nonstandard atmospheric temperatures except that low temperatures need to be considered for terrain clearance purposes.

NOTE.—Standard temperature at sea level is 15° C (59° F). The temperature gradient from sea level is 2° C (3.5° F) per 1000 feet. Pilots should apply corrections for static pressure systems and/or instruments, if appreciable errors exist.

2. Experience gained with the present volume of operations has led to the adoption of a standard altimeter setting for flights operating in the higher altitudes.

A standard setting eliminates altitude conflicts caused by altimeter settings derived from different geographical sources. In addition, it eliminates station barometer errors and some of the altimeter instrument errors. The following outlines the current procedures in effect:

a. The cruising altitude or flight level of aircraft shall be maintained by reference to an altimeter which shall be set, when operating:

(1) Below 18,000 feet MSL—to the current reported altimeter setting of a station along the route and within 100 nautical miles of the aircraft; or, if there is no station within this area, to the current reported altimeter setting of an appropriate available station. In the case of an aircraft not equipped with a radio, set to the elevation of the departure airport or use an appropriate altimeter setting available prior to departure.

(2) At or above 18,000 feet MSL—to 29.92'' Hg (standard setting).

NOTE.—The lowest usable flight level is determined by the atmospheric pressure in the area of operation, as shown in the following table:

| Altimeter Setting (Current Reported) | Lowest usable flight level |
|---|---|
| 29.92 or higher | 180 |
| 29.91 to 29.42 | 185 |
| 29.41 to 28.92 | 190 |
| 28.91 to 28.42 | 195 |
| 28.41 to 27.92 | 200 |

(3) Where the minimum altitude, as prescribed in FAR Parts 91.79 and 91.119, is above 18,000 feet M.S.L. the lowest usable flight level shall be the flight level equivalent of the minimum altitude plus the number of feet specified in the following table:

| Altimeter Setting | Correction Factor |
|---|---|
| 29.92 or higher | None |
| 29.91 to 29.42 | 500 feet |
| 29.41 to 28.92 | 1000 feet |
| 28.91 to 28.42 | 1500 feet |
| 28.41 to 27.92 | 2000 feet |
| 27.91 to 27.42 | 2500 feet |

NOTE.—For example, the minimum safe altitude of a route is 19,000 feet MSL and the altitimeter setting is reported between 29.92 and 29.42 Hg., the lowest usable flight level will be 195, which is the flight level equivalent of 19,500 feet M.S.L. (minimum altitude plus 500 feet).

## PILOTS AUTOMATIC TELEPHONE WEATHER ANSWERING SERVICE (PATWAS)

At some locations the numbers of pilots requiring flight weather briefings are too numerous for person-to-person briefings. To assist in this important service, recorded weather briefings are available at several locations. This service is called Pilots Automatic Telephone Weather Answering Service (PATWAS). The recorded telephone briefing includes a weather forecast which emphasizes expected weather up to about 12 hours in advance. Forecasts given around 6 p.m. may be an 18 hour forecast, with emphasis on the outlook for the next morning. The PATWAS locations are found in Part 2 of the AIM under the FSS-CS/T Weather Service Telephone Numbers section.

# KEY TO AVIATION WEATHER REPORTS . . . . . . .

| LOCATION IDENTIFIER AND TYPE OF REPORT* | SKY AND CEILING | VISIBILITY WEATHER AND OBSTRUCTION TO VISION | SEA-LEVEL PRESSURE | TEMPERATURE AND DEW POINT | WIND | ALTIMETER SETTING | RUNWAY VISUAL RANGE | CODED PIREPS |
|---|---|---|---|---|---|---|---|---|
| MKC | 15SCT M25OVC | 1R-K | 132 | /58/56 | /1807 | /993/ | R04LVR20V40 | /UA OVC 55 |

## SKY AND CEILING

Sky cover contractions are in ascending order. Figures preceding contractions are heights in hundreds of feet above station.

Sky cover contractions are:

CLR Clear: Less than .1 sky cover.
SCT Scattered: .1 to .5 sky cover.
BKN Broken: .6 to .9 sky cover.
OVC Overcast: More than .9 sky cover.

— Thin (When prefixed to the above symbols.)
–X Partial obscuration: .1 to less than 1.0 sky hidden by precipitation or obstruction to vision (bases at surface).
X Obscuration: 1.0 sky hidden by precipitation or obstruction to vision (bases at surface).

Letter preceding height of layer identifies ceiling layer and indicates how ceiling height was obtained. Thus:

| E | Estimated height | V | Immediately following numerical values indicates a variable ceiling |
|---|---|---|---|
| M | Measured | | |
| W | Indefinite | | |

## VISIBILITY

Reported in statute miles and fractions. (V Variable)

## WEATHER AND OBSTRUCTION TO VISION SYMBOLS

| A | Hail | IC | Ice crystals | S | Snow |
|---|---|---|---|---|---|
| BD | Blowing dust | IF | Ice fog | SG | Snow grains |
| BN | Blowing sand | IP | Ice pellets | SP | Snow pellets |
| BS | Blowing snow | IPW | Ice pellet showers | SW | Snow showers |
| D | Dust | K | Smoke | T | Thunderstorms |
| F | Fog | L | Drizzle | T+ | Severe thunderstorm |
| GF | Ground fog | R | Rain | ZL | Freezing drizzle |
| H | Haze | RW | Rain showers | ZR | Freezing rain |

Precipitation intensities are indicated thus:
Light: (no sign) Moderate: + Heavy

## WIND

Direction in tens of degrees from true north, speed in knots. 0000 indicates calm. G indicates gusty. Peak speed of gusts follows G or Q when gusts or squall are reported. The contraction WSHFT followed by GMT time group in remarks indicates windshift and its time of occurrence. (Knots X 1.15=statute mi/hr.)

EXAMPLES: 3627=360 Degrees, 27 knots; 3627G40=360 Degrees, 27 knots Peak speed in gusts 40 knots.

## ALTIMETER SETTING

The first figure of the actual altimeter setting is always omitted from the report.

## RUNWAY VISUAL RANGE (RVR)

RVR is reported from some stations. Extreme values during 10 minutes prior to observation are given in hundreds of feet. Runway identification precedes RVR report.

## CODED PIREPS

Pilot reports of clouds not visible from ground are coded with ASL height data preceding and/or following sky cover contraction to indicate cloud bases and/or tops, respectively. UA precedes all PIREPS.

## DECODED REPORT

Kansas City: Record observation, 1500 feet scattered clouds, measured ceiling 2500 feet overcast, visibility 1 mile, light rain, smoke, sea-level pressure 1013.2 millibars, temperature 58°F, dewpoint 56°F, wind 180° at 7 knots, altimeter setting 29.93 inches, Runway 04 left, visual range 2000 feet variable to 4000 feet. Pilot reports top of overcast 5500 feet.

## *TYPE OF REPORT

The omission of type-of-report data identifies a scheduled record observation for the hour specified in the sequence heading. An out-of-sequence, special observation is identified by the letters "SP," following station identification and a 24-hour clock time group, e.g., "PIT SP 1715—X M10VC." A special report indicates a significant change in one or more elements.

# KEY TO AVIATION WEATHER FORECASTS

**TERMINAL FORECASTS** contain information for specific airports on expected ceiling, cloud heights, cloud amounts, visibility, weather and obstructions to vision and surface wind. They are issued 3 times/day and are valid for 24 hours. The last six hours of each forecast are covered by a categorical statement indicating whether VFR, MVFR, IFR or LIFR conditions are expected. Terminal forecasts will be written in the following form:

CEILING: Identified by the letter "C."
CLOUD HEIGHTS: In hundreds of feet above the station (ground).
CLOUD LAYERS: Stated in ascending order of height.
VISIBILITY: In statute miles but omitted if over 6 miles.
WEATHER AND OBSTRUCTION TO VISION: Standard weather and obstruction to vision symbols are used
SURFACE WIND: In tens of degrees and knots; omitted when less than 10

## EXAMPLE OF TERMINAL FORECAST

DCA 221010—DCA Forecast 22nd day of month—valid time 10Z—10Z.
19SCT C18BKN 55W— 3415G25 OCNL C8 X1SW—Scattered clouds at 1900 feet, ceiling 1800 feet broken, visibility 5 miles, ing 1800 feet becoming, surface wind 340 degrees 15 knots Gusts to 25 knots, occasional ceiling 8 hundred feet sky ob-

12Z C50BKN 3312G22—At 12Z becoming ceiling 5000 feet broken, surface wind 330 degrees 12 knots Gusts to 22.
04Z MVFR Last 6 hours of FT after 04Z marginal VFR due to ceiling.

**AREA FORECASTS** are 18-hour aviation outlook prepared 2 times/day giving general description of cloud cover, weather and frontal conditions for an area the size of several states Heights of cloud tops, and icing are referenced ABOVE SEA LEVEL (ASL); ceiling heights, ABOVE GROUND LEVEL (AGL): bases of cloud layers are AGL unless indicated. Each SIGMET or AIRMET affecting an FA area will also serve to amend the Area Forecast.

**SIGMET or AIRMET** messages warn airmen in flight of potentially hazardous weather such as squall lines, thunderstorms, fog, icing, and turbulence. SIGMET concerns severe and extreme conditions of importance to all aircraft. AIRMET concerns less severe conditions which may be hazardous to some aircraft or to relatively inexperienced pilots. Both are broadcast by FAA on NAVAID voice channels.

**WINDS AND TEMPERATURES ALOFT (FD) FORECASTS** are 12-hour forecasts of wind direction (nearest 10° true N) and speed (knots) for selected flight levels. Temperatures aloft (°C) are included for the 3-, 6- and 9-foot levels.

## EXAMPLES OF WINDS AND TEMPERATURES ALOFT (FD) FORECASTS:

FD WBC 121745
BASED ON 121200Z DATA
VALID 130000Z FOR USE 1800-0300Z. TEMPS NEG ABV 24000

| FT | 3000 | 6000 | 9000 | 12000 | 18000 | 24000 | 30000 | 34000 | 39000 |
|---|---|---|---|---|---|---|---|---|---|
| BOS | | 3425-07 | 3420-11 | 3421-16 | 3516-27 | 3512-38 | 311649 | 292451 | 283451 |
| JFK | 3127 | | | | | | | | |
| | 3026 | 3327-08 | 3324-12 | 3322-16 | 3120-27 | 2923-38 | 284248 | 285150 | 285749 |

At 6000 feet ASL over JFK wind from 330° at 27 knots and temperature minus 8°C

**TWEB (CONTINUOUS TRANSCRIBED WEATHER BROADCAST)**—Individual route forecasts covering a 25 nautical mile zone either side of a course line for the route. By requesting a specific route number, detailed en route weather for a 12- to 18-hour period plus a synopsis can be obtained.

**PILOTS . . .** report in-flight weather to nearest FSS. The latest surface weather reports are available by phone at the nearest pilot weather briefing office by calling at H | 10.

As the pilot gains experience, he (or she) is able to select the most accurate data available and integrate it into a system of navigation that best fits the existing flight conditions. A knowledge of weather conditions that may be expected in flight is provided pilots by the numerous weather stations. Since weather is a prime factor affecting any flight, it is important that the pilot have a thorough understanding of weather information and the services available.

## WEATHER CHARTS

The pioneer in weather charting, the Air Weather Service (AWS) assembles weather information from all regions of the world to provide the U. S. Air Force with a world-wide forecasting service. This weather information is collected at regular and frequent intervals from thousands of observing stations. The Air Weather Service has comparatively few stations; therefore, it depends upon civilian weather services for data concerning

(NMC), AWS Weather Centrals and Forecast Centers, only the most important information such as wind speed and direction, temperature, dew point, and existing weather is included.

The weather analysis depicted on the surface chart in figure 6-2 illustrates:

- Surface frontal position
- Pressure system centers
- Precipitation areas and type
- Isobars (lines connecting points of equal pressure)

Weather which is hazardous to flight is indicated in red symbols.

### Surface Prognostic Charts

Surface prognostic (forecast) charts are prepared by the NMC, AWS Weather Centrals and Forecast Centers. They are transmitted via facsimile network and teletype bulletins. These charts can also be prepared by the forecaster when circumstances require. Figure 6-3 shows a prog-

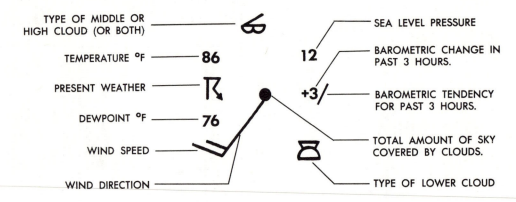

*Figure 6-1    Plotted Data Around Station Circle on Facsimile Surface Chart*

North America. Ships, aircraft, and stations of other U. S. military services also furnish information. The surface and upper air data observed by stations throughout the world are collected and plotted on surface and constant pressure charts.

### Surface Charts

The station circle illustrated in figure 6-1 is used on the facsimile surface charts. The facsimile surface chart (weather map) is received every three hours; from these charts, forecasters obtain a picture of conditions existing at the time of the observations.

When charts are prepared for facsimile transmission by the National Meteorological Center

nostic chart which depicts the expected position and orientation of fronts, pressure systems, cloud pattern, and other areas of weather significant to flying operations. Prognostic charts are valid for 12 to 72 hours after the time of preparation; most mission planning is based on a 12- or 18-hour prognostic chart.

After careful study of current and prognostic charts, the forecaster relies on his training, experience, and judgment to make local and operational forecasts.

### Constant Pressure Charts

Upper air data for selected standard pressure levels are plotted on constant pressure charts and analyzed. This information is obtained by upper air soundings supplemented by aircraft inflight reports (AIREP). The aircraft report may be the only information available to the forecaster for overwater areas or in areas where there are minimum reporting stations.

The standard pressure levels, for which constant pressure charts (CPC) are constructed and transmitted on facsimile, are shown in figure 6-4 . A

charts and surface charts furnishes the user with:
• Wind direction and speed at specific levels
• Temperature and dewpoint depression (temperature dewpoint spread)
• "D" value and expected drift
• Intensity, speed, and direction of movement of frontal and pressure systems
• Amount, type, and intensity of cloud forms and precipitation areas

*Figure 6-2   Surface Chart Prepared in Weather Station*

typical constant pressure chart for the 300-mb level is illustrated in figure 6-5 . These charts are prepared from 0000Z and 1200Z observations. The CPC analysis illustrates:
• Contour lines (lines of equal true altitude)
• Isotherms (lines of equal temperature)
• Isotachs (lines of equal wind speed)
• Height centers (highs and lows)
• Region of maximum wind

Current facsimile constant pressure charts are prepared with computer inputs. All of the five fields of data listed above will not appear on any *given* chart but can be determined from any *series* of charts for all standard levels.

The constant pressure charts, together with surface charts and other charts and diagrams, present a three-dimensional picture of the atmosphere. The combined information from constant pressure

• Areas of thunderstorms.

From these charts there are derived three general rules which the navigator can safely use for flight planning purposes. These are:
1. The winds blow parallel to the contour lines.
2. The speed of the wind is proportional to the spacing of the contour lines; the closer the contour lines, the stronger the winds.
3. Wind blows clockwise around a high and counterclockwise around a low.

**Constant Pressure Prognostic Charts**

Constant pressure prognostic charts are prepared by the National Meteorological Center, AWS Weather Centrals and Forecast Centers; they are transmitted to field weather stations via facsimile network. A prognostic chart for the 300-mb level is illustrated in figure 6-6 .

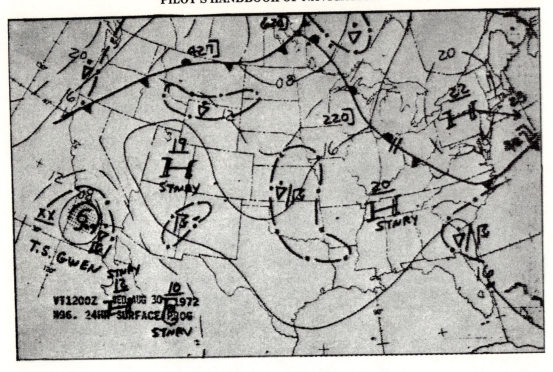

Figure 6-3. Surface Prognostic (Forecast) Chart

| Pressure Altitude | | Temperature | Pressure |
|---|---|---|---|
| (meters) | (feet) | (°C) | (mb) |
| 16,180 | 53,083 | −57 | 100 |
| 11,784 | 38,662 | −57 | 200 |
| 9,164 | 30,065 | −46 | 300 |
| 5,574 | 18,289 | −21 | 500 |
| 3,012 | 9,882 | − 5 | 700 |
| 1,457 | 4,781 | + 6 | 850 |

Mean Sea Level (59°F) 15°C; (29.92" Hg); 1013.25 mb.

Figure 6-4. Standard Pressure Levels

These prognostic charts indicate:
• Forecast position and orientation of contours.
• Forecast position of trough lines, and isotachs.
• Forecast position of circulation centers.

NOTE: Remember prognostic charts represent weather conditions anticipated at a specific time, not average weather conditions over a period of time.

### Winds Aloft Charts

Winds aloft charts are prepared four times daily from data obtained from upper air observations at 0000Z, 0600Z, 1200Z, and 1800Z. The information collected and plotted on winds aloft charts contains winds for selected levels in the troposphere and stratosphere. Some typical winds aloft charts are shown in figure 6-7.

NOTE: Winds aloft charts do not contain forecast winds, they contain actual winds which can be up to 12 hours old.

Despite the fact that winds from these charts are not necessarily current, they are important to aircrews for computing headings, altitudes, groundspeeds, and time enroute. The detachment forecaster can provide valuable guidance or assistance in determining the representability of these winds.

### Summary

With world-wide coverage and various facilities, the Air Weather Service provides vital weather information to aircrews throughout the USAF. It is the job of the navigator to correctly interpret the information provided and use it to best advantage. The foregoing discussion is an introduction to the charts most often used. It is not a complete coverage of the subject. Some of the weather charts discussed in this chapter may not be displayed in all weather stations, but are available upon request.

## WEATHER REPORTS AND SYMBOLS

### Surface Observations

Surface weather observations are made hourly by Air Weather Service observers or National Weather Service personnel. When a weather ele-

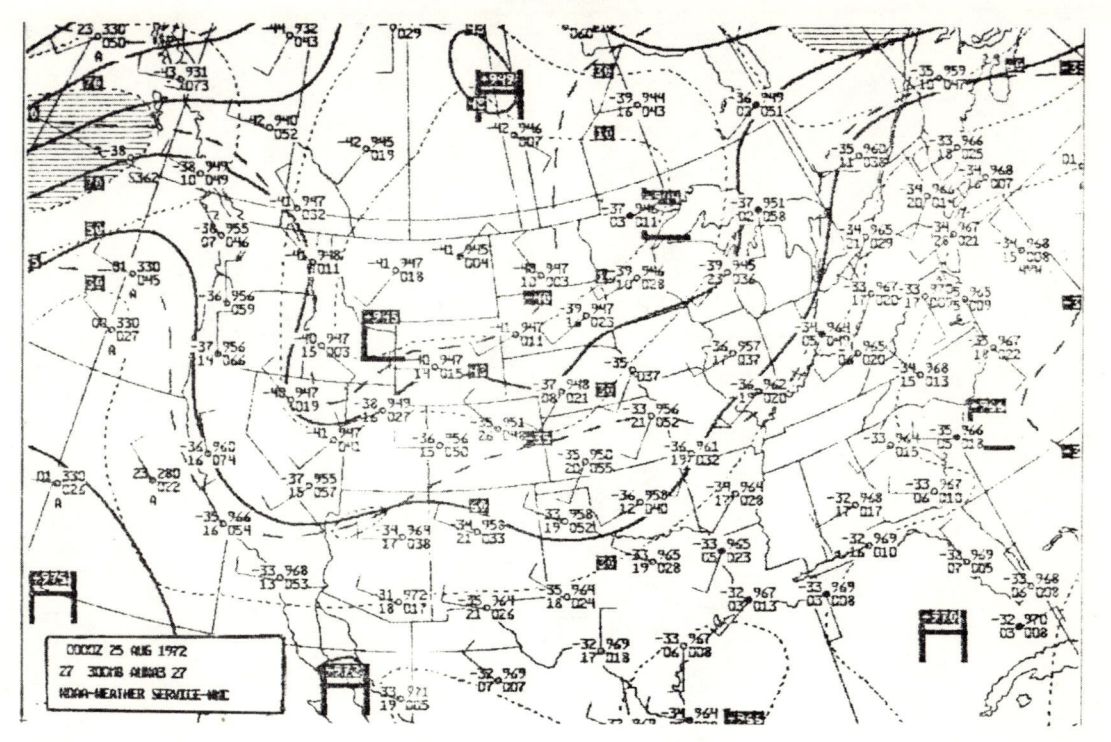

Figure 6-5. 300-mb Constant Pressure Chart

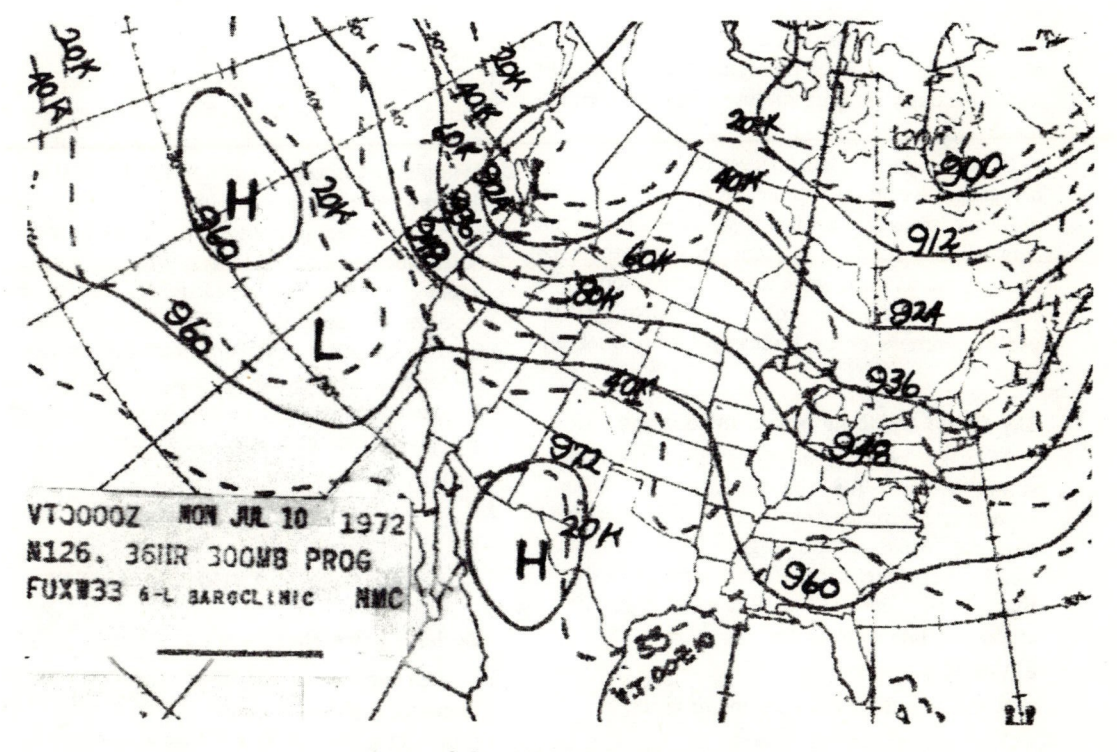

Figure 6-6. 300-mb Prognostic Chart

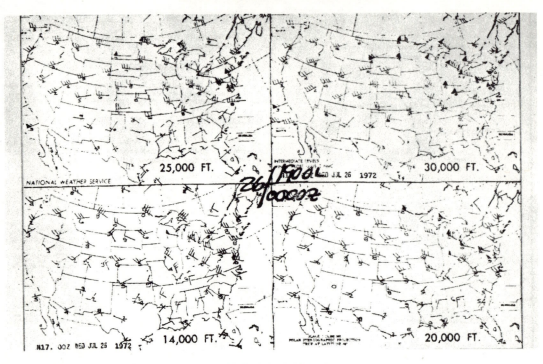

*Figure 6-7. Winds Aloft Charts*

ment changes significantly, a special observation is taken. Automatic and continuous observations of such elements as the ceiling, visibility, wind, pressure, temperature, and dewpoint are made by weather instruments. These observations are placed in the hands of the using agencies almost instantaneously through the use of modern equipment.

Observations of vital interest to crew members are called Aviation Weather Reports. These reports are transmitted over a world-wide teletype network and received by individual weather stations in the aviation weather reporting code. The reports are collected in sequence and displayed for use by aircrews or by forecasters who brief aircrews.

### Sky Cover Symbols and Contractions

The sky cover symbols and contractions shown in figure 6-8 and the keys to weather reports and forecasts on page 73 are an international weather language. They provide weather personnel with information in a format that is easily understood. A typical report includes the following items:

- Visibility
- Weather and/or obstructions to vision
- Temperature and dewpoint
- Wind

- Altimeter setting

### PILOT-TO-FORECASTER SERVICE

Air Weather Service operates a pilot-to-forecaster facility at many bases. These facilities enable airborne crew members to communicate with a forecaster. It is the best and fastest method of obtaining weather information for aircraft in flight. Aircrews use this service to obtain terminal forecasts, altimeter settings, current winds aloft, or other weather data. This also provides a means of passing a pilot report (PIREP), when actual weather encountered differs significantly from the forecast weather. Pilot-to-forecaster service has the highest priority in the weather station except forecaster duties associated with aircraft emergencies or emergency war orders.

### WEATHER FOR FLIGHT PLANNING

#### Gathering Weather Data

Weather is an extremely important factor in planning any flight mission. These are the steps to follow in gathering weather data:

*Step 1.* Know exactly what weather information is needed. This normally consists of, but is not

| Summation Amount of Sky Cover | | Symbol | Contraction | Remarks |
|---|---|---|---|---|
| 1/10 to less than 10/10 surface-based obscuring phenomena | | -X | -X | No height assigned this condition. Vertical visibility is not completely restricted. |
| 10/10 surface-based obscuring phenomena | | X | X | Always preceded by a vertical visibility (height) value. Height value preceding this symbol is normally prefixed with the ceiling designator W. |
| Clear | | ◯ | CLR | This symbol (contraction) is not used in combination with any other. |
| Less than 1/10 thru 5/10, half or more thin | Only used to report layers aloft | -◐ | -SCT | Height values preceding these symbols (contractions) are never designated as ceiling layers. Note that "less than 1/10" relates to a trace of clouds. A trace of obscuring phenomena aloft is not reportable as sky cover. |
| Less than 1/10 thru 5/10, more than half opaque | | ◐ | SCT | |
| 6/10 thru less than 10/10, half or more thin | | -◑ | -BKN | |
| 6/10 thru less than 10/10, more than half opaque | | ◑ | BKN | Height value preceding this symbol (contraction) is prefixed with a ceiling designator (M or E), provided a lower ceiling layer is not present. |
| 10/10, half or more thin | | -⊕ | -OVC | Height value preceding this symbol (contraction) is never designated as a ceiling layer. |
| 10/10, more than half opaque | | ⊕ | OVC | Height value preceding this symbol (contraction) is prefixed with a ceiling designator (M or E), provided a lower broken ceiling layer is not present. This symbol (contraction) is used in combination with lower overcast layers only when such layers are classified as thin. |

*Figure 6-8*
*Sky Cover Symbols and Contractions*

necessarily limited to, the weather and winds to expect enroute, the weather at destination and alternate destinations, and the local weather for the time of takeoff and climb-out.

*Step 2.* Inform the weather forecaster of aircraft type, estimated time of departure (ETD), proposed route and flight altitude, estimated time enroute (ETE), and any additional information that will help him visualize the flight. The more information he is given, the better he is able to provide data pertinent to the flight.

*Step 3.* When the weather briefing is completed, insure that the following information is complete:

WEATHER FOR TAKEOFF AND CLIMB

- Surface temperature and pressure altitude (or density altitude)
- Surface winds
- Bases and tops of cloud layers
- Visibility
- Precipitation
- Freezing level
- Climb winds

FORECAST WEATHER ENROUTE

- Bases, tops, type, and amount of each cloud layer
- Visibility at flight altitude
- Type, location, intensity, and direction and speed of frontal movements
- Freezing levels
- Temperatures and winds at flight altitude
- Areas of hazardous weather (thunderstorms, hail, icing, and turbulence)
- Areas of good weather (for use in event of an emergency landing enroute).

FORECAST WEATHER FOR DESTINATION AND ALTERNATES

- Bases, tops, type, and amount of cloud layers
- Visibility
- Weather and obstructions to vision
- Freezing level
- Surface wind speed and direction
- Forecast altimeter setting

Insure that all essential weather elements are included in the briefing and request clarification or additional information concerning any weather data about which there are doubts.

*NOTE: If the time of departure is delayed longer than 1½ hours after the time of the briefing, the weather must be revalidated.*

In addition to forecaster briefings at the weather stations, some bases use closed-circuit television to brief aircrews. This gives the aircrews a visual weather briefing without going to the weather station. In addition, a small self-briefing weather display is often available to allow the crew to flight plan before getting the final weather briefing.

## Applying Weather Data

Getting all the facts and applying them correctly is essential in flight planning. Departure weather can be the deciding factor when an emergency arises soon after takeoff. In some cases, the weather may deteriorate rapidly shortly after takeoff. The navigator must know the winds up

```
ATC ROUTE-OVERLAND EAST
THIS CFP VALID FOR DEPARTURES 3 HOURS EITHER SIDE OF ETD DISPLAYED

AFGWC TIME PHASED ROUTE FORECAST                          ROUTE PRIORITY-17
DEPARTURE POINT KMHR
LAT 3834N  LONG 12118W DEPT ALT    1000FT  ETD 71 NOV 95 1600Z
NO COLUMNAR GHD GWD
NAVIGATION PARAMETERS ARE BASED ON LEG MIDPOINTS
ACCUMULATED TIMES ADJUSTED FOR 1 PCT ROUTE DEVIATIONS
LOCID    LAT    LONG ALT TTT DDVVV    WF THD MHD   GS  TAS TIM ATIM
A      3833N 12033W  90  13 28001     1 092 075   161  160  13 0013
B      3911N 12016W 150   0 27024     6 013 356   196  190  12 0025
C      3932N 12016W 150  -8 27035    -8 345 327   182  190   7 0032
D      4027N 11730W 150  -9 28037    30 062.044   220  190  39 0111
E      4111N 11500W 150 -10 29043    32 062 045   222  190  33 0144
F      4158N 11200W 160 -13 29051    35 062 046   225  190  39 0223
G      4111N 11500W 160 -14 29052   -42 264 247   148  190  59 0322
H      4027N 11730W 160 -12 29046   -38 259 242   152  190  49 0411
I      3932N 12018W 160 -11 28039   -34 254 237   156  190  55 0506
J      3650N 12054W 160 -11 27035   -21 224 206   169  190  18 0524
K      3841N 12106W 160 -10 27035   -27 234 217   163  190   5 0529
 MHR   3834N 12118W 160 -10 27034   -29 240 223   161  190   5 0534
ETP SUMMARY        FWF -10, WF1  22, WF2 -36, TIME 0230, DIST  517
OVERALL TO KMHR   AWF  -10   TDIS   981   ATIM 0534  ETA 2134Z
```

*Figure 6-9 Computer Flight Plan*

to flight altitude in order to compute the distance flown and the fuel consumed during the climb to altitude.

A knowledge of the clouds at flight altitude gives the aircrew an idea of the areas of possible precipitation, icing, turbulence, and other hazards to flight. If the navigator has this knowledge, it increases his ability to make the proper operational decision in the event of an enroute emergency.

### Computer Flight Plans

One of the newer and more sophisticated aids for the navigator is the Computer Flight Plan (CFP). The computer simulates the response of an aircraft to the environmental conditions likely to occur during a given flight. If given track, altitude, true airspeed, and time of departure, the computer may then determine the location of the aircraft and its groundspeed, heading, wind factor, remaining fuel load, and many other factors for any specified point along the route.

The CFP is used extensively by Air Force's Strategic Air Command and Military Airlift Command on their long-haul flights. The use of this aid has materially reduced flight planning and forecaster preparation times with a substantial improvement in overall accuracy in many cases. During the winter months, however, errors in individual leg winds and associated wind factors may

be large, but when averaged over several legs, the predicted wind factor will normally be very close to the actual value.

CFPs are most useful and accurate for the long flight. To obtain a Computer Flight Plan, the navigator should contact the local weather forecaster preferably 11 to 24 hours before takeoff. If the route is a non-standard route. the forecaster will require specific flight information: Takeoff time, track (latitude and longitude and ICAO identifiers if appropriate and desired) altitude, and true airspeed. A wide variety of navigational options and print-out formats are available and may be requested. An example of a CFP is illustrated in figure 6-9.

### SUMMARY

Aviation weather reports provide vital information to the aircrews when planning a flight. These reports can be used to obtain data pertinent to all phases of flight. Data obtained during weather briefing may be rapidly updated through use of the pilot-to-forecaster service.

It is essential that the navigator thoroughly understand all the factors of weather information that are required to plan a flight adequately. Getting all the facts and correctly interpreting them leads to successful completion of each phase of the flight.

# 7 GOOD OPERATING PRACTICES

It should be remembered that adherence to air traffic rules does not eliminate the need for good judgment on the part of the pilot. Compliance with the following Good Operating Practices will greatly enhance the safety of every flight.

## ALERTNESS

Be alert at all times, especially when the weather is good. Most pilots pay attention to business when they are operating in full IFR weather conditions, but strangely, air collisions almost invariably have occurred under ideal weather conditions. Unlimited visibility appears to encourage a sense of security which is not at all justified. Considerable information of value may be obtained by listening to advisories being issued in the terminal area, even though controller workload may prevent a pilot from obtaining individual service.

## JUDGMENT IN VFR FLIGHT

Use reasonable restraint in exercising the prerogative of VFR flight, especially in terminal areas. The weather minimums and distances from clouds are minimums. Giving yourself a greater margin in specific instances is just good judgment.

Conducting a VFR operation in a Control Zone when the official visibility is 3 or 4 miles is not prohibited, but good judgment would dictate that you keep out of the approach area.

It has always been recognized that precipitation reduces forward visibility. Consequently, although again it may be perfectly legal to cancel your IFR flight plan at any time you can proceed VFR, it is good practice, when precipitation is occuring, to continue IFR operation into a terminal area until you are reasonably close to your destination.

In conducting simulated instrument flights, be sure that the weather is good enough to compensate for the restricted visibility of the safety pilot and your greater concentration on your flight instruments. Give yourself a little greater margin when your flight plan lies in or near a busy airway or close to an airport.

## USE OF CLEARING PROCEDURES

**1. Before Takeoff.** Prior to taxiing onto a runway or landing area in preparation for takeoff, pilots should scan the approach areas for possible landing traffic, executing appropriate clearing maneuvers to provide him a clear view of the approach areas.

**2. Climbs and Descents.** During climbs and descents in flight conditions which permit visual detection of other traffic, pilots should execute gentle banks, left and right at a frequency which permits continuous visual scanning of the airspace about them.

**3. Straight and level.** Sustained periods of straight and level flight in conditions which permit visual detection of other traffic should be broken at intervals with appropriate clearing procedures to provide effective visual scanning.

**4. Traffic pattern.** Entries into traffic patterns while descending create specific collision hazards and should be avoided.

**5. Traffic at VOR sites.** All operators should emphasize the need for sustained vigilance in the vicinity of VORs and airway intersections due to the convergence of traffic.

**6. Training operations.** Operators of pilot training programs are urged to adopt the following practices:

**a.** Pilots undergoing flight instruction at all levels should be requested to verbalize clearing procedures (call out, "clear" left, right, above or below) to instill and sustain the habit of vigilance during maneuvering.

**b.** High-wing airplane, momentarily raise the wing in the direction of the intended turn and look.

**c.** Low-wing airplane, momentarily lower the wing in the direction of the intended turn and look.

**d.** Appropriate clearing procedures should precede the execution of all turns including chandelles, lazy eights, stalls, slow flight, climbs, straight and level, spins and other combination maneuvers.

## GIVING WAY

If you think another aircraft is too close to you, give way instead of waiting for the other pilot to respect the right-of-way to which you may be entitled. It is a lot safer to pursue the right-of-way angle after you have completed your flight.

## USE OF FEDERAL AIRWAYS

Pilots not operating on an IFR flight plan, and when in level cruising flight, are cautioned to conform with VFR cruising altitudes appropriate to direction of flight. During climb or descent, pilots are encouraged to fly to the right side of the center line of the radial forming the airway in order to avoid IFR and VFR cruising traffic operating along the center line of the airway.

## FOLLOW IFR PROCEDURES EVEN WHEN OPERATING VFR

1. To maintain IFR proficiency, pilots are urged to practice IFR procedures whenever possible, even when operating VFR. Some suggested practices include:

**a.** Obtain a complete preflight and weather briefing. Check the NOTAMS.

**b.** File a flight plan. This is an excellent low cost insurance policy. The cost is the time it takes to fill it out. The insurance includes the knowledge that someone will be looking for you if you become overdue at destination.

**c.** Use current charts.

**d.** Use the navigation aids. Practice maintaining a good course—keep the needle centered.

**e.** Maintain a constant altitude (appropriate for the direction of flight).

**f.** Estimate enroute position times.

**g.** Make accurate and frequent position reports to the FSS's along your route of flight.

2. Simulated IFR flight is recommended (under the hood), however, pilots are cautioned to review and adhere to the requirements specified in FAR 91.21 before such flight.

## VFR AT NIGHT

When flying VFR at night, in addition to the altitude appropriate for the direction of flight, pilots should maintain an altitude which is at or above the minimum enroute altitude as shown on charts. This is especially true in mountainous terrain, where there is usually very little ground reference. Don't depend on your being able to see those built-up rocks or TV towers in time to miss them.

## FLIGHT OUTSIDE THE UNITED STATES AND U.S. TERRITORIES

## AVOID FLIGHT BENEATH UNMANNED BALLOONS

The majority of unmanned free balloons currently being operated have, extending below them, either a suspension device to which the payload or instrument package is attached, or a trailing wire antenna, or both.

## FISH AND WILDLIFE SERVICE REGULATION

The Fish and Wildlife Service has the following regulation in effect governing the flight of aircraft on and over wildlife areas:

"The unauthorized operation of aircraft at low altitudes over, or the unauthorized landing of aircraft on a wildlife refuge area is prohibited, except in the event of emergency."

The Fish and Wildlife Service requests that pilots maintain a minimum altitude of 1,000 feet above the terrain of a wildlife refuge area.

## PARACHUTE JUMP AIRCRAFT OPERATIONS

Pilots of aircraft engaged in parachute jump operations are reminded that all reported altitudes must be with reference to mean sea level, or flight level as appropriate, to enable ATC to provide meaningful traffic information.

## STUDENT PILOTS RADIO IDENTIFICATION

The FAA desires to help the student pilot in acquiring sufficient practical experience in the environment in which he will be required to operate. To receive additional assistance while operating in areas of concentrated air traffic, a student pilot need only identify himself as a student pilot during his initial call to an FAA radio facility. For instance, "Dayton Tower, this is Fleetwing 1234, Student Pilot, over." This special identification will alert FAA air traffic control personnel and enable them to provide the student pilot with such extra assistance and consideration as he may need. This procedure is not mandatory.

## OPERATION OF AIRCRAFT ROTATING BEACON

1. There have been several incidents in which small aircraft were overturned or damaged by prop/jet blast forces from taxing large aircraft. A small aircraft taxiing behind any large aircraft with its engines operating could meet with the same results. In the interest of preventing ground upsets and injuries to ground personnel due to prop/jet engine blast forces, the FAA has recommended to air carriers/commercial operators that they establish procedures for the operation of the aircraft rotating beacon any time the engines are in operation.

2. General aviation pilots utilizing aircraft equipped with rotating beacons are also encouraged to participate in this program and operate the beacon any time the aircraft engines are in operation as an alert to other aircraft and ground personnel that prop/jet engine blast forces may be present. Caution must be exercised by all personnel not to rely solely on the rotating beacon as an indication that aircraft engines are in operation, since participation in this program is voluntary.

## MOUNTAIN FLYING

Your first experience of flying over mountainous terrain (particularly if most of your flight time has been over the flatlands of the midwest) could be a never-to-be-forgotten nightmare if proper planning is not done and if you are not aware of the potential hazards awaiting. Those familiar section lines are not present in the mountains; those flat, level fields for forced landings are practically non-existent; abrupt changes in wind direction and velocity occur; severe updrafts and downdrafts are common, particularly near or above abrupt changes of terrain such as cliffs or rugged areas; even the clouds look different and can build up with startling rapidity. Mountain flying need not be hazardous if you follow the recommendations below:

1. File a flight plan. Plan your route to avoid topography which would prevent a safe forced landing. The route should be over populated areas and well known mountain passes. Sufficient altitude should be maintained to permit gliding to a safe landing in the event of engine failure.

2. Don't fly a light aircraft when the winds aloft, at your proposed altitude, exceed 35 miles per hour. Expect the winds to be of much greater velocity over mountain passes than reported a few miles from them. Approach mountain passes with as much altitude as possible. Downdrafts of from 1500 to 2000 feet per minute are not uncommon on the leeward side.

3. Don't fly near or above abrupt changes in terrain. Severe turbulence can be expected, especially in high wind conditions.

4. Some canyons run into a dead-end. Don't fly so far up a canyon that you get trapped. ALWAYS BE ABLE TO MAKE A 180 DEGREE TURN!

5. Plan your trip for the early morning hours. As a rule, the air starts to get bad at about 10 a.m., and grows steadily worse until around 4 p.m., then gradually improves until dark Mountain flying at night in a single engine light aircraft is asking for trouble.

6. When landing at a high altitude field, the same indicated airspeed should be used as at low elevation fields. *Remember:* that due to the less dense air at altitude, this same indicated airspeed actually results in a higher true airspeed, a faster landing speed, and more important, a longer landing distance. During gusty wind conditions which often prevail at high altude fields, a power approach and power landing is recommended. Additionally, due to the faster groundspeed, your takeoff distance will increase considerably over that required at low altitudes.

7. *Effects of Density Altitude.* Performance figures in the aircraft owner's handbook for length of takeoff run, horsepower, rate of climb, etc., are generally based on standard atmosphere conditions (59° F, pressure 29.92 inches of mercury) at sea level. However, inexperienced pilots as well as experienced pilots may run into trouble when they encounter an altogether different set of conditions. This is particularly true in hot weather and at higher elevations. Aircraft operations at altitudes above sea level and at higher than standard temperatures are commonplace in mountainous areas. Such operations quite often result in a drastic reduction of aircraft performance capabilities because of the changing air density. Density altitude is a measure of air density. It is not to be confused with pressure

altitude—true altitude or absolute altitude. It is not to be used as a height reference, but as a determining criteria in the performance capability of an aircraft. Air density decreases with altitude. As air density decreases, density altitude increases. The further effects of high temperature and high humidity are cumulative, resulting in an increasing high density altitude condition. High density altitude reduces all aircraft performance parameters. To the pilot, this means that— the normal horsepower output is reduced, propeller efficiency is reduced and a higher true airspeed is required to sustain the aircraft throughout its operating parameters. It means an increase in runway length requirements for takeoff and landings, and a decreased rate of climb. (Note.—A turbo-charged aircraft engine provides some slight advantage in that it provides sea level horsepower up to a specified altitude above sea level.) An average small airplane, for example, requiring 1,000 feet for takeoff at sea level under standard atmospheric conditions will require a takeoff run of approximately 2,000 at an operational altitude of 5,000 feet.

NOTE.—All flight service stations will compute the current density altitude upon request.

## ASK FOR ASSISTANCE

"I'm not lost, I just don't know for sure where I am, but a familiar landmark will show up soon." "I *think* I have enough fuel to get there." "I think it will be smoother if I go above these clouds, there are bound to be some holes I can get down through when I get near home." "I'd look pretty silly if I asked for help and then found out I didn't really need it." The first time one of these thoughts pop into your mind, *it is time* to ask for assistance. Do not wait until the situation has deteriorated into an emergency before letting ATC know of your predicament. A little embarrassment is better than a big accident!

## TO CONTACT AN FSS

Flight Service Stations are allocated frequencies for different functions, for Airport Advisory Service the pilot should contact the FSS on 123.6 MHz, for example. Other FSS frequencies are listed in Part 3 of the AIM. If you are in doubt as to what frequency to use to contact an FSS, transmit on 122.1 MHz and advise them of the frequency you are receiving on. (See Microphone Technique in Chapter 3.)

## ALTIMETER ERRORS

1. The importance of frequently obtaining current altimeter settings can not be overemphasized. If you do not reset your altimeter when flying *from* an area of high pressure or high temperatures *into* an area of low temperatures or low pressure, *your aircraft will be closer to the surface than the altimeter indicates.* An inch error on the altimeter equals 1,000 feet of altitude. To quote an old saw: "GOING FROM A HIGH TO A LOW, LOOK OUT BELOW."

2. A reverse situation—without resetting the altimeter when going from a low temperature or pressure area into a high temperature or high pressure area, the aircraft will be higher than the altimeter indicates.

3. The possible result of the situation in 1. above, is obvious, particularly if operating at the minimum altitude. In situation 2. above, the result may not be as spectacular, but consider an instrument approach: If your altimeter is in error you may still be on instruments when reaching the minimum altitude (as indicated on the altimeter), whereas you might have been in the clear and able to complete the approach if the altimeter setting was correct. FAR 91.81 defines current altimeter setting practices.

## AIRCRAFT CHECKLISTS

Most Owners' Manuals contain recommended checklists for the particular type of aircraft. As checklists vary with different types of aircraft and equipments, it is not practical to recommend a complete set of checklists to cover all aircraft. Additions to the procedures in your Owners' Manual may better fill your personal preference or requirements. However, it is most important that checklists be designed to include all of the items in the Owners' Manual. By using these lists for every flight, the possibility of overlooking important items will be lessened.

## VFR IN A CONGESTED AREA?—LISTEN!

As pointed out in the item TURBOJETS IN TERMINAL AREAS—KEEP-'EM-HIGH in Chapter 3, "A high percentage of near midair collisions occur below 8,000 feet AGL and within 30 miles of an airport ─────." When operating VFR in these highly congested areas, whether you intend to land at an airport within the area or are just flying through, it is recommended that extra vigilance be maintained and that you monitor an appropriate control frequency. Normally the appropriate frequency is an approach control frequency. By such monitoring action you can "get the picture" of the traffic in your area. When the approach controller has radar, traffic advisories may be given to VFR pilots who requests them, subject to the provisions included in RADAR TRAFFIC INFORMATION SERVICE in Chapter 3.

## IFR APPROACH TO A NONTOWER/NON FSS AIRPORT

AIRPORT ADVISORIES AT NONTOWER AIRPORTS in Chapter 4 outlines the practices, frequencies to use, and when to report your position and intentions when operating on or in the vicinity of airports not served by a tower or FSS. All pilots should monitor the appropriate frequency and make these reports, *including the pilot making an IFR approach to the airport.*

When making an IFR approach to an airport not served by a tower or FSS, after the ATC controller advises "CHANGE TO ADVISORY FREQUENCY APPROVED" you should broadcast your intentions, including the type of approach being executed, your position, and when over the final approach fix inbound. Continue to monitor the appropriate frequency (UNICOM, etc.) for reports from other pilots.

## OPERATION LIGHTS ON

FAA has initiated a voluntary pilot safety program, "Operation Lights On" to enhance the "see-and-be-seen" concept of averting collisions both in the air and on the ground, and to reduce bird strikes. All pilots are encouraged to turn on their anti-collision lights any time the engine(s) are running day or night. All pilots are further encouraged to turn on their landing lights when operating within 10 miles of any airport (day and night), in conditions of reduced visibility and in areas where flocks of birds may be expected, i.e., coastal areas, lake areas, swamp areas, around refuse dumps. etc.

Although turning on aircraft lights does enhance the "see-and-be-seen" concept, pilots should not become complacent about keeping a sharp lookout for other aircraft. Not all aircraft are equipped with lights and some pilots may not have their lights turned on. The aircraft manufacturers' recommendations for operation of landing lights and electrical systems should be observed.

**Conformance with the above Good Operating Practices is not as simple as it sounds. It requires constant attention to innumerable details and a deep-seated realization that additional hazards incident to flight operations exist whenever vigilance is relaxed.**

# NAVIGATION AIDS 8

## GENERAL

Various types of air navigation aids are in use today, each serving a special purpose in our system of air navigation.

These aids have varied owners and operators namely: the Federal Aviation Administration, the military services, private organizations; and individual states and foreign governments.

The Federal Aviation Administration has the statutory authority to establish, operate, and maintain air navigation facilities and to prescribe standards for the operation of any of these aids which are used by both civil and military aircraft for instrument flight in federally controlled airspace. These aids are tabulated in the Airport/Facility Directory by State in Part 3 of the *Airman's Information Manual*.

A brief description of these aids follows. Also, a composite table of normal usable altitudes and distances appears in Class of VOR/VORTAC/TACAN.

## NON-DIRECTIONAL RADIO BEACON (NDB)

1. A low or medium-frequency radio beacon transmits nondirectional signals whereby the pilot of an aircraft equipped with a loop antenna can determine his bearing and "home" on the station. These facilities normally operate in the frequency band of 200 to 415 kHz and transmit a continuous carrier with 1,020-cycle modulation keyed to provide identification except during voice transmission.

2. When a radio beacon is used in conjunction with the Instrument Landing System markers, it is called a Compass Locator.

3. All radio beacons except the compass locators transmit a continuous three-letter identification in code except during voice transmissions. Compass locators transmit a continuous two-letter identification in code. The first and second letters of the three-letter location identifier are assigned to the front course outer marker compass locator (LOM), and the second and third letters are assigned to the front course middle marker compass locator (LMM).

**Example:**

ATLANTA, ATL, LOM-AT, LMM-TL.

4. Voice transmissions are made on radio beacons unless the letter "W" (without voice) is included in the class designator (HW).

5. Radio beacons are subject to disturbances that result in ADF needle deviations, signal fades and interference from distant station during night operations. Pilots are cautioned to be on the alert for these vagaries.

## VHF OMNIDIRECTIONAL RANGE (VOR)

1. VOR's operate within the 108.0–117.95 MHz frequency band and have a power output necessary to provide coverage within their assigned operational service volume. The equipment is VHF, thus, it is subject to line-of-sight restriction, and its range varies proportionally to the altitude of the receiving equipment. There is some "spill over," however, and reception at an altitude of 1000 feet is about 40 to 45 miles. This distance increases with altitude.

2. There is voice transmission on the VOR frequency available over the VOR's.

3. The effectiveness of the VOR depends upon proper use and adjustment of both ground and airborne equipment.

a. **Accuracy:** The accuracy of course alignment of the VOR is excellent, being generally plus or minus 1°.

b. **Roughness:** On some VORs, minor course roughness may be observed, evidenced by course needle or brief flag alarm activity (some receivers are more subject to these irregularities than others). At a few stations, usually in mountainous terrain, the pilot may occasionally observe a brief course needle oscillation, similar to the indication of "approaching station." Pilots flying over unfamiliar routes are cautioned to be on the alert for these vagaries, and in particular, to use the "to-from" indicator to determine positive station passage.

(1) Certain propeller RPM settings can cause the VOR Course Deviation Indicator to fluctuate as much as ±6°. Slight changes to the RPM setting will normally smooth out this roughness. Helicopter rotor speeds may also cause VOR course disturbances. Pilots are urged to check for this propeller modulation phenomenon prior to reporting a VOR station or aircraft equipment for unsatisfactory operation.

4. The only positive method of identifying a VOR is by its Morse Code identification or by the recorded automatic voice identification which is always indicated by use of the word "VOR" following the range's name. Reliance on determining the identification of an omnirange should never be placed on listening to voice transmissions by the Flight Service Station (FSS) (or approach control facility) involved. Many FSS remotely operate several omniranges which have different names from each other and in some cases none have the name of the "parent" FSS. (During periods of maintenance the coded identification is removed. See MAINTENANCE OF FAA NAVAIDS.)

5. Voice identification has been added to numerous VHF omniranges. The transmission consists of a voice announcement, "AIRVILLE VOR" alternating with the usual Morse Code identification.

## VOR RECEIVER CHECK

1. Periodic VOR receiver calibration is most important. If a receiver's Automatic Gain Control or modulation circuit deteriorates, it is possible for it to display acceptable accuracy and sensitivity close in to the VOR or VOT and display out-of-tolerance readings when located at greater distances where weaker signal areas exist. The likelihood of this deterioration varies between receivers, and is generally considered a function of time. The best assurance of having an accurate receiver is periodic calibration. Yearly intervals are recommended at which time an authorized repair facility should recalibrate the receiver to the manufacturer's specifications.

2. Part 91.25 of the Federal Aviation Regulations provides for certain VOR equipment accuracy checks prior to flight under instrument flight rules. To comply with this requirement and to ensure satisfactory operation of the airborne system, the FAA has provided pilots with the following means of checking VOR receiver accuracy: (1) FAA VOR test facility (VOT) or a radiated test signal from an appropriately rated radio repair station, (2) certified airborne check points, and (3) certified check points on the airport surface.

a. The FAA VOR test facility (VOT) transmits a test signal for VOR receivers which provides users of VOR a convenient and accurate means to determine the operational status of their receivers. The facility is designed to provide a means of checking the accuracy of a VOR receiver while the aircraft is on the ground.

The radiated test signal is used by tuning the receiver to the published frequency of the test facility. With the Course Deviation Indicator (CDI) centered the omnibearing selector should read 0° with the to-from indication being "from" or the omnibearing selector should read 180° with the to-from indication reading "to". Should the VOR receiver operate an RMI (Radio Magnetic Indicator), it will indicate 180° on any OBS setting when using the VOT. Two means of identification are used with the VOR radiated test signal. In some cases a continuous series of dots is used while in others a continuous 1020 hertz tone will identify the test signal. Information concerning an individual test signal can be obtained from the local Flight Service Station.

**b.** A radiated VOR test signal from an appropriately rated radio repair station serves the same purpose as an FAA VOR signal and the check is made in much the same manner with the following differences: (1) the frequency normally approved by the FCC is 108.0 MHz the repair stations are not permitted to radiate the VOR test signal continuously, consequently the owner/operator must make arrangements with the repair station to have the test signal transmitted. This service is not provided by all radio repair stations, the aircraft owner/operator must determine which repair station in his local area does provide this service. A representative of the repair station must make an entry into the aircraft logbook or other permanent record certifying to the radial accuracy which was transmitted and the date of transmission. The owner/operator or representative of the repair station may accomplish the necessary checks in the aircraft and make a logbook entry stating the results of such checks. It will be necessary to verify with the appropriate repair station the test radial being transmitted and whether you should get a "to" or "from" indication.

**c.** Airborne and ground check points consist of certified radials that should be received at specific points on the airport surface, or over specific landmarks while airborne in the immediate vicinity of the airport.

**d.** Should an error in excess of ±4° be indicated through use of a ground check, or ±6° using the airborne check, IFR flight shall not be attempted without first correcting the source of the error. CAUTION: no correction other than the "correction card" figures supplied by the manufacturer should be applied in making these VOR receiver checks.

**e.** The list of airborne check points and ground check points is published in Part 3. VOTs are included with the Airport/Facility Directory in Part 3 *Airman's Information Manual*.

**f.** If dual system VOR (units independent of each other except for the antenna) is installed in the aircraft, the person checking the equipment may check one system against the other. He shall turn both systems to the same VOR ground facility and note the indicated bearing to that station. The maximum permissible variations between the two indicated bearings is 4°.

## TACTICAL AIR NAVIGATION (TACAN)

**1.** For reasons peculiar to military or naval operations (unusual siting conditions, the pitching and rolling of a naval vessel, etc.) the civil VOR–DME system of air navigation was considered unsuitable for military or naval use. A new navigational system, Tactical Air Navigation (TACAN), was therefore developed by the military and naval forces to more readily lend itself to military and naval requirements. As a result, the FAA has been in the process of integrating TACAN facilities with the civil VOR–DME program. Although the theoretical, or technical principles of operation of TACAN equipment are quite different from those of VOR–DME

facilities, the end result, as far as the navigating pilot is concerned, is the same. These integrated facilities are called VORTAC's.

**2.** TACAN ground equipment consists of either a fixed or mobile transmitting unit. The airborne unit in conjunction with the ground unit reduces the transmitted signal to a visual presentation of both azimuth and distance information. TACAN is a pulse system and operates in the UHF band of frequencies. Its use requires TACAN airborne equipment and does not operate through conventional VOR equipment.

## VHF OMNIDIRECTIONAL RANGE/TACTICAL AIR NAVIGATION (VORTAC)

**1.** VORTAC is a facility consisting of two components, VOR and TACAN, which provides three individual services: VOR azimuth, TACAN azimuth and TACAN distance (DME) at one site. Although consisting of more than one component, incorporating more than one operating frequency, and using more than one antenna system, a VORTAC is considered to be a unified navigational aid. Both components of a VORTAC are envisioned as operating simultaneously and providing the three services at all times.

**2.** Transmitted signals of VOR and TACAN are each identified by three-letter code transmission and are interlocked so that pilots using VOR azimuth with TACAN distance can be assured that both signals being received are definitely from the same ground station. The frequency channels of the VOR and the TACAN at each VORTAC facility are "paired" in accordance with a national plan to simplify airborne operation.

## DISTANCE MEASURING EQUIPMENT (DME)

**1.** In the operation of DME, paired pulses at a specific spacing are sent out from the aircraft (this is the interrogation) and are received at the ground station. The ground station (transponder) then transmits paired pulses back to the aircraft at the same pulse spacing but on a different frequency. The time required for the round trip of this signal exchange is measured in the airborne DME unit and is translated into distance (Nautical Miles) from the aircraft to the ground station.

**2.** Operating on the line-of-sight principle, DME furnishes distance information with a very high degree of accuracy. Reliable signals may be received at distances up to 199 NM at line-of-sight altitude with an accuracy of better than ½ mile or 3% of the distance, whichever is greater. Distance information received from DME equipment is SLANT RANGE distance and not actual horizontal distance.

**3.** DME operates on frequencies in the UHF spectrum between 962 MHz and 1213 MHz. Aircraft equipped with TACAN equipment will receive distance information from a VORTAC automatically, while aircraft equipped with VOR must have a separate DME airborne unit.

**4.** VOR/DME, VORTAC, ILS/DME, and LOC/DME navigation facilities established by the FAA provide course and distance information from colocated components under a frequency pairing plan. Aircraft receiving equipment which provides for automatic DME selection assures reception of azimuth and distance information from a common source whenever designated VOR/DME, VORTAC, ILS/DME, and LOC/DME are selected.

**5.** Due to the limited number of available frequencies, assignment of paired frequencies has been required for certain military noncolocated VOR and TACAN facilities which serve the same area but which may be separated by distances up to a few miles. The military is presently undergoing a program to colocate 'VOR and TACAN facilities or to assign nonpaired frequencies to those facilities that cannot be colocated.

6. VOR/DME, VORTAC, ILS/DME, and LOC/DME facilities are identified by synchronized identifications which are transmitted on a time share basis. The VOR or localizer portion of the facility is identified by a coded tone modulated at 1020 Hz or by a combination of code and voice. The TACAN or DME is identified by a coded tone modulated at 1350 Hz. The DME or TACAN coded identification is transmitted one time for each four times that the VOR coded identification is transmitted. Where VOR code and voice identification is used, the DME or TACAN code identification is transmitted once for each three times that the VOR identification is transmitted. (At some VORTAC facilities, the VOR and DME coded identification is supplemented with VOR voice identification transmitting continuously in the background.) When either the VOR or the DME is inoperative, it is important to recognize which identifier is retained for the operative facility. A single coded identification transmitting once every 37½ seconds indicates that the DME only is operative. The absence of the single coded identification every 37½ seconds indicates that the VOR or localizer only is operative.

7. Aircraft receiving equipment which provides for automatic DME selection assures reception of azimuth and distance information from a common source whenever designated VOR/DME, VORTAC and ILS/DME navigation facilities are selected. Pilots are cautioned to disregard any distance displays from automatically selected DME equipment whenever VOR or ILS facilities, which do not have the DME feature installed, are being used for position determination.

## CLASS OF NAVAIDS

VOR, VORTAC, and TACAN aids are classed according to their operational use. There are three classes.

T   (Terminal)
L   (Low altitude)
H   (High altitude)

The normal service range for the T, L, and H class aids is included in the following table. Certain operational requirements make it necessary to use some of these aids at greater service ranges than are listed in the table. Extended range is made possible through flight inspection determinations. Some aids also have lesser service range due to location, terrain, frequency protection, etc. Restrictions to service range are listed in Part 3 of the *Airman's Information Manual*.

### VOR/VORTAC/TACAN NAVAIDS
#### Normal Usable Altitudes and Radius Distances

| Class | Altitudes | Distance (miles) |
|---|---|---|
| T | 12,000' and below | 25 |
| L | Below 18,000' | 40 |
| H | Below 18,000' | 40 |
| H | Within the conterminous 48 states only, between 14,500' and 17,999' | 100 |
| H | 18,000' — FL 450 | 130 |
| H | Above FL 450 | 100 |

### NON-DIRECTIONAL RADIO BEACON (NDB)
#### Usable Radius Distances for all Altitudes

| Class | Power (watts) | Distance (miles) |
|---|---|---|
| Compass Locator | Under 25 | 15 |
| MH | Under 50 | 25 |
| H | 50 — 1999 | *50 |
| HH | 2000 or more | 75 |

* Service range of individual facilities may be less than 50 miles. See Restrictions to Enroute Navigation Aids, Part 3 of the *Airman's Information Manual*.

## MARKER BEACON

1. Marker beacons serve to identify a particular location in space along an airway or on the approach to an instrument runway. This is done by means of a 75-MHz transmitter which transmits a directional signal to be received by aircraft flying overhead. These markers are generally used in conjunction with enroute navaids and the Instrument Landing Systems as point designators.

2. The class FM fan markers are used to provide a positive identification of positions at definite points along the airways. The transmitters have a power output of approximately 100 watts. Two types of antenna array are used with class FM fan markers. The first type used, and generally referred to as the standard type, produces an elliptical-shaped pattern, which at an elevation of 1000 feet above the station is about four miles wide and 12 miles long. At 10,000 feet the pattern widens to about 12 miles wide and 35 miles long.

3. The second array produces a dumbbell or bone-shaped pattern, which, at the "handle" is about three miles wide at 1000 feet. The boneshaped marker is preferred at approach control locations where "timed" approaches are used.

4. The class LFM or low-powered fan markers have a rated power output of 5 watts. The antenna array produces a circular pattern which appears elongated at right angles to the airway due to the directional characteristics of the aircraft receiving antenna.

5. The Station Location, or Z-Marker, was developed to meet the need for a positive position indicator for aircraft operating under instrument flying conditions to show the pilot when he was passing directly over a Low Frequency navigational aid. The marker consists of a 5-watt transmitter and a directional antenna array which is located on the range plot between the towers or the loop antennas.

6. ILS marker beacon information is included under "ILS."

## INSTRUMENT LANDING SYSTEM (ILS)

### 1. GENERAL

a. The instrument landing system is designed to provide an approach path for exact alignment and descent of an aircraft on final approach to a runway.

b. The ground equipment consists of two highly directional transmitting systems and, along the approach, three (or fewer) marker beacons. The directional transmitters are known as the localizer and glide slope transmitters.

c. The system may be divided functionally into three parts:

Guidance information—localizer, glide slope
Range information—marker beacons
Visual information—approach lights, touchdown and centerline lights, runway lights

d. Compass locators located at the outer marker or middle marker may be substituted for these marker beacons. DME when specified in the procedure may be substituted for the outer marker.

**e.** At some locations a complete ILS system has been installed on each end of a runway; on the approach end of runway 4 and the approach end of runway 22 for example. When such is the case, the ILS systems are not in service simultaneously during IFR weather conditions. In other words, if instrument approaches are being conducted to runway 4, the ILS system on the opposite end of runway 4, runway 22, will be out of service.

## 2. LOCALIZER

**a.** The localizer transmitter, operating on one of the forty ILS channels within the frequency range of 108.10 MHz to 111.95 MHz, emits signals which provide the pilot with course guidance to the runway centerline.

**b.** The approach course of the localizer, which is used with other functional parts, e.g., glide slope, marker beacons, etc., is called the front course. The localizer signal emitted from the transmitter at the far end of the runway is adjusted to produce an angular width between 3° and 6°, as necessary, to provide a linear width of approximately 700' at the runway approach threshold.

**c.** The course line along the extended centerline of a runway, in the opposite direction to "b" above is called the back course. CAUTION—unless your aircraft's ILS equipment includes reverse sensing capability, when flying inbound on the back course it is necessary to steer the aircraft in the direction opposite of the needle deflection on the airborne instrument when making corrections from off-course to on-course. This "flying away from the needle" is also required when flying outbound on the front course of the localizer. DO NOT UTILIZE BACK COURSE SIGNALS for approach unless a BACK COURSE APPROACH PROCEDURE has been published for the particular runway and is authorized by ATC.

**d.** Identification is in International Morse Code and consists of a three-letter identifier preceded by the letter I (●●) transmitted on the localizer frequency.

**Example:** I–DIA

**e.** The localizer provides course guidance throughout the descent path to the runway threshold from a distance of 18 NM from the antenna between an altitude of 1000' above the highest terrain along the course line and 4500' above the elevation of the antenna site. Proper off-course indications are provided throughout the following angular areas of the operational service volume:

1. To 10° either side of the course along a radius of 18 NM from the antenna, and

2. From 10°–35° either side of the course along a radius of 10 NM.

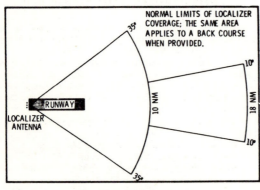

**Figure** 8-1

**f.** Proper off-course indications are generally not provided between 35–90° either side of the localizer course. Therefore, instrument indications of possible courses in the area from 35° to 90° off-course should be disregarded.

All unrestricted localizer facilities provide reliable course guidance information within the areas described in 2.e., above.

**g.** Momentary localizer flag activity and course aberrations may be observed when other aircraft cross over the localizer antenna or are in a position to affect the radiated signal.

**h.** The Atlantic City, Atlanta, and San Francisco Category III localizers include a far field monitor which will automatically remove the localizer from service when the course signal is displaced 10 microampres (approximately 20 feet at the runway threshold) or more from the runway centerline for a period of approximately 70 seconds or longer. The 70 second delay is necessary to prevent unnecessary localizer shutdowns for temporary conditions. These temporary conditions may be caused by taxiing aircraft, by aircraft overflight or other conditions between the localizer antenna and the far field monitor (normally located at the middle marker).

**i.** Approximately 65 seconds after the far field monitor detects a continuing 10 microamperes or more displacement of the course it will transmit simultaneously 1900 and 2100 Hertz audible alert tones on the localizer frequency. This feature is to advise pilots that the localizer will shutdown 5 seconds after the alert tones are first transmitted. Out of tolerance signals detected by the main internal monitor at the localizer will shutdown the facility immediately without warning.

**j.** The localizer-type directional aid (LDA) is of comparable utility and accuracy to a localizer but is not part of a complete ILS. The LDA course usually provides a more precise approach course than the similar SDF installation, which may have a course width of 6° or 12°. The LDA is not aligned with the runway. Straight-in minima may be published where alignment does not exceed 30 degrees between the course and runway. Circling minima only are published where this alignment exceeds 30 degrees.

## 3. GLIDE SLOPE

**a.** The UHF glide slope transmitter, operating on one of the forty ILS channels within the frequency range 329.15 MHz, to 335.00 MHz radiates its signals primarily in the direction of the localizer front course. Normally, a glide slope transmitter is not installed with the intent of radiating signals toward the localizer back course; however, there are a few runways at which an additional glide slope transmitter is installed to radiate signals primarily directed toward the localizer back course to provide vertical guidance. The two glide slope transmitters will operate on the same channel but are interlocked to avoid simultaneous radiation to support either the front course or the back course, but not both at the same time. Approach and landing charts for the runways which have glide slopes on the localizer back course will be depicted accordingly.

*Caution:* Spurious glide slope signals may exist in the area of the localizer back course approach which can cause the glide slope flag alarm to disappear and present unreliable glide slope information. Disregard all glide slope signal indications when making a localizer back course approach unless a glide slope is specified on the approach and landing chart.

**b.** The glide slope transmitter is located between 750' and 1250' from the approach end of the runway (down the runway) and offset 400–600' from the runway centerline. It transmits a glide path beam 1.4° wide. The term "glide path" means that portion of the glide slope that intersects the localizer.

**c.** The glide path projection angle is normally adjusted to 3 degrees above horizontal so that it intersects the middle marker at about 200 feet and the outer marker at about 1400 feet above the runway elevation. The glide slope is normally usable to the distance of 10 NM. However, at some locations, the glide slope has been

Figure 8-2

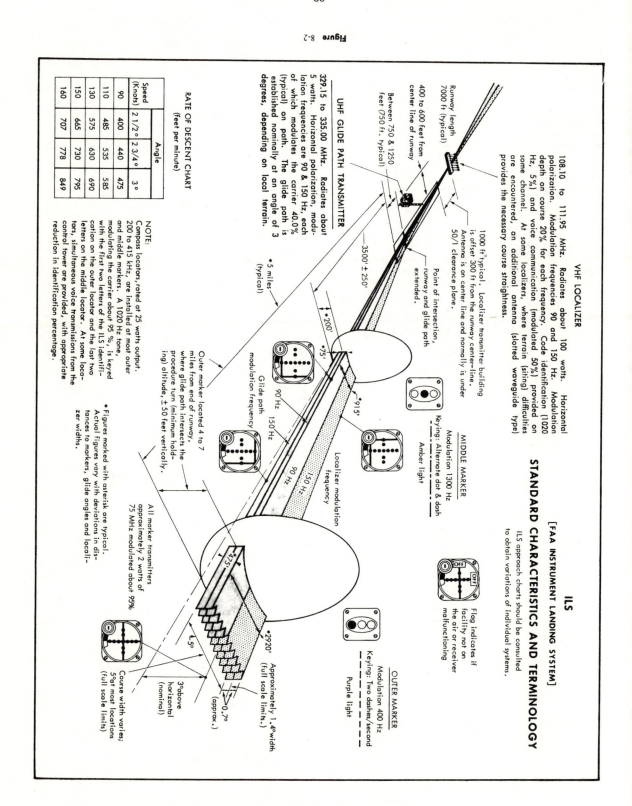

# ILS
### [FAA INSTRUMENT LANDING SYSTEM]
## STANDARD CHARACTERISTICS AND TERMINOLOGY

ILS approach charts should be consulted to obtain variations of individual systems.

### VHF LOCALIZER

108.10 to 111.95 MHz. Radiates about 100 watts. Horizontal polarization. Modulation frequencies 90 and 150 Hz. Modulation depth on course 20% for each frequency. Code identification (1020 Hz, 5%) and voice communication (modulated 50%) provided on some channel. At some localizers, where terrain (siting) difficulties are encountered, an additional antenna (slotted waveguide type) provides the necessary course straightness.

### UHF GLIDE PATH TRANSMITTER

329.15 to 335.00 MHz. Radiates about 5 watts. Horizontal polarization, modulation frequencies are 90 & 150 Hz, each of which modulates the carrier 40.0% (typical) on path. The glide path is established nominally at an angle of 3 degrees, depending on local terrain.

### RATE OF DESCENT CHART
(feet per minute)

| Speed (Knots) | Angle 2 1/2° | 2 3/4° | 3° |
|---|---|---|---|
| 90 | 400 | 440 | 475 |
| 110 | 485 | 535 | 585 |
| 130 | 575 | 630 | 690 |
| 150 | 665 | 730 | 795 |
| 160 | 707 | 778 | 849 |

### NOTE:

Compass locators, rated at 25 watts output, 200 to 415 kHz, are installed at most outer and middle markers. A 1020 Hz tone, modulating the carrier about 95%, is keyed with the first two letters of the ILS identification on the outer locator and the last two letters on the middle locator. At some locators, simultaneous voice transmissions from the control tower are provided, with appropriate reduction in identification percentage.

Runway length 7000 ft (typical)

400 to 600 feet from center line of runway

Between 750 & 1250 feet (750 ft. typical)

1000 ft. typical. Localizer transmitter building is offset 300 ft from the runway center-line. Antenna is on center line and normally is under 50/1 clearance plane.

Point of intersection, runway and glide path extended.

3500' ± 250'

*5 miles (typical)

*200'

*75'

*915'

Glide path modulation frequency

Localizer modulation frequency

90 Hz
150 Hz
90 Hz
150 Hz

MIDDLE MARKER
Modulation 1300 Hz
Keying: Alternate dot & dash
Amber light

OUTER MARKER
Modulation 400 Hz
Keying: Two dashes/second
Purple light

Flag indicates if facility not on the air or receiver malfunctioning

Outer marker located 4 to 7 miles from end of runway, where glide path intersects the procedure turn (minimum holding) altitude, ± 50 feet vertically.

All marker transmitters approximately 2 watts of 75 MHz modulated about 95%

*2920'

Approximately 1.4° width (full scale limits)

*475'

5°

0.7° (approx.)

3° above horizontal (nominal)

Course width varies; 5° at most locations (full scale limits)

*Figures marked with asterisk are typical. Actual figures vary with deviations in distances to markers, glide angles and localizer widths.

certified for an extended service volume which exceeds 10 NM.

d. In addition to the desired glide path, false course and reversal in sensing will occur at vertical angles considerably greater than the usable path. The proper use of the glide slope requires that the pilot maintain alertness as the glide path interception is approached and interpret correctly the "fly-up" and "fly-down" instrument indications to avoid the possibility of attempting to follow one of the higher angle courses. Provided that procedures are correctly followed and pilots are properly indoctrinated in glide path instrumentation, the fact that these high angle courses exist should cause no difficulty in the glide path navigation.

e. Every effort should be made to remain on the indicated glide path (reference: FAR 91.87(d) (3) ). Extreme caution should be exercised to avoid deviations below the glide path so that the predetermined obstacle/terrain clearance provided by an ILS instrument approach procedure is maintained.

f. A glide slope facility provides a path which flares from 18–27 feet above the runway. Therefore, the glide path should not be expected to provide guidance completely to a touchdown point on the runway.

### 4. MARKER BEACON

a. ILS marker beacons have a rated power output of 3 watts or less and an antenna array designed to produce an elliptical pattern with dimensions, at 1000 feet above the antenna, of approximately 2400 feet in width and 4200 feet in length. Airborne marker beacon receivers with a selective sensitivity feature should always be operated in the "low" sensitivity position for proper reception of ILS marker beacons.

b. Ordinarily, there are two marker beacons associated with an instrument landing system; the outer marker and middle marker. However, some locations may employ a third marker beacon to indicate the point at which the decision height should occur when used with a Category II ILS.

c. The outer marker (OM) normally indicates a position at which an aircraft at the appropriate altitude on the localizer course will intercept the ILS glide path. The OM is modulated at 400 Hz and identified with continuous dashes at the rate of two dashes per second.

d. The middle marker (MM) indicates a position at which an aircraft is approximately 3500 feet from the landing threshold. This will also be the position at which an aircraft on the glide path will be at an altitude of approximately 200 feet above the elevation of the touchdown zone. The MM is modulated at 1300 Hz and identified with alternate dots and dashes keyed at the rate of 95 dot/dash combinations per minute.

e. The inner marker (IM), where installed, will indicate a point at which an aircraft is at a designated decision height (DH) on the glide path between the middle marker and landing threshold. The IM is modulated at 3000 Hz and identified with continuous dots keyed at the rate of six dots per second and a white marker beacon light.

f. A back course marker, where installed, normally indicates the ILS back course final approach fix where approach descent is commenced. The back course marker is modulated at 3000 Hz and identified with two dots at a rate of 72 to 95 two-dot combinations per minute and a white marker beacon light.

### 5. COMPASS LOCATOR

a. Compass locator transmitters are often situated at the middle and outer marker sites. The transmitters have a power of less than 25 watts, a range of at least 15 miles and operate between 200 and 415 kHz. At some locations, higher-powered radio beacons, up to 400 watts, are used as outer marker compass locators. These generally carry Transcribed Weather Broadcast information.

b. Compass locators transmit two-letter identification groups. The outer locator transmits the first two letters of the localizer identification group, and the middle locator transmits the last two letters of the localizer identification group.

### 6. ILS MINIMUMS

ILS Minimums with all components operative normally establish a DH (Decision Height MSL) with a HAT (Height Above Touchdown) of 200 feet, and a visibility of one-half statute mile. Refer to Inoperative Component Table (FAR 91.117c) for adjustment to minimums when airborne equipment is inoperative or not used or when a component of the ILS is reported out of service.

### 7. ILS FREQUENCY

The following frequency pairs have been allocated for ILS. Although all of these forty frequency pairs are not being utilized, they will be in the future.

**ILS**

| Localizer mHz | Glide Slope mHz | Localizer mHz | Glide Slope mHz |
|---|---|---|---|
| 108.10 | 334.70 | 110.1 | 334.40 |
| 108.15 | 334.55 | 110.15 | 334.25 |
| 108.3 | 334.10 | 110.3 | 335.00 |
| 108.35 | 333.95 | 110.35 | 334.85 |
| 108.5 | 329.90 | 110.5 | 329.60 |
| 108.55 | 329.75 | 110.55 | 329.45 |
| 108.7 | 330.50 | 110.70 | 330.20 |
| 108.75 | 330.35 | 110.75 | 330.05 |
| 108.9 | 329.30 | 110.90 | 330.80 |
| 108.95 | 329.15 | 110.95 | 330.65 |
| 109.1 | 331.40 | 111.10 | 331.70 |
| 109.15 | 331.25 | 111.15 | 331.55 |
| 109.3 | 332.00 | 111.30 | 332.30 |
| 109.35 | 331.85 | 111.35 | 332.15 |
| 109.50 | 332.60 | 111.50 | 332.9 |
| 109.55 | 332.45 | 111.55 | 332.75 |
| 109.70 | 333.20 | 111.70 | 333.5 |
| 109.75 | 333.05 | 111.75 | 333.35 |
| 109.90 | 333.80 | 111.90 | 331.1 |
| 109.95 | 333.65 | 111.95 | 330.95 |

**Figure 8-3**

## SIMPLIFIED DIRECTIONAL FACILITY (SDF)

The Simplified Directional Facility provides a final approach course which is similar to that of the ILS localizer described in this chapter. A clear understanding of the ILS localizer and the additional factors listed below completely describe the operational characteristics and use of the SDF.

1. The SDF transmits signals within the range of 108.10 MHz to 111.95 MHz. It provides no glide slope information.

2. For the pilot, the approach techniques and procedures used in the performance of an SDF instrument approach are essentially identical to those employed in executing a standard no-glide-slope localizer approach except that the SDF course may not be aligned with the runway and the course may be wider, resulting in less precision.

3. Usable off-course indications are limited to 35° either side of the course centerline. Instrument indications in the areas between 35° and 90° are not controlled and should be disregarded.

4. The SDF antenna may be offset from the runway centerline. Because of this, the angle of convergence between the final approach course and the runway bearing should be determined by reference to the instrument approach procedure chart. This angle is generally not more than 3°. However, it should be noted that inasmuch as the approach course originates at the antenna site, an approach which is continued beyond the runway threshold will lead the aircraft to the SDF offset position rather than along the runway centerline.

**5.** The SDF signal emitted from the transmitter is fixed at either 6° or 12° as necessary to provide maximum flyability and optimum course quality.

**6.** Identification consist of a three letter identifier transmitted on the SDF frequency. Example: SAN, ETT, etc. The appropriate instrument approach chart will indicate the identifier used at a particular airport.

## INTERIM STANDARD MICROWAVE LANDING SYSTEM (ISMLS)

**1.** The ISMLS is designed to provide approach information similar to the ILS for an aircraft on final approach to a runway. The system provides both lateral and vertical guidance which is displayed on a conventional course deviation indicator or approach horizon. Operational performance and coverage areas are also similar to the ILS as defined; AIM, Part I, ILS, para. 2.e.1.2.

**2.** ISMLS operates in the C-band microwave frequency range (about 5000 MHz) so the signal will not be received by unmodified VHF/UHF ILS receivers. Aircraft utilizing ISMLS must be equipped with a C-band receiving antenna in addition to other special equipment mentioned below. The receiving aperture of the C-band antenna limits reception of the signal to an angle of about 50° from the inbound course. Therefore, an aircraft so equipped will not receive the ISMLS signal until flying a magnetic heading within 50° either side of the inbound course. Because of this, ISMLS procedures are designed to preclude use of the ISMLS signal until the aircraft is in position for the final approach. Transition to the ISMLS, holding and procedure turns at the ISMLS facility must be predicated on other navigation aids such as NDB, VOR, etc. Once established on the approach course inbound, the system can be flown identical to ILS. No back course is provided.

**3.** The Interim Standard Microwave Landing System consists of the following basic components:
    **a.** C-Band (5000 MHz–5030 MHz) localizer.
    **b.** C-Band (5220 MHz–5250 MHz) glide path.
    **c.** VHF marker beacons (75 MHz).
    **d.** A VHF/UHF ILS receiver modified to be capable of receiving the ISMLS signals.
    **e.** C-Band antenna.
    **f.** Converter unit.
    **g.** A Microwave/ILS Mode Control.

**4.** The identification consists of a three letter Morse Code identifier preceded by the Morse Code for "M" (−−). Example: "M-STP." The "M" will distinguish this system from ILS which is preceded by the letter "I."

**5.** Approaches published in conjunction with the ISMLS will be identified as "MLS Rwy —— (Interim)." The frequency displayed on the ISMLS approach chart will be a VHF frequency. ISMLS frequencies are tuned by setting the receiver to the listed VHF frequencies. When the ISMLS mode is selected, receivers modified to accept ISMLS signals will not receive the VHF/UHF frequency but a paired LOC/GS C-Band frequency that will be processed by the receiver.
**CAUTION:** Aircraft not equipped for ISMLS operation should not attempt to fly ISMLS procedures.

## MAINTENANCE OF FAA NAVAIDS

**1.** During periods of routine or emergency maintenance, the coded identification (or code and voice, where applicable) will be removed from certain FAA navaids; namely, ILS localizers, VHF ranges, NDB's, compass locators and 75 MHz marker beacons. The removal of identification serves as warning to pilots that the facility has been officially taken over by "Maintenance" for tune-up or repair and may be unreliable even though on the air intermittently or constantly.

## NAVAIDS WITH VOICE

**1.** Voice equipped en route radio navigational aids are under the operational control of an FAA Flight Service Station (FSS), or an approach control facility. Most are remotely operated.

**2.** Unless otherwise noted on the chart, all radio navigation aids operate continuously except during interruptions for voice transmissions on the same frequencies where simultaneous transmission is not available, and during shutdowns for maintenance purposes. Hours of operation of those facilities not operating continuously are annotated on the charts.

## SIMULTANEOUS VOICE TRANSMISSIONS FROM A SINGLE FSS LOCATION

**1.** At several FAA facilities, simultaneous voice transmissions are made from a single FSS location. For example, the New York FSS controls the transmitters at Hampton and Riverhead VOR's.

**2.** To avoid confusion, pilots are urged to use the name of the VOR to which they are listening when a reply on the VOR voice channel is desired. For example, call "RIVERHEAD RADIO" and the call will be answered by the New York FSS, "THIS IS NEW YORK RADIO."

## USER REPORTS ON NAVAID PERFORMANCE

**1.** Users of the National Airspace System can render valuable assistance in the early correction of navaid malfunctions by reporting their observations of undesirable performance. Although the navaid is monitored by electronic detectors within the equipment or in the field of radiation near the antenna, adverse effects of electronic interference, new obstructions or changes in terrain near the navaid can exist without detection by the ground monitors. Some of the characteristics of malfunction or deteriorating performance which should be reported are: Erratic course or bearing indications; intermittent, or full, flag alarm; garbled, missing or obviously improper coded identification; poor quality communications reception; or, in the case of frequency interference, an audible hum or tone accompanying radio communications or navaid identification.

Reports should identify the navaid, location of the aircraft, time of the observation and type of aircraft and describe the condition observed; the type of receivers in use will also be useful information. The FAA has established procedures to respond to user reports which can be made in any of the following ways:

    **a.** Report by radio communication to the controlling Air Route Traffic Control Center, Control Tower, or Flight Service Station. This method will provide the most timely resolution of the reported condition.

    **b.** Report by telephone to the nearest FAA facility.

    **c.** Report by FAA Form 8000-7, Safety Improvement Report, a self addressed, postage-paid card designed with this purpose in mind. A supply of these cards may be found at FAA Flight Service Stations, General Aviation District Offices and General Aviation Fixed Base Operations.

**2.** In complex aircraft radio installations involving more than one receiver, there are many combinations of possible interference between units. This interference can cause either erroneous navigation indications, or complete or partial blanking out of the communications. Pilots should be familiar enough with the radio installation of particular airplanes they fly to recognize this type of interference.

## LORAN

Loran-C has been selected by the Federal Government as the radio-navigation system for the Coastal Confluence Region, Harbors and Estuaries of the United States. Loran-C, with its high accuracy and availability, will meet the requirements of all users in the Coastal Confluence Region while providing the nation the capability of wider system application, including over-land use, in the future. Loran-C is a companion to the Omega system, which is being implemented for long-range, worldwide usage beyond the U.S. Coastal Confluence Region.

During the major transition while the Loran-C system is being constructed, a primary concern is the future of our present Loran-A system. To alleviate much of the difficulty in this major shift of systems, the U.S. Coast Guard will continue operation of the Loran-A system for a period of two years subsequent to completion of each phase of the Loran-C system. Thus the schedule for Loran-A termination in the Coastal Confluence Region is:

| Area | Date of Termination |
|---|---|
| Hawaiian Islands | 1 July 1979 |
| Aleutian Islands | 1 July 1979 |
| Gulf of Alaska | 1 July 1979 |
| U.S. West Coast | 1 July 1979 |
| Gulf of Mexico | 1 July 1980 |
| U.S. East Coast | 1 July 1980 |
| Caribbean | 1 July 1980 |

## CONSOLAN

1. Consolan is a long-range navigation aid operating in the low/medium frequency band. One station operational in the U.S.—San Francisco (SFI) 192 kHz.

2. A receiver equipped with CW (BFO) position without AVC should be used. When a loop antenna is used, best results are obtained by placing the loop on the maximum signal position.

3. A range of 1000–1400 nautical miles or more is to be expected. Greatest range is found over water. The range depends on the receiver, noise, ground conductivity, ionospheric conditions, frequency, and transmitter power. With experience, operators can obtain accurate bearings through considerable noise. Narrow band receivers provide a high operational range. The systems are not usable within 25 miles of the station.

## VHF/UHF DIRECTION FINDER

1. The VHF/UHF Direction Finder (VHF/UHF/DF) is one of the Common System equipments that helps the pilot without his being aware of its operation. The VHF/UHF/DF is a ground-based radio receiver used by the operator of the ground station where it is located.

2. The equipment consists of a directional antenna system, a VHF and a UHF radio receiver. At a radar-equipped tower or center, the cathode-ray tube indications may be superimposed on the radarscope.

3. The VHF/UHF/DF display indicates the magnetic direction of the aircraft from the station each time the aircraft transmits. Where DF equipment is tied into radar, a strobe of light is flashed from the center of the radarscope in the direction of the transmitting aircraft.

4. DF equipment is of particular value in locating lost aircraft and in helping to identify aircraft on radar. (See VHF/UHF Direction Finding Instrument Approach Procedure in Chapter *23*)

## RADAR

**1. Capabilities.**

**a.** A method whereby radio waves are transmitted into the air and are then received when they have been reflected by an object in the path of the beam. *Range* is determined by measuring the time it takes (at the speed of light) for the radio wave to go out to the object and then return to the receiving antenna. The *direction* of a detected object from a radar site is determined by the position of the rotating antenna when the reflected portion of the radio wave is received.

**b.** More reliable maintenance and improved equipment have reduced radar system failures to a negligible factor. Most facilities actually have some components duplicated—one operating and another which immediately takes over when a malfunction occurs to the primary component.

**2. Limitations.**

**a.** It is very important for the aviation community to recognize the fact that there are limitations to radar service and that ATC controllers may not always be able to issue traffic advisories concerning aircraft which are not under ATC control and cannot be seen on radar.

(1) The characteristics of radio waves are such that they normally travel in a continuous straight line unless they are:

(a) "Bent" by abnormal atmospheric phenomena such as temperature inversions;

(b) Reflected or attenuated by dense objects such as heavy clouds, precipitation, ground obstacles, mountains, etc.; or

(c) Screened by high terrain features.

(2) The bending of radar pulses, often called anomalous propagation or ducting, may cause many extraneous blips to appear on the radar operator's display if the beam has not been bent toward the ground and may decrease the detection range if the wave is bent upward. It is difficult to solve the effects of anomalous propagation, but using beacon radar and electronically eliminating stationary and slow moving targets by a method called moving target indicator (MTI) usually negate the problem.

(3) Radar energy that strikes dense objects will be reflected and displayed on the operator's scope thereby blocking out aircraft at the same range and greatly weakening or completely eliminating the display of targets at a greater range. Again, radar beacon and MTI are very effectively used to combat ground clutter and weather phenomena, and a method of circularly polarizing the radar beam will eliminate some weather returns. A negative characteristic of MTI is that an aircraft flying a speed that coincides with the canceling signal of the MTI (tangential or "blind" speed) may not be displayed to the radar controller.

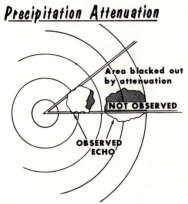

### *Precipitation Attenuation*

The nearby target absorbs and scatters so much of the out-going and returning energy that the radar does not detect the distant target.

**Figure** 8-4

(4) Relatively low altitude aircraft will not be seen if they are screened by mountains or are below the radar beam due to earth curvature. The only solution to screening is the installation of strategically placed multiple radars which has been done in some areas.

(5) There are several other factors which affect radar control. The amount of reflective surface of an aircraft will determine the size of the radar return. Therefore, a small light airplane or a sleek jet fighter will be more difficult to see on radar than a large commercial jet or military bomber. Here again, the use of radar beacon is invaluable if the aircraft is equipped with an airborne transponder. Altitude information must be obtained by radio from the pilot since present en route and airport surveillance radars are only two dimensional (range and azimuth). However, at several ATC facilities beacon Mode C is used to electronically display altitude information to the controller from appropriately equipped aircraft.

(6) The controllers' ability to advise a pilot flying on instruments or in visual conditions of his proximity to another aircraft will be limited if the unknown aircraft is not observed on radar, if no flight plan information is available, or if the volume of traffic and workload prevent his issuing traffic information. First priority is given to establishing vertical, lateral, or longitudinal separation between aircraft flying IFR under the control of ATC.

3. FAA radar units operate continuously at the locations shown in Part 3 of the Airman's Information Manual, in the Airport/Facility Directory, and their services are available to all pilots, both civil and military. Contact the associated FAA control tower or ARTCC on any frequency guarded for initial instructions, or in an emergency, any FAA facility for information on the nearest radar service.

4. Radar Traffic Information Service Procedures, Expanded Radar Service for Arriving and Departing Flights in Terminal Areas Procedures, and Terminal Radar Service Procedures are found in Chapter *23*.

## AIR TRAFFIC CONTROL RADAR BEACON SYSTEM (ATCRBS)

1. The Air Traffic Control Radar Beacon System (ATCRBS), sometimes referred to as secondary surveillance radar, consists of three main components:

**a. Interrogator.** Primary radar relies on a signal being transmitted from the radar antenna site and for this signal to be reflected or "bounced back" from an object (such as an aircraft). This reflected signal is then displayed as a "target" on the controller's radarscope. In the ATCRBS, the *Interrogator*, a ground based radar beacon transmitter-receiver, scans in synchronism with the primary radar and transmits discrete radio signals which repetitiously requests all transponders, on the mode being used, to reply. The replies received are then mixed with the primary returns and both are displayed on the same radarscope.

**b. Transponder.** This airborne radar beacon transmitter-receiver automatically receives the signals from the interrogator and selectively replies with a specific pulse group (code) only to those interrogations being received on the mode to which it is set. These replies are independent of, and much stronger than a primary radar return.

**c. Radarscope.** The radarscope used by the controller displays returns from both the primary radar system and the ATCRBS. These returns, called targets, are what the controller refers to in the control and separation of traffic.

2. The job of identifying and maintaining identification of primary radar targets is a long and tedious task for the controller. Some of the advantages of ATCRBS over primary radar are:

**a.** Reinforcement of radar targets.

**b.** Rapid target identification.

**c.** Unique display of selected codes.

3. A part of the ATCRBS ground equipment is the decoder. This equipment enables the controller to assign discrete transponder codes to each aircraft under his control. Normally only one code will be assigned for the entire flight. Assignments are made by the ARTCC computer on the basis of the National Beacon Code Allocation Plan. The equipment is also designed to receive Mode C altitude information from the aircraft. Refer to figures 8-5 and 8-6 with explanatory legends for an illustration of the target symbology depicted on radar scopes in the NAS Stage A (enroute), the ARTS III (terminal) systems, and other non-automated (broadband) radar systems.

4. It should be emphasized that aircraft transponders greatly improve the effectiveness of radar systems. (See Transponder Operation in Chapter *23*)

## SURVEILLANCE RADAR

1. Surveillance radars are divided into two general categories: Airport Surveillance Radar and Air Route Surveillance Radar. Airport Surveillance Radar (ASR) is designed to provide relatively short range coverage in the general vicinity of an airport and to serve as an expeditious means of handling terminal area traffic through observation of precise aircraft locations on a radarscope. The ASR can also be used as an instrument approach aid. Air Route Surveillance Radar (ARSR) is a long-range radar system designed primarily to provide a display of aircraft locations over large areas.

2. Surveillance radars scan through 360° of azimuth and present target information on a radar display located in a tower or center. This information is used independently or in conjunction with other navigational aids in the control of air traffic.

## PRECISION RADAR

1. Precision approach radar is designed to be used as a *landing aid*, rather than an aid for sequencing and spacing aircraft. PAR equipment may be used as a primary landing aid, or it may be used to monitor other types of approaches. It is designed to display *range, azimuth* and *elevation* information.

2. Two antennas are used in the PAR array, one scanning a vertical plane, and the other scanning horizontally. Since the range is limited to 10 miles, azimuth to 20 degrees, and elevation to 7 degrees, only the final approach area is covered. Each scope is divided into two parts. The upper half presents altitude and distance information, and the lower half presents azimuth and distance.

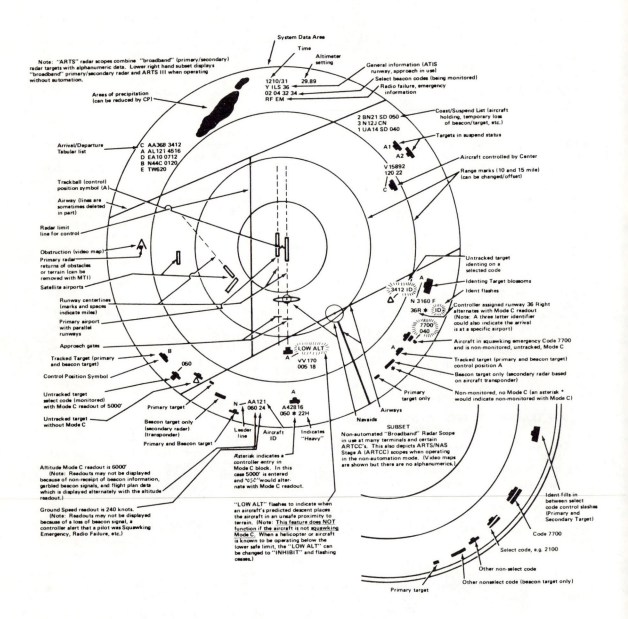

FIGURE 8-5 ARTS III Radar Scope with Alphanumeric Data. Note: A number of radar terminals do not have ARTS equipment. Those facilities and certain ARTCC's outside the contiguous US would have radar displays similar to the lower right hand subset. ARTS facilities and NAS Stage A ARTCC's, when operating in the non-automation mode would also have similar displays and certain services based on automation may not be available.

Figure 8-6. NAS Stage A Controllers Plan View Display. This figure illustrates the controller's radar scope (PVD) when operating in the full automation (RDP) mode, which is normally 20 hours per day. (Note: When not in automation mode, the display is similar to the broadband mode shown in Figure 1-5. Certain ARTCC's outside the contiguous U.S. also operate in "broadband" mode.)

**Target Symbols**

| | | |
|---|---|---|
| 1 | Uncorrelated primary radar target | + |
| 2 | *Correlated primary radar target | X |
| 3 | Uncorrelated beacon target | / |
| 4 | Correlated beacon target | \ |
| 5 | Identifying beacon target | ≡ |

(*Correlated means the association of radar data with the computer projected track of an identified aircraft)

**Position Symbols**

| | | |
|---|---|---|
| 6 | Free track (No flight plan tracking) | △ |
| 7 | Flat track (flight plan tracking) | ◇ |
| 8 | Coast (Beacon target lost) | # |
| 9 | Present Position Hold | ⊠ |

**Data Block Information**

10 *Aircraft identification

11 *Assigned Altitude FL280, mode C altitude same or within ±200' of assgnd altitude

12 *Computer ID #191, Handoff is to Sector 33 (0-33) would mean handoff accepted) ("Nr's 10, 11, 12 constitute a "full data block.")

13 Assigned altitude 17,000', aircraft is climbing, mode C readout was 14,300 when last beacon interrogation was received

14 Leader line connecting target symbol and data block

15 Track velocity and direction vector line (Projected ahead of target)

16 Assigned altitude 7000, aircraft is descending, last mode C readout (or last reported altitude was 100', above FL230

17 Transponder code shows in full data block only when different than assigned code

18 Aircraft is 300' above assigned altitude

19 Reported altitude (No mode C readout) same as assigned. An "N" would indicate no reported altitude)

20 Transponder set on emergency code 7700 (EMRG flashes to attract attention)

21 Transponder code 1200 (VFR) with no mode C

22 Code 1200 (VFR) with mode C and last altitude readout

23 Transponder set on Radio Failure code 7600, (RDOF flashes)

24 Computer ID #228, CST indicates target is in Coast status

25 Assigned altitude FL290, transponder code (These two items constitute a "limited data block.")

**Other symbols**

26 Navigational Aid

27 Airway or jet route

28 Outline of weather returns based on primary radar (See Chapter 4, ARTCC Radar Weather Display. H's represent areas of high density precipitation which might be thunderstorms. Radial lines indicate lower density precipitation)

29 Obstruction

30 Airports Major: □ , Small: ⌐

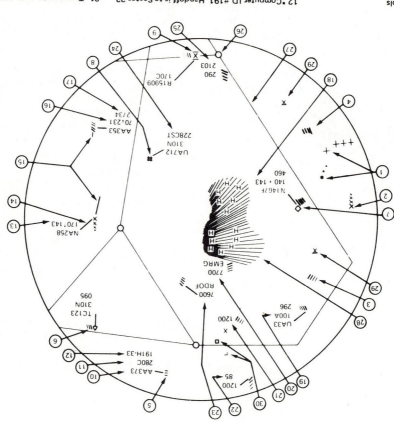

# 9 RADIO PRINCIPLES

## General

Radio communication and radio navigation usually are necessary during navigation flights. The pilot should be familiar with radio principles and the capabilities and employment of aircraft radio equipment to successfully navigate.

## Wave Transmission

According to the wave theory of radiation, sound, light, and electrical energy are transmitted by waves.

*a. Wave.* Energy traveling through a substance or space by vibrations or impulse moves in waves. *For example,* when a stone is dropped into a pond, the energy of motion from the stone causes ripples on the water surface. The ripples (waves of energy) travel outward from the place where the stone struck the water, but the water itself does not move outward. The rise and fall above and below the normal undisturbed water level can be graphed as a curved line.

*b. Cycle.* A cycle is an *alternation* of a wave from a specific amplitude through a complete series of movement back to the same amplitude (*e* below); i.e., one complete wave vibration. A cycle (fig. 9–1) is represented by the portion of the wave from A to E, from B to F, from C to G, or between any other two points encompassing exactly one complete amplitude variation. *For example,* a cork floating on calm water is subjected to cyclic wave movement when a stone is dropped into the water. One wave cycle occurs as the cork (1) rises from the calm water level (normal position) up to the wave crest, (2) drops back to normal, (3) falls into the wave trough, and (4) rises back to normal. A cycle is also completed when the cork moves from the crest of the wave down to the normal position, falls into the wave trough, rises back to normal, and continues rising to the top of the wave crest. Thus a *cycle* is any complete sequence of amplitude variation in a repetitive series of wave movements.

*c. Frequency.* The frequency of a wave is measured by the number of cycles completed in 1 second. If two cycles are completed in 1 second, the wave frequency is two cycles per second. Since the number of cycles per second (*cps*) runs into high figures, radio frequencies are commonly expressed as kilocycles (1,000 cycles) or megacycles (1,000,000 cycles). The "per second" time period is dropped but understood.

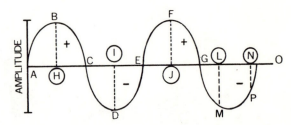

*Figure 9–1. Wave representation (alternating current).*

*d. Wavelength.* The linear distance of a cycle is known as the *wavelength.* In figure 9–1 the wavelength from A to E can be expressed in meters, feet, miles, or any other suitable linear measurement.

*e. Amplitude.* The *amplitude* of a wave is its magnitude measured from a specific reference level. In figure 9–1 the peak wave amplitude is represented by the lines BH, ID, or FJ; other amplitudes are shown as lines LM and NP. All representations were measured in linear distance above (+) or below (−) reference line AO.

## Dc and Ac Current

An electrical current flows by the movement of electrons through a conductor. Direct current (dc) flows in only one direction. An alternating current (ac) flows in one direction for a time and then flows in the opposite direction for the same length of time, with a continuous movement. An alternating current (fig. 9–1) can be represented as a continuous flow of electrons with half of each cycle being negative and the other half being positive.

## Radio Waves

An electrical current builds up a magnetic field around the conductor through which it flows. When alternating current flows through a wire, the magnetic field around the wire alternately builds up and collapses. An alternating current of high frequency is used to generate radio waves which are emitted during the build-up and collapse of the magnetic field

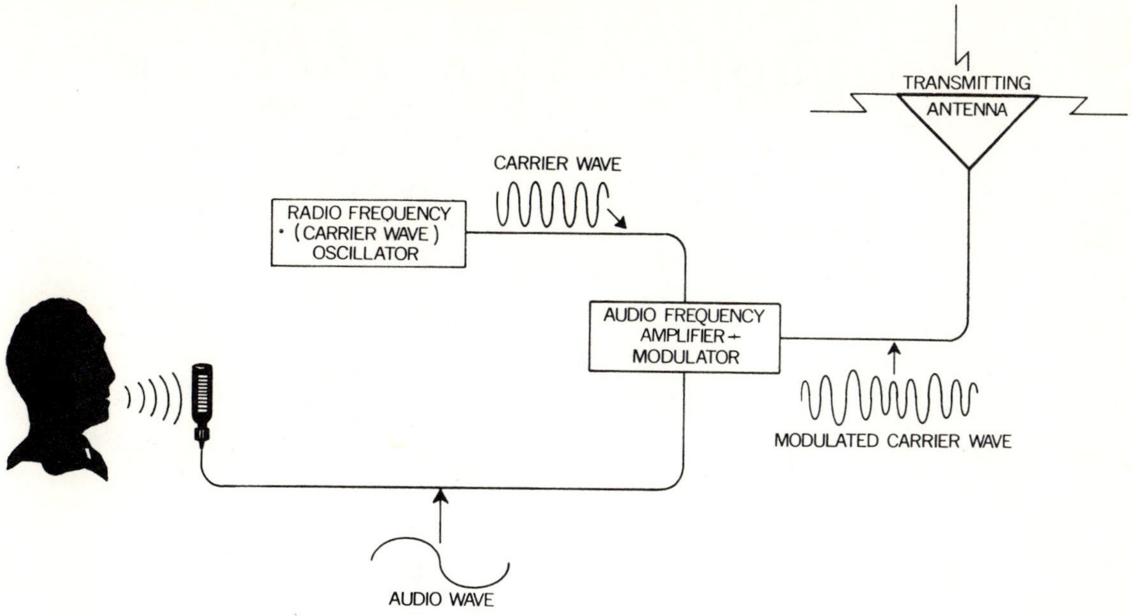

*Figure 9–2. Radiotelephone transmitter schematic.*

around a conductor (the antenna). Radio frequencies extend from 10 kilocycles to above 300,000 megacycles.

## Principles of the Transmitter

*a. Generating and Transmitting a Radio Signal.* Fundamentally, a radio signal is transmitted by generating an alternating electric current of the desired frequency and connecting it to an antenna suitable for radiating that particular wavelength (fig. 9–2). The current frequency is determined by the number of times per second that the alternating current changes direction of flow in the antenna.

*b. Altering the Radiated Signal.* To transmit intelligible data, the radiated carrier wave (A, fig. 9–3) is altered in some manner and these alterations are decoded at the receiver. Code is transmitted either by interrupting the carrier wave (B fig. 9–3) into a series of dots and dashes or by modulating the carrier wave with another steady tone (C, fig. 9–3) which is interrupted to produce the desired code. Voice is transmitted by molding or modulating the carrier wave signal with audio wave transmissions (D, fig. 9–3) generated through the radio microphone. The combined carrier wave and audio wave appear as in E, figure 9–3 if the

audio wave is superimposed by amplitude modulation (AM) of the carrier wave. If the audio wave is superimposed by frequency modulation (FM), the combined wave will appear as shown in F of figure 9–3. Both AM and FM are used in transmitting voice to Army aircraft, but AM is the most common method. A modulated signal is commonly called a modulated carrier wave (mcw) when either voice or tone signals are used in the modulation process (C, E, and F of fig. 9–3 represent modulated carrier waves). Figure 9–2 illustrates voice modulation of a carrier wave.

## Principles of the Receiver

*a. Tuning.* Radio waves induce minute electrical currents in receiving antennas. This process is the same as inducing an alternating current in one conductor by placing it near another conductor carrying alternating current. The method of selecting the desired signal from the many induced signals is called *tuning.* The tuning circuit in the receiver is adjusted to *resonance* with the frequency of the desired signal; other frequencies are rejected by the tuning circuit. The selected frequency is unintelligible at this receiving stage since it is

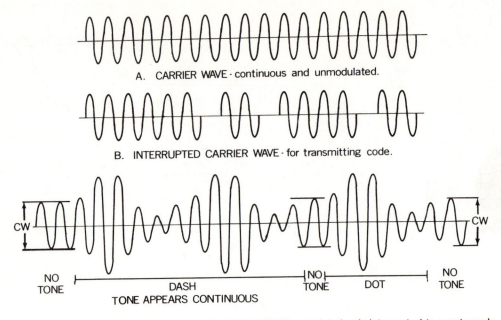

A.  CARRIER WAVE - continuous and unmodulated.

B.  INTERRUPTED CARRIER WAVE - for transmitting code.

C. CONTINUOUS WAVE MODULATED WITH TONE SIGNAL - modulation is interrupted to create code.

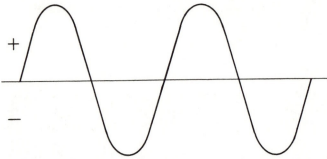

D.  AUDIO WAVE - continuous audible tone signal. When superimposed on carrier wave, the positive 180°
increases the carrier wave amplitude and the negative 180° decreases the carrier wave amplitude
(E below).

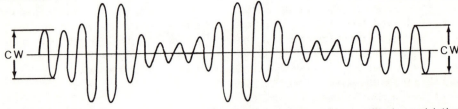

E.  CARRIER WAVE (A above) with audio wave (D above) superimposed by amplitude modulation (AM).

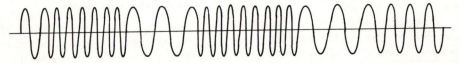

F.  CARRIER WAVE (A above) with audio wave (D above) superimposed by frequency modulation (FM).

*Figure 9–3. Radio waves.*

still a combination of the radio wave (carrier wave) and the audio wave (voice).

*d. Demodulating.* Another stage of the receiver called a demodulator or detector is used to separate the audio wave from the carrier wave. The audio portion of the wave is amplified and used to vibrate the diaphragm of the headset or a speaker. This vibrating surface causes audible sound waves, reproducing those which entered the microphone at the transmitter (fig. 9–2).

*Note.* Since the frequency of a carrier wave is above the audible sound range, a beat frequency oscillator (BFO) (para 11–3*d*) is used to convert coded interrupted carrier wave signals into audible, intelligible sound.

(6) Ultrahigh frequency (UHF), 300 to 3,000 mc.

(7) Superhigh frequency (SHF), 3,000 to 30,000 mc.

(8) Extremely high frequency (EHF), 30,000 to 300,000 mc.

*Note.* Although the Federal Communications Commission (FCC) designates 300 to 3,000 mc as UHF, the UHF radio frequency band in aviation communication begins at 200 mc.

## Low Frequency Radio Wave Propagation (Nondirectional)

A radio wave leaves the transmitting antenna in all directions. That portion of the radiated

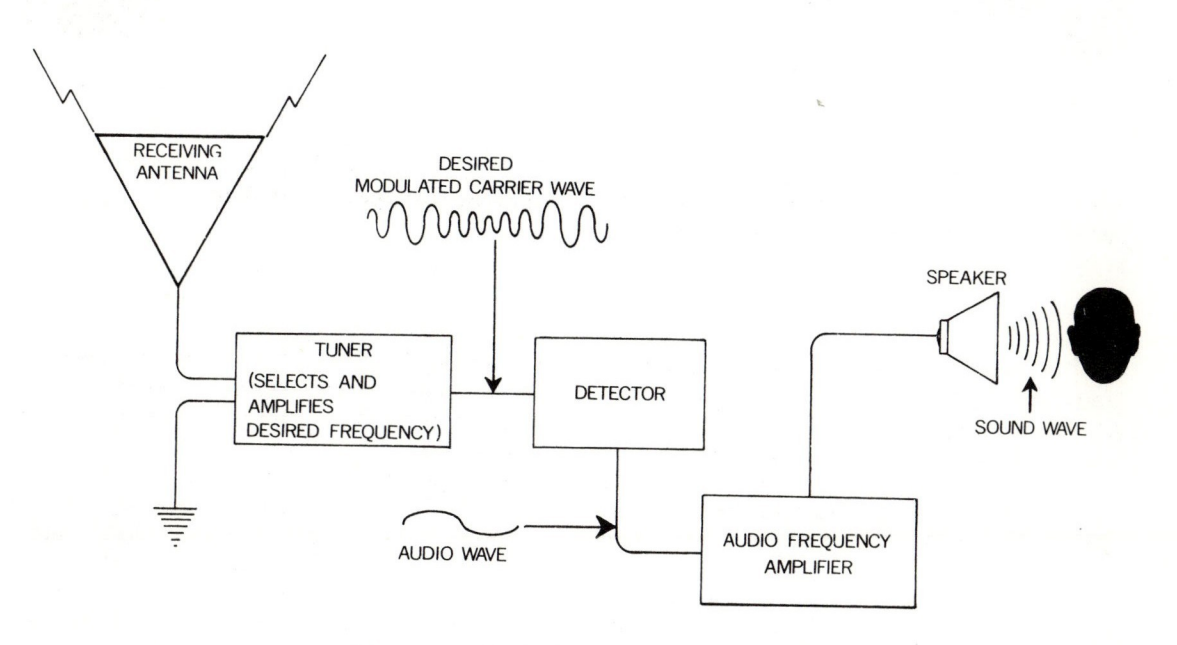

*Figure 9–4. Radiotelephone receiver schematic.*

## Classification of Frequencies

*a. Audio Frequency (AF).* Twenty to 20,000 cps (cycles per second).

*b. Radio Frequency (RF).*

(1) Very low frequency (VLF), 10 to 30 kc (kilocycles).

(2) Low frequency (LF), 30 to 300 kc.

(3) Medium frequency (MF), 300 to 3,000 kc.

(4) High frequency (HF), 3,000 to 30,000 kc.

(5) Very high frequency (VHF), 30 to 300 mc (megacycles).

wave following the ground is called the *ground wave* (fig. 9–5). The ground wave is conducted along the earth until its energy is absorbed (depleted by the attenuation process.

The remainder of the radiated energy is called the *shy wave* (fig. 9–5). The sky wave is radiated into space and would be lost were it not for the refracting layers in the atmosphere. These layers are in the region of the atmosphere called the *ionosphere* (region where air is ionized by radiation of the sun). The refracting effect on the waves returns them to earth and permits signals to be received at distant points. The effect on reception distance is determined by the height and density of the ionosphere and by the angle at

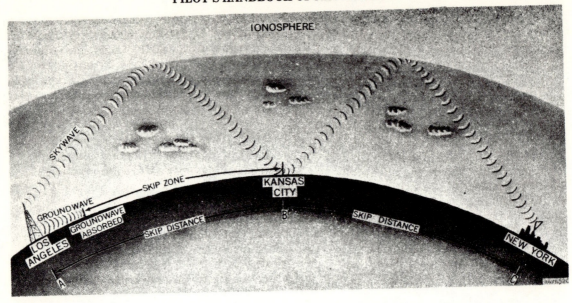

*Figure 9–5. Radio wave transmission.*

which the radiated wave strikes the ionosphere. The ionosphere varies in height and density with the seasons, time of day, and latitude.

### Skip Distance

The distance between the transmitting antenna and the point where the sky wave first returns to the ground is called the *skip distance* (AB, fig. 9–5). By extension, this term also includes the distance between each surface reflection point in multihop transmission (BC, fig. 9–5). The distance between the point where the ground wave can no longer be received and the point where the sky wave is returned is called the *skip zone*. Since solar radiation changes the position and density of the ionosphere, a great change in skip distance occurs at dawn and dusk, causing the fading of signals to be more prevalent than usual.

### Effect of All Matter on Radiation

All matter within the universe has a varying degree of conductivity or resistance to radio waves. The earth itself acts as the greatest resistor to radio waves. The part of the radiated energy that travels near the ground induces a voltage in the ground that subtracts energy from the wave. Therefore, the ground wave is attenuated (decreased in strength) as its distance from the antenna increases. The molecules of air, water, and dust in the atmosphere and matter at the earth's surface—such as trees buildings, and mineral deposits—also absorb radiation energy in varying amounts.

### Effect of Static Upon Low and Medium Frequency Reception

Static disturbance is either manmade or natural interference. Manmade interference is caused, for example, by an ordinary electric razor. Each small spark, whether originating at a spark plug, contact point, or brushes of an electric motor, is a source of radiation. All frequencies from 0 to approximately 50 mc are transmitted from each spark and, consequently, add their energy to any radio reception within this frequency range. Natural static may be divided into two types. Interference which originates from natural sources away from the aircraft is called *atmospheric static*. Interference caused by electrostatic discharges from the aircraft is called *precipitaion* or *canopy static*.

### General Nature of High Frequency Propagation (3,000 Kc to 30 Mc)

The attenuation of the ground wave at frequencies above approximately 3,000 kc is so great as to render the ground wave of little use for communication except at very short distances. The sky wave must be used, and since it reflects back and forth from sky to ground, communication can be maintained over long distance (12,000 miles, for example). Frequencies between LF and VHF produce the greatest radio transmission range between points on the earth because they are refracted by layers of the ionosphere and follow the curvature of the earth. The range of low frequencies (LF) is reduced by attenuation and atmospheric absorption, and VHF or higher frequencies penetrate the ionosphere and escape to outer space.

## General Nature of Very High Frequency (VHF) and Ultra High Frequency (UHF) Propagation (30 to 3,000 Mc)

Practically no ground wave propagation occurs at frequencies above approximately 30 mc. Ordinarily there is little refraction from the ionosphere, so that communication is possible only if the transmitting and receiving antennas are raised far enough above the earth's surface to allow the use of a direct wave. This type of radiation is known as "line-of-sight" transmission. Thus, VHF/UHF communication is dependent upon the position of the receiver in relation to the transmitter. When using airborne VHF/UHF equipment, it is of utmost importance for the aviator to be aware of the factors limiting his communication range.

## Range of VHF and UHF Transmission

The range of VHF and UHF transmission is limited primarily by the altitude of the aircraft and the power of the station. Both VHF and UHF are line-of-sight transmission, and at 1,000 feet above level terrain are usable for approximately 39 nautical miles. At higher altitudes, VHF and UHF transmissions can be received at greater distances as indicated below.

| Altitude above ground station (feet) | Reception distance (nautical miles) |
|---|---|
| 1,000 | 39 |
| 3,000 | 69 |
| 5,000 | 87 |
| 10,000 | 122 |
| 15,000 | 152 |
| 20,000 | 174 |

# 10 VHF OMNIDIRECTIONAL RANGE SYSTEM (VOR)

## Section I. COMPONENTS AND OPERATION

### General

The VHF omnidirectional range system (VOR) is the major navigational system used by Army aviators in the United States. The VOR is a VHF facility which eliminates atmospheric static problems and provides the aviator with 360 usable courses to or from the station. The terms *omni* and *VOR* are often used interchangeably.

### Transmitter and Receiver Fundamentals

The VOR transmitter emits two signals—the

magnetic north. At all other points around the station, there is a definite difference between the signals (fig. 10–1). Receivers in the aircraft detect the phase difference and present the information to the aviator either by centering an indicator needle representing an on-course position or by deflecting the needle to the right or left of center representing an off-course position. The signals may also be fed to a compass-type indicator to show direction to the transmitting station.

### Army Receivers

Standard VOR receiver set is the ARN–30.

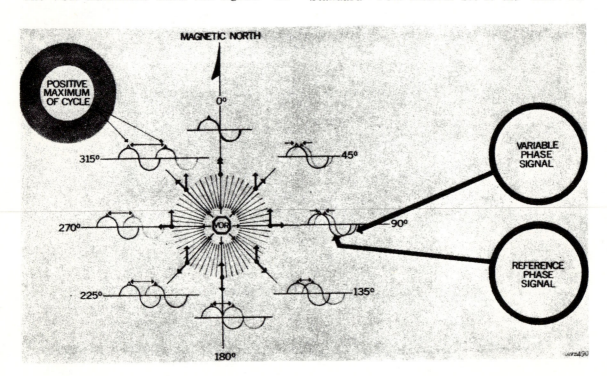

*Figure 10–1. VOR signal phase differences.*

*variable signal* and the *reference signal*. The variable signal is transmitted in only one direction at any given time; however, the direction of its transmission varies so rapidly that the signal appears to be a continuous signal rotating clockwise around the station at approximately 1,800 rpm. Receivers which are correctly tuned to a station will receive both signals. The two signals are in phase *only* at

The control panel for the ARN–30A is illustrated in figure 10–2. This set can receive VHF signals from 108 mega Hertz (MHz) to 136 MHz. The lower part of this band (108 to 118 MHz) is used for reception of voice and navigational signals from VOR stations and the localizer transmitter of the instrument landing system (ILS). Since VOR stations do not operate above 118 MHz, the upper part of the receiver band is used only for voice reception of various radio facilities operating on these channels. Some Army aircraft are equipped with the ARN–30D receiver. This receiver is basically the same as the ARN–30A but has a different tuning mechanism (fig. 10–3). The ARN–30A control panel has a continuous tuning crank, whereas the ARN–30D has two digital tuning knobs. The left knob changes the whole mega Hertz frequency and the right knob changes the decimal values. The tuned frequency appears in the window on the receiver control panel. This digital tuning feature of the ARN–30D is far more efficient and convenient in flight, especially when the aviator's attention must be devoted to several things simultaneously. Another type receiver (ARN–30E) is similar to the ARN–30D but is coupled with a glideslope receiver to allow simultaneous tuning of the ILS localizer frequency and the glideslope. The frequency span for the ARN–30D and –30E is 108.00 to 126.90 MHz. The glideslope feature of the ARN–30E is discussed in chapter 15.

## VOR Course Indicator Components (ID–453)

The VOR navigation signal is displayed to the aviator in most Army aircraft on an instrument called the *course indicator* (ID–453) (fig. 10–4).

*a. Components.* This instrument consists of—

(1) A course selector (bearing selector) (A and A¹, fig. 10–4).

(2) A course deviation indicator (CDI) commonly called the *needle*) (B, fig. 10–4).

(3) A to–from indicator (sense indicator) (C, fig. 10–4).

(4) A glideslope deviation indicator (D, fig. 10–4).

*Note.* The glideslope indicator is not shown in other figures of this chapter because it is not used for the procedures discussed herein.

(5) A VOR receiver warning flag (localizer warning flag) (E, fig. 10–4).

(6) A glideslope receiver warning flag (F, fig. 10–4).

*b. Course Selector.* The aviator operates the course selector manually with the course selector knob labeled ARC (fig. 10–4). The arrow end of the course selector (A, fig. 10–4) is set to any course the aviator desires; the course reciprocal then appears opposite the course arrow under the ball (A¹, fig. 10–4). The to–from indicator (sense indicator) will respond automatically to any course the aviator selects. The to–from indicator (C, fig. 10–4) senses the position of the aircraft with respect to the station. It indicates *TO* any time a selected course (if flown) would bring the aircraft closer to the station. *For example*, if an aircraft is located southeast of a station and the aviator selects a course of 270°, the to–from indicator

*Figure 10–2. ARN–30A receiver control panel.*

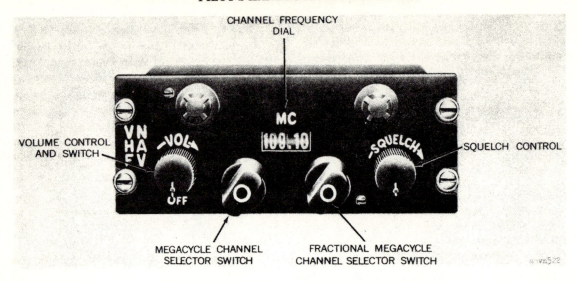

*Figure 10–3. ARN–30D or –30E receiver control panel.*

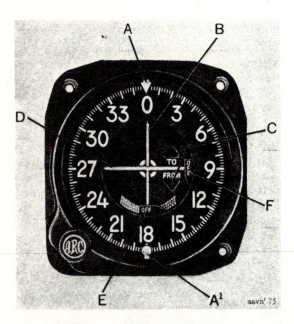

*Figure 10–4. ID–453 course indicator.*

centered regardless of the aircraft's heading.

(2) *Full-scale deflection.* If the aircraft is off the selected course by 10° or more (dashed radials, fig. 10–6), the needle deflects full-scale to one side (aircraft D, fig. 10–6). The indicator face is graduated in 2° increments, with the edge of the small center circle representing 2° and each dot (alined horizontally) representing 2°.

d. *Warning Flags.* The VOR receiver warning flag (localizer warning flag) is at the bottom of the course deviation indicator. This flag which is labeled *OFF* (E, fig. 10–4) will be visible any time the signal being received is too weak to operate the receiver properly. This flag must be completely out of sight while the receiver is being used for navigation. The glideslope receiver warning flag (F, fig. 10–4) is discussed in chapter 15.

immediately responds with a *TO* indication. If the aircraft flies the 270° course, it will move closer *to* the station (fig. 10–5).

c. *Deviation Indicator.* The course deviation indicator (needle) (B, fig. 10–4) will indicate whether the aircraft is presently located on the selected course or its reciprocal indicated under the ball. Figure 10–6 shows course deviation indicator readings with respect to a selected course of 300°.

(1) *Centered.* When the aircraft is actually located on the selected course (aircraft A, B, and C, fig. 10–6), the deviation needle is

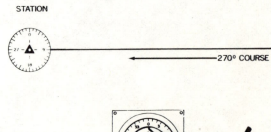

*Figure 10–5. Course selector operation.*

104

## Tuning

*a. ARN–30A.* To tune the VOR receiver with the ARN–30A receiver control panel (fig. 10–2) —

(1) Turn the combined off–on switch and volume control (labeled VOL/OFF, fig. 10–2) clockwise.

(2) Place the selector switch (labeled OMNI/VAR–LOC, fig. 10–2) in the OMNI position. (The VAR–LOC position is discussed in ch. 15.)

(3) Determine the frequency to be used and "crank in" this frequency on the dial (MC, fig. 10–2) with the tuning crank.

(4) When the dial (MC, fig. 10–2) is set to the correct position, listen in the headset for the correct station identification. Increase the volume temporarily if necessary. After identification is completed, the volume may be reduced to the desired level.

(5) Observe the behavior of the VOR warning flag (*OFF* flag) on the course indicator (E, fig. 10–4). This flag will drop out of sight as the correct frequency is tuned and the signal from the station is reliably received.

*b. ARN–30D or –30E.* Tuning with the ARN–30D or –30E receiver (fig. 10–3) is similar to the ARN–30A except for *a*(2) and (3) above. There is no OMNI/VAR–LOC select switch, and the frequency is set with the digital tuning knobs on the control panel. The knob on the left controls the whole-frequency digits and the knob on the right is used to set the tenths and hundredths of mega Hertz. Other procedures involved in tuning are the same as described for the ARN–30A.

*Note.* The squelch knob is used to reduce background noise in the receiver headset.

*c. Signal Reception.*

(1) Because of VHF radio characteristics (ch. 9), the signal from a station may not be received adequately if the aircraft is on the ground in a blind spot, too far away from the station, or at an altitude below the reliable reception altitude. Station monitors normally detect irregularities in the transmission of the signal from the station.

(2) If the station is transmitting a signal known to be unreliable, the station identification is removed from the transmission. Consequently, if the station identification is not received while tuning, even though other reactions are normal, the signal should be considered unreliable for navigation. However, the voice capability of the station at this time may still be normal and usable.

## Section II. FLIGHT PROCEDURES WITH THE ID–453

## Orientation

*a. Courses and Radials.* The term *course* may be used to refer to any desired or intended track into or away from a VOR station. The aviator selects such courses with the course selector. The term *radial* refers to a course emanating from a VOR station. On navigation charts, courses are published as directions outbound from the VOR stations (radials). It is frequently convenient to refer to an aircraft's position in terms of the radial on which it is located; *for example,* figure 10–7 shows three aircraft on the 090° radial. Aircraft A is on the 090° radial following a 270° course inbound to the station. Aircraft B is on the 090° radial following a 090° course outbound from the station. Aircraft C is crossing the 090° radial flying a heading of 320°.

*b. Orientation Procedure.* Orientation using the VOR indicator (ID–453) is accomplished by determining the radial on which the aircraft is located; i.e., determining the direction to the station. To accomplish the orientation procedure—

(1) Tune and identify the station.

(2) Rotate the course selector until the to-from indicator reads *TO* and the deviation indicator (needle) is centered.

(3) Read the direction to the station under the course arrow when the needle centers with a *TO* indication. The radial on which the aircraft is located also can be read under the ball (A, fig. 10–7).

*Note.* The procedures in (2) above can be altered to result in a *FROM* reading if desired. If altered in this manner, the radial on which the aircraft is located appears under the course arrow and the direction to the station appears under the ball (B and C, fig. 10–7).

## Tracking

*Tracking* is the process of maintaining a specific course into or away from a station; i.e., to "make good" an intended track. Figure 10–8 shows an aircraft following a 0° course inbound to the VOR station.

*a. Position A.* In position A, the aircraft is on course; the needle is centered.

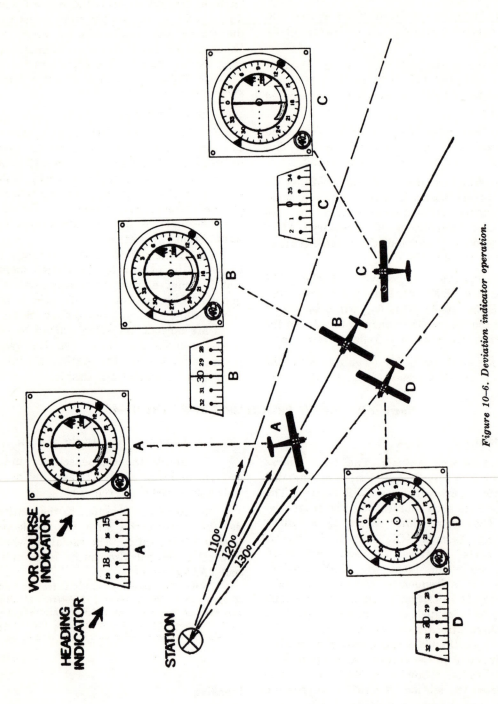

Figure 10–6. Deviation indicator operation.

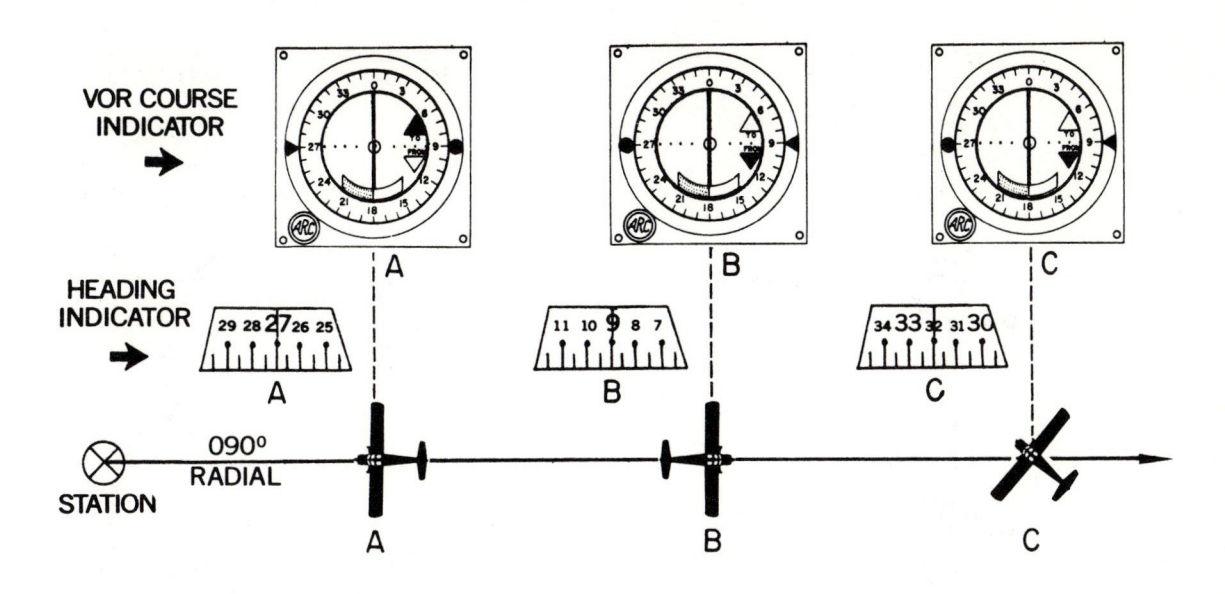

*Figure 10–7. Aircraft positions described by radial and heading.*

*b. Position B.* In position B, the aircraft has been blown off course approximately 5°. The crosswind is from the left and the deviation indicator is deflected to the left. To return to the intended track, the aviator must correct his heading to the left. The standard correction under normal wind conditions is 20° for aircraft with airspeed above 90 kt and 30° for aircraft operating below 90 kt.

*c. Position C.* In position C, the aviator is applying the standard correction.

*d. Position D.* In position D, the aircraft has returned to the intended track—the needle has recentered. At this time if the aviator maintains his present heading, he will fly through the intended track. If he returns to the original course heading, he will be blown off course again.

*e. Trial Drift Correction.* To avoid both situations, the aviator corrects the heading (turning toward the intended track; by half the amount of the initial correction; i.e., he turns toward the track 10° (15° if flying a slow helicopter). This results in the first trial drift correction for the crosswind. This drift correction may later prove to be either correct, too small, or too large.

(1) *Correction too small.* If the first trial drift correction is too small (wind is stronger than anticipated), the aircraft is again blown off course 5° from point E to point F (fig. 10–9). The aviator corrects his heading (G, fig. 10–9) to intercept the intended track at 20° or 30°, whichever is appropriate to his aircraft. He reintercepts the intended track at point H. He corrects his heading (I, fig. 10–9) by turning toward the intended track 5° (a 10° turn is' used for slow aircraft). This bracketing procedure can be repeated if necessary.

(2) *Correction too large.* If the first trial drift correction is too large (wind not quite as strong as expected), the aircraft will fly off course upwind. In figure 10–10, the aircraft is overcorrecting at point U and flies off course into the wind at point V. The aviator returns to the intended track by turning parallel to the track (point W) to let the wind blow the aircraft back on course at point X. When back on course, the aviator applies a 5° drift correction (not as large as the initial correction) into the wind (point Y) ; or a 10° drift correction is applied for slow aircraft. This systematic approach to establishing the proper drift correction is usually the most effective method of quickly establishing the proper heading. An aviator *could* track by "following the needle"; *for example,* with left needle deflection—turning left, and for right needle deflection—turning right; but this method, sometimes called "chasing the needle," normally wastes time, fuel, and effort.

(3) *Correction for unusually strong wind.* On some occasions, unusually strong winds will prevent the aircraft from returning to the intended track even when a 20° or 30° correction is used. If, after applying a 20° or 30° correc-

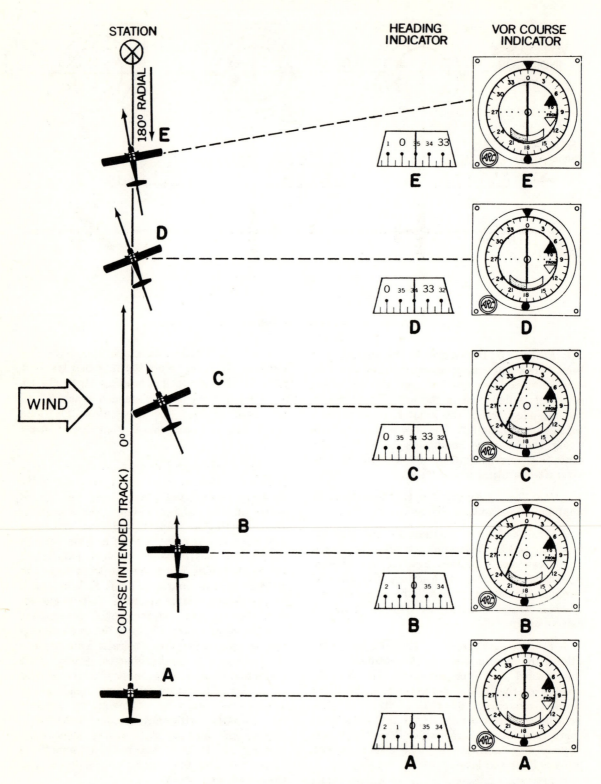

Figure 10-8. Tracking inbound (VOR).

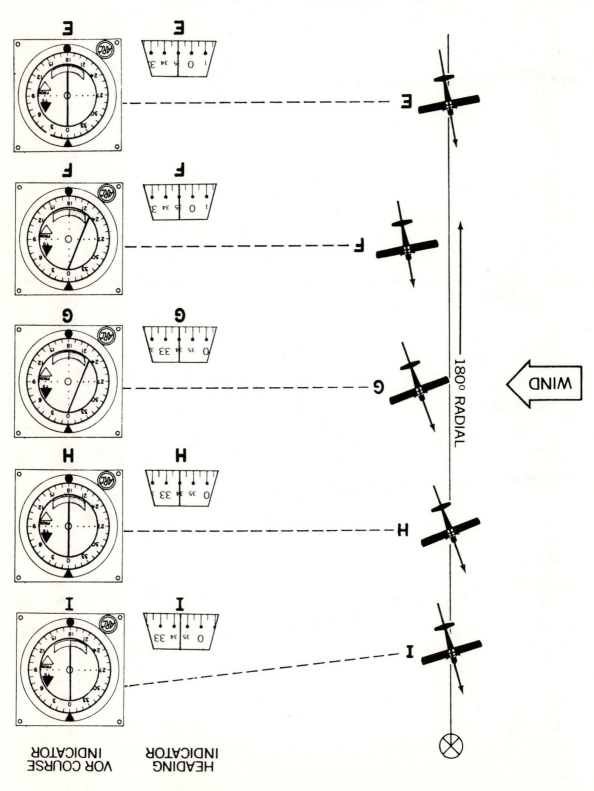

Figure 10-9. Trial drift correction too small (VOR).

VHF OMNIDIRECTIONAL RANGE SYSTEM (VOR)

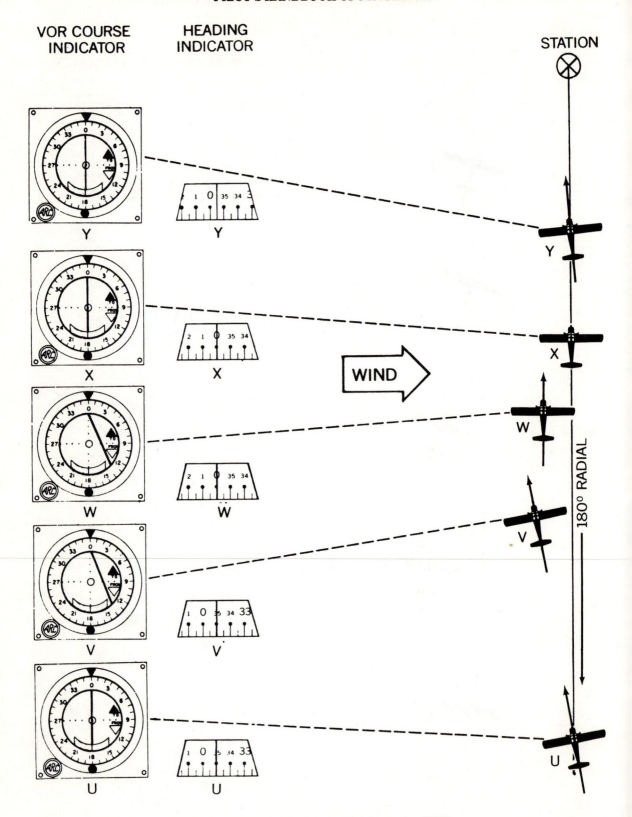

Figure 10-10. Trial drift correction too large (VOR).

tion, the aviator sees that the needle is still not recentering in a reasonable amount of time, he may have to apply a correction of 40° or more to return to the course. The pilot must assume that if 40° is required to return to the intended track, approximately half of the correction (in this example 20°) will be required to stay on course.

*Note.* Tracking procedures discussed here are for guidance only; in flight, these procedures are refined by the pilot to suit specific flight conditions.

### Station Passage

Recognition of station passage is very important because aviators use VOR stations to fix their exact position. These stations are also used as holding points for air traffic control and are often the destination point of an IFR flight to be used during the instrument approach to the airfield. Station passage is determined as follows:

*a.* Since the pilot usually knows his approximate arrival time over a station, he watches the clock and, as this time approaches, observes the to–from indicator reaction.

(1) While inbound to the station, the indicator will read *TO*.

(2) As the aircraft passes over the station, the to–from indicator will fluctuate momentarily, then drop to *FROM*. The time that this occurs is station passage time. While flying over the station, the aviator may also notice fluctuations of the deviation needle and the momentary appearance of the warning flag.

*b.* If the pilot intends to continue his flight along the same course, he continues to track outbound. The only indicator change is the reversal of the to–from indicator. If there is a course change (fig. 10–11), the pilot resets the course selector and turns the aircraft to the new heading.

*c.* Figure 10–12 illustrates another important consideration when the to–from indicator reading changes. The aircraft is flying toward the station (point A) but is *not* inbound on the selected course. The aircraft continues on the same heading and flies past the station (point B). At the time the aircraft is abeam the station, the to–from indicator will change to read *FROM*. This *FROM* reading will remain in the indicator as the aircraft flies away from the station (point C).

### Position Fixing

*a.* The Victor (V) airways system (see *note*

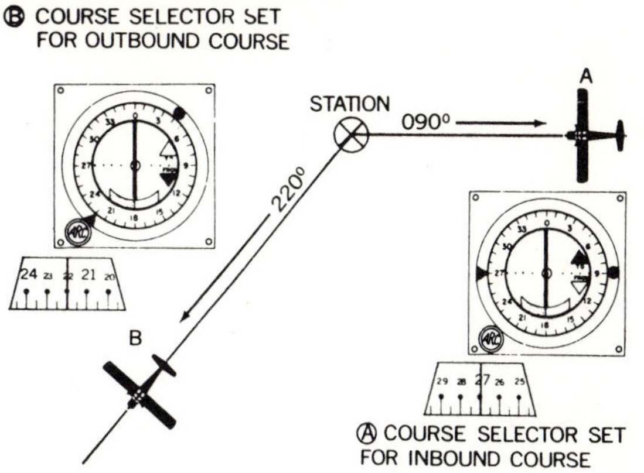

*Figure 10–11. Course selector reset to track outbound on a different course.*

below) is based upon the operation of several hundred VOR stations and has, in addition to the stations themselves, numerous other flight checkpoints (intersections). An *intersection* is a point where two or more radials from different VOR stations intersect. Checkpoints can be established at these intersections for position fixing. The procedure for fixing position over intersections by using one VOR receiver (course selector reading *FROM*) is illustrated in figure 10–13.

*Note.* Airways are routes through navigable airspace on which air route traffic control is maintained over en route IFR traffic. Airways are labeled along their centerlines point-to-point, and are comprised of the airspace charted on either side of the designated centerline and within the designated upper and lower limits. Airways established with VOR facilities in the low altitude route structure (position determined by radials from VOR stations) are called Victor airways and are labeled with a V and a number; e.g., V-111. The north-south airways have odd numbers and the east-west airways even numbers. Airways in the high altitude route structure are labeled with a J and a number, and are called jet routes.

(1) The aircraft proceeds outbound (W, fig. 10–13) from station A with the receiver tuned to station A. During this outbound flight, the aviator establishes the correct heading for remaining on the course (090°).

(2) After establishing the desired heading to remain on a 090° course, the aviator tunes and identifies station B.

(3) The 130° radial from station B crosses the 090° radial from station A to establish the intersection (open triangle symbol). The pilot sets the course selector on 130° and, since this is the radial from the station, the to–from indicator will read *FROM* (X, fig. 10–13).

(4) At the time the aircraft is exactly over the intersection (Y, fig. 10–13), the deviation indicator will center since the aircraft's

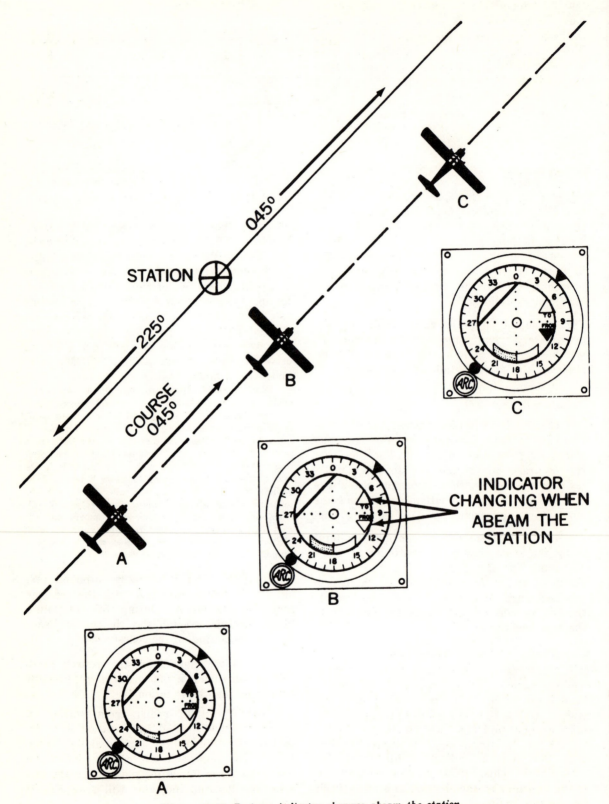

Figure 10-12. To-from indicator changes abeam the station.

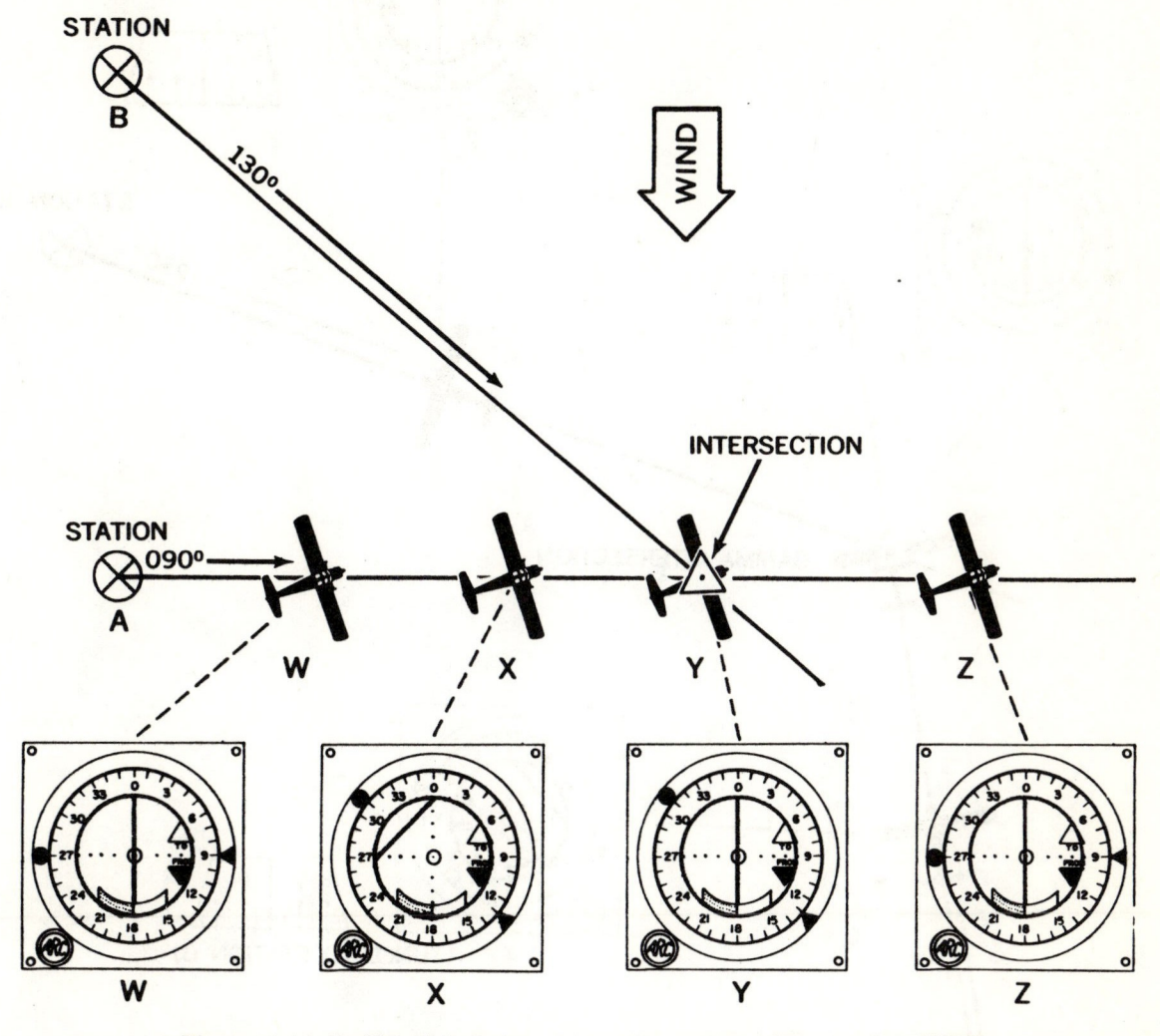

*Figure 10–13. Position fixing at an intersection, course selector reading FROM.*

position will then be on the 130° radial from station B. Having fixed his position over the intersection, the pilot then resets the course selector to 090°, retunes and identifies station A, and checks for the centering of the needle to remain on course outbound from station A (Z, fig. 10–13).

*b.* In performing the procedure discussed in *a* above, it is important that the pilot be able to interpret the direction of needle deflection. In the situation described in figure 10–13, the needle is deflected to the left while the aircraft

is at point X. Prior to arrival at the intersection, the needle will be deflected to the same side on which the station is located if the course selector has been set on the published *radial* which causes the to–from indicator to read *FROM.*

*c.* It may be convenient or necessary to fix an intersection by setting the course selector for a *TO* reading, as illustrated in figure 10–14. In flying from station A to station B, the aviator will turn inbound to station B when the aircraft arrives over the Gamma intersection.

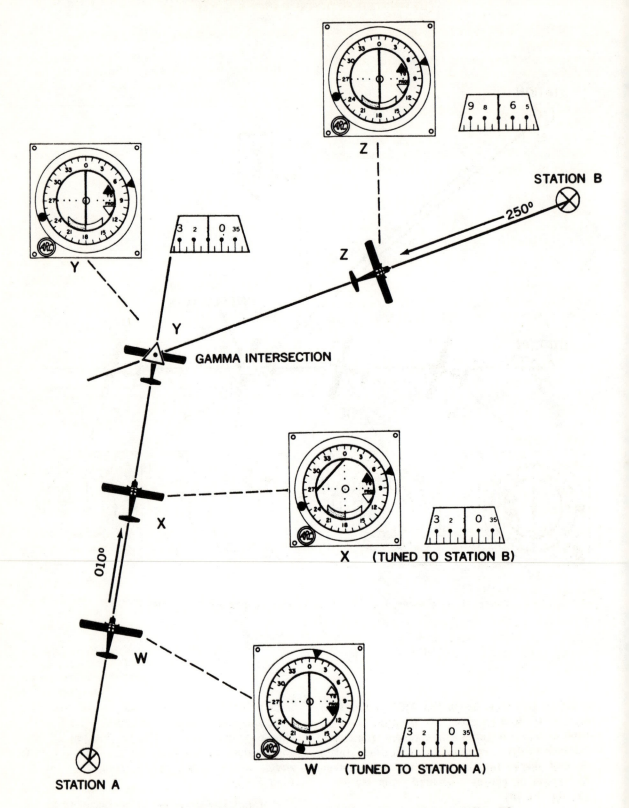

Figure 10–14. *Position fixing at an intersection, course selector reading TO.*

(1) The pilot departs station A tracking outbound, with the course selector set on 010°, the needle centered (on course), and the to–from indicator reading *FROM* (W, fig. 10–14).

(2) Prior to reaching the Gamma intersection, the pilot tunes and identifies station B (X, fig. 10–14). The published radial for station B is 250°, but the pilot knows that 250° is the direction *outbound* from B. Since he desires to go inbound, he sets the course selector to 070°, the reciprocal of 250°. The resultant reading on the to–from indicator is *TO* because a course of 070° will take the aircraft *to* station B. Station B is to the aviator's right from point X to Gamma intersection, but the needle deflects to the left. Since the course selector is set on the reciprocal of the published radial to produce the *TO* reading in the to–from indicator, the needle deflects to the side opposite the station. (Compare with the deflection described in *b* above.)

The needle centers when the aircraft arrives over Gamma intersection (Y, fig. 10–14), and remains centered inbound to station B (Z, fig. 10–14).

## Track Interception

It is occasionally necessary to intercept a desired track some distance from the position of the aircraft. Several procedures can be applied, as explained below.

*a. From a Known Position—45° or 90° Interception.* At A in figure 10–15, the aircraft is flying a heading of 350° while crossing the 200° radial. The flight has been cleared to intercept Victor airway 13 (V–13), which is the 180° radial from the station. Since the present position (radial) with respect to the radial to be intercepted is known, proceed as follows.

(1) Determine the direction of turn— right or left.

(2) Select a heading which will intercept the desired track at an angle of 45°. In this case, since the desired track inbound to the station is 0°, a heading of 045° would intercept the track at a 45° angle.

*Note.* The standard interception angle is 45°; however, others may be used. If ATC requests the pilot to "expedite," a 90° interception angle should be used.

(3) Turn to the selected heading and set the course selector on the desired inbound track—0°.

(4) Turn to the inbound heading of 0° when the needle centers (B, fig. 10–15); i.e.,

when the aircraft has reached the track.

(5) Procedure (2) above can be changed to intercept the track at a 90° angle (C, fig. 10–15) if necessary to reach the track in the least possible time.

*b. From a Known Position—Double-the-Angle Interception.* The double-the-angle method of intercepting a desired track from a known position consists of the following procedures (fig. 10–16):

(1) Determine the angular difference between the radial on which the aircraft is presently located and the radial which represents the desired track. At A of figure 10–16 the aircraft is on the 150° radial and the aviator wishes to intercept the inbound course of 0° to the station. (This is the 180° radial, so the angular difference is 30°.)

(2) Double the angular difference and this will give a desirable interception angle. In this case, the interception angle will be 60°.

*Note.* When using this procedure, initial interception angles of less than 20° are usually not practical. Also, an interception angle of 90° is the maximum; thus, an angle greater than 45° would not be doubled.

(3) Select the heading which will cause the aircraft to intercept the desired track at the desired interception angle. In this case, a heading of 300° will intercept the inbound track of 0° at the desired 60° angle.

(4) Turn the aircraft to the selected heading and reset the course selector for the inbound track of 0°.

(5) At the time the needle centers (B, fig. 10–16), the aircraft has reached the track and the pilot turns inbound to the station.

*Note.* When using this technique, the leg flown to intercept (from point A to point B) is equal in length to the leg remaining to the station (from point B to the station). Consequently, the time required to fly the interception leg is the approximate time remaining to fly to the station, from the time that interception occurs.

*c. Leading the Needle.* If the pilot waits until the needle is fully centered to turn to the track heading, he runs the risk of overshooting the track; if he turns to the track heading too soon, he will roll out short of the track. For use during initial track interception or reinterception, he must perfect the technique of leading the needle. The movement rate of the deviation needle and the size (degree) of the interception angle are good indicators of how much the aviator should lead the needle just prior to the actual track interception.

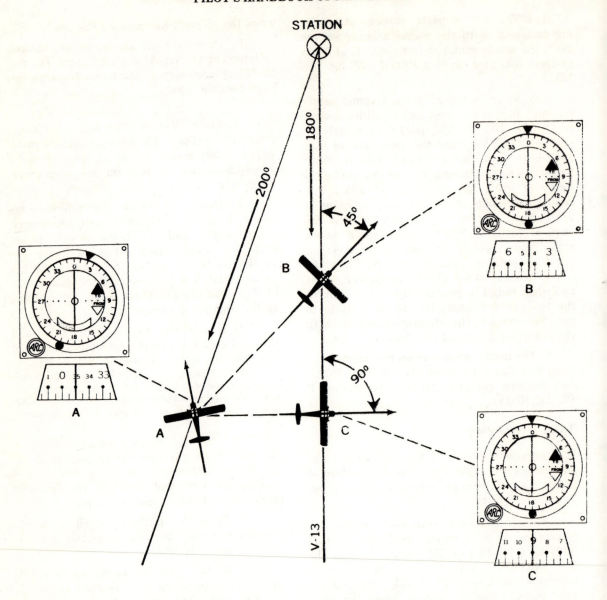

*Figure 10–15. Track interception at 45° or 90° from a known position (VOR).*

*d. Interception of a Given Track From an Unknown Position.* A pilot may be directed to intercept a specific track, and at a time when he is uncertain of the track location. He may not know if it is to his right, left, front, or rear. One simple method of quickly orienting the aircraft with respect to a desired track is as follows:

(1) From the present heading of the aircraft, turn the shortest way to a heading which is parallel to the desired track. (The station has previously been tuned and identified.)

(2) While turning ((1) above), set the course selector to the desired track.

(3) After rolling out of the turn ((1)

above), observe the deflection of the needle. The track lies to the same side as the needle deflection. The to–from indicator will now indicate if the station is ahead or behind the aircraft.

(4) Turn toward the track (needle), to a heading which will intercept the track at an appropriate angle.

## Estimating Time and Distance to a Station

*a.* In most situations, an aviator will be flying in regions where two VOR stations, or one VOR station and some other type of station, are within reception distance. The pilot can,

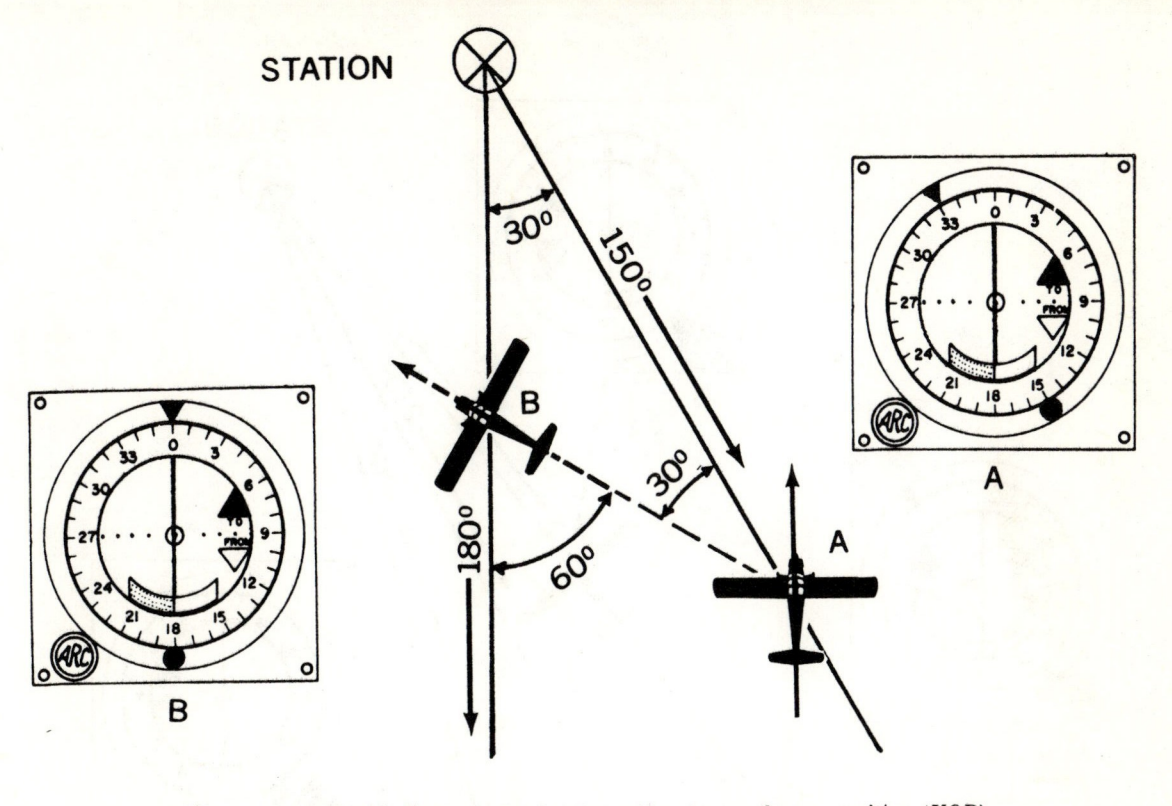

*Figure 10-16. Double-the-angle track interception from a known position (VOR).*

if necessary, fix his position and estimate the time and distance to either station by determining the bearing to each.

*b.* In some isolated cases, it may be necessary to estimate the time or distance to a station by using the signal from a single station.

*c.* A different method is illustrated in figure 10-17. The aircraft is inbound to the station on the 200° radial. To estimate the time and distance to this station, proceed as follows (fig. 10-17):

(1) Turn the aircraft through 80° (left in fig. 10-17).

(2) Move the course selector 10° (from 020° at point A to 030° at point B) to a known radial ahead of the aircraft.

(3) Wait for the needle to recenter and take a time check (e.g., 1412:50).

(4) Move the course selector an additional 10° (from 30° at point B to 40° at point C).

(5) Wait for the needle to recenter and take a second time check (e.g., 1414:55, or 2 minutes and 5 seconds elapsed during the 10°

bearing change).

(6) Turn inbound to the station (D) and estimate the time to the station by applying the following formula (data taken from situation in fig. 10-17):

$$\text{Time Remaining to Station} = \frac{\text{Minutes Flown} \times 60}{\text{Degree Change}} = \frac{\text{Seconds Flown}}{\text{Degree Change}}$$

$$\text{Time Remaining to Station} = \frac{2 \ 1/12 \ \text{Minutes} \times 60}{10°} = \frac{125 \ \text{seconds}}{10°} = 12.5 \ \text{Minutes}.$$

*Note.* If aircraft is turned 80° right ((1) above), move course selector from 020° to 010° in (2) above and from 010° to 0° in (4) above.

(7) The approximate distance to the station may be estimated by using the following formula:

$$\text{Distance to Station} = \frac{\text{True Airspeed} \times \text{Minutes Flown}}{\text{Degree Change}}$$

(8) Substituting the data from figure 10-17 (assume the TAS is 120 knots)—

$$\text{Distance to Station} = \frac{120 \times 2 \ 1/12}{10} = 25 \ \text{Nautical Miles}.$$

*Note.* Seconds must be changed to fractional or decimal parts of a minute.

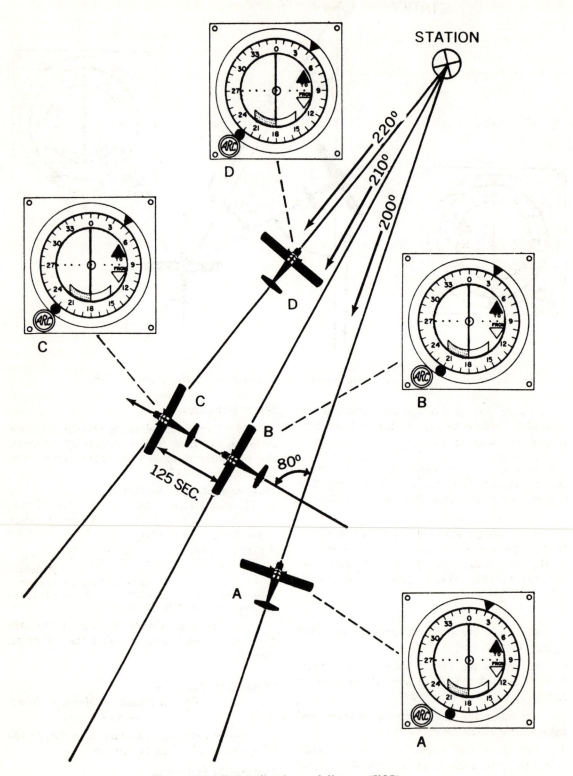

Figure 10–17. *Estimating time and distance (VOR).*

*d.* The following limiting factors should be kept in mind when applying the above time and distance formulas:

(1) They are based on the assumption that a 1° angle is 1 mile wide 60 miles away from the vertex. This is an approximation.

(2) They do not take into account adverse wind conditions that may cause groundspeeds to vary considerably on headings which differ by 90°.

(3) To determine time-distance required, the aircraft must turn so that it will fly abeam the station during the time required for the aircraft to fly through a 10° change in the course selector reading.

(4) The bearing change (change in the course selector setting) may be allowed to vary from 5° to 15°. Ten degrees is used as a mathematical convenience.

## Section III. RECEIVER CHECKS

### General

VOR receiver sand their associated indicators (e.g., ID–453) must be checked periodically for accuracy. The specific requirement for performing these checks is contained in Federal Air Regulations. There are several types of checks which can be performed to insure equipment accuracy. In performing these checks, current data for designated station frequencies, specific VOR radials, and station identification are contained in current navigational publications.

### Radiated Test Signal

Equipment installed at many airports transmits a continuous test signal receivable at any point on the airport. Although designed primarily as a ground test system, this equipment is also usable at relatively low altitudes in flight over the airport. The procedure for using the radiated test signal (VOT) to check receivers is to—

*a.* Tune the designated frequency.

*b.* Listen for the proper identification; i.e., either a continuous series of dots or a continuous 1,020-cycle tone.

*c.* Check for the disappearance of the *OFF* flag.

*d.* Set the course arrow to either 180° or 0°.

*e.* Check the reaction of the to–from indicator. If the course arrow is set on 180°, the indicator should read *TO.* If the course arrow is set on 0°, the indicator should read *FROM.*

*f.* Check the deviation needle. It should be centered. If the needle is not centered, rotate the course selector until the needle centers. If the course selector does not have to be rotated more than 4° in order to center the needle, then the equipment is within tolerance. If the

needle will not center within a 4° tolerance, the equipment is unreliable for instrument flight.

### Other Ground Checks

Not all airports have equipment for radiated test signals (VOT). However, many airports have VOR stations situated nearby from which selected radials can be used for checkpoints. Data pertaining to the certified ground check radial is published in current navigation publications. In the illustration (fig. 10–18), the 120° radial from a station passes directly over the end of runway 27. An exact spot is marked on this runway. The aircraft is taxied to this spot and the receiver check is performed in the following manner:

*a.* Tune the designated frequency of the station.

*b.* Listen for the correct station identification.

*c.* Check for the disappearance of the *OFF* flag.

*d.* Set the course arrow on the specific radial for the check.

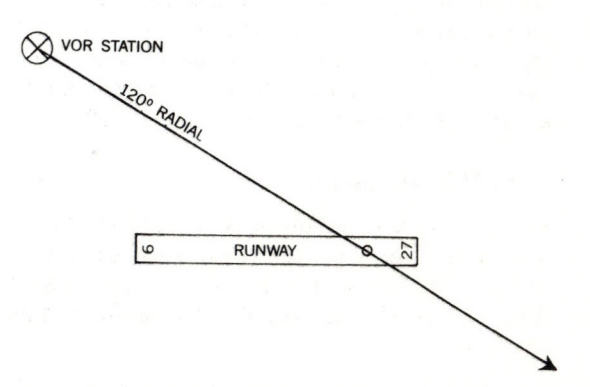

*Figure 10–18. Ground receiver check (VOR).*

*e.* Check the reaction of the to–from indicator. If the correct radial is set under the course arrow, it should indicate *FROM*.

*f.* Check the deviation needle for centered position. Plus or minus 4° tolerance is allowed on the course selector setting for centering the needle. If movement of the course selector within 4° of the published radial will cause the deviation needle to center, the equipment is usable. Equipment that does not meet these tolerance limits is unreliable for instrument flight.

## Airborne Check

*a.* At airports where radiated test signals (VOT's) or other ground check radials have not been established, an airborne check radial may exist. Airborne checks are performed like ground checks except that an air checkpoint is specified instead of a designated spot at the airport. *For example,* if a prominent water tower exists within a few miles of the VOR station, a certain radial can be selected which passes over this tower. As the aircraft flies over the tower, the accuracy of the equipment can be checked. A published airborne check over the tower may appear in navigational publications as: "DALLAS, TEXAS (Love Field)—213°, over striped water tower on Loop 12 (Highway) approximately 2.6 miles east-northeast of Love Field."

*b.* To perform the airborne check—

(1) Tune and identify the Dallas, Texas VOR.

(2) Set the course arrow on 213° and check for a *FROM* reading.

(3) Fly over the water tower described.

(4) When over the water tower, check the deviation needle for centered position. If the needle is within 6° of center, or if a course selector movement of 6° or less from the published radial will cause the needle to center, the equipment is within tolerance. Equipment that does not meet these tolerance limits should not be used for instrument flight.

## Dual VOR Receivers

If an aircraft is equipped with dual receivers, one receiver may be checked against the other. If receivers are within 4° of each other, both may be considered reliable. To perform this check—

*a.* Tune and identify one VOR station with both VOR receivers.

*b.* Using dual course indicators (ID–453 or equivalent), rotate the course selectors of each until the needles are centered.

*c.* Check to determine that the to–from indicators on each instrument are in agreement.

*d.* Check the course arrow readings. These readings must be within 4° of each other. If receivers do not meet these limits, one or both are unreliable. Each will have to be checked independently to determine if one is within allowable tolerance.

## Unpublished Receiver Check

An aviator in a location where no receiver checks are published may establish a checkpoint from a nearby VOR station. To accomplish an unpublished receiver check—

*a.* Select a VOR radial that lies along the centerline of an established VOR airway.

*b.* Select a prominent ground point along the selected radial preferably more than 20 miles from the VOR ground facility, and maneuver the aircraft directly over the checkpoint at a reasonably low altitude.

*c.* Note the VOR bearing indicated by the receiver when over the ground point. (The maximum permissible variation between the published radial and the indicated bearing is 6°.)

## Needle Sensitivity

At the same time that the VOR receiver is checked for accuracy, the indicator can be checked for needle sensitivity. The face of the indicator (ID–453) (fig. 10–19) is graduated in 2° intervals. Moving from center to either side, the edge of the small circle is 2° and each dot (alined horizontally) represents 2°. When the needle is fully deflected to one side, the aircraft is off the selected course by at least 10°. Consequently, if the receiver is checked for a centered needle with the course selector set on a given radial—for example 140° (A, fig. 10–19)—a full swing of the needle can be checked by setting the course selector on 130° (B, fig. 10–19) and then on 150° (C, fig. 10–19). If the needle swings full scale to each side as the course selector is displaced ±10° from a published check radial, the needle reaction is correct. Full scale needle swing can vary between approximately 8° and 12° when the receiver is within the reliable accuracy tolerances.

*Note.* The sensitivity tolerance of the needle is different if the receiver is used in conjunction with the ILS (ch. 15).

*Warning:* **Pertinent information for VOR receiver checks should be verified from current navigational publications and regulations. This information is subject to change.**

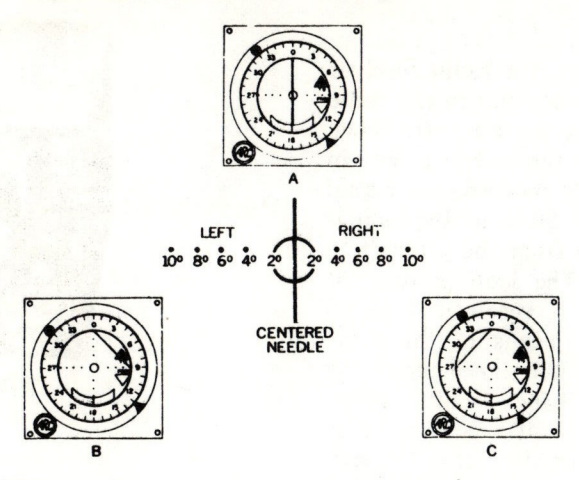

*Figure 10–19. Needle sensitivity check.*

## Section IV.   VOR STATION CLASSIFICATION

### Classification by General Types

The general types of VOR stations are—

  *a.* T (Terminal).

  *b.* L (Low Altitude).

  *c.* H (High Altitude).

  *d.* VORTAC is a dual facility consisting of a VOR and a TACAN facility. TACAN (ch. 20) is a UHF system of navigation similar to VOR combined with a distance measuring capability. The VOR and TACAN are collocated and have some common components, but either facility may be used independently.

### Classification by Reception Capabilities

Stations are also classified by their interference-free reception capabilities with respect to distance and altitude. This classification is the basis for establishing the interference-free reception range of transmitter frequencies. The following data shows station classification with maximum reliable distances and altitudes:

| Classification | Altitudes | Distances (nm) |
|---|---|---|
| T | 12,000 ft. and below | 25 |
| L | Below 18,000 ft. | 40 |
| H | Below 18,000 ft. | 40 |
| H | 14,500 ft–17,999 ft. | 100* |
| H | 18,000 ft–FL 450 | 130 |
| H | Above FL 450 | 100 |

*Applicable only east of 105° meridian within the contiguous 48 states.

*Note.* Classification of stations is subject to change. Current operational publications should be consulted for latest information.

# 11 RADIO DIRECTION FINDER

## Section I. COMPONENTS AND OPERATION

### General

The *radio direction finder* is a radio receiver set used to determine the bearing to the radio transmitting station from the aircraft. When a loop antenna of a radio receiver is placed in a radio transmitter signal pattern, the signal is heard *except* when the plane of the loop is perpendicular to the line from the aircraft to the navigation facility. The loop position at which no audible signal is received is called the *null*. Navigation by means of the radio direction finder uses the null for determining direction to the transmitting facility. Automatic direction finding (ADF) equipment in Army aircraft is constructed so that the aviator can either manually rotate the loop and its indicator needle as he listens for a null or operate the receiver in the automatic position and read the instrument indicator as the loop automatically determines the null position. A bearing to the transmitting facility can be read from the indicator face by either method.

*Note.* The bearing to the beacon measured clockwise from the aircraft heading is the *relative bearing* (the bearing is relative to the aircraft heading). Relative bearing can be converted to a magnetic bearing, or the indicator can be set so as to read the magnetic bearing directly.

### Components of the ARN–59 Automatic Direction Finder (ADF)

The ARN–59 is the basic automatic direction finder receiver set found in most Army aircraft. The components of this set (fig. 11–1 ① and ②) are—

a. The receiver, which is an LF/MF radio receiver.

b. The loop antenna, which is directional and consists of an iron-core loop housed in a sealed container.

c. The sensing antenna which is a length of wire running parallel to the fuselage of the aircraft or enclosed within a nonmetallic cover.

d. The control panel, which is used for tuning the receiver and for selecting the mode of operation.

e. The radio compass indicator (also called

**RECEIVER**

**LOOP ANTENNA**

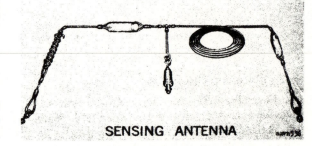

**SENSING ANTENNA**

*Figure 11–1①. ARN–59 components (part 1 of 2)*

the ADF indicator), which consists of a compass card, an azimuth indicator (pointer), and a card rotation knob (labeled VAR). The indicator is used to determine the bearing to the radio navigation facility.

*Note.* Another type of indicator (the radio magnetic indicator, ch. 12) has a compass card that moves automatically to show the aircraft heading.

### Receiver Operation

a. *Frequency Range.* The ARN–59 can receive signals transmitted between .19 MHz

122

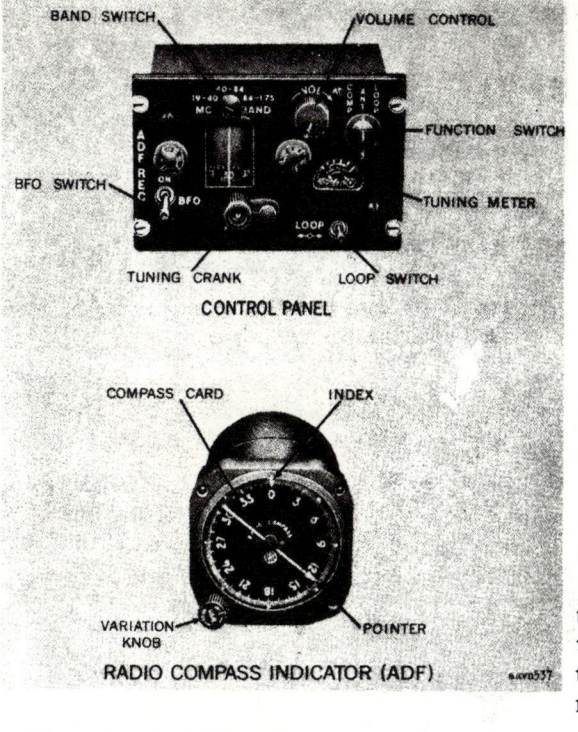

*Figure 11–1②. ARN–59 components (part 2 of 2).*

(190 kHz) and 1.75 MHz (1,750 kHz). To facilitate rapid tuning, this range is divided into three separate bands on the control panel from .19 MHz to .40 MHz, from .40 MHz to .84 MHz and from .84 MHz to 1.75 MHz.

*b. Receivable Facilities.*

(1) *Homing beacon transmitter.* The primary radio navigational transmitting facility used with the ARN–59 receiver is the *radio beacon.* The LF/MF radio beacon is also called a nondirectional beacon, homing beacon, compass locator, or a homer, and is classified by power output, as follows:

| Radio class | Power output (watts) | Usable distance (statute miles) |
|---|---|---|
| L | Less than 25 | 15 |
| MH | 25 to less than 50 | 25 |
| H | 50 to less than 2,000 | 50 |
| HH | 2,000 or more | 75 |

(2) *Other transmitters.* In addition to radio beacon transmitters, other facilities usable with the ARN–59 receiver are commercial broadcasting stations, radio marine beacons, and low or medium frequency transmitters operating in control towers or in other communications installations.

*c. Tuning.* Inaccurate tuning to a radio frequency may cause the azimuth indicator to fluctuate, making accurate interpretations of bearing difficult. The procedure for accurate tuning is as follows (refer to the control panel, fig. 11–1 ②):

(1) Turn on the set by rotating the VOL (volume) switch to the right (clockwise) and adjust the volume to the desired level.

(2) Place the function selector knob in the COMP (compass) position.

(3) Select the correct radio frequency band. For example, if the transmitting beacon broadcasts on 310 kHz (.31 MHz), set the band selector on .19 MHz to .40 MHz.

(4) Operate the tuning crank until the desired radio frequency appears on the dial. At this time the beacon identification code should be audible. Make adjustments with the crank until the strongest signal is heard. Make further slight adjustments with the crank to obtain the maximum deflection on the tuning meter.

(5) Adjust volume to desirable level.

*d. Beat Frequency Oscillator (BFO).*

(1) *Function.* Certain types of radio beacons transmit an interrupted but unmodulated radio carrier wave which is inaudible unless the BFO switch is turned on. The beat frequency oscillator converts the inaudible keyed (interrupted) carrier wave into an audible, intelligible sound. (On some ADF receivers the BFO switch is labeled CW.) Beacons transmitting an unmodulated carrier wave signal are common in Europe and other oversea regions. The overseas navigation publications indicate these beacons with the classification "A–1 emission."

(2) *Tuning.* While tuning with the BFO switch, a continuous monotone is heard until the beacon frequency is approximately tuned. At this point, the monotone fades to a null. A peak tone exists on each side of the null, but the strongest signal is received from the peak on the lower frequency side of the null (as indicated in the dial window). After the set is tuned, the BFO switch is turned off for normal operation in the COMP position. For operation of the set in the LOOP position (sec. III), the BFO switch is left on so that the signal can be heard.

## Section II. AUTOMATIC DIRECTION FINDER FLIGHT PROCEDURES

### Orientation

*a. ADF Indicator (Radio Compass).* The radio compass indicator (fig. 11–1 ②) usually is used with the ADF receiver to determine direction to the transmitting facility. The compass card (indicator face) can be rotated manually with the VAR knob on the indicator head to facilitate position fixing on airways. Normally, however, the card is set with 0° at the index (the small mark at the top center of the indicator head which represents the nose of the aircraft).

*b. Relative Bearing.* Standard ADF flight procedures are based upon setting 0° on the compass card under the index representing the nose of the aircraft. Therefore, the bearing read directly from the ADF indicator is measured from 0°. For this reason, the bearing is called the *relative bearing,* and is the bearing to the facility as measured clockwise from the nose of the aircraft. Although measured clockwise, the relative bearing may be read from the ADF indicator as degrees left or right of the nose; i.e., a relative bearing of 270° clock-

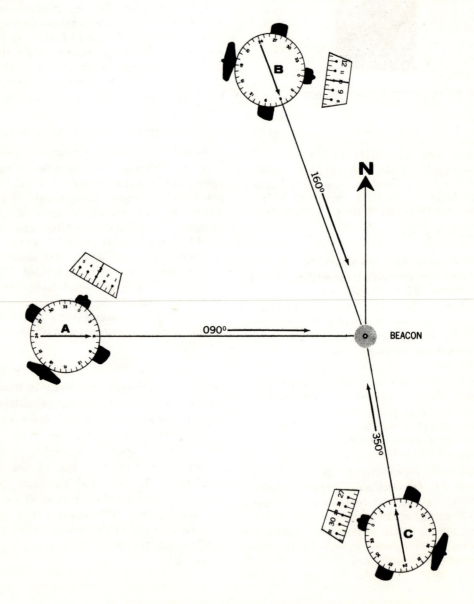

*Figure 11–2. Relative bearing of 060° on three different aircraft headings.*

wise is equivalent to a bearing of 90° left of the nose.

(1) In figure 11–2, the relative bearing to the beacon is 060° regardless of the aircraft heading; i.e., the beacon is 60° right of the actual aircraft heading. To determine the magnetic direction to the beacon, add 60° to the magnetic heading; i.e., at point A, figure 11–2, the magnetic heading is 030°, the beacon is 60° right; the magnetic direction to the beacon is 090°. The magnetic directions at points B and C in figure 11–2 are 160° and 350°, respectively. Once the direction to the beacon is known, the aviator can orient his aircraft with respect to this beacon.

(2) Figure 11–3 depicts a beacon 330° right of the aircraft heading (060°). In this situation it is more practical to interpret the indication as 30° left of the nose, although the relative bearing is 330° right of the nose. Since the magnetic heading of the aircraft is 060°, the magnetic direction to the beacon is computed as 030° by subtracting 30° (left of the nose) from the magnetic heading of 060°.

c. *Summary*. The procedure for ADF orientation is summarized as follows:

(1) Tune and identify the beacon.

(2) From the ADF indicator, determine the number of degrees left or right between the beacon and the aircraft heading.

(3) Add the number of degrees right to the heading or subtract the number of degrees left from the heading to determine the magnetic course to the beacon.

## Homing

Flying toward a beacon by keeping the nose of the aircraft oriented toward the beacon at all times is called homing (fig. 11–4); i.e., maintaining a constant relative bearing of 0°. After the pilot determines the direction to the beacon (para 11–4), he homes to the beacon by turning toward it until the azimuth indicator reads 0° (A, fig. 11–4). If the azimuth indicator drifts off zero, the aviator turns the aircraft toward the indicator (the beacon) until the indicator returns to zero. With a crosswind, the path flown (*track*) will curve (B to C, fig. 11–4). As the aircraft approaches the beacon, the azimuth indicator tends to fluctuate because of the strong signal. The aviator should estimate his close proximity to the transmitting beacons and not "chase the needle." Arrival at the beacon and beacon

passage is indicated when the needle reverses (to 180° in D, fig. 11–4).

## Tracking

*Tracking* is the technique of flying a definite path (track) into or away from a beacon. The correction for wind must be determined and applied to remain on track. The method for

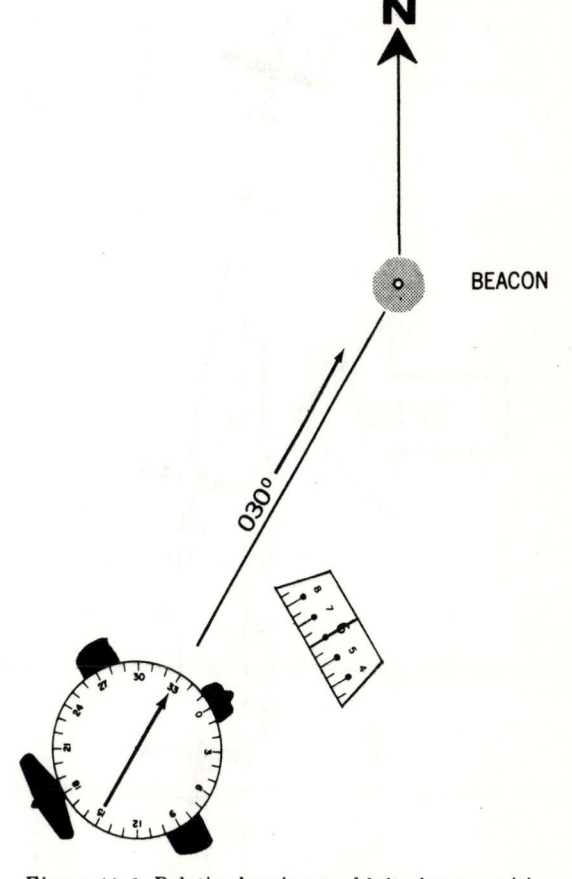

*Figure 11–3. Relative bearing read left of nose position.*

applying drift correction varies slightly for slow-speed aircraft because with slower airspeeds the wind correction must be greater. Both the usual tracking technique and the technique used on slow aircraft are illustrated in figure 11–5.

a. *Tracking Inbound—Aircraft Above 90 Knots* ((A), fig. 11–5).

(1) *Point A*. Aircraft is inbound on a course of 350°, heading is 350°, and azimuth indicator reads 0°

(2) *Point B.* Crosswind from the left has caused the aircraft to drift off course to the right. The ADF indicator reads 355°, so the intended track is 5° to the left. Normally, a correction in heading to return to course is applied when the aircraft moves 5° or more off course. However, a 5° deflection of the azimuth indicator is not significant close to the beacon.

(3) *Point C.* The aviator applies 20° of correction to return to the desired track. The heading is 330°. This standard 20° correction will return the aircraft to the desired track unless the wind is exceptionally strong. After turning to the 330° heading, the azimuth indicator reads 015°, which indicates that the beacon is 15° to the right of the aircraft.

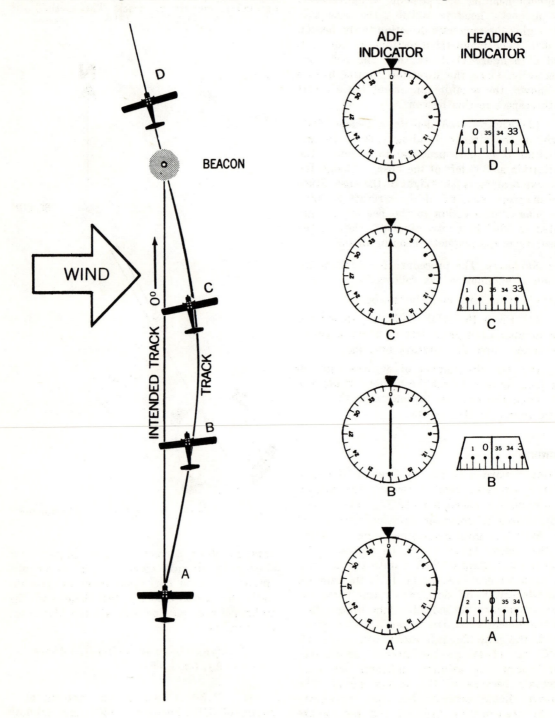

*Figure 11–4. Homing (ADF).*

(4) *Point D.* The aircraft has returned to the desired track and the azimuth indicator shows the beacon 20° right of the nose.

(5) *Point E.* Once on track, a new heading of 340° is used, which incorporates a 10° trial drift correction to compensate for the crosswind from the left. The azimuth indicator reads 010°, indicating the applied drift correction. From this point to the beacon, the aircraft will remain on track if the azimuth reading and heading do not change.

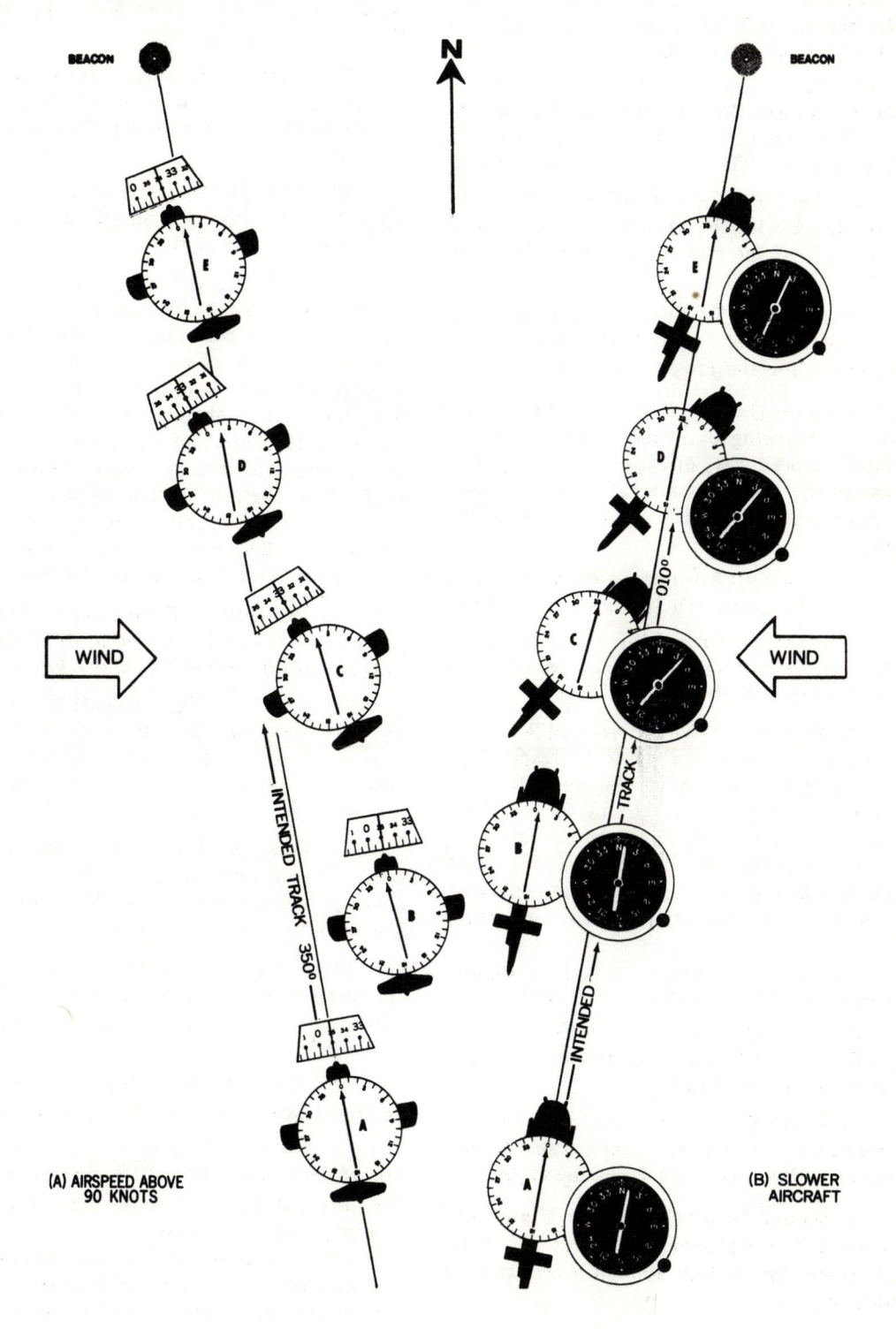

Figure 11-5. Tracking inbound (ADF).

*b. Tracking Inbound—Aircraft Below 90 Knots* ((B), fig. 11–5).

(1) *Point A.* Aircraft is inbound on a course of 010°, heading is 010° and azimuth indicator reads 0°.

(2) *Point B.* Crosswind from the right drifts the aircraft off course to the left; azimuth indicator reads 005°.

(3) *Point C.* A correction of 30° is applied to return the aircraft to the desired track. New heading is 040° and the azimuth indicator reads 335° (25° left of the nose).

(4) *Point D.* The aircraft returns to the desired track with the azimuth indicator reading 30° left of the nose (a relative bearing of 330°).

(5) *Point E.* The new heading of 025° incorporates a 15° trial drift correction to compensate for the crosswind from the right.

*c. Tracking Outbound* (fig. 11–6). Procedures for tracking outbound are the same as inbound procedures discussed in *a* and *b* above except that the aircraft tail (azimuth indicator reading of 180°) is the reference point.

(1) *Point A.* Aircraft is outbound on a course of 350°; azimuth indicator reads 180°.

(2) *Point B.* Crosswind from the right drifts the aircraft off course to the left; azimuth indicator reads 175°.

(3) *Point C.* The 20° correction is applied upwind to return the aircraft to track. This increases the deviation of the azimuth indicator to the right of the tail by 20°. The indicator does not swing through the tail position as it does through the nose position when tracking inbound (*a*(3) above). The new heading is 010° and the azimuth indicator reads 155°.

Note. The 30° correction is used for tracking outbound in aircraft below 90 knots (*b* above).

(4) *Point D.* The aircraft returns to the desired track with the azimuth reading 160° (station 20° right of tail).

(5) *Point E.* The new heading of 0° incorporates a 10° trial drift correction to compensate for the crosswind from the right.

*d. Inadequate Initial Corrections.* Figure 11–7 illustrates procedures to correct for wind drift when the initial applied correction is inadequate.

(1) *Point A.* An aircraft has reintercepted the desired track and is holding a heading of 100° to maintain a track of 090°, allowing a 10° drift correction for the crosswind from the right.

(2) *Point B.* The wind is stronger than the aviator anticipated and the aircraft has drifted off the desired track; i.e., the relative bearing has changed from 350° (10° left) to 355° (5° left).

(3) *Point C.* The aviator turns to a heading of 110° (azimuth indicator reads 345°) to return the aircraft to the desired track at point D.

(4) *Point D.* The aircraft returns to the desired track with the azimuth indicator reading 340° (20° left of the nose).

(5) *Point E.* After returning to track, the aviator turns to a heading of 105° which allows 15° drift correction for the wind. If an unusually strong wind prevents the aircraft from returning to track by using a 20° correction, the aviator should try a 40° correction. Upon reinterception of the track by using the 40° correction, the azimuth indicator will read 40° to the left or right of the nose. Upon successful track reinterception, the aviator should try a 20° drift correction to remain on the desired track in the strong wind condi-

*e. Overcorrection.* Figure 11–8 illustrates procedures to correct for wind drift when the initial applied correction is too great.

(1) *Point A.* The aircraft is holding a 15° left correction to remain on an outbound course of 270°. The aviator is using a 15° correction because of his slow airspeed (*b* above).

(2) *Point B.* The aircraft has flown off track in an upwind direction. The relative bearing has changed from 195° (15° left of the tail) to 190° (10° left of the tail).

(3) *Point C.* To return to track, the aviator turns parallel to the desired track (heading 270°) and allows the wind to blow him back on track.

(4) *Point D.* The aircraft is back on the desired track with a relative bearing of 180°

(5) *Point E.* The aviator selects a heading of 260° (10° left drift correction) since the original 15° left drift correction (255° heading) was too great.

Note. The system of trial-and-error drift correction described above (bracketing) illustrates the tracking principle, but in practice each aviator must refine the bracketing procedure to satisfy his particular flight technique.

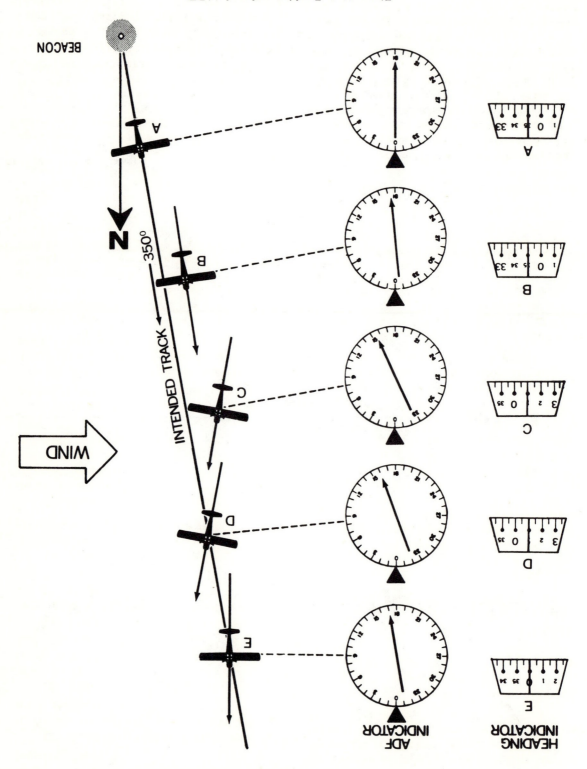

Figure 11-6. Tracking outbound (ADF).

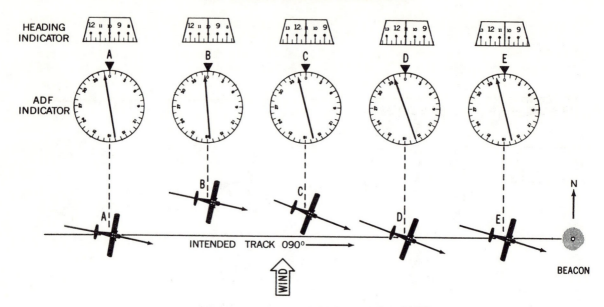

*Figure 11-7. Inadequate initial correction (ADF).*

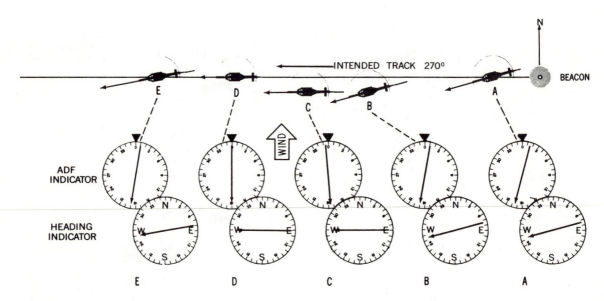

*Figure 11-8. Overcorrection (ADF).*

## Position Fixing (Fixes)

*a. Position Fixing at Airway Reporting Points.* If an aircraft has flown a known track from a beacon, the track can be represented by a course line or airway on a navigation chart. At any point along this course, the aviator can take a radio bearing to a second beacon with the ADF receiver and plot this bearing as a second line on the chart. The intersection of these two chart lines fixes the approximate location of the aircraft. This method of position fixing can be used in flight training without drawing any lines on the chart; *for example*—

(1) In figure 11-9, an aircraft is tracking inbound to beacon A on a track of 090°. The aircraft has been maintaining the track in a no-wind condition. Ahead of the aircraft is a reporting point (the triangle) at which the aviator must fix his position and report to air traffic control. A check of the navigation chart shows the direction to beacon B is 120° from the reporting point. Since the direction to beacon B is 120° from the reporting point. Since the direction to beacon A is 090°, there will be a 30° difference between the heading to beacon A (090°) and the radio bearing to beacon B over the reporting point (120°);

130

i.e., at the time the aircraft arrives over the mandatory reporting point, beacon B will be 30° to the right of the nose of the aircarft. If the aviator tunes beacon B on the ADF receiver, he will be over the reporting point when the azimuth indicator shows a relative bearing of 30°. Prior to arriving over the repointing point, the relative bearing on the indicator will be less than 30°.

(2) In a similar situation (fig. 11-10), the aircraft is holding a heading of 080° to track inbound on 090° (applied drift correction of 10° left). When the aircraft reaches the reporting point, the relative bearing on the azimuth indicator will be 040° (the difference between the heading and the bearing to beacon B).

(3) Figure 11-11 illustrates three other situations involving the principles discussed in (1) and (2) above, but with changes in the position of the beacon with respect to the aircraft. At the reporting point (point B), figure 11-11① depicts the instrument readings when the beacon is 45° left of the nose; figure 11-11② depicts the beacon 40° left of the tail; and figure 11-11③ depicts the beacon 60° right of the tail.

*Note.* 10° left drift correction.

(4) The position of the aircraft over a reporting point may also be fixed by rotating the compass card with the VAR knob so that the aircraft heading is at the index (nose position). As illustrated in figure 11-12, the aircraft will be over the intersection (point B) when the azimuth indicator shows the published bearing (120°) to beacon C. This method eliminates the need for computing the angle between the heading and the station (relative bearing) ((1), (2), and (3) above).

*b. Position Fixing When Lost or Disoriented.* If a pilot is lost or disoriented, he can quickly determine his position on a navigational chart when there are at least two LF/MF radio beacons within reception distance of the azimuth receiver. The relative bearing to one bacon is determined by orientation and plotted on the chart as a line of position (direction from the beacon to the aircraft). The relative bearing to a second beacon is determined by the same method and plotted. The intersection of the two lines is the approximate location of the aircraft. An angular difference of 30° between the two lines of position usually is large enough for a fix, but the closer this angle is to 90° the more accurate is the fix. The relative bearing to a third beacon may be used to form a more accurate fix.

## Track Interception

*a. From a Known Position.* The procedure for intercepting an inbound track is illustrated in figure 11-13①. The pilot is cleared to the beacon via the 080° magnetic course. From his present inbound track of 050° (point A) he knows that he must turn left to intercept the new track at the desired angle of 45°. In this situation, the pilot turns left to a heading of 035° and flies to intercept the desired course (080°) at a 45° angle (point B). The desired interception is accomplished when the azimuth indicator shows a relative bearing of 46° (right of the nose). The pilot then turns inbound to a heading of 080° and proceeds to the beacon. (Although 45° is only one of many angles that could be used to intercept the course, it is a standard interception angle considered. In another inbound course interception situation (point D, fig. 11-13②), the relative bearing is 315° (45° left of the nose) when the beacon is to the left of the nose of the

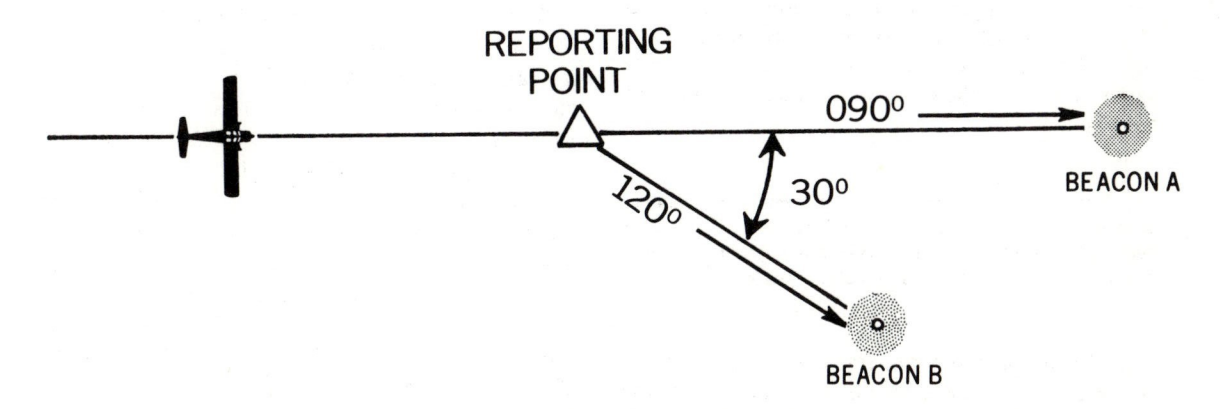

*Figure 11-9. Position fixing under no-wind condition (ADF).*

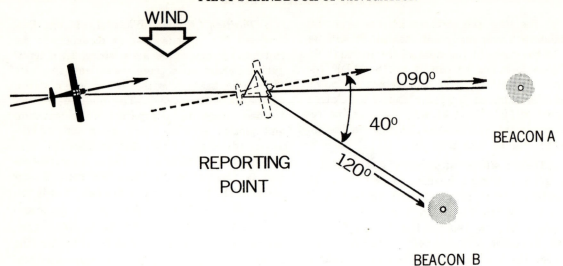

Figure 11-10. *Position fixing with drift correction (ADF).*

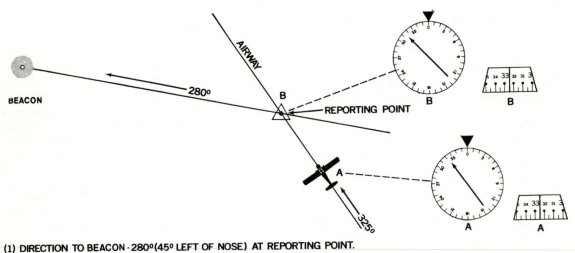

(1) DIRECTION TO BEACON-280°(45° LEFT OF NOSE) AT REPORTING POINT.

Figure 11-11①. *Typical position fixes (ADF).*

aircraft at the time of interception. When intercepting an outbound course (fig. 11-13⑤), the 45° interception angle is shown off the tail (180°) as relative bearings of 135° or 225°. When ATC requests the pilot to "expedite" the interception, the pilot is to intercept the desired course as soon as possible. Therefore, he turns to intercept the course at a 90° angle, following the shortest flightpath to the course rather than using the standard 45° interception angle (fig. 11-13④).

*Note.* In all interception problems, the relative bearing read on the azimuth indicator at the time of course interception (from either the nose or the tail) must equal the desired interception angle.

*b. From an Unknown Position.* If a pilot is directed to track inbound on a specific course but is uncertain of the exact location of the track, the procedures depicted in figure 11-14 may be used. The aircraft at point A is inbound to beacon X on a course of 010° according to the flight instruments. ATC directs the pilot | to track inbound on a course of 075° to beacon Y. Although he is uncertain of his present position with respect to beacon Y, he can orient himself by first turning parallel to the desired course (075°) while tuning his azimuth receiver to beacon Y. After paralleling the desired course (point B), if he observes the azimuth indicator is deflected to the right, he knows that the beacon and course are located to the right of the aircraft; if to the left, the beacon and course are to the left. In figure 11-14 at point B, the course is to the right. The pilot turns to a heading of 120° to intercept the desired course at the standard 45° angle (point C).

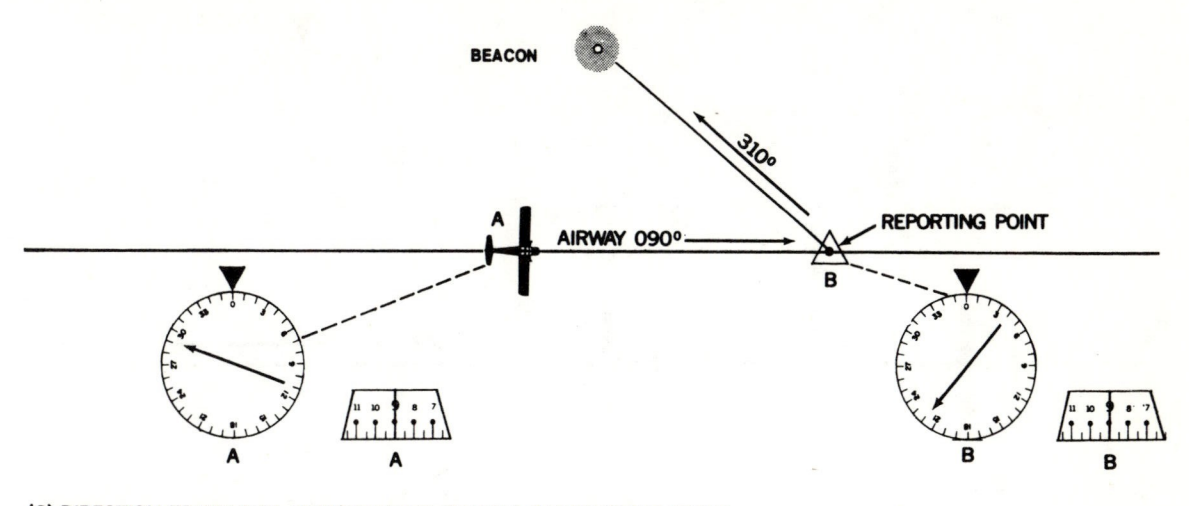

(2) DIRECTION TO BEACON - 310° (40° LEFT OF TAIL) AT REPORTING POINT.

*Figure 11–11②. Typical position fixes (ADF) (part 2 of 3).*

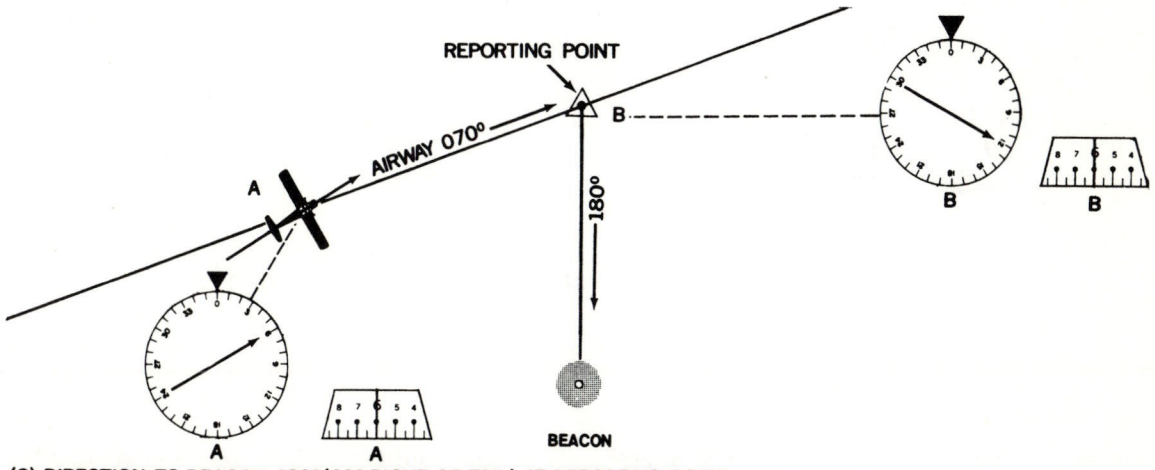

(3) DIRECTION TO BEACON - 180° (60° RIGHT OF TAIL) AT REPORTING POINT.

*Figure 11–11③. Typical position fixes (ADF) (part 3 of 3).*

*c. Double-the-Angle Method.* This procedure (fig. 11–15) for intercepting a course is as follows:

(1) With the azimuth receiver tuned to beacon Z, determine the angular difference between the aircraft and the desired course. If necessary, turn parallel to the course to determine the angular difference. When the present position of the aircraft and the desired course are known, a turn to parallel may not be required.

(2) Double the angle derived in (1) above and use the product as an interception angle. At point X of figure 11–15, the angular difference of 30° produces an interception angle of 60°.

(3) Turn to a heading which will give the desired interception angle. A heading of 135° intercepts the inbound course (point Y, fig. 11–15) of 075° at the desired 60° angle.

(4) The aircraft is at the point of interception when the azimuth indicator reads 60° left of the nose (relative bearing of 300°, fig. 11–15).

(5) The triangle XYZ (fig. 11–15) shows that the leg flown to intercept from point X to point Y is equal in distance to the remaining leg from Y to Z; therefore, the time required

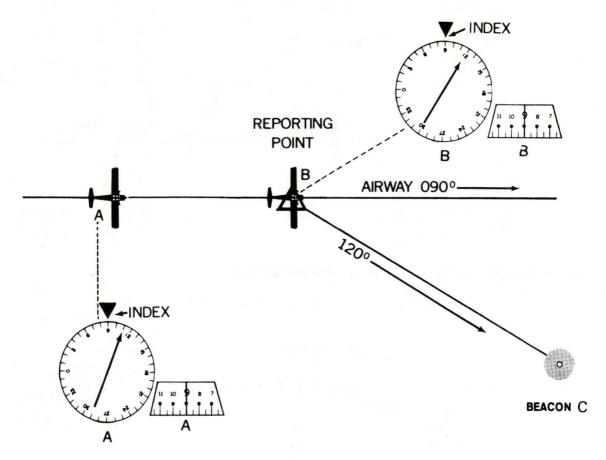

Figure 11–12. Position fixing with heading set under index (ADF).

to fly the interception from X to Y is approximately equal to the time required to fly the remaining leg from Y to Z (disregarding wind conditions). An advantage of the double-the-angle method is that it enables the aviator to estimate his flying time to the station. This method will not work successfully when the angular distance between the aircraft and the desired course is greater than 45° because doubling such angles exceeds 90°. Unless the interception is effected very close to the beacon, the selected interception angle should never be less than 20° because the wind may prevent course interception in a reasonable time.

RADIO DIRECTION FINDER

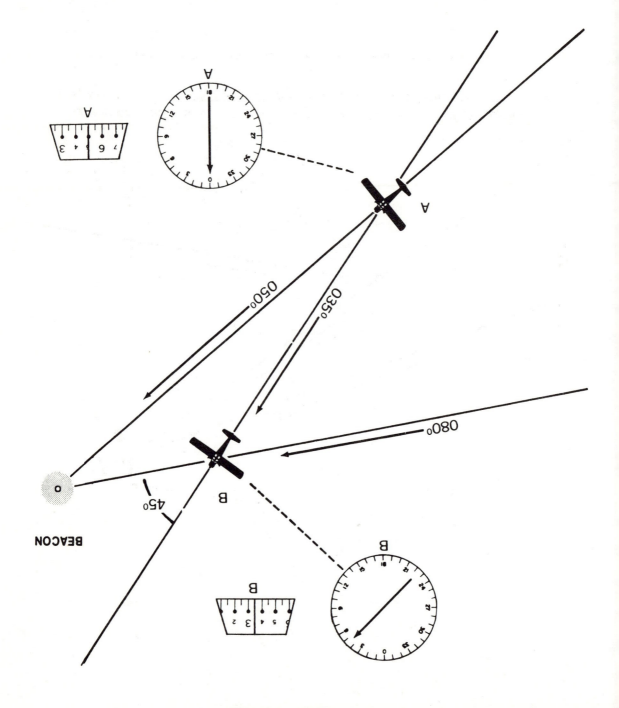

(1) INBOUND WITH BEACON 45° RIGHT OF NOSE AT TIME OF TRACK INTERCEPTION. HEADING (035°) PLUS INTERCEPTION ANGLE (045°) EQUALS DESIRED INBOUND TRACK (080°).

*Figure 11-13⓪. Track interception from a known position (ADF)  (part 1 of 4).*

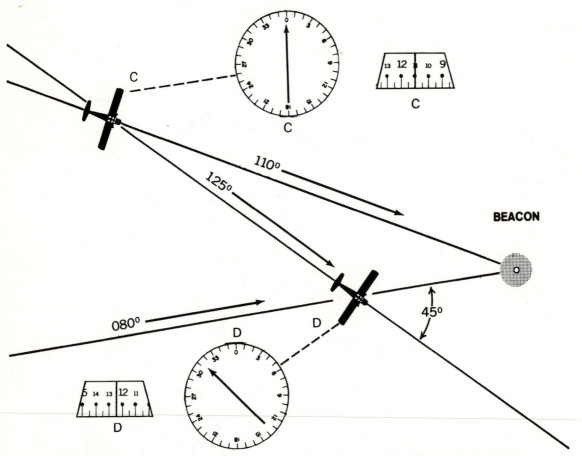

(2) INBOUND WITH BEACON 45° LEFT OF NOSE AT TIME OF TRACK INTERCEPTION.
HEADING (125°) MINUS INTERCEPTION ANGLE (045°) EQUALS DESIRED INBOUND TRACK.

*Figure 11–13②. Track interception from a known position (ADF) (part 2 of 4).*

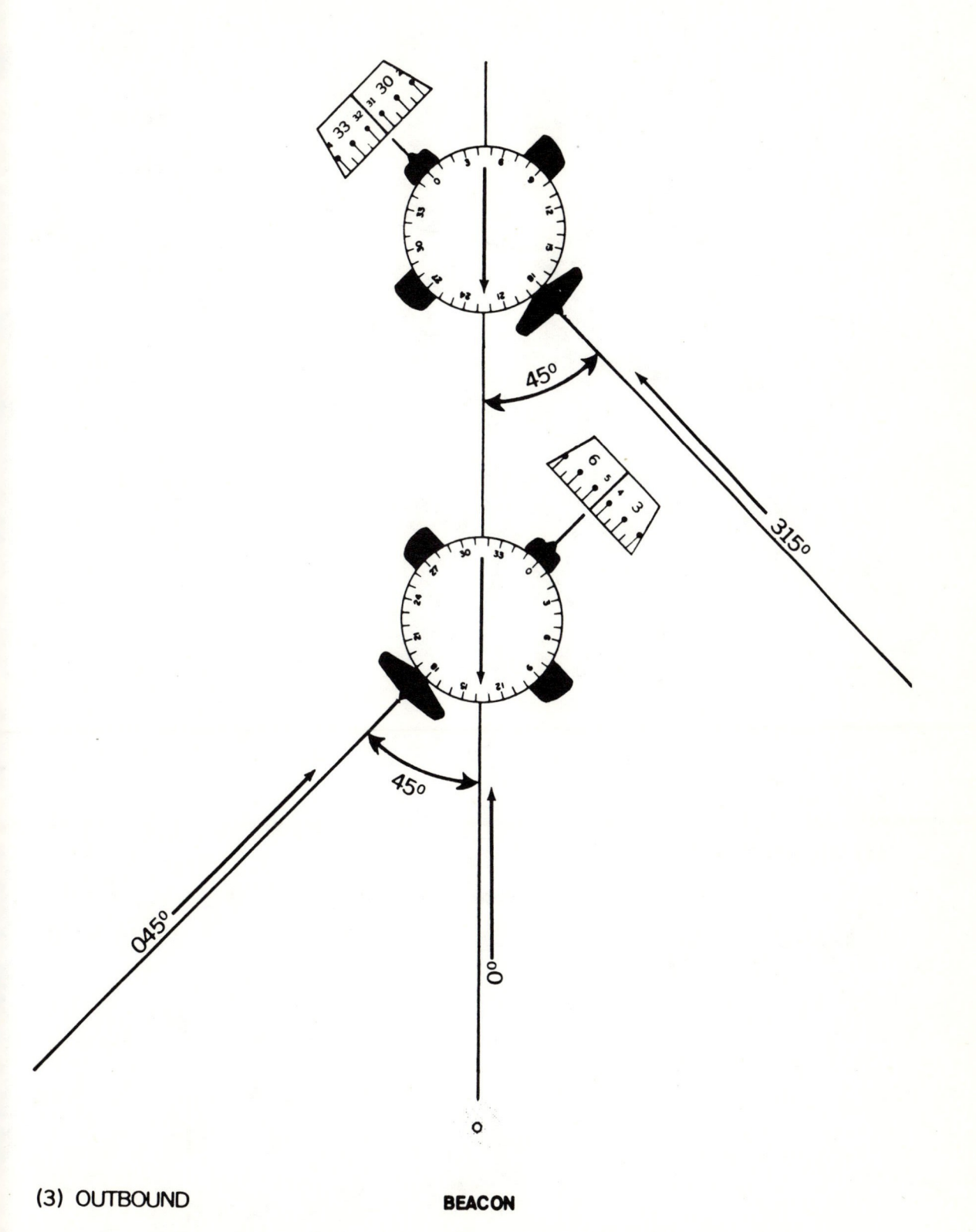

(3) OUTBOUND                                          BEACON

*Figure 11-13③. Track interception from a known position (ADF) (part 3 of 4).*

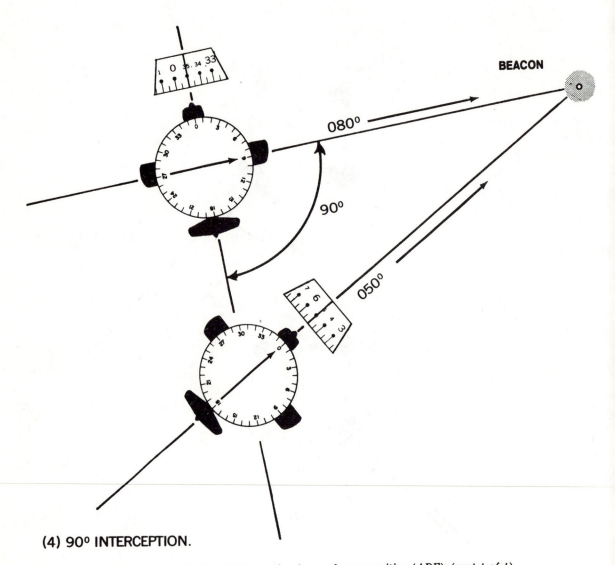

**(4) 90° INTERCEPTION.**

*Figure 11–13④. Track interception from a known position (ADF) (part 4 of 4).*

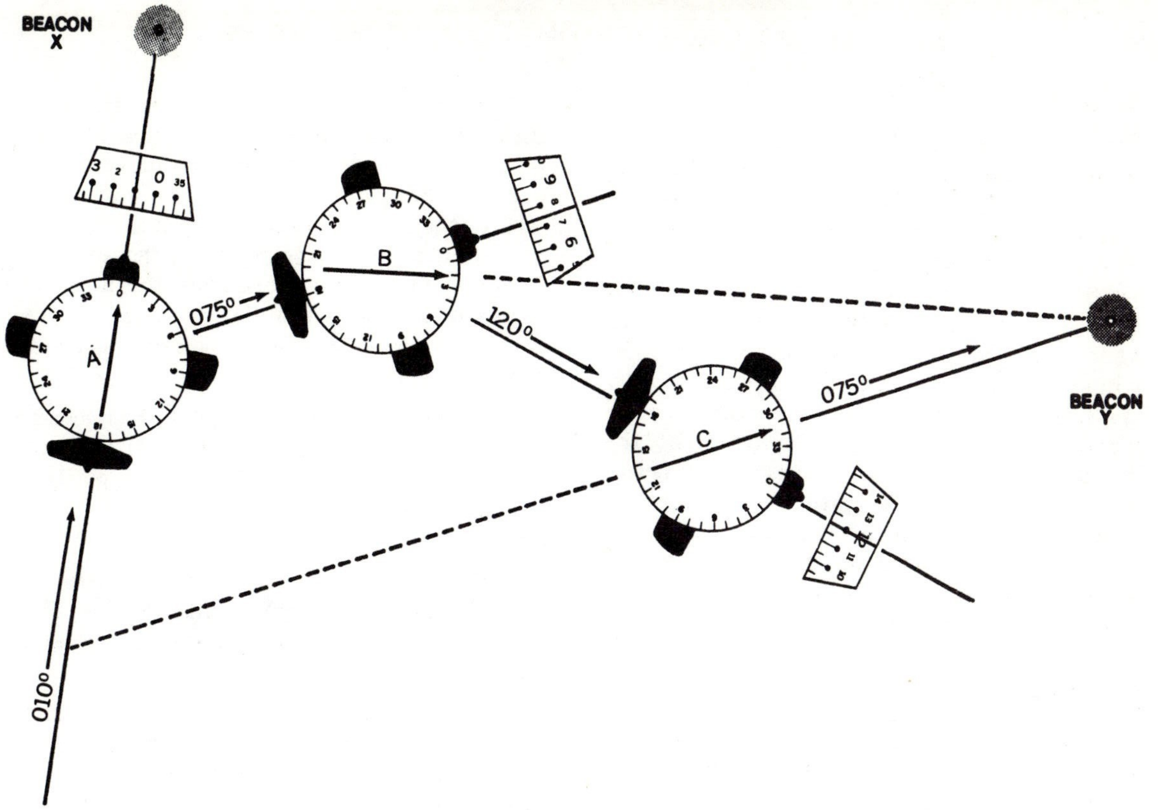

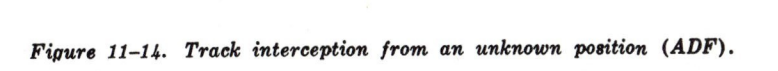

Figure 11-14. Track interception from an unknown position (ADF).

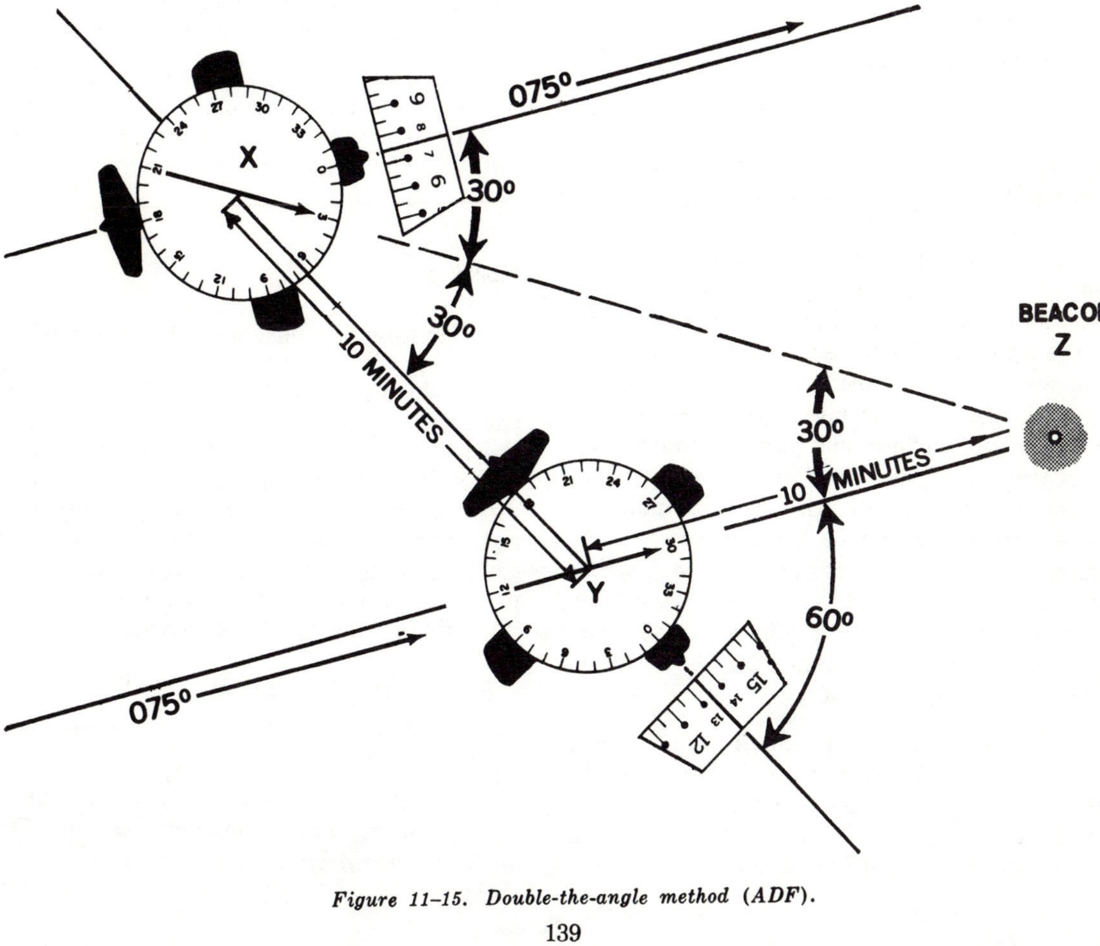

Figure 11-15. Double-the-angle method (ADF).

## Section III.  MANUAL (LOOP) OPERATION OF THE ARN-59

### General

The radio compass (ARN-59) may be operated manually for navigational use with the selector switch in the LOOP position. Manual operation may be necessary when the signal or indicator readings received in the COMP position are unreliable. Navigational procedures are the same in the LOOP position as when using the COMP position; however, the azimuth indicator is positioned manually by the loop drive switch to locate the null by sound. If the switch is moved to the right, the indicator arrow moves to the right (clockwise); if the switch is moved to the left, the indicator moves to the left (counterclockwise). An aural null (minimum reception) results when the plane of the loop antenna is perpendicular to a line from the beacon.

### Orientation

Orientation procedures used in determining direction to the transmitting beacon are explained below.

a. Tune and identify the beacon in the ANT position.

b. Move the selector switch to the LOOP position.

c. Move the loop drive switch and listen to the signal. At some point the signal will fade; this is the null position. As the aviator rotates the azimuth indicator, the signal will build on each side of the null position. Ideally, the null should be no more than 5° wide on the face of the azimuth indicator. For example, if the signal begins to fade when the indicator reaches 120° and immediately builds up again at 125°, the null is reasonably narrow. Normally, a well defined null can be obtained by increasing the volume. After the null is definitely located, the azimuth indicator indicator points toward the beacon but the indication is ambiguous; i.e., the correct relative bearing may be at either end of the indicator. For example, when the arrow points to 120°, the relative bearing to the station may be at either 120° or its reciprocal 300°.

d. To resolve the ambiguity, rotate the loop manually until the azimuth indicator points to 090° or 270°. Turn the aircraft right or left until the signal, which increased in strength as the azimuth indicator was rotated, again fades to a null. The beacon is then either to

the left or to the right of the aircraft. Maintain a constant heading until the aircraft flies out of the null; depress the loop drive switch again and relocate the null. If the loop is rotated to the right to relocate the null, the beacon is to the right (clockwise) of the aircraft; if rotated to the left to relocate the null, the beacon is to the left of the aircraft. Procedures for resolving ambiguity are illustrated in figure 11-16.

(1) At point A, the pilot has located an aural null on a heading of 270°, with the azimuth indicator in the 0° to 180° position. The beacon is either directly ahead of or behind the aircraft.

(2) At point B, the pilot rotates the azimuth indicator (loop) to the 090° to 270° (wingtip) position, which causes the signal to rebuild.

(3) At point C, the aircraft is turned until the null reappears at the wingtip position (heading 0°), indicating the beacon is either to the left or right of the aircraft. The pilot flies this heading for a short time and the signal rebuilds. This indicates that the aircraft has flown out of the wingtip null.

(4) The null is relocated at point D by rotating the azimuth indicator (loop) clockwise to 110°. Therefore, the beacon is at point X, 110° right of the aircraft heading of 0°.

### Homing

a. Homing to a beacon with the ARN-59 set operating in the LOOP position is accomplished by first locating the beacon and then turning the aircraft until the null is on the nose position. If the aircraft drifts out of the null position, the pilot determines the direction of drift by rotating the loop left or right to relocate the null. The pilot then turns the aircraft until the null is again on the nose position and repeats the procedure until in the immediate vicinity of the beacon.

b. To determine arrival over the beacon—

(1) Estimate the time of arrival accurately.

(2) Prior o arrival, set the azimuth indicator on the wingtip position (090° to 270°) with the loop drive switch to receive a strong signal.

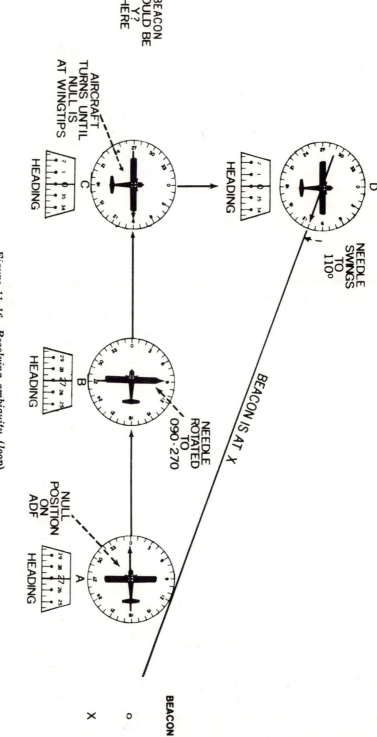

Figure 11–16. Resolving ambiguity (loop).

(3) As the aircraft flies over the beacon (or abeam the beacon), a sharp null of short duration will be detected.

## Tracking

Aural null tracking procedures are identical to those used for ADF tracking except that the pilot must manually rotate the loop to determine the null position and relative bearing. In tracking toward or away from a beacon (null at 0° or 180°), drift is indicated by movement of the null from the nose or tail position. *For example*, if the aircraft drifts off course to the left, a 20° correction may be made to reintercept the course. Procedures are as follows:

a. Using the heading indicator, turn 20° right to intercept the course. Set the indicator (loop) to 20° left of the nose.

b. Continue on the same heading until the null reappears at this new setting (20° left of nose). Aircraft is back on desired course.

c. Turn back toward original heading by 10° and relocate the null (10° left of nose).

d. If original correction of 10° is excessive or inadequate, make additional corrections of 5°.

## Time and Distance

In remote areas, the pilot may, if necessary, estimate the time or distance to a radio beacon with the ADF receiver by following the VOR procedure described in Homing. The bearing change during flight is observed on the azimuth indicator. Time and distance estimates with the ADF are less reliable than with the use of VOR.

## Section IV.  DIRECTION FINDING (DF) STATIONS

### General

A wide network of DF stations exists in this country and overseas. Many control towers are equipped with DF radios operating in the VHF and UHF frequency bands. When navigational radios are inoperative or the aviator is lost, a nearby DF beacon can be requested to provide the aviator with a DF steer.

### Station Capabilities

DF stations provide steers for aircraft on selected VHF and UHF channels. Equipment is available for operation during normal working hours at most stations and usually on a 5-minute notice at other times. (Consult current navigation publications for this data.) This equipment has an effective line-of-sight range of approximately 100 miles, although greater distances can be obtained when the signal strength of the aircraft's radio transmitter and its altitude are ideal. The emergency frequencies of 121.5 and 243.0 MHz can be used when an emergency occurs.

### Typical DF Steer

The following is a typical DF homing procedure:

a. Pilot tunes a specific DF frequency using the ARC–55 or equivalent radio.

b. Pilot transmits (for example), "Cairns tower, 75088, requests practice steer (or emergency steer) over."

c. DF station replies, "75088, Cairns tower, transmit for steer, over."

d. Pilot replies, "Army 088, Roger." (Pilot then depresses the microphone button for about 10 seconds.)

e. DF station replies, "088, steer 180° to Cairns, over."

f. Pilot acknowledges, "088, Roger."

g. Subsequent corrections to the DF steer can be provided as necessary.

## Components

A radio magnetic indicator (RMI) (fig. 12–1) consists of a compass card, a heading index, and two bearing pointers. Indicator readings enable the aviator to determine simultaneously the present magnetic heading of the aircraft, the direction to and from the navigation facility to which the number 1 pointer receiver is tuned, and the direction to and from the navigation facility to which the number 2 pointer receiver is tuned.

*a. Compass Card.* The compass card is actuated by the aircraft's compass system; it is a remote indicator for a compass system. When working properly, the card shows the correct magnetic heading under the index at all times.

*Note.* Preflight cross-check of compass card with magnetic compass should always be made.

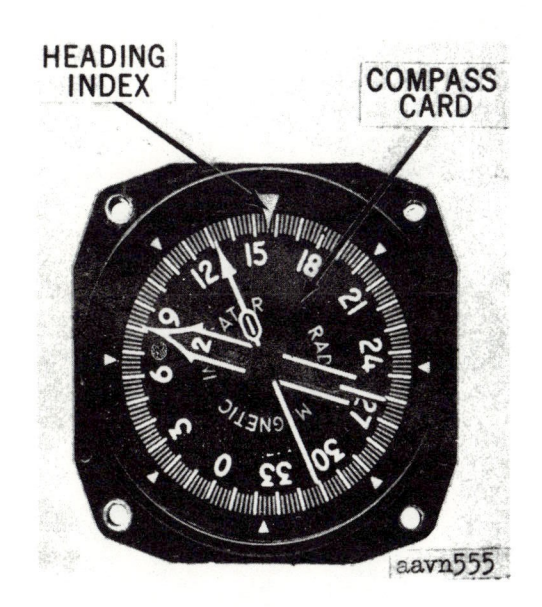

*Figure 12–1. RMI components.*

*b. Number 1 and 2 Bearing Pointers.* These two bearing pointers are actuated by either VOR, Tactical Air Navigation (TACAN), or ADF radio receivers.

*Caution:* **Bearing pointers will not function in relation to the instrument landing system (ILS).**

Each bearing pointer when coupled to a navigation receiver, will indicate the direction to the navigation facility being received. Based on the number and type receivers installed, there are several coupling arrangements possible. Information on coupling arrangements for specific aircraft types and models can be found in each aircraft operator's manual. In this chapter, reference to *facility number 1* and *facility number 2* applies equally to the VOR, TACAN, or the LF/MF facility. RMI flight procedures are the same for either type facility (except during a compass system failure.

## Orientation

*a.* Using the RMI, a pilot can orient himself instantly with respect to a facility. When the receiver is properly tuned to a facility, the head of the bearing pointer indicates the magnetic direction to the facility. The tail of the bearing pointer shows the radial on which the aircraft is located. The readings on the RMI face ((A), fig. 12–2) show an aircraft on a heading of 280° located on the 120° radial from facility number 1 and the 240° radial of facility number 2. The aviator uses a navigation chart to visualize a specific radial from a facility. (B) of figure 12–2 presents graphically the readings shown on the RMI face in (A) of figure 12–2. The orientation consists of two procedures:

(1) Tune and identify a facility.

## (A) RMI READINGS.

(2) Under the tail of the appropriate bearing pointer, read the radial on which the aircraft is located.

*Note.* For convenience, the concept of a radial used for VOR flight procedures (ch 10) may be applied to ADF flight procedures. That is, a radial may be considered as a magnetic bearing *from* the facility.

*b.* To fly directly to a beacon, the pilot turns the aircraft toward the bearing point. As the aircraft turns, both the compass card and the pointer rotate. When the head of the pointer reaches the heading index, the aircraft is headed directly toward the beacon. The RMI, when operating properly, always shows the magnetic direction to a navigation beacon.

## Tracking

*a. Inbound.* When tracking inbound with the RMI, the desired magnetic course to a beacon should appear under the appropriate bearing pointer. Figure 12–3 illustrates a tracking sequence using the same tracking techniques discussed in chapters 10 and 11.

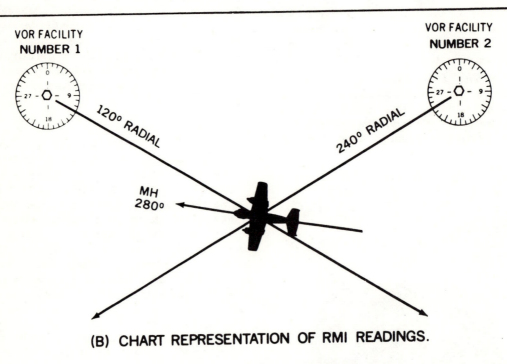

(B) CHART REPRESENTATION OF RMI READINGS.

*Figure 12-2. Orientation (RMI).*

**RADIO MAGNETIC INDICATOR (RMI)**

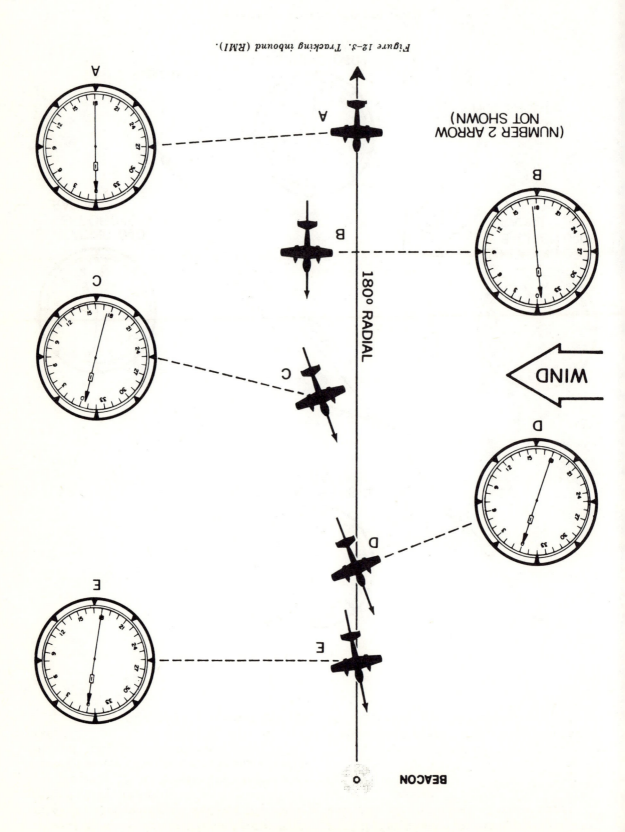

*Figure 12-3. Tracking inbound (RMI).*

(1) *Point A.* Aircraft is on course with no drift correction. Heading is 0°, RMI indicates 0° (180° radial).

(2) *Point B.* Left crosswind blows aircraft off course 5° to the right. Heading is still 0°, but RMI is now indicating 355° (175° radial).

(3) *Point C.* Aircraft turns left 20° to-

ward the intended course (heading 340°). This heading is held until back on course (point D), at which time the RMI indicates 0° (180° radial).

(4) *Point E.* After returning to intended track, aircraft turns 10° right to a heading of 350°, allowing 10° drift correction for wind

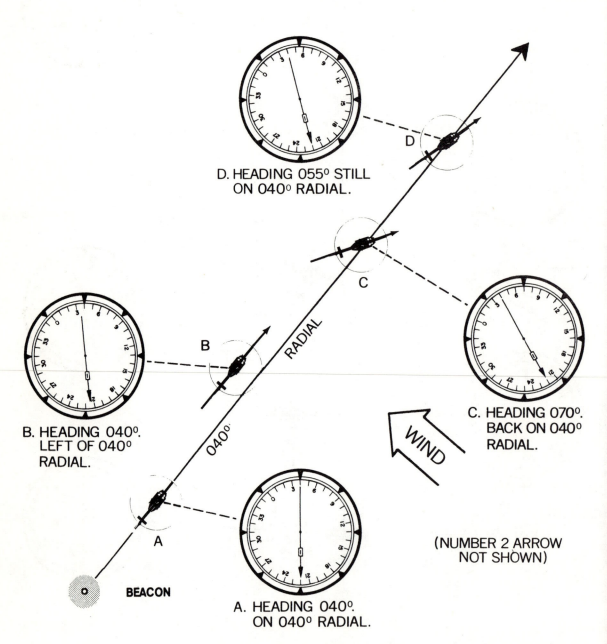

D. HEADING 055° STILL
ON 040° RADIAL.

B. HEADING 040°.
LEFT OF 040°
RADIAL.

C. HEADING 070°.
BACK ON 040°
RADIAL.

WIND

BEACON

(NUMBER 2 ARROW
NOT SHOWN)

A. HEADING 040°.
ON 040° RADIAL.

*Figure 12-4. Tracking outbound (RMI).*

effect. RMI still indicates the course of 0° (180° radial).

*b. Outbound.* While tracking outbound, the pilot observes the tail of the bearing pointer to determine whether he is on the correct radial or has drifted off course. When using either type of facility, the pilot can consider the tail of the RMI bearing pointer as indicating the radial on which the aircraft is located. Figure 12–4 illustrates an outbound tracking sequence employing the initial correction of 30° used for slower aircraft (chs. 10 and 11).

### Position Fixing

*a.* (A) of figure 12–5 illustrates a navigation chart presentation of a VOR intersection as it normally is published. Intersections using VOR radials are published with the radial passing through the intersection. Arrival over these intersections is easily identified with the RMI. (B) of figure 12–5 illustrates the RMI reading using the VOR intersections as a position fix.

(1) The number 1 bearing pointer is pointing to the VOR facility ahead of the aircraft (facility 1). This facility is being used

to maintain the airway track.

(2) The aircraft heading necessary to maintain this airway track is read under the RMI index.

(3) The number 2 bearing pointer is pointing to an off-airways facility (facility 2) used to fix the intersection. Since the navigation chart shows the radial from the facility, this radial (305°) appears under the tail of the number 2 bearing pointer when the aircraft arrives at the intersection.

*Note.* If both facility number 1 and number 2 are VOR facilities, and if the aircraft is equipped with only one VOR receiver, the pilot maintains the airway track by flying a predetermined heading and tunes the receiver to the off-airways facility to fix the intersection.

*b.* Figure 12–6 illustrates a situation similar to the one in figure 12–5, but the intersection is based on an LF/MF facility. Therefore, the bearing from the intersection to the facility, rather than a radial (direction from the facility), is shown. The bearing pointer is read to determine the magnetic course to the facility. The aircraft is maintaining a heading of 305° to make good a track of 295°. Regardless of the aircraft heading, arrival at the intersection

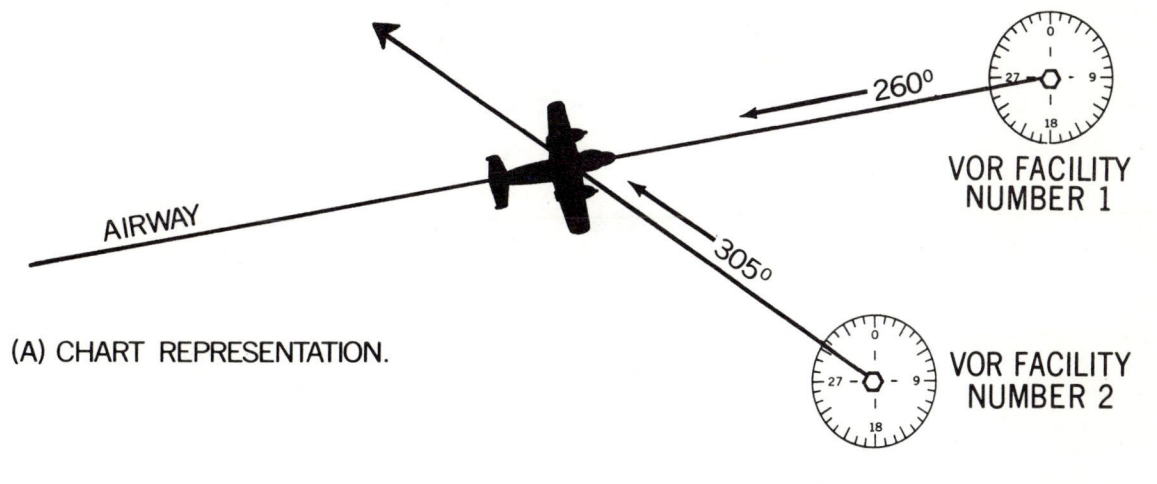

(A) CHART REPRESENTATION.

VOR FACILITY NUMBER 1

VOR FACILITY NUMBER 2

AIRWAY

260°

305°

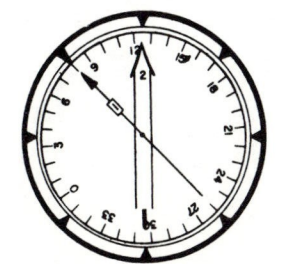

(B) RMI READINGS.

*Figure 12–5. Position fixing using VOR or TAGAN intersection (RMI).*

occurs when the number 2 bearing pointer points to the desired bearing (235° in fig. 12–6).

## Track Interception

*a.* If pilot, using the RMI, is required to intercept a track other than the one he is presently following or crossing, he visualizes the position of the desired track with reference to the radial on which the aircraft is presently located.

*b.* A of figure 12–7 shows the RMI indication with the aircraft located on the 300° radial on a heading of 020°. If the pilot is cleared to Alpha via the 270° radial, he will—

(1) Determine his present location from the RMI face (300° radial).

(2) Determine the location of the desired radial (270°). The 270° radial is counterclockwise from 300° and south of the aircraft's present position.

(3) Determine a heading which will in-

tercept the desired radial at a reasonable angle while flying outbound. Using a standard interception angle of 45°, he will turn to a heading of 225° and maintain the desired heading until interception (B, fig. 12–7). Interception occurs when the RMI bearing pointer (tail end) indicates the desired radial (270°) (C, fig. 12–7).

*Note.* Surrounding the face of the RMI compass card are tick marks at 45° intervals. These marks assist the pilot in determining 45° and 90° interceptions (inbound or outbound), and also aid in determining when the aircraft is abeam a station.

## Compass Failure

*a.* If the aircraft's compass system fails, the heading information shown under the RMI index will be unreliable. If the compass card stops and remains fixed on a given heading, subsequent heading changes will not be reflected by the card. Another type of failure causes the compass card to drift (float). In either type failure, the pilot must rely on supplementary directional gyros or the mag-

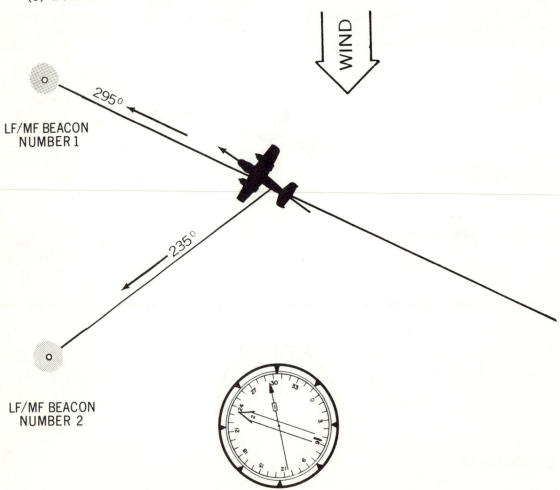

*Figure 12–6. Position fixing using ADF facility (RMI).*

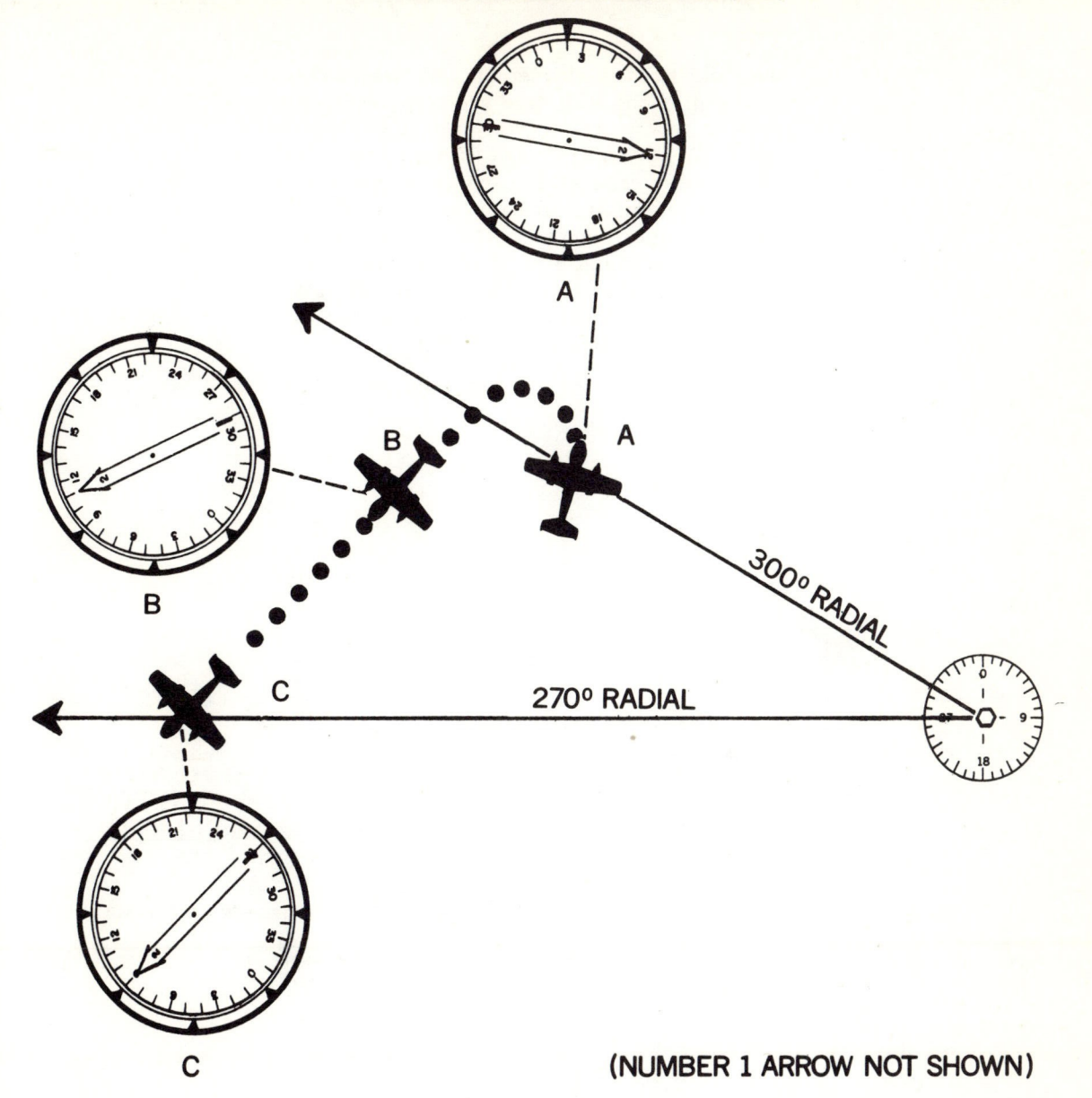

(NUMBER 1 ARROW NOT SHOWN)

*Figure 12–7. RMI readings during track interception.*

netic compass (depending on type of equipment installed in the aircraft) for heading information.

*b.* The RMI bearing pointers will continue to operate regardless of compass failure as long as the radio signals and instrument relay circuits are operating properly. However, the bearing pointer reaction when coupled with a VOR receiver is different from the reaction when coupled with an ADF receiver.

(1) If either the bearing pointer number 1 or bearing pointer number 2 is coupled with a VOR or TACAN receiver, the bearing pointer continues to indicate the correct magnetic bearing to the VOR or TACAN facility, but an incorrect heading appears under the index. The aviator can continue to read the correct radial under the tail of the bearing pointer and the correct direction to the facility under the head of the bearing pointer (fig. 12–8). The magnetic compass can be used to determine the correct heading.

(2) If either bearing pointer is coupled with the ADF receiver, the bearing pointer will point to the LF/MF beacon and indicate a bearing relative to the heading under the index (30° right of heading in fig. 12–9). The magnetic compass can be used to determine the correct heading. If the compass failure results in a fixed card (*a* above), the aviator can continue to read the ADF bearing pointer using relative bearing procedures similar to

those discussed in chapter 11. The tick marks surrounding the RMI face are very useful in showing relative bearings of 045°, 090°, 135°, 180°, 225°, 270°, 315°, and 0° (fig. 12–9). If the compass card is drifting as a result of compass failure, it will be of little use as an ADF reference; however, the tick marks at 45° intervals can still be of considerable help. In addition, the pilot can turn the aircraft as necessary to position the bearing pointer at the index and home to the facility

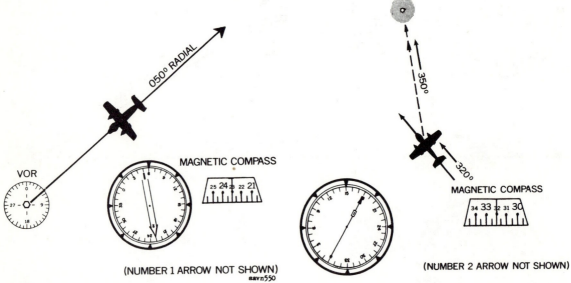

*Figure 12–8. RMI compass failure (receiver tuned to VOR station).*

*Figure 12–9. RMI compass failure (receiver tuned to LF/MF station).*

# INTRODUCTION TO INSTRUMENT 13
## APPROACH PROCEDURES

## Section I. INSTRUMENT APPROACHES

### Purpose

Instrument approaches are designed to assist the aviator in landing during low ceiling and low visibility conditions by—

*a.* Allowing movement from en route courses and altitudes to a position and altitude at which the final descent on a final approach course can be started.

*b.* Providing for safe descent on the final approach course with accurate directional guidance.

*c.* Guiding the aircraft down on the approach path to a minimum altitude from which a safe landing can be made if the aviator has visual reference to the runway.

### Instrument Approach Segments

An instrument approach procedure may have five separate segments (fig. 13–1). These include transition, initial, intermediate, final, and missed approach segments. In addition, an area for circling the airport under visual conditions shall be considered. The approach segments begin and end at designated fixes; however, under some circumstances certain segments may begin at specified points where no fixes are available. The fixes are named to coincide with the associated segment. *For example,* the intermediate segment begins at the intermediate fix (IF) and ends at the final approach fix (FAF). The segments are discussed in the same order in this chapter that the pilot would fly them in a complete procedure (i.e., from a transition or initial, through an intermediate, to a final approach). Only those segments which are required by local conditions need be included in a procedure. The design of the approach should blend all segments to provide an orderly maneuvering pattern appropriate to the local area with regard to airspace requirements.

*a. Transition Segment.* The transition segment, when required, is used to designate course and distance from a fix in the en route structure to the initial approach fix (IAF). Only those transitions which provide an operational advantage shall be established and published.

*b. Initial Approach Segment.* In the initial approach, the aircraft has departed the en route or transitional phase of flight and is maneuvering to enter the intermediate segment. An initial approach may be made along an arc, radial, course, heading, radar vector, or any combination thereof. Procedure turns, shuttle descents, and high altitude teardrop penetrations are also initial segments.

*c. Intermediate Approach Segment.* This is the segment which blends the initial approach segment into the final approach segment. It is the segment in which aircraft configuration, speed, and positioning adjustments are made for entry into the final approach segment. The intermediate segment begins at the IF or point, and ends at the FAF. There are two types of intermediate segments—the radial or course and the arc.

*d. Final Approach Segment.* This is the segment in which alinement and descent for landing are accomplished. The final approach segment considered for obstruction clearance begins at the final approach fix, or point, and ends at the missed approach point. Final approach may be made to a runway for a straight-in landing, or to an airport for a circling approach.

*e. Missed Approach Segment.* A missed approach segment begins at the missed point and provides obstruction clearance and course guidance to a fix for holding or return to the en route structure. The missed approach point specified in the approach procedure may be the point of intersection of an electronic glidepath with a decision height (DH) or minimum descent altitude (MDA), a navigation facility, a fix, or a specified distance from the final approach fix.

*Note.* The altitudes used in figures 13–2 through 13–20 are examples only. Actual altitudes authorized on instrument approaches are published in navigation publications; they vary considerably for each installation.

### Straight-In Approaches

In figure 13–2, an aircraft is cleared to Smithdale VOR (final approach fix for a VOR Runway 9 approach to Smithdale Airport) via the Robinsville VOR 135° radial, Jones intersection, direct to Smithdale VOR maintaining 3,000 feet. Over the Robinsville VOR IAF, the aircraft is cleared to "descend and maintain

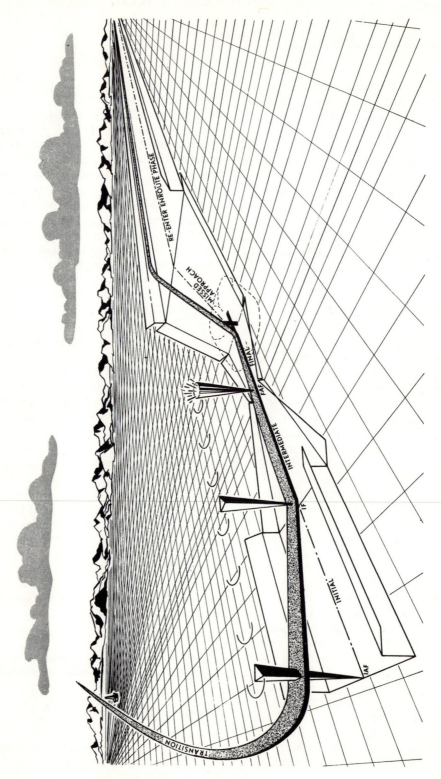

*Figure 13–1. Segments of an approach procedure.*

2,000." The aircraft descends to 2,000 feet while flying between Robinsville VOR and Jones intersection (the initial approach segment) on the Robinsville VOR 135° radial. The term "NO PT" (no procedure turn) appears on the course between Jones intersection and Smithdale VOR indicating that when cleared for "straight-in VOR Runway 9 approach" the pilot shall not fly a procedure turn. (See next para. for an approach using a procedure turn.) Over Jones intersection, the aircraft is cleared for "straight-in VOR Runway 9 approach." The aircraft descends to 1,300 feet between Jones intersection and Smithdale VOR (intermediate approach segment) on the 270° radial. After passing Smithdale VOR (FAF), inbound to the airport on the 090° radial, the aircraft begins its final descent to minimum descent altitude published on the approach chart. During the descent from 1,300 feet to landing minimums, the pilot should establish visual contact with the runway environment and be in a position to complete the visual landing. If visual contact with the runway environment is not made or cannot be maintained by the time the aircraft has reached the missed approach point, a missed approach procedure will be executed. Straight-in approaches are authorized to be commenced (1) from over a fix depicted as "NO PT" or "FINAL," (2) on the final approach course within specified distances from the final approach fix, or (3) when on an intercept heading of the final approach course specified by a radar controller.

*Note.* The term *straight-in approach* as used in this paragraph refers to an instrument approach procedure that does not include a procedure turn. It should not be confused with *straight-in landing.* An aircraft may execute a *straight-in approach* to a specified runway and then circle to another runway for landing. Circling minimums will be applied in this case. Straight-in landing minimums apply when published on the approach chart and a landing is made to a runway alined within 30° of the final approach course.

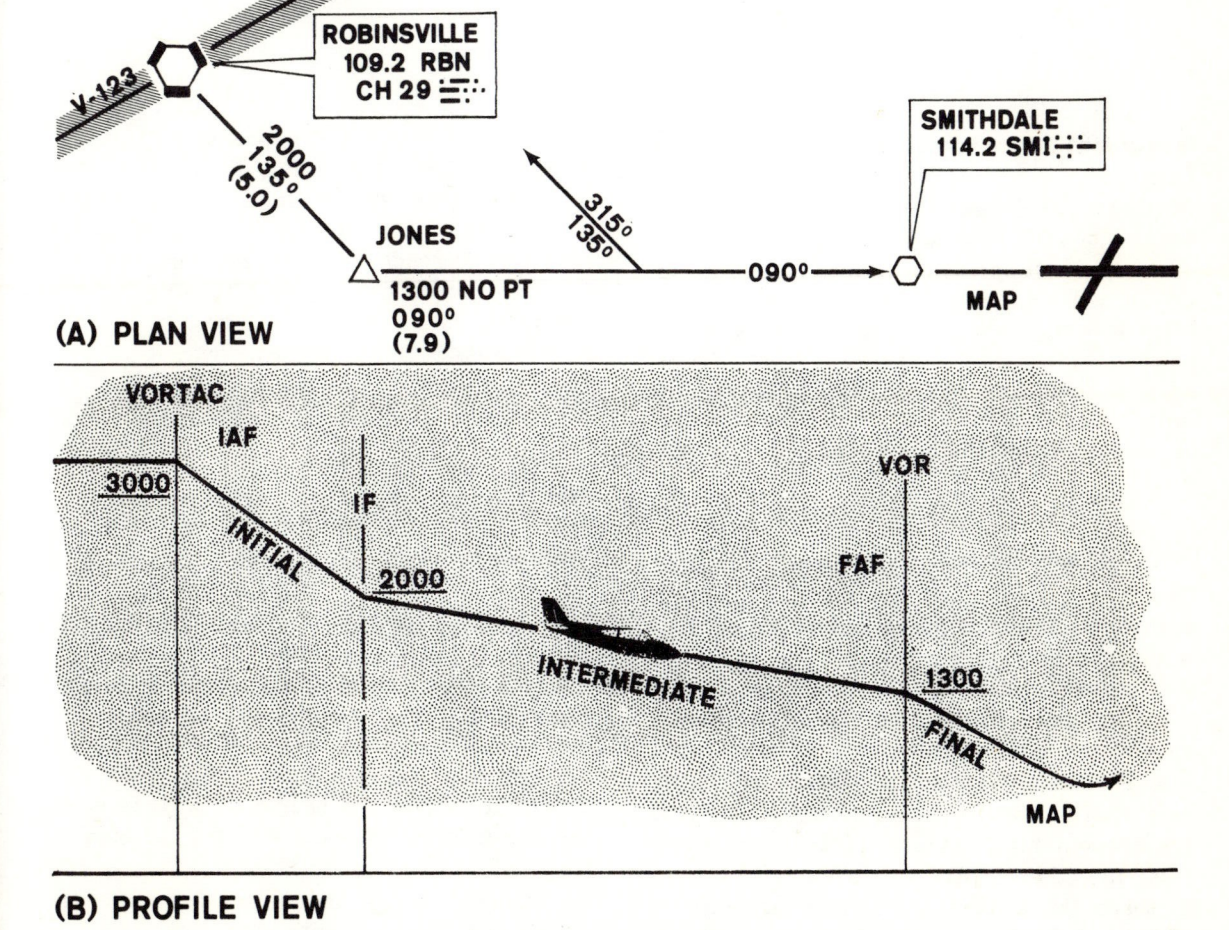

(A) PLAN VIEW

(B) PROFILE VIEW

*Figure 13-2. Straight-in approach.*

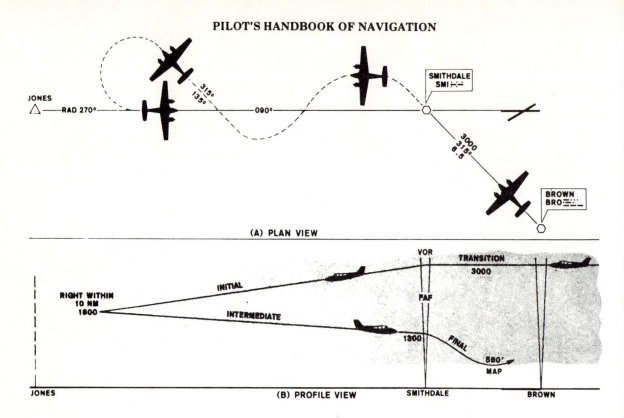

*Figure 13-3. Approach with a procedure turn.*

## Instrument Approach With Procedure Turn

In figure 13-3, an aircraft is approaching Smithdale VOR from a direction not suited for a straight-in approach. In this case, the approach will require a procedure turn. The aircraft will maintain 3,000 feet (or assigned altitude) during the transition from Brown VOR to Smithdale VOR ((A), fig. 13-3). The initial approach begins when the aircraft crosses Smithdale VOR outbound. As the aircraft flies outbound for a procedure turn, it descends to 1,800 feet ((B), fig. 13-3). After completion of procedure turn (sec. III), the aircraft begins the intermediate approach segment and descends to 1,300 feet. The final approach segment begins after passing Smithdale VOR (FAF) inbound, and descent to landing minimums is commenced.

*Note.* Format, symbols, and abbreviations on figure 13-3 parallel those used in current navigation publications.

## Section II. TRANSITIONS

### Definition and Purpose

The term *transition* as applied to instrument approaches refers to the procedure whereby an aircraft departs one en route facility or fix and proceeds along a specified course to a nearby initial approach facility or fix. Figure 13-4 shows three facilities in a terminal area: Robinsville (RBN), Brown (BRO), and Smithdale (SMI). Two of these (RBN and BRO) are not suitably located to serve as approach aids; the other (SMI) is conveniently located to serve the airport. Air traffic arriving at RBN and BRO must transition to the SMI VOR to make an instrument approach to the airport.

### Publication

Information on course, distance, and minimum altitude, which is necessary for the aviator to execute a transition is published on instrument approach charts where an operation advantage is provided. Figure 13-4 shows an area with several published transitions from VOR facilities and intersections to the SMI VOR approach facility for the nearby airport. In each case, the information published for the transition consists of—

*a.* Course, with the magnetic direction printed and indicated with an arrow.

*b.* Distance, shown to the nearest tenth of a mile.

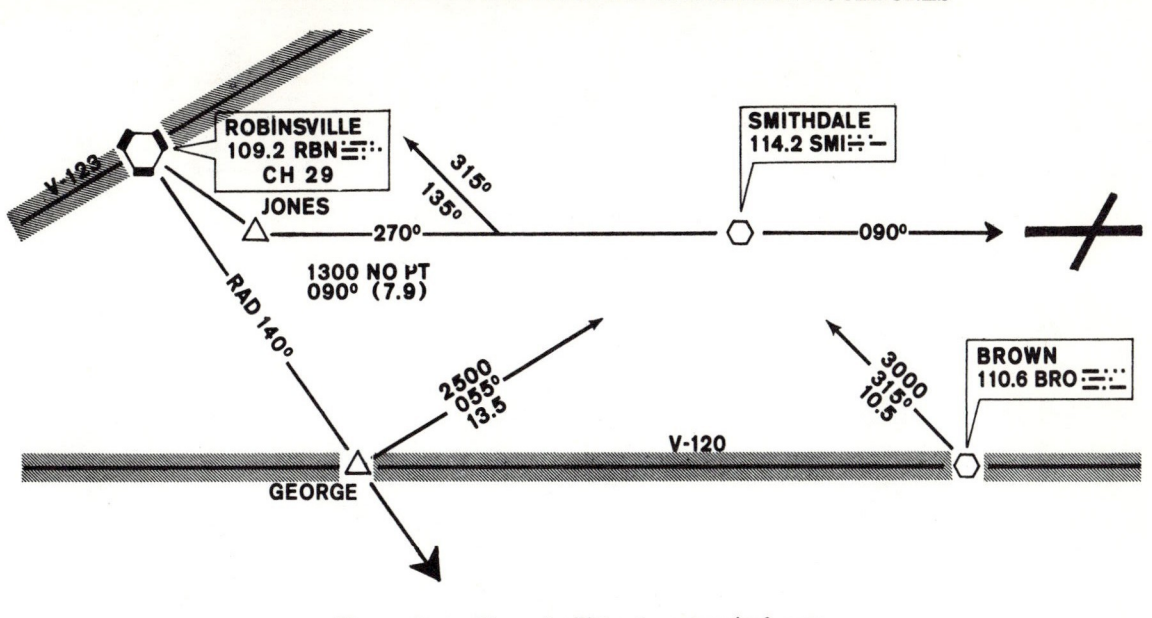

*Figure 13-4. Three facilities in a terminal area.*

*c.* Minimum authorized altitude, which is based upon a standard obstruction clearance of 1,000 feet above obstacles within 4 nautical miles of the transition course. NO PT, or the word FINAL, is printed following the minimum authorized altitude (at Jones intersection on fig. 13-4) to indicate that aircraft approaching Smithdale from Jones intersection will be cleared for a straight-in approach.

### Execution

A transition is executed in accordance with the clearance delivered by the air traffic controller. This clearance may authorize either of the following:

*a.* Transition clearance without authority to execute an approach, is given when the controller is unable to give the actual approach clearance (because of traffic). The clearance for a transition in this situation will include a specific route and altitude assignment and, when necessary, holding instructions with an expected approach clearance time.

*b.* Transition to the initial approach fix with clearance to execute the approach (e.g., "Army 12345 cleared to Smithdale via Brown 315° radial, cleared for VOR approach . . ."). In this situation, the recommended procedure is to descend to the published minimum altitude authorized for the transition, unless the procedure turn altitude is higher.

## Section III. PROCEDURE TURNS

### Purpose

A *procedure turn* is a maneuver which allows the aviator to—

*a.* Reverse flight direction.

*b.* Descend from initial approach (transition) altitude or last assigned altitude to a specified procedure turn altitude from which descent for final approach is begun.

*c.* Intercept the inbound course at a sufficient distance away from the approach fix to aline the aircraft for the final approach.

### Typical Patterns

Typical procedure turn flight patterns are illustrated in (A) and (B) of figure 13-5. A description is given for each illustration (*a* and *b* below).

*a. 45° Turn From Nonprocedure Turn Side* ((A), fig. 13-5).

(1) In this situation the aircraft flies outbound to point A either on the reciprocal of the approach course or south of the course.

(2) At point A the aircraft turns right to the procedure turn heading published on the approach chart, 315°. The aircraft then flies 40 seconds from the approach course to point B. The aviator may adjust the time to compensate for known headwinds or tailwinds.

(3) At point B the aircraft turns left to

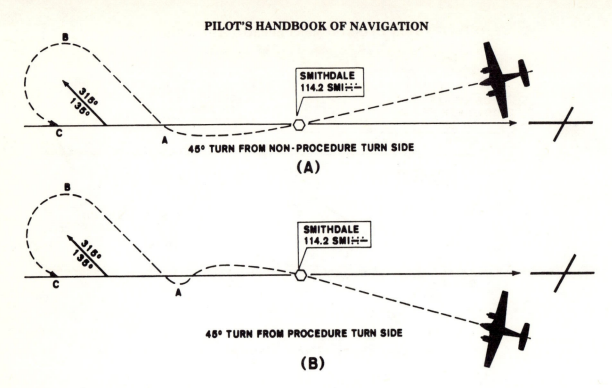

*Figure 13–5. 45° turns.*

intercept the approach course at point C and flies inbound to the final approach fix.

*b. 45° Turn From Procedure Turn Side* ((B), fig. 13–5).

(1) In this situation the aircraft flies outbound to point A north of the approach course.

(2) At point A the aircraft turns left to intercept the approach course. After intercepting the approach course, the aircraft turns right to the procedure turn heading published on the approach chart, 315°. The aircraft then continues the procedure as discussed in *a* above.

*c. Teardrop Turn.* A teardrop turn (fig. 13–6) may be executed at the aviator's discretion in lieu of the more conventional 45° type turn.

(1) Upon arrival over approach fix, the aviator follows a track outbound not to exceed 30° of the reciprocal of the approach course.

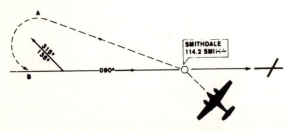

*Figure 13–6. Teardrop turn.*

(2) At point A, which is about 1 minute from the approach fix, the aviator turns inbound to intercept the approach course at point B.

*Note.* These procedures are applied using turns in the opposite direction when the procedure turn is on the other side of the approach course.

## Procedure Turn Area

*a.* Figure 13–7 shows the airspace area within which obstruction clearance is provided for procedure turn maneuvering.

*b.* The limiting distance for procedure turns is published on the profile view of approach charts ((B), fig. 13–3). It is normally 10 nautical miles. Variations from normal will be clearly depicted on the approach charts.

*c.* In flying outbound from the approach fix to execute the procedure turn, the aviator normally flies a minimum of 1 minute. This outbound leg may be extended, if necessary, to lose additional altitude or compensate for adverse wind effects. However, in no event may the distance outbound from the station exceed the one published on the approach chart.

## Obstruction Clearance—Minimum Altitude

The published procedure turn altitude will provide a minimum of 1,000 feet of clearance in the primary area (fig. 13–7). In the secondary

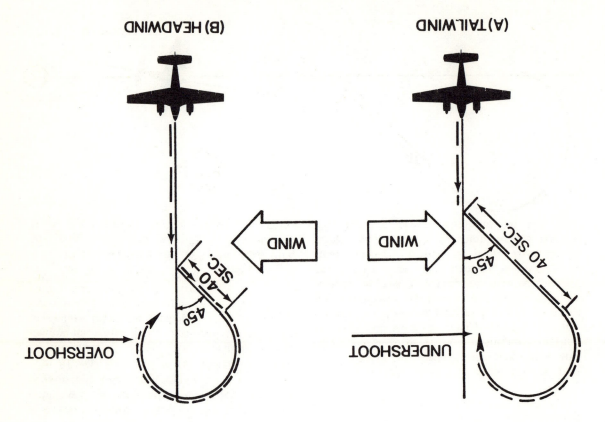

Figure 13-8. Improper procedure turn patterns caused by wind effects.

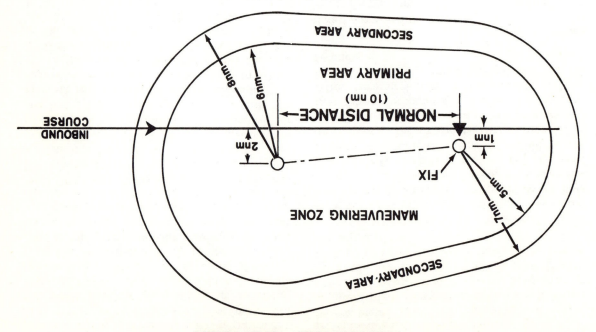

Figure 13-7. Procedure turn area.

# PROCEDURE TURN AREA

## INTRODUCTION TO INSTRUMENT APPROACH PROCEDURES

area, 500 feet of obstruction clearance shall be provided at the inner edge, tapering uniformly to zero feet at the outer edge. The pilot will not descend below this published minimum ((B), fig. 13–3). In flying outbound from the approach fix, the aviator normally descends from the transition altitude to the procedure turn altitude. This descent may vary between several thousand feet and a few hundred feet – or there may be no descent if the transition altitude is the same as the procedure turn altitude. The rate of descent is a matter of aviator judgment; however, it should not exceed a maximum safe rate at which the aviator has complete control of the aircraft. A descent rate of 500 feet per minute is recommended for the last 1,000 feet of altitude change. If the aircraft has not arrived at the minimum procedure turn altitude at the time the turn starts, the descent is continued during the turn until the minimum altitude is reached. If the initial approach altitude is unusually high, it may be necessary to lose the excessive altitude in a holding pattern.

## Turning Rate

The procedure turn is made at the standard rate of 3° per second; however, the bank angle should not exceed 30°. In aircraft equipped with an integrated flight system which uses a steering pointer (ch. 13), the turn is executed with a centered steering pointer (approximately a 25° bank angle).

*a. Flying Outbound to the Procedure Turn.* If the aircraft encounters unusually strong headwinds while flying outbound to execute a procedure turn, the outbound time should be extended to provide ample time inbound for course alinement of the aircraft.

*b. The 45° Procedure Turn.*

(1) In executing the 45° procedure turn, the aviator will normally fly for 40 seconds from the approach course on the procedure turn heading. This timing is calculated so that the subsequent turn to the inbound course will be completed when the final approach course is intercepted. However, the 40 seconds flying time must be adjusted if adverse headwinds or tailwinds are encountered. Figure 13–8 illustrates the results when suitable adjustments are not made.

(2) Adjustment to the 40 seconds flying time is based upon the known or estimated drift correction required to fly the track outbound. An allowance of 1 second for each degree of drift correction used on the outbound leg should be applied to the 40 seconds flown on the leg of the procedure turn. Figure 13–9 shows the aircraft holding a 10° drift correction flying outbound for the procedure turn. After turning left 45°, the aircraft will be headed into the wind and will fly for 50 seconds.

## Missed Approaches

If the instrument approach and landing cannot be completed successfully, the aviator executes a *missed approach* procedure. This procedure is published on the approach chart and normally is supplemented by further instructions and clearances from the controller.

*Note.* Format, symbols, and abbreviations in figure 13–10 parallel those used in current navigation publications.

*a. Procedure.* The typical procedure normally directs the aircraft to proceed on a specified course to or from a designated facility, and to climb to a specified minimum altitude. Figure 13–10 shows plan and profile views of an instrument approach procedure, with the missed approach procedure printed beneath the profile view. The aircraft is making the final approach on the 090° radial from the facility and is unable to complete the landing. The aviator begins climbing on the same

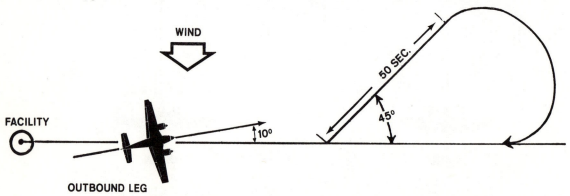

*Figure 13–9. Adjusting procedure turn for wind effects.*

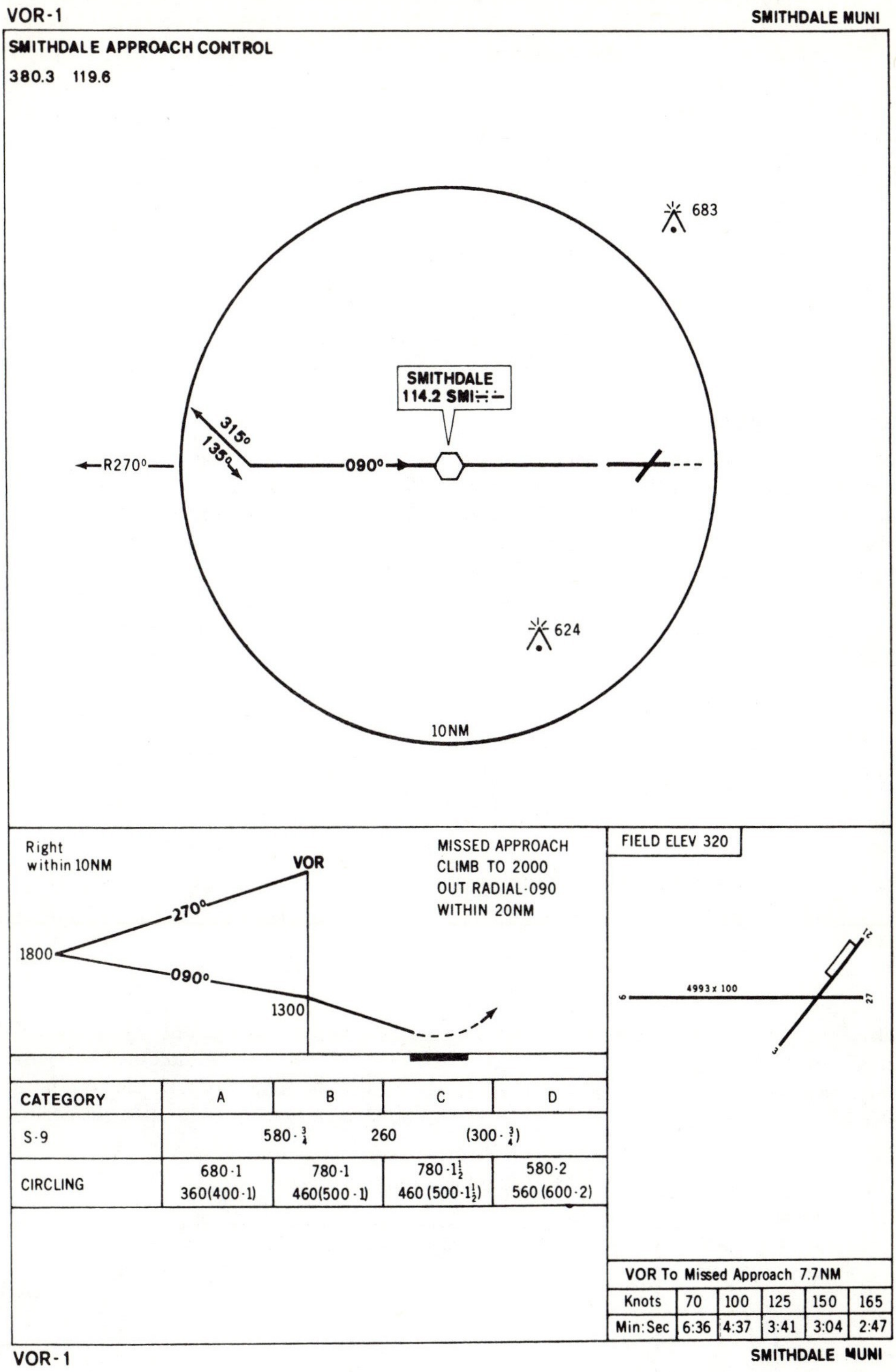

VOR-1                                                                 SMITHDALE MUNI

**SMITHDALE APPROACH CONTROL**

**380.3   119.6**

```
                                        ☀ 683

                              ┌──────────────┐
                              │  SMITHDALE   │
                              │ 114.2 SMI ═ ═│
                              └──────┬───────┘
         315°                        ⬡
         135°
  ←R270°────────────────090°──────────────────┼ ─ ─ ─ ─

                                        ☀ 624

                              10NM
```

| | | | FIELD ELEV 320 |
|---|---|---|---|

Right within 10NM                    VOR         MISSED APPROACH
                                                 CLIMB TO 2000
        270°                                     OUT RADIAL·090
                                                 WITHIN 20NM
1800

        090°

               1300                                    4993 x 100

| CATEGORY | A | B | C | D |
|----------|---|---|---|---|
| S-9 | 580 · $\frac{3}{4}$ | 260 | | (300 · $\frac{3}{4}$) |
| CIRCLING | 680·1 360(400·1) | 780·1 460(500·1) | 780·1$\frac{1}{2}$ 460 (500·1$\frac{1}{2}$) | 580·2 560 (600·2) |

**VOR To Missed Approach 7.7NM**

| Knots | 70 | 100 | 125 | 150 | 165 |
|-------|----|----|-----|-----|-----|
| Min:Sec | 6:36 | 4:37 | 3:41 | 3:04 | 2:47 |

VOR-1                                                                 SMITHDALE MUNI

*Figure 13–10.  Plan and profile views of an instrument approach procedure.*

course, climbing to 2,000 feet indicated on the altimeter (1,680 feet above the airport). The procedure is based upon the use of the 090° radial of Smithdale VOR; the missed approach altitude guarantees adequate obstacle clearance provided the pilot begins the climb at the missed approach point and follows the published procedure.

*b. Report.* The pilot must report a missed

approach to the controller as soon as practical after he starts the procedure. The report states the time the missed approach began and includes a specific request for further clearance. The pilot may request clearance to execute another approach (if feasible), or he may request clearance to an alternate airport if conditions warrant it.

## Section IV.  HOLDING

*Warning:* The information and procedures pertaining to holding in this section are subject to change. Pilots are cautioned to consult current operational publications for changes to procedures discussed in this chapter.

### Definition and Purpose

Because of heavy traffic conditions en route or at busy air terminals, air traffic controllers occasionally instruct pilots to *hold. Holding* is the procedure used to delay an aircraft at a definite position and assigned altitude. In some instances the pilot is directed to climb or descend to a newly assigned altitude in the holding pattern.

### Holding Pattern Configuration

The *standard* holding pattern consists of *right* turns (fig. 13–11), the *nonstandard* holding pattern of *left* turns.

### Timing

*a.* The initial outbound leg of a holding pattern at or below 14,000 feet mean sea level is flown for 1 minute. Above 14,000 feet mean sea level, the leg is flown for 1½ minutes.

*b.* Subsequent outbound legs are adjusted (depending on the wind) so that the inbound leg is 1 minute at and below 14,000 feet mean sea level and 1½ minutes above 14,000 feet mean sea level. *For example:*

(1) A helicopter flying a true airspeed of 75 knots experiences a 30-knot headwind on the outbound leg and a 30-knot tailwind on the inbound leg. The following tabular data shows the comparative times flown on the outbound and inbound legs to compensate for this wind (no allowance is made for drift during the inbound turn):

| Outbound time | Inbound time |
| --- | --- |
| 1 minute | 26 seconds |
| 1 minute and 20 seconds | 34.5 seconds |
| 1 minute and 40 seconds | 43 seconds |
| 2 minutes | 51.5 seconds |
| 3 minutes | 1 minute and 17 seconds |

(2) In this example, therefore, the aviator must fly approximately 2 minutes and 20 seconds on the outbound leg to achieve the desired 1-minute flying time on the inbound leg.

*c.* Outbound timing (fig. 13–12) begins over or abeam the holding fix, whichever occurs later. The position abeam the station can be determined by reference to a certain radial

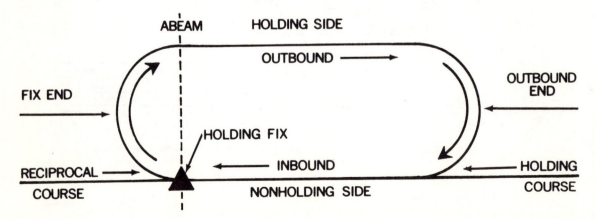

*Figure 13–11. Standard holding pattern.*

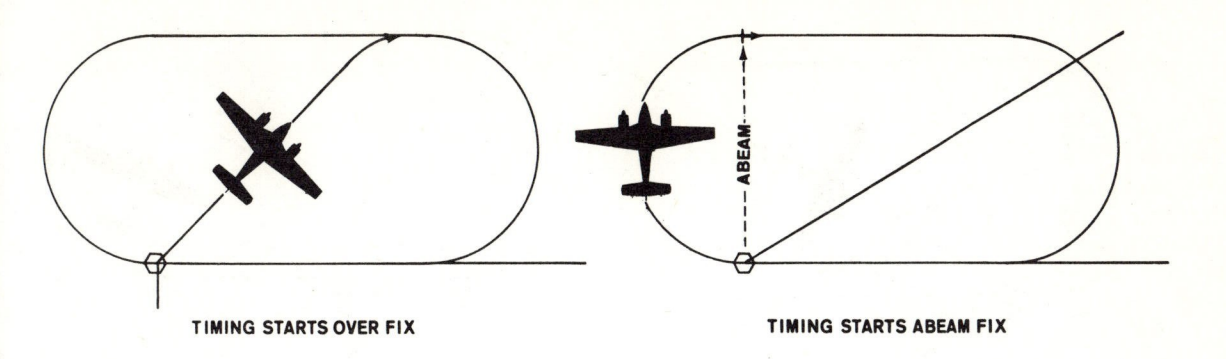

TIMING STARTS OVER FIX          TIMING STARTS ABEAM FIX

*Figure 13–12. Outbound timing.*

(VOR ) or bearing (ADF). If the abeam position cannot be determined, the outbound timing is started when the turn to the outbound is completed.

*Note.* TACAN (DME) holding is subject to the same entry and holding procedures except that distances (nautical miles) are used in lieu of time values.

### Airspeeds

Maximum indicated airspeed (IAS) allowed for holding is 175 knots for all propeller-driven aircraft. Different speeds are allowed for civil and military turbojet aircraft, depending on holding altitude and aircraft category, as listed in current navigation publications. Aircraft oprating en route at normal cruise airspeeds higher than the maximum authorized for holding are required to reduce airspeed within 3 minutes before reaching a holding fix.

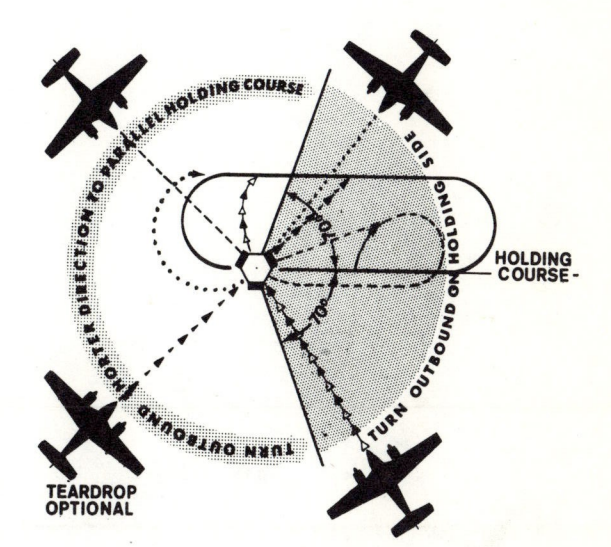

*Figure 13–13. Holding pattern entry.*

### Turns

Turns made during entry and while holding will be at the rate of 3° per second (not to exceed a 30° angle of bank), or with the steering pointer of an integrated flight system .

### Holding Pattern Entry

The aircraft is considered to be in the holding pattern at the time of initial passage of the holding fix. The aircraft *heading* at initial holding fix passage determines the direction of turn to enter the holding pattern (fig. 13–13).

 a. If the aircraft heading is within 70° of the inbound holding course (fig. 13–14), turn outbound in the same direction as the holding pattern to parallel the holding course.

 b. If the aircraft heading is not within 70° of the inbound holding course (fig. 13–15), turn outbound in the shorter direction to parallel the holding course. If this places the aircraft on the nonholding side at completion of the outbound leg, turn toward the holding side.

 c. The teardrop entry (fig. 13–16), may be used at the pilot's discretion when entering the holding pattern on a heading conveniently aligned with the teardrop course. The teardrop course is 30° or less outbound from the holding course and on the holding side. Course interception is not mandatory prior to turning inbound.

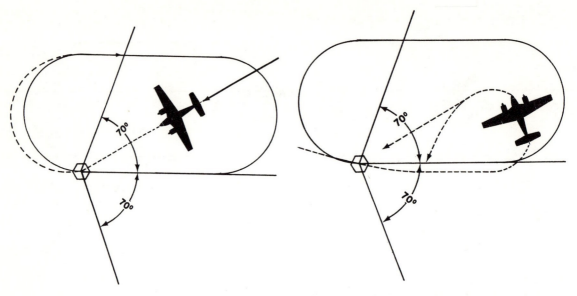

Figure 13-14. *Holding pattern entry, direct entry.*   Figure 13-15. *Holding pattern entry, parallel outbound.*

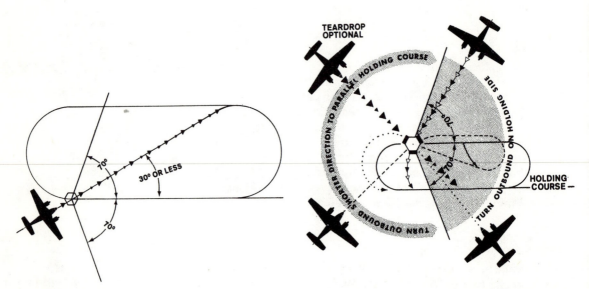

Figure 13-16. *Holding pattern entry, teardrop.*   Figure 13-17. *Nonstandard holding pattern entry.*

*d.* When turning inbound the first time, proceed directly to the fix or intercept and fly the holding course. If the holding pattern is nonstandard (left turns), the entry patterns are reversed (fig. 13-17).

### Departing the Holding Pattern

When cleared by the controller to leave the holding fix, the pilot normally departs the pattern from over the fix. An exception to this occurs when the controller specifically states ".... cleared from your present position. ... If the controller has specified a departure time, the pilot must adjust the holding pattern so that the aircraft is over the holding fix ready to depart at the specified time. If an aircraft is holding on the published final approach course at an approach fix

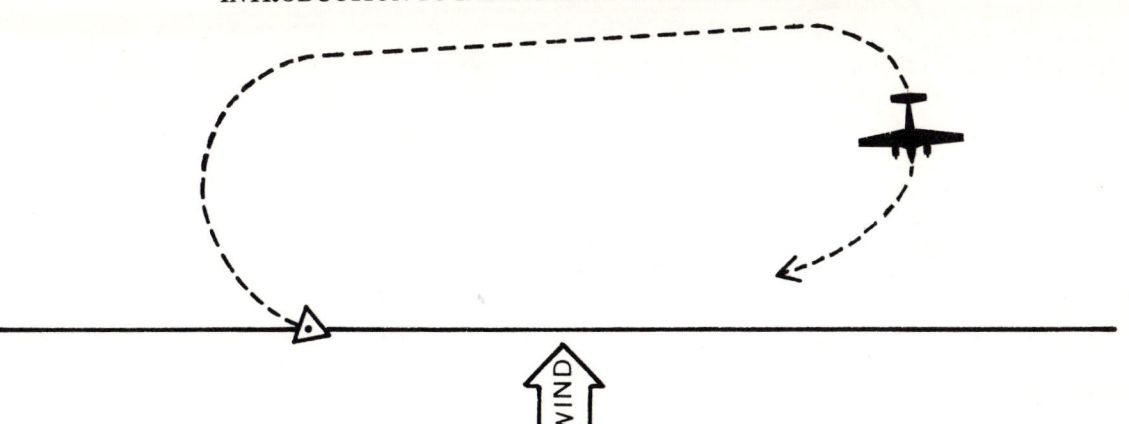

## (A) RIGHT CROSSWIND OUTBOUND

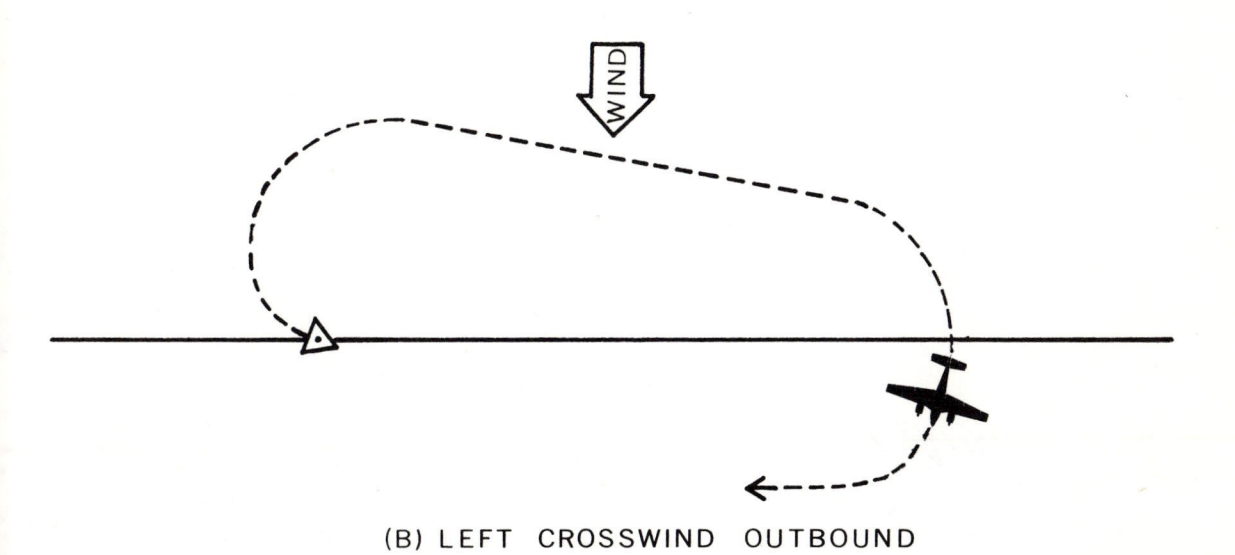

## (B) LEFT CROSSWIND OUTBOUND

*Figure 13–18. Wind effect on holding pattern flight.*

and receives clearance for the approach, the pilot normally begins the final approach from the holding pattern without executing the conventional procedure turn.

*Note.* At some locations, beginning the final approach from the holding pattern may be prohibited by notes published on the approach chart.

### Drift Correction in the Holding Pattern

*a.* If no attempt is made to correct for adverse affects of crosswinds while holding, the aircraft will follow a wide arc in one turn and a tight arc in the opposite turn (fig. 13–18).

*b.* If the same amount of drift correction is flown for both inbound and outbound legs (but applied in opposite directions), the outbound leg will parallel the inbound leg; however, the turns will still be wide and tight, respectively. Since the aviator has little control over the aircraft's track while turning, he must adjust the track of the outbound leg to avoid turning short of or overshooting the inbound leg.

*c.* For holding pattern drift correction—

(1) Determine the correction necessary to maintain the track inbound.

(2) While flying the outbound leg, double the inbound correction and apply it in the op-

posite direction; or if the inbound correction is over 10°, use an outbound correction of 10° plus the inbound correction.

*Note.* This guide must be adjusted to fit each situation. Analysis of the initial inbound turn (overshooting or undershooting) should be used as a basis for a subsequent adjustment.

*d.* Two examples of applied drift correction are—

(1) In (A) of figure 13–19, the inbound correction is 5° right; therefore, the correction used outbound is 10° left.

(2) In (B) of figure 13–19, the inbound correction is 15° left; therefore, the correction used outbound is 25° right.

### Holding Clearances and Reports

*a.* When delivering an ATC clearance for holding with a previously assigned altitude, the controller is required to give the following information, in the order shown:

(1) Direction to hold from holding point.

(2) The holding fix.

(3) Radial, course, magnetic bearing, or airway number which constitutes the holding course.

(4) Outbound leg length in nautical miles if DME is used.

(5) Time to expect further clearance (EFC time) or time to expect approach clearance (EAC time).

*b.* If the clearance is for nonstandard holding (left turns), the controller must state "Left turns" after giving the information in *a*(3) or (4) above.

*Note.* When a new altitude assignment is made as part of the holding clearance, the controller will specify the altitude.

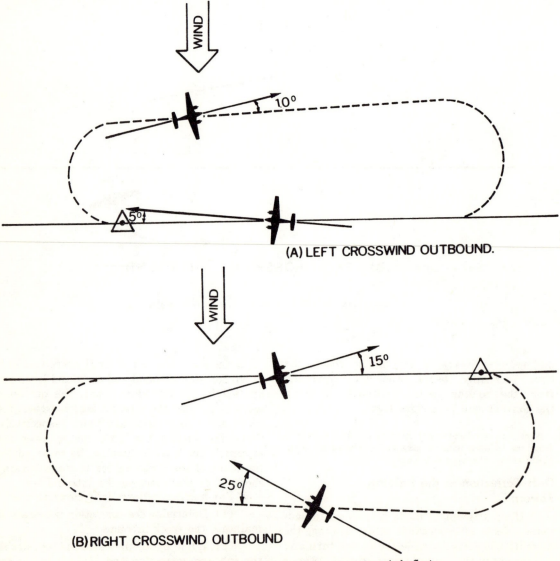

**(A) LEFT CROSSWIND OUTBOUND.**

**(B) RIGHT CROSSWIND OUTBOUND**

*Figure 13–19. Adjusting holding pattern for wind effect.*

*c.* Typical clearances are—

(1) "Hold south of the Ajax VOR on the 190 radial; expect approach clearance at 30" ((A), fig. 13–20).

(2) "Hold northeast of Red Cliff intersection on V–21, left turns, expect further clearance at 50" ((B), fig. 13–20).

*d.* Pilots are required to report when arriving over and when departing the holding fix. The arrival report normally will include identification, position, time, and altitude. The report when departing the holding fix normally includes identification and time of departure.

## Stacking

*Stacking* is the procedure when two or more aircraft are holding, one above the other, on the same fix. As the lower aircraft leaves the stack to complete its approach, the aircraft above it is cleared to the next lower holding altitude. This clearance is given after the aviator of the approaching aircraft has reported that he is vacating his altitude and is leaving the radio facility inbound. The second aircraft is cleared for an approach when the first aircraft is sighted by the tower and when the tower considers that a normal, safe landing will be accomplished. The length of time an aircraft is required to hold in a stack depends upon the time required by the aircraft in the lower positions to land. Since the delay may be of considerable duration, the pilot should fly at an airspeed and power setting which will provide fuel economy but still permit adequate aircraft control.

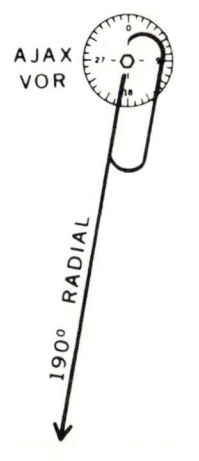

**(A) HOLDING AT FACILITY**

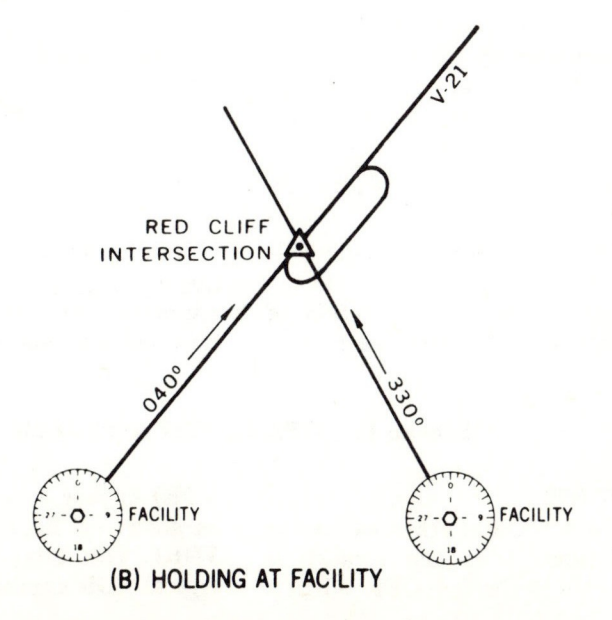

**(B) HOLDING AT FACILITY**

*Figure 13–20. Holding patterns flown for typical clearances.*

# 14 VOR AND ADF APPROACHES

## Section I. APPROACH CHARTS

### General

The separate VOR and ADF approach charts published by Federal agencies and private companies contain complete information on current instrument approach procedures at specific airfields. The format of all these charts is basically the same. Therefore, once the aviator has studies one type chart and its legend, he is usually able to use other types effectively.

### Typical VOR Approach Chart

*a. General.* The Cairns AAF VOR runway 6 approach chart (fig. 14–1) is typical of those found in current navigational publications. Its format and general data presentation are a guide for the pilot. For detailed explanation of approach chart symbols, consult FLIP instrument approach procedure charts and their legends printed with each volume.

*b. Explanatory Data for Figure 14–1.* The following numbered items apply to the same numbers shown on figure 14–1.

(1) Chart heading includes type of approach (VOR, ADF, ILS, etc.) and to which runway; and name of airport, city, and State.

(2) Communications data includes frequencies for approach control, tower, ground control, etc.; and type of radar (PAR/ASR) or other service available.

(3) Transition data from outer fixes includes minimum altitude (2,000 ft.), direction (230°), and distance (27.1 nm).

(4) Minimum sector altitudes within 25 nautical miles.

(5) Approach facility location with call-out to show frequency, identifier, and code.

(6) Procedure track shows direction of procedure turn with 45° off-course bearings.

(7) Missed approach track with description listed in profile.

(8) Procedure turn data: direction of turn (right), limiting distance (10 nm), and minimum altitude (1,700 ft).

(9) Straight-in (rnwy 6) minima data by aircraft category (aircraft category based on stall speed and maximum gross weight).

(10) Circling minima data by aircraft category.

(11) Minimum Descent Altitude (MDA) shown in feet above mean sea level (msl). This is the lowest altitude to which descent is authorized until airport or runway environment is in sight.

*Note.* On precision approaches (approaches with glideslope information), this value is referred to as a Decision Height (DH).

(12) Visibility values are expressed as Runway Visual Range (RVR), Prevailing Visibility (PV), or Runway Visibility (RV). RVR is shown in feet; e.g., 24 equals 2,400 feet. PV and RV are shown in miles and fractions thereof; e.g., 1½ equals 1½ miles.

(13) Height above touchdown (HAT) indicates height of MDA (or DH on precision approaches) above the runway elevation in the touchdown zone (first 3,000 feet) of runway for *straight-in* approaches.

(14) Height above airport (HAA) indicates height of MDA above airport elevation for *circling* approaches.

(15) Ceiling and visibility value for military use in accordance with directives of each service.

**(16) Airport diagram to show field elevation, runway location with dimensions, approach lights information, and also direction and distance from related facility.**

## Section II. TYPICAL VOR APPROACH

### VOR Station Location

VOR stations used in VOR approaches may be located some distance from the airport as shown in figure 13–10 (called OFF Airport VOR) or may be located on or near the airport as shown in figure 14–1 (called ON Airport VOR). The latter type is used to depict the typical VOR approach procedure.

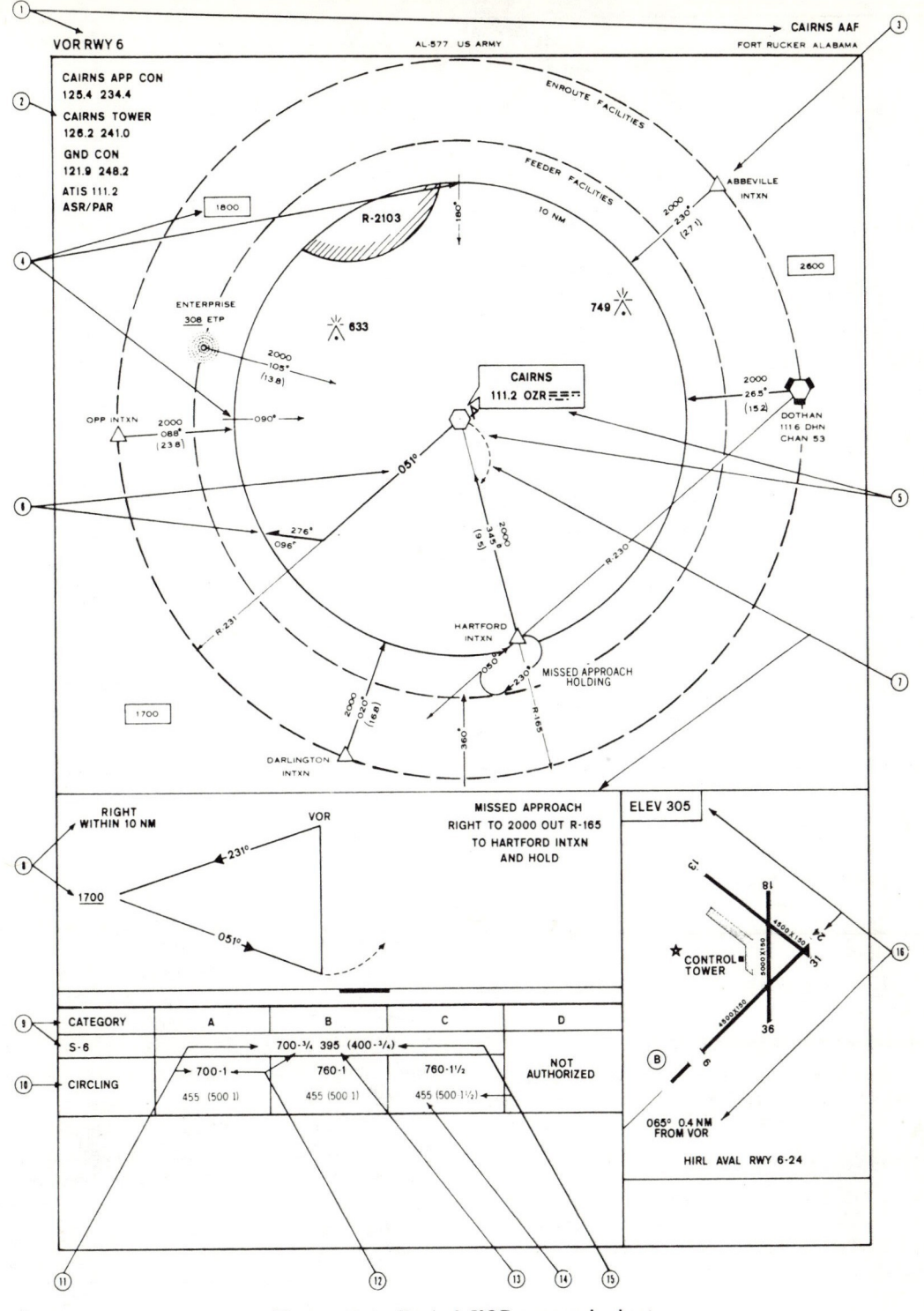

*Figure 14-1. Typical VOR approach chart.*

## Initial Contact and Arrival

A pilot is flying eastbound on V–241 at 5,000 feet with Cairns AAF as his destination (fig. 14–2). In compliance with ATC instructions, he establishes radio contact with Cairns approach control over Darlington intersection. Cairns approach control clears him to the Cairns VOR from over Hartford intersection, with clearance to hold southwest of Cairns VOR on the 231° radial left turns. He is cleared to descend and maintain 4,000 feet and given an expected approach time of 25 minutes past the hour. Upon arrival at Cairns VOR at 4,000 feet, the aviator—

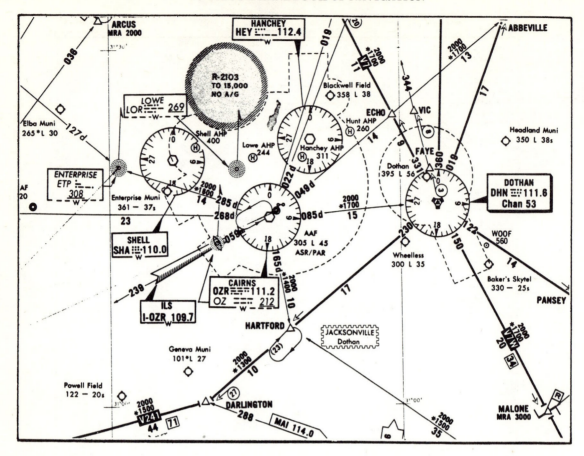

*Figure 14-2. Typical VOR station locator.*

*a.* Checks the time.

*b.* Turns outbound to enter the holding pattern.

*c.* Reduces airspeed to prescribed holding speed.

*d.* Reports station passage.

*Note.* Actions *a* through *c* above are performed almost simultaneously. The report is not made until after station passage.

## VOR Holding

*a.* Initial passage of Cairns VOR occurs when the TO–FROM indicator reverses readings (*TO* to *FROM*). The pilot then turns outbound to a heading of 231° (fig. 14–3) to enter the holding pattern. Use of the course selector and deviation needle to track outbound during the entry procedure is optional, but this procedure will aid the aviator in orienting himself with respect to the VOR station and to the holding radial. He may either set the course selector on the desired outbound radial or fly a heading outbound with the course selector set for tracking inbound on the holding course.

*b.* After flying 1 minute on the outbound heading of the entry leg, the pilot turns left to intercept the holding course inbound (051°, fig. 14–3). Prior to turning, the course selector is set on 051° and the TO–FROM indicator reads *TO*. The pilot should adjust the inbound turn as he monitors the course indicator **and/or the RMI to intercept the desired inbound course.**

*c.* During the initial inbound leg of the holding course, the pilot should determine (1) the drift correction necessary to remain on the desired track, and (2) the time flown on the inbound leg. Subsequent outbound legs of the holding pattern are adjusted so that each inbound leg requires 1 minute. Drift corrections in the holding pattern are discussed in chapter 13.

*d.* After flying over the VOR facility, the pilot makes a 180° turn to the outbound heading of the holding course. Timing for the outbound leg should begin when the aircraft is abeam the station. One accurate method for determining his position abeam the station is by rotating the course selector 90° to fix the aircraft position abeam the station (fig. 14–4). This technique permits the pilot to time the outbound leg accurately from a position abeam the station.

(1) *Point A*. During the left turn outbound, the pilot rotates the course selector 90° to the left (reading of 321°), thereby enabling him to fix his position abeam the station. During the turn, the deviation needle deflects full left.

(2) *Point B*. Needle centers abeam the station. Outbound timing begins at this time.

(3) *Point C*. After passing point B, course selector is reset to 051° to intercept the holding course inbound. The needle deflects to the side away from the holding course during the outbound portion of the holding pattern.

(4) *Point D*. Needle centers as aircraft turns inbound and intercepts the holding course.

*e*. Another method of accurately determining position abeam the station is by using RMI indications (ch. 12).

*f*. When holding at a fix where methods described in *d* and *e* above cannot be used, the pilot should begin timing the outbound leg immediately after rolling out of the 180° standard rate turn.

## Descent While Holding

*a*. The pilot is holding at 4,000 feet over Cairns VOR. The approach chart ((8), fig. 14–1) shows the minimum procedure turn altitude for the VOR approach to the field as 1,700 feet. As lower air traffic departs the holding pattern, the controller clears the aviator to descend to a lower holding altitude. In this situation, the clearance is received to 3,000 feet. The pilot continues the established holding pattern and establishes a 500-foot-per-minute rate of descent. When the pilot reports leaving 4,000 feet, the controller can assign this holding altitude to another aircraft. (The 500-foot-per-minute rate of descent is used for 1,000-foot descents in the holding pattern.)

*b*. If the aircraft had been at a higher altitude (e.g., 9,000 feet), and were cleared to a low altitude (e.g., 3,000 feet), the pilot could have established the maximum rate of descent at which he could still fully control the aircraft. He could have used this steep rate to within 1,000 feet above the newly assigned holding altitude; he would then reduce to a rate not to exceed 500-feet-per-minute for the last 1,000 feet of descent.

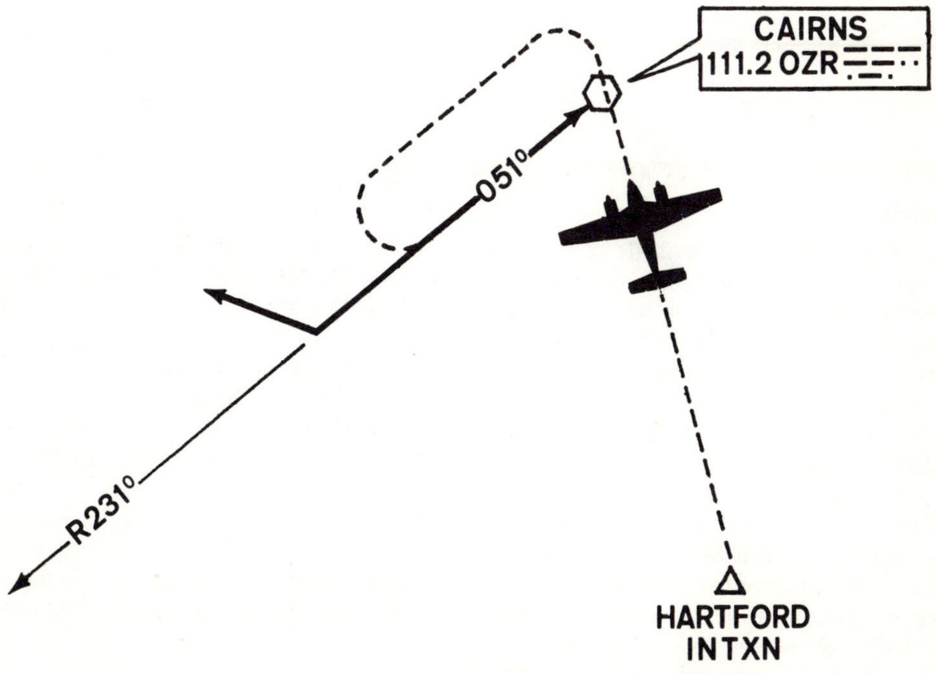

*Figure 14–3. Holding pattern entry.*

169

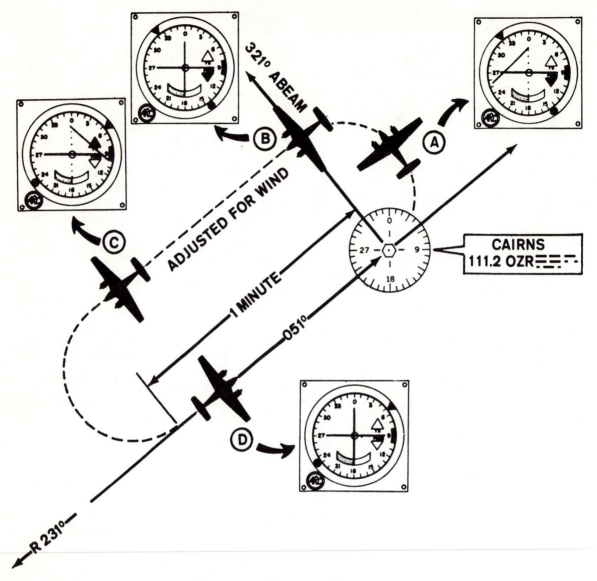

*Figure 14-4. Flying a holding pattern.*

## The Approach

a. The pilot has been advised of his expected approach clearance time (fig. 14-1). As air traffic conditions change, the controller revises the expected approach clearance time and advises the pilot accordingly. When the pilot is cleared for the approach, he may immediately begin the descent from the 3,000-foot holding altitude to the 1,700-foot procedure turn altitude, regardless of his position in the holding pattern. The final turn inbound from the holding pattern serves as the procedure turn, so the aviator could extend the outbound leg to lose altitude if necessary, provided he does not exceed the 10 miles ((8), fig. 14-1) prior to turning inbound. Since this approach is an ON Airport VOR approach, the final segment of the approach begins with completion of the procedure turn.

b. After intercepting the final approach course inbound, the pilot may descend from the authorized procedure turn altitude to the minimum descent altitude (MDA) authorized for the approach (700 feet; (11), fig. 14-1).

c. Descent below the 700-foot MDA is not authorized until the pilot establishes visual contact with the runway environment and can reasonably expect to maintain visual contact throughout the landing. In making an ON-Airport VOR approach, the VOR is the missed approach point. Should the pilot not make visual contact by the time he reaches the VOR, he would execute a missed approach.

## Landing

Clearance to land on a specified runway is issued when the pilot is on final approach. If the pilot is cleared to land on runway 6 (fig. 14-1), he will land straight in. Clearance to land on other runways will require circling to land. In most cases, the MDA and visibility minimums are higher for circling-type landings. In this case, the visibility requirement is higher but the MDA remains the same (700-1; (11), fig. 14-1).

## Missed Approach

If, for any reason. the landing is not accomplished, the pilot executes the missed approach procedure. To accomplish the procedure as specified in figure 14-1, the pilot

*a.* Adjusts power and attitude, as necessary, to begin an immediate climb.

*b.* Turns right to intercept the 165° radial

of Cairns VOR.

*c.* Sets the course selector to 165°. (This results in a *FROM* indication and a right needle deflection on the course indicator.)

*d.* Reports a missed approach to the controller and requests further clearance, either for another approach or to his alternate airport, as appropriate. (If he requests clearance to the alternate, flight plan data must be given to the controller.)

*e.* Checks for centered needle at the 165° radial.

*f.* Continues climb to missed approach altitude (2,000 feet).

*g.* Complies with subsequent ATC instructions.

## Section III.  TYPICAL ADF (NONDIRECTIONAL BEACON) APPROACH

### General

ADF approach charts are similar in appearance and format to VOR approach charts. The approach procedures are essentially the same as those for VOR. The significant difference between VOR and ADF approaches is the aircraft instrumentation; i.e., the ADF procedure is accomplished by reading bearings from the

ADF indicator or the RMI (see RMI techniques, ch. 12).

*Note.* Symbols, abbreviations, and format in figures parallel those used in current navigation publications.

### Transition

Figure 14-5 shows the ADF transition procedure and instrument readings during the tran-

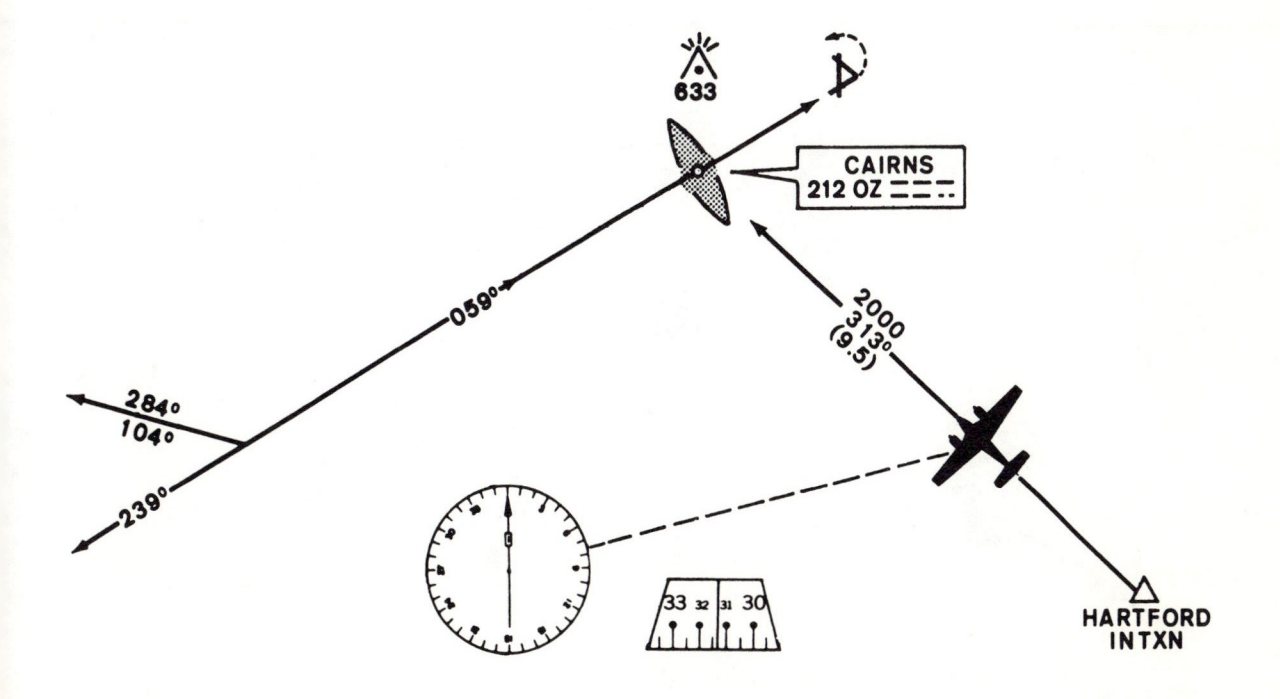

*Figure 14-5. ADF transition.*

sition. The aircraft is making a transition from Hartford intersection to Cairns RBn (OZ) to execute an ADF approach. The pilot has been cleared to hold at OZ and await an approach clearance.

## Holding Entry

Figure 14–6 shows the ADF holding pattern entry procedure and instrument readings in the entry pattern. The pilot arrives at OZ and accomplishes the normal entry into a left-hand holding pattern by turning outbound on the holding side.

*a. Point A.* After passing OZ inbound from Hartford intersection, the pilot observes station passage (indicator arrow reverses from 0° to 180°) and turns left.

*b. Point B.* The aircraft is turning inbound to the station; the relative bearing should be approximately 315° (45° left of nose) when on the intercept heading of 104°.

*c. Point C.* The aircraft is inbound to the station with a relative bearing of 0° on a heading of 059°. The pilot checks the inbound drift correction and flight time accurately for adjusting subsequent outbound legs.

## Holding

Figure 14–7 shows the holding pattern with aircraft locations shown at various points in the pattern.

*a. Point A.* After passing the beacon inbound the aviator begins the left turn.

*b. Point B.* He completes the turn to the outbound heading (239°); the ADF indicator passes the wingtip position (270° relative bearing), and the pilot checks the time.

*c. Point C.* Having flown the required outbound time (based on inbound time), the pilot begins a left turn to reintercept the inbound course.

*d. Point D.* The pilot again establishes the aircraft inbound and rechecks timing for further refinement.

## Descent

Figure 14–8 shows the ADF descent procedure from the holding pattern. At point A, the pilot receives the ADF approach clearance and immediately begins descent to procedure turn altitudes while in the holding pattern. The 180° inbound turn serves as the procedure turn.

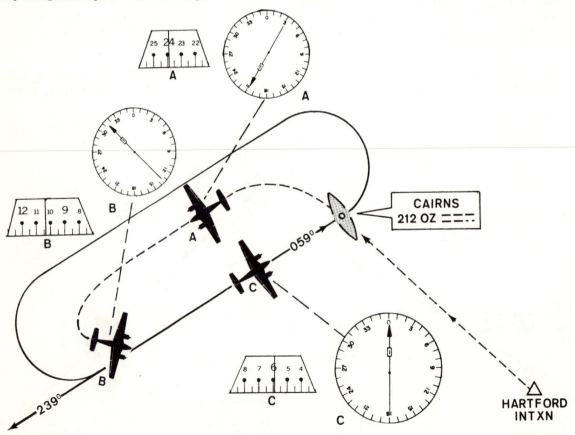

*Figure 14–6. ADF holding pattern entry.*

## Intermediate And Final Approach

*a.* Figure 14–9 shows the procedure and instrument readings during the intermediate and final segment of the approach. After completing the procedure turn, the pilot intercepts the 059° inbound course and continues at the minimum altitude authorized (2,000 ft) until passing the beacon (OZ).

*b.* The pilot determines station passage, notes the time, begins descent to the minimum descent altitude and reports to the controller.

*c.* In this approach, the facility (OZ) is located OFF Airport and the pilot will continue to fly toward the field at the MDA for a time based on the distance to field and the estimated groundspeed (fig. 14–9).

*d.* If the pilot establishes visual contact during the time from facility to field, he completes the landing; if not, he executes the missed approach.

## Section IV.  HELICOPTER ADF FIGURE EIGHT APPROACH

### Holding

The tactical holding procedure is a 4-minute figure eight pattern, composed of four 15-second legs (dashed line, fig. 14–10) and two 90-second 270° turns (solid curved lines, fig. 14–10).

*a.* In the pattern, airspeed is limited to 60 knots and standard rate turns are used.

*b.* The holding axis (fig. 14-10) is the reference line for the figure eight holding and letdown pattern. Its direction (to nearest 10°) is furnished by the person in charge of beacon emplacement.

*c.* The figure eight holding pattern in a no-wind condition—

(1) Limits the flightpath to an equal ground distance left and right of the holding axis.

(2) Limits the flightpath to an equal ground distance along the axis measured from the beacon.

(3) Allows the pilot to identify his position over the beacon twice in each 4-minute circuit of the pattern.

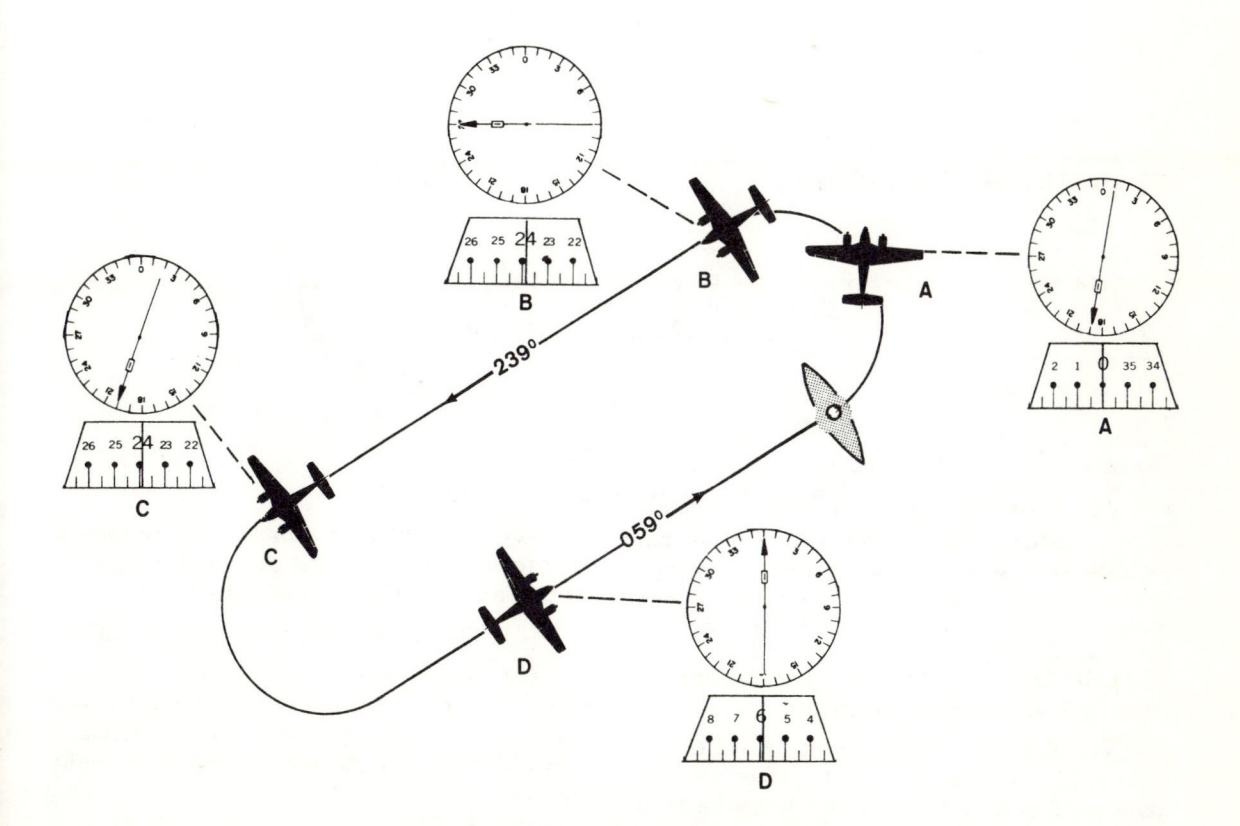

*Figure 14-7.  ADF holding pattern procedure.*

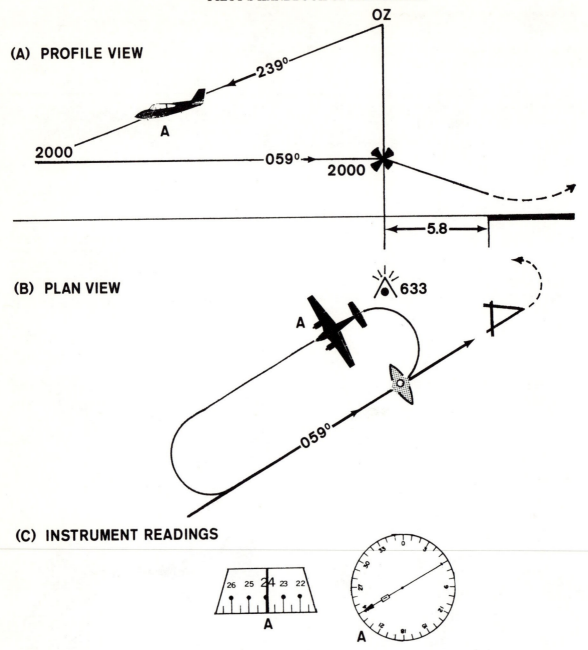

*Figure 14–8. ADF descent.*

(4) Allows the pilot to hold by homing.

*Caution:* **These procedures become unreliable when surface winds exceed 20 knots.**

Note. Distances shown for the figure eight patterns (fig. 14–10) are based on 60 knots airspeed (no wind).

d. Holding in the figure eight pattern by homing is accomplished as follows:

(1) Referring to the holding pattern in figure 14–10, assume pilot , after beacon passage, flies on a heading 45° outbound from the holding axis (in this case a heading of 45°) for 15 seconds and then begins a standard rate

270° left turn.

(2) The pilot continues the standard rate turn to the left until the ADF needle indicates a relative bearing of zero.

(3) The pilot then homes to the station.

(4) Upon crossing the beacon, the pilot establishes the desired 135° outbound heading for 15 seconds and then begins a standard rate 270° right turn to continue the figure eight pattern.

(5) A strong north wind could cause him to make a continual turn to the left (dashed

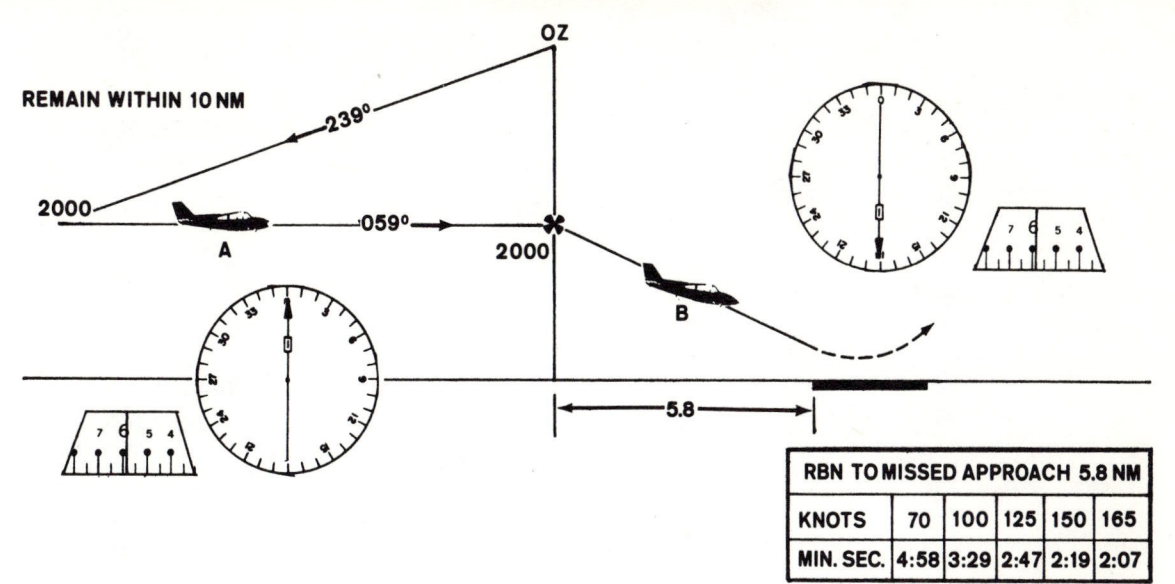

Figure 14-9. ADF final inbound.

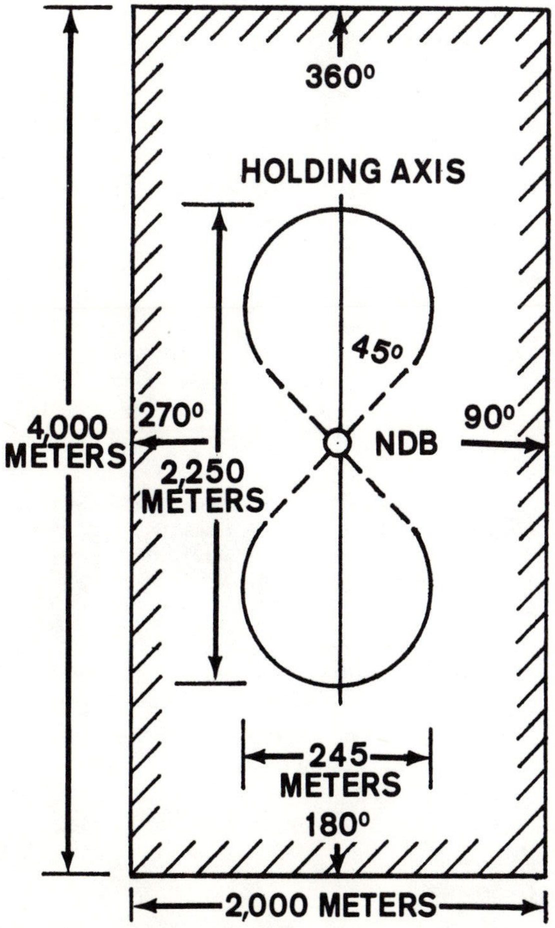

| RBN TO MISSED APPROACH 5.8 NM | | | | | |
|---|---|---|---|---|---|
| KNOTS | 70 | 100 | 125 | 150 | 165 |
| MIN. SEC. | 4:58 | 3:29 | 2:47 | 2:19 | 2:07 |

Figure 14-10. Helicopter figure eight holding pattern.

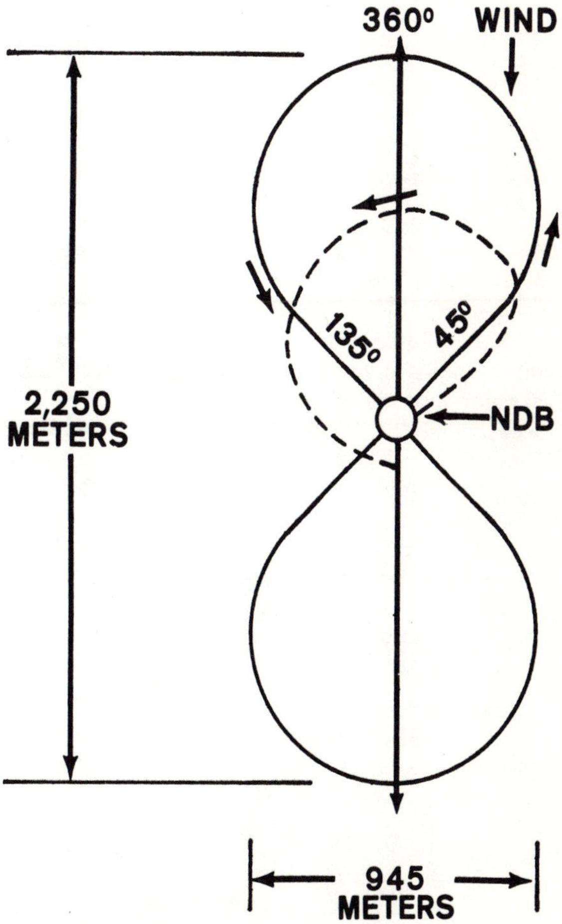

Figure 14-11. Wind effect on holding.

line, fig. 14–11) in excess of heading 090° in order to establish the ADF reading at zero. In this case, the pilot would stop the turn at heading 090° as a limiting point for this type procedure. Once the aircraft reaches heading 090°, the pilot maintains heading 090° until abeam the station, at which time he initiates a right turn and completes the figure eight by homing back to the beacon.

*e.* Entry into the tactical ADF figure eight pattern is accomplished upon initial beacon passage by turning to the nearest 15-second leg outbound; i.e., the nearest outbound heading in relation to the helicopter heading at the time of beacon passage.

## Descent

Tactical ADF letdowns are accomplished while holding in the figure eight pattern.

## Approach

The tactical ADF approach is a continuation of the figure eight holding and letdown pattern. The pilot descends in the figure eight pattern to the established *low station altitude* of 700 feet above minimum descent altitude rounded to the nearest 100 feet msl. He maintains low station altitude until beacon passage into the downwind portion of the figure eight, at which time he descends to the minimum descent altitude. He executes a landing or missed approach at the beacon. *For example:*

*a.* Assume the pilot is inbound to the beacon (as shown in fig. 14–12) on a heading of 135° at 1,300 feet indicated altitude. Assuming the station altitude to be 400 feet msl with a 40-foot obstacle in the area, the 200-foot minimum added to 440 feet would give a minimum descent altitude of 640 feet msl. Another 700 feet added for low station altitude would be 1,340 feet msl and, rounded to the nearest 100 feet, would give a low station altitude of 1,300 feet msl.

*b.* Upon crossing low station (altitude 1,300 ft. indicated) on a heading of 135° outbound (downwind portion, fig. 14–12) at 60 knots minute rate of descent and continues the figure eight approach pattern.

*c.* He normally arrives at minimum descent altitude approximately 90 to 105 seconds after low station passage (point A, fig. 14–12). His altitude at approach minimums is approximately 200 feet above obstacles (640 ft. indicated).

*d.* With approximately 15 to 30 seconds remaining between minimum descent altitude and the beacon, the pilot continues his turn until the relative bearing is 0°.

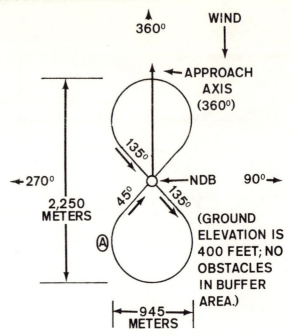

*Figure 14–12. Helicopter figure eight approach altitude.*

*e.* The copilot should be able to see the ground when the helicopter reaches minimum descent altitude. If he does, he takes over the helicopter control, continues to the station, and lands. However, if visual ground contact is not established, the pilot maintains minimum descent altitude until over or abeam the beacon where he executes a missed approach.

## Missed Approach

*a.* The normal missed approach procedure is accomplished by tracking from the beacon on the approach axis heading (360°, fig. 14–12) while maintaining a normal climb airspeed until reaching a safe terrain-clearance altitude.

*Note.* During the flight planning phase, the aviator must determine the rate of climb that will provide safe obstacle clearance on a straight-out missed approach. If this rate of climb is impracticable, the aviator uses the figure eight pattern missed approach.

*b.* If high terrain or the combat situation beyond the buffer area along the approach axis prevents a normal missed approach (the straight climb explained in *a* above), the aviator remains in the figure eight pattern and climbs at 60 knots maximum indicated airspeed until reaching a safe terrain-clearance altitude.

## Departures

Departures from beacon sites are accomplished in the same manner as missed approaches.

# INSTRUMENT LANDING SYSTEM 15

## Introduction

The instrument landing system (ILS) is a complex array of radio and light navigation aids. It is the most efficient system in widespread use for safe landing under extremely low ceiling and visibility conditions. Its effectiveness as an approach aid is matched by radar, but the preferred system at most major air terminals is the instrument landing system supplemented by radar. More advanced systems have been undergoing tests for several years, but several factors have prevented placing these systems in an operational status.

## Ground Components

*a. Required Components.* For a complete ILS to be commissioned operational at an airport, the following ground components must be installed and operating within specified tolerances as determined by flight checks:

(1) Localizer transmitter

(2) Glideslope transmitter

(3) Outer marker beacon

(4) Middle marker beacon

(5) Approach lights

   *Note.* Airports use several types of transmitting equipment, but the design differences are relatively minor.

*b. Supplementary Components.* The ILS is frequently supplemented by installing one or more of the following approach aids:

(1) *Compass locators*

(2) *Transmissometers.* This device "looks" electronically down the instrument runway in the landing direction and either determines the runway visibility by reference to ordinary runway lights or computes the runway visual range (RVR) by reference to high-intensity runway lights.

(3) *Surveillance and precision radar systems*

(4) *Distance measuring equipment* (*DME*). This aid, although normally installed at VOR, Tactical Air Navigation (TACAN), and VOR–TAC sites, is occasionally installed at the site of the ILS localizer. With proper airborne receiving equipment, the aviator can read the distance to or from the transmitter at all times.

(5) *Visual approach slope indicator* (*VASI*). This aid provides by visual reference the same information that the glideslope unit of the ILS provides electronically. It provides a visual light path within the approach zone which the aviator can use for descent guidance during an approach to a landing. The basic principle of the VASI is that of color differentiation between red and white. The light units are arranged so that the aviator during approach will see the following colors:

(*a*) Above glideslope—all white lights.

(*b*) On glideslope—red above white lights (combination).

(*c*) Below glideslope—all red lights.

   *Note.* Course guidance during the VASI-guided approach is obtained by reference to the runway lights.

(6) *Condenser-discharge sequenced flashing light system.* This system consists of a series of brilliant blue-white bursts of light flashing in sequence along the approach path in the approach light system. This creates the illusion of one high-intensity light moving rapidly down the approach path toward the runway touchdown point.

(7) *High-intensity runway lights.* Special lights of high brilliancy are used to outline the sides of the active runway.

(8) *In-runway lighting aids.* These include—

(*a*) *Touchdown zone lighting*—two rows of transverse light bars placed symmetrically about the runway centerline in the runway touchdown zone.

(*b*) *Runway centerline lighting*—bidirectional lights recessed along the centerline of the runway.

(*c*) *Taxiway turnoff lights*—flush lights defining the curved path of aircraft travel from the runway centerline to a point on the taxiway.

(9) *Runway end identifier lights* (*REIL*). These flashing light pairs provide rapid positive identification of the approach end of a particular runway.

   *Note.* Consult current navigation publications to determine the exact supplementary ILS components available for specific airfields and airports.

## Localizer

*a. Location and Signal Pattern* (fig. 15–1). The localizer transmitter is located near the end of the primary instrument runway opposite the approach end. It produces two signal patterns which overlap along the runway centerline and extend in both directions from the transmitter. One side of the signal pattern is referred to as the *blue* sector, the other as the *yellow* sector. The "beam" produced by the overlap of the sectors is usually from 4° to 5° wide. The portion of the beam extending from the transmitter toward the outer marker (OM) (fig. 15–1), is called the *front course*. The sectors are arranged so that, when flying inbound toward the runway on the front course, the blue sector is to the right of the aircraft and the yellow sector to the left. While flying inbound on the *back course* (extending from the transmitter to the left (fig. 15–1)), the blue sector is to the left of the aircraft and the yellow sector is to the right. Both the front course and the back course may be approved for instrument approaches; however, only the front course will be equipped with a glideslope and associated marker beacons and lighting aids. (Some major airports are equipped with two complete ILS installations, thus providing a front course for each end of the runway. In these cases, however, only one ILS will be operated at a time.)

*b. Transmission Frequencies.*

(1) Localizer transmitters are assigned 1 of 20 VHF channels, from 108.1 MHz to 111.9 MHz, as indicated below.

(2) Only frequencies ending with an odd digit are assigned as localizer frequencies. Those within the same frequency band ending with even digits are assigned to VOR stations.

| Sequence or channel number | Localizer MHz | Glide-slope MHz |
|---|---|---|
| 1 | 110.3 | 335.0 |
| 2 | 109.9 | 333.8 |
| 3 | 109.5 | 332.6 |
| 4 | 110.1 | 334.4 |
| 5 | 109.7 | 333.2 |
| 6 | 109.3 | 332.0 |
| 7 | 109.1 | 331.4 |
| 8 | 110.9 | 330.8 |
| 9 | 110.7 | 330.2 |
| 10 | 110.5 | 329.6 |
| 11 | 108.1 | 334.7 |
| 12 | 108.3 | 334.1 |
| 13 | 108.5 | 329.9 |
| 14 | 108.7 | 330.5 |
| 15 | 108.9 | 329.3 |
| 16 | 111.1 | 331.7 |
| 17 | 111.3 | 332.3 |
| 18 | 111.5 | 332.9 |
| 19 | 111.7 | 333.5 |
| 20 | 111.9 | 331.1 |

The use of this frequency band provides the typical reception advantages of VHF, but also imposes the usual line-of-sight limitation.

Since localizers and VOR stations have overlapping frequency bands and their principle of operation is similar, the same VHF navigation receiver is used with both types of stations.

*c. Tuning.* To tune a localizer transmitter with the ARN–30A receiver, the selector switch (OMNI-VAR LOC) on the control panel (fig. 15–2) is placed in the VAR LOC position. This changes the functioning of the receiver to make it operate by signals from a localizer. In this function the course selector and the to–from indicator are disconnected from the circuit. Also the needle sensitivity of the deviation indicator changes from a total of 20° for VOR use to a total of 4° to 5° for localizer use. Failure to place the selector

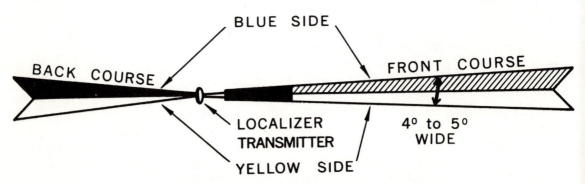

*Figure 15–1. Sample ILS localizer location and signal pattern.*

switch in the correct position will give unreliable results on the course indicator. The ARN-30D and ARN-30E receiver control panels (fig. 15-2) provide digital tuning of the station and eliminate the OMNI-VAR LOC selector switch. On these models the change in function to localizer is automatic when a localizer frequency is selected.

*d. Flight Checks.* Flight checks of the localizer will insure that it has a usable distance of at least 25 miles in a sector extending 30° on each side of the course line, from 1,000 feet above the highest terrain on the localizer course to 6,250 feet above site elevation. When reception at greater altitudes and distances is needed, appropriate flight checks must be made and the facility approved for use as required.

*e. Identification and Voice.* Localizer transmitters identify themselves continuously in code with the three-letter identifier assigned to the airport; however, the basic identification is preceded by the letter I; e.g., IOZR identifies the Cairns AAF localizer. The localizer channel usually is capable of transmit-

ting voice and is frequently used for transmitting air traffic control instructions.

*f. Signal Failure.* If there is a malfunction in the receiver or if, for some other reason, the localizer signal is not received, the *OFF* flag will appear in the ID-453 course indicator at the bottom of the vertical needle. Automatic ground monitors are situated to provide a continuous check on ground equipment. If a failure or malfunction is detected, either the standby equipment is turned on or the broadcast of the signal is terminated. Alarm systems immediately indicate functional trouble to controllers at the airport.

*g. Localizer Tracking.* The ID-453 course indicator has blue and yellow markings on its face. The colors are arranged so that when the aircraft is proceeding inbound on the front course or outbound on the back course, the needle indications are directional. *For example,* if the aircraft flies into the blue sector (fig. 15-3), the needle is deflected to the blue (left), indicating that a left turn is required to return to the localizer track. However, if the aircraft is flying inbound on the back

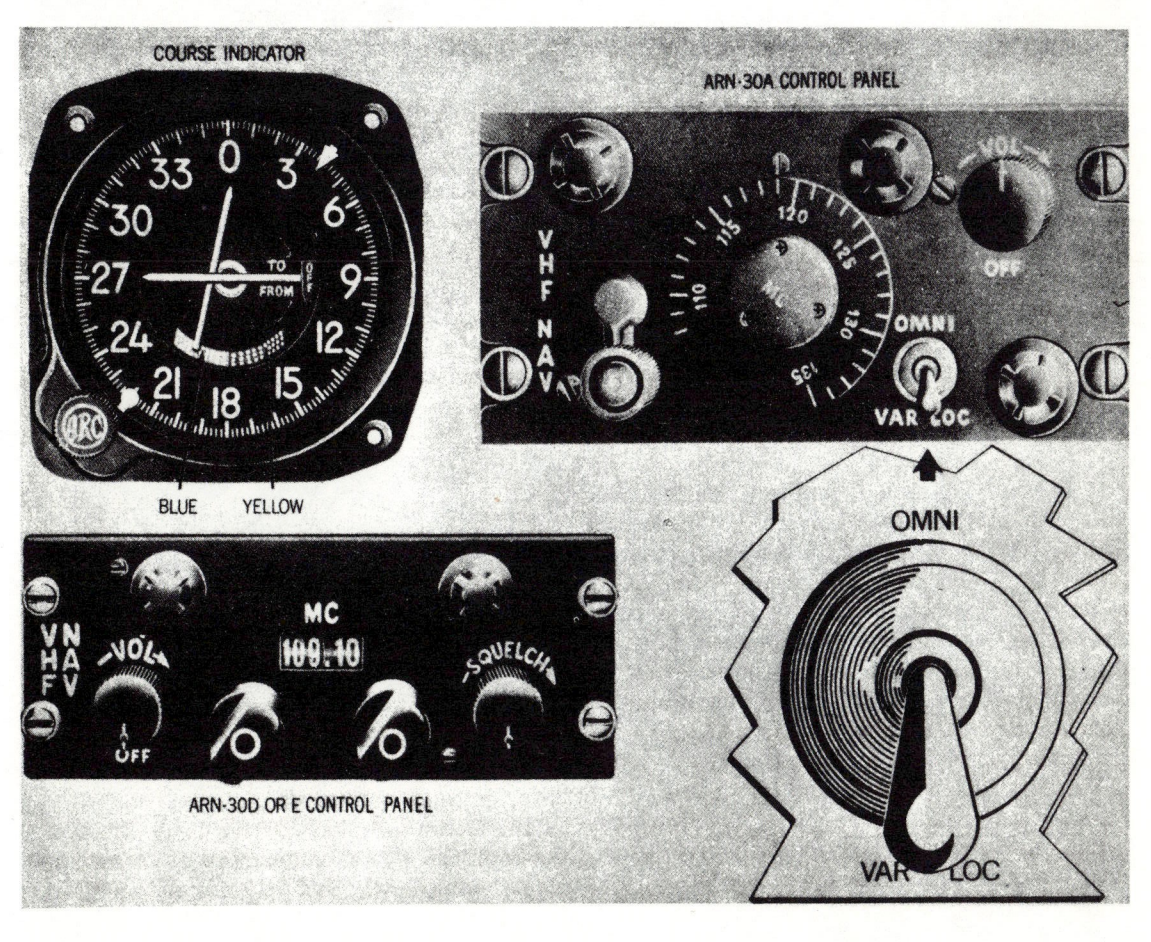

*Figure 15-2. ID-453 course deviation indicator and control panels.*

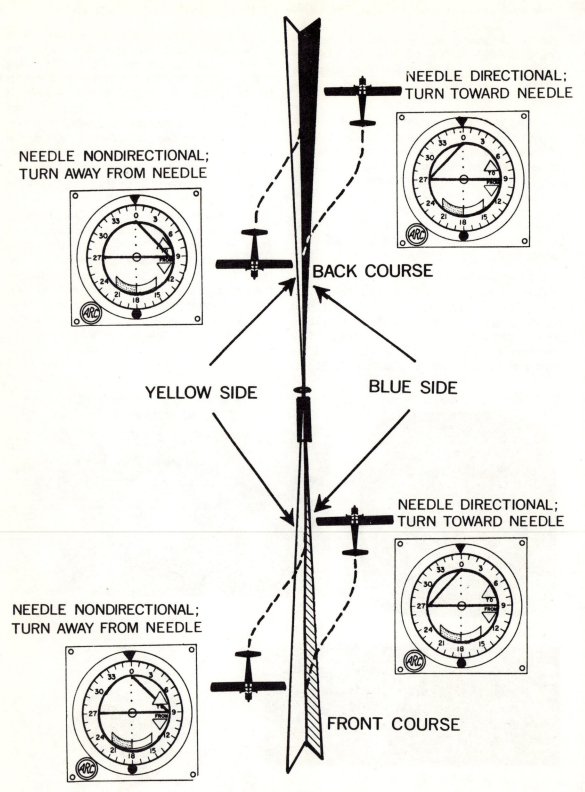

Figure 15-3. Localizer tracking.

course or outbound on the front course, the needle is no longer directional *although the color indications are correct. For example,* if the aircraft flies into the yellow sector (fig. 15–3), the needle will indicate yellow (right), but the aircraft must turn left (away from the needle) to return to the localizer track. The amount of correction required depends on the distance between the aircraft and the transmitter. Corrections of 20° or more to return to the localizer may be necessary if the aircraft is following a track 25 miles away from the airport. However, during the latter part of final approach, corrections larger than 2° to 5° are seldom required to remain on the localizer track.

**Glideslope**

a. *Transmitter Location.* The glideslope transmitter is located near the touchdown end of the runway, but is placed to one side so it will not cause a final approach hazard. Figure 15–4 shows a typical installation. These transmitters are located to produce a glideslope only on the front course of the localizer.

b. *Signal Pattern.*

(1) The normal elevation of the glideslope is 2½°, but this may be raised to 3° to provide adequate obstacle clearance. The absolute elevation of the 2½° glideslope above the base line at various distances is as follows:

| ¼ mile | 65.5' |
|--------|-------|
| ½ mile | 131' |
| 1 mile | 262' |
| 2 miles | 525' |
| 4½ miles | 1,180' |

(2) The thickness of the glideslope may vary but is normally 1.4°. Therefore, the glideslope needle is very sensitive.

(3) The indications of the needle are always directional; i.e., if the aircraft is below

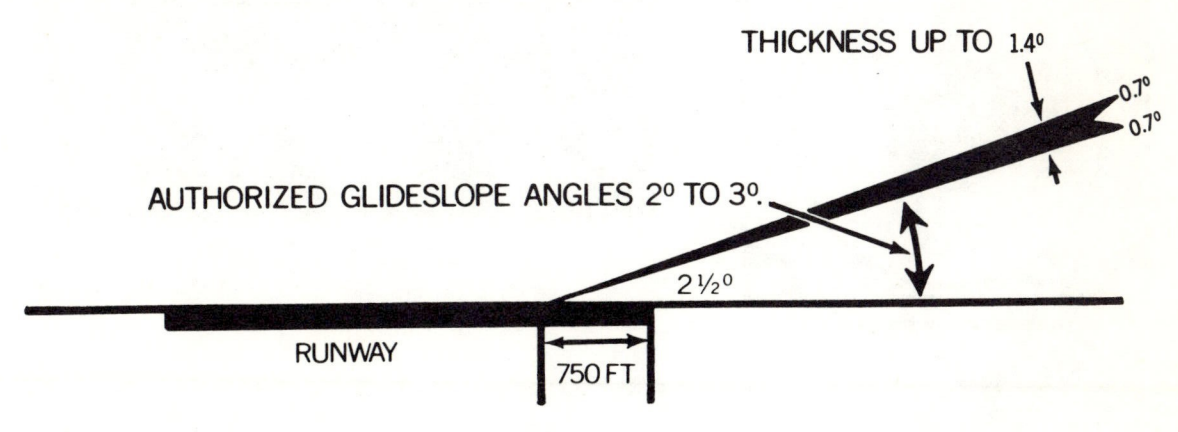

Figure 15–4. Typical glideslope transmitter installation.

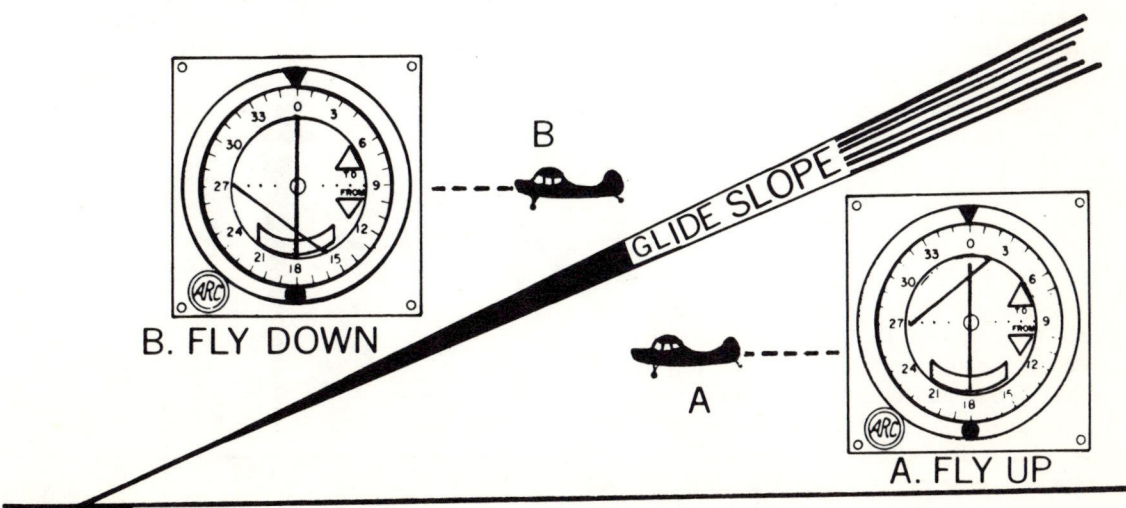

Figure 15–5. Directional aspect of glideslope indicator.

the glideslope, the needle is deflected to the "fly up" position (fig. 15–5). Although the needle reactions are referred to as "fly up" and "fly down," the actual corrections are made by the coordinated use of aircraft pitch controls and power settings to adjust the attitude and rate of descent of the aircraft.

*c Tuning.* Since glideslope and localizer frequencies are paired in the predesignated combinations, tuning of the glideslope transmitter is accomplished automatically and simultaneously with the tuning of the localizer on certain receiver sets. The Army receiver control panel ARN–30E provides for this arrangement. When the frequen-

## Marker Beacons and Compass Locators

*a. Marker Beacons.* A marker beacon is a radio facility capable of transmitting a signal in a vertical direction only. Its signal is received only while flying directly over the facility (fig. 15–7). The primary purpose of the marker beacon is to provide the aviator with a definite radio position fix. The horizontal cross section of the vertical radiation pattern of a marker beacon resembles either an ellipse or a bone (fig. 15–8). The type used with ILS is the elliptical pattern. It is quite narrow so that an aircraft will pass through the

(B) ARN-30A RECEIVER CONTROL PANEL.

(A) GLIDESLOPE CONTROL PANEL.

*Figure 15–6. Coordinated tuning of glideslope receiver and VOR receiver (ARN–30A).*

cy of 108.7 MHz is tuned on the ARN-30E, the corresponding frequency of 330.5 MHz is automatically tuned on the glideslope receiver. On some aircraft there is a separate control panel for the glideslope receiver. However, instead of being graduated with the actual glideslope frequencies ranging from 329.6 MHz to 335.0 MHz, the control panel is graduated with the corresponding localizer channel. Therefore, to tune both receivers the pilot would set 108.7 MHz on each control panel (fig. 15-6).

*d. Signal Reception.* Prior to tuning the glideslope receiver, the glideslope *OFF* flag is visible at the right end of the horizontal needle of the ID–453 course indicator. When the flag disappears, the signal is being reliably received. At any time the signal fails, the *OFF* flag will reappear. Interception and tracking of the glidescope signal is discussed in Approach paragraph later in this chapter.

pattern rapidly, thereby insuring the accuracy of the fix. Since all marker beacons transmit on a frequency of 75 MHz, the receiver is preset to a 75 MHz frequency to receive signals from any beacon. The marker beacon signal is modulated with a coded (or continuous) audio frequency for identification purposes.

*Figure 15–7. Marker beacon signal pattern (vertical cross section).*

# FAN SHAPED

# BONE SHAPED

*Figure 15–8. Marker beacon signal pattern (horizontal cross section).*

The marker beacon receiver is arranged so that the signal can be either heard in the headset or seen as a marker beacon light on the aircraft's instrument panel, or both.

*b. Outer Marker.* On a standard front course ILS approach, the outer marker is normally the final approach fix and is located on the localizer course, usually at a distance of 4 to 7 miles from the end of the runway. The ideal distance is approximately 4.5 miles (fig. 15–9). The outer marker transmits a signal of continuous dashes modulated at an audio frequency of 400 cps (low pitched). On final approach the aircraft will intercept the glide-slope at or near the outer marker (fig. 15–9).

*Note.* On instrument approach charts, the altitude of the glide-slope at the outer marker will be shown.

*c. Middle Marker.* The middle marker is located approximately 3,500 feet from the approach end of the runway (fig. 15–9). The decision height (DH) is normally reached at or near this point.

*d. Compass Locators.* A compass locator is a nondirectional homing beacon. It is used with the ADF receiver and is a supplementary component of the ILS. Two normally are used—one at the outer marker site and another at the middle marker site (fig. 15–10). The combination of a marker beacon and compass locator at the outer marker site is referred to in navigation publications as an *LOM;* i.e., *L*—compass locator and *OM*—outer marker. *LMM* refers to a compass locator and middle marker beacon situated together at the middle marker site. Compass locators are low-powered facilities which normally have a reliable range of 15 miles. These locators are used in two ways: (1) they aid the aviator in transitioning to the outer marker from a fix or facility in the area, and (2) the compass locator feature may be used as a substitute for either marker beacon if the marker beacon or marker beacon receiver fails. Compass locators transmit identification in code. The outer compass locator transmits the first and second letters of the basic airport identification (e.g., the outer compass locator at Cairns AAF (OZR) would transmit OZ as its identification). The middle compass locator (when installed) transmits the second and third letters of the basic airport identification.

## Approach Lights

Several different approach lighting systems are used at airports as an integral part of the ILS. Current navigation publications contain characteristic features of specific systems; e.g., approach lights and high-intensity approach lights. However, all approach lighting

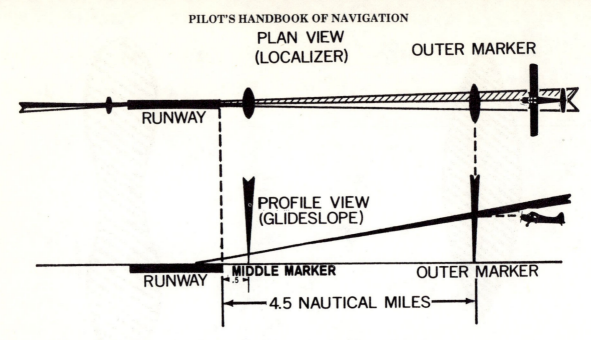

Figure 15–9. Outer and middle marker locations and signal pattern.

systems are designed for the same purpose—to provide the aviator with a visual aid in completing the last phase of his approach. The electronic aids are designed to guide the aircraft to reasonably low ceiling and visibility minimums; e.g., 200 feet and ½ mile. The lighting aids assist the aviator in establishing visual contact with the runway and in com-

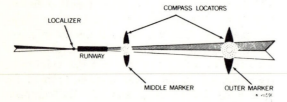

Figure 15–10. Compass locator positions.

pleting the approach visually. Approach lights are situated in the final approach path within the last ¼ to ½ mile of the final approach. When the approach lighting system fails, the ceiling and visibility minimums for the instrument approach are normally increased. (Consult approach charts and their legends for specific airport approach lighting data.)

**Transition**

Figure 15–11 shows Cairns AAF with the localizer course and LOM in relation to the airways network, and the related facilities in the area. The sequence of arrival in the area, transition to the ILS, and the ILS approach (without the use of radar) might typically proceed as follows:

*a.* Aircraft 64264 is inbound to Hartford intersection (the clearance limit) from the southwest on airway V–241. The pilot has been instructed to establish contact with Cairns approach control over Darlington intersection:

(1) Pilot: "Cairns approach control, 64264—estimating Hartford intersection at 30 at 3,000, over."

(2) Approach control: "Roger 64264, you are cleared from over Hartford intersection to the Cairns LOM direct; maintain 3,000 feet. Report Hartford intersection. Current weather: measured ceiling 300, visibility ¾, light rain, fog, altimeter 29.98."

(3) Pilot : " 64264, Roger, out."

*b.* The pilot analyzes the clearance and observes that he has not been cleared to hold. The controller is expecting to clear him for the approach with no delay. He also notes that he has been cleared to continue at 3,000 feet and, by referring to the approach chart (fig. 15–12), that the procedure turn altitude is 2,000 feet. The approach chart also shows the distance and direction from Hartford intersection to the LOM (9.5 miles and 313°). The pilot knows he will be within range of the compass locator signal before arriving at the Hartford intersection.

*c.* The pilot tunes the ADF receiver to the outer compass locator (212 KHz) and receives the coded identification (OZ). The aircraft (Army 64264) is equipped with only one VHF navigation receiver (ARN–30A)

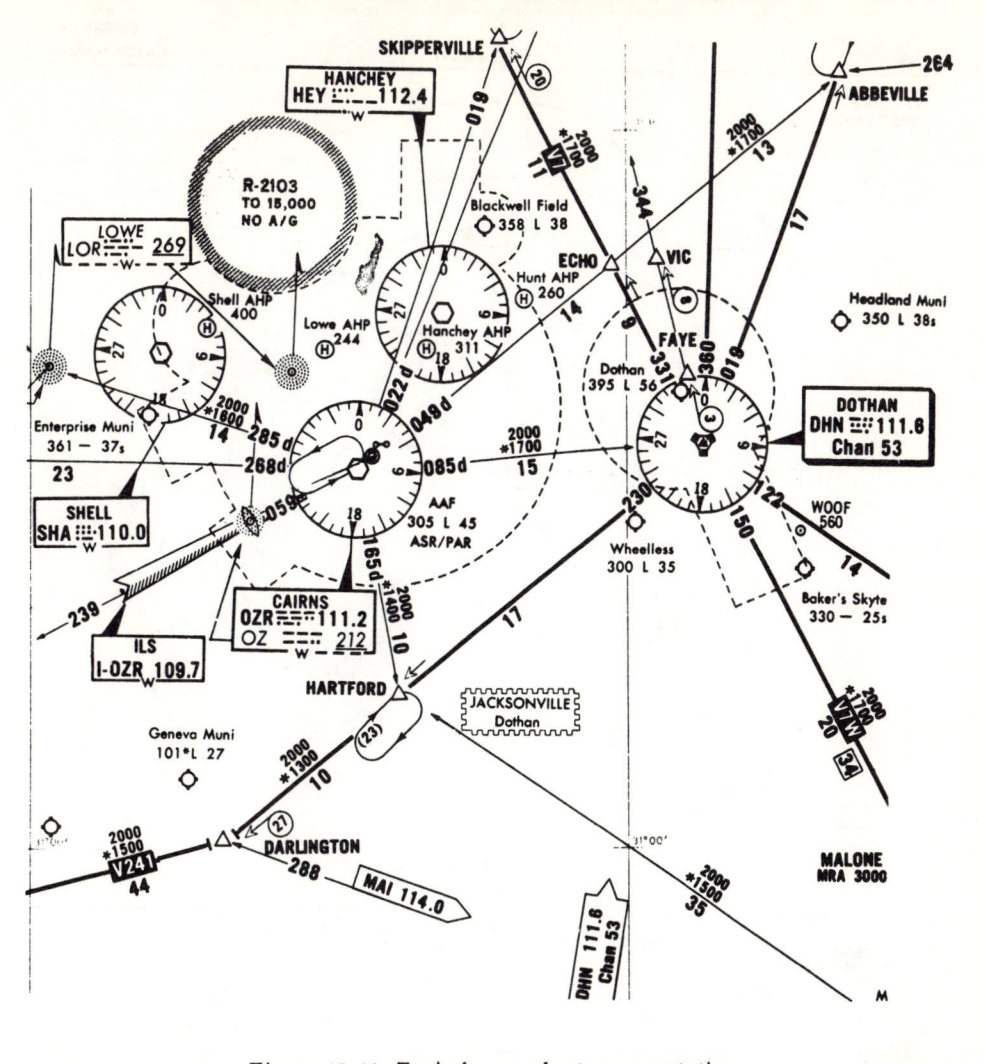

*Figure 15–11. Typical area chart representation.*

which remains tuned to Dothan VOR until near Hartford intersection. This aircraft has a glideslope receiver and, although the pilot is not in a position to use it yet, he tunes the glideslope receiver by setting the 109.7 MHz frequency in the window. The actual glideslope frequency of 333.2 MHz is paired with the localizer channel of 109.7 MHz. The glideslope warning flag will not disappear until the aircraft is in position to receive a reliable signal from the glideslope transmitter.

*d.* Upon arrival at Hartford intersection, the pilot turns to 313° and reports—

(1)    Pilot : " 64264 Hartford intersection 30. . . ."

(2) Approach control: "Roger 264, you are cleared for an ILS approach to runway 06, wind zero five zero degrees at 6 kt, over."

(3)    Pilot : " 264, Roger, out."

*e.* At this point the pilot performs each of the following actions:

(1)    Cross-checks heading (313°) and ADF relative bearing (0°) to verify that aircraft is proceeding directly to the LOM.

(2)    Tunes VHF navigation receiver (ARN–30A) to 109.7 MHz, setting the selector switch to VAR LOC position. Verifies identification IOZR, checks for disappearance of *OFF* flag, and observes that the needle is deflected to blue side of instrument.

(3)    Checks minimum authorized transition altitude (2,000 feet (fig. 15–12)), begins descent to 2,000 feet, and reports leaving 3,000 feet. This is the recommended procedure because the pilot is already cleared for the approach and can begin losing altitude inbound to the LOM.

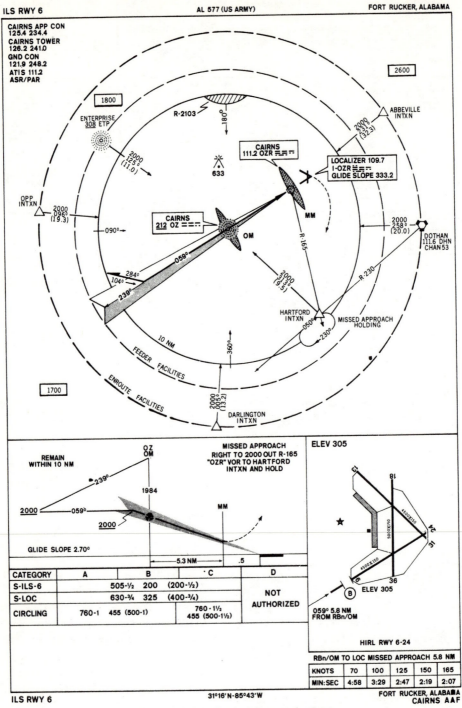

*Figure 15–12. Typical approach chart.*

(4) Checks the indications of the localizer needle. It should be in the blue sector (full left deflection) and should move rapidly when the aircraft gets within 2½° (5° localizer) of the localizer course.

(5) Cross-checks power switch on marker beacon receiver to verify that it is turned on with appropriate volume setting. The marker beacon signal (continuous dashes—low pitched) will be received as the aircraft flies over the LOM.

(6) Cross-checks the glideslope warning flag. As the aircraft flies into a good reception position with the receiver tuned, the flag will disappear

(7) Checks the instrument approach

chart, noting the direction, limiting distance, and minimum altitude for the procedure turn. (The pilot has previously studied the approach and is familiar with it.) In this situation, the procedure turn is to the right, within 10 miles of the LOM, and at a minimum altitude of 2,000 feet. Also, the pilot has previously determined that his minimums for this approach are based on a prevailing visibility value of ½ mile and, since the current weather is 300 feet and ¾ mile, he does not expect any difficulty in completing the approach visually.

*f.* After flying the 9.5 miles from Hartford intersection, the pilot detects passage of the LOM by reversal of the ADF bearing pointer and simultaneous reception of the marker beacon signal (light and tone). Also, the localizer needle has moved through the center position to the yellow sector.

**Approach**

With the transition to the ILS complete, the pilot r flies outbound for the procedure turn.

*a.* During this phase the pilot —

(1) Turns to the outbound heading (239°) and checks the position of the localizer needle and ADF indicator. The color indication of the needle is correct, but while flying outbound on the front course, track corrections are made by turning away from the needle. If the needle deflects to the yellow (right), the pilot turns left. The ADF bearing pointer will be approximately on the tail (180°) but will deflect if the aircraft flies off track. If the aircraft drifts into the yellow sector while maintaining the outbound heading of the front course, the ADF will show a relative bearing greater than 180° (left of tail).

(2) Flies outbound approximately 1 minute. The actual limiting distance (usually 10 miles) is shown on the profile of the instrument approach chart.

(3) Since the descent to 2,000 feet has been completed, he maintains 2,000 feet.

(4) Executes the procedure turn by turning (in this case) to the right 45° to a heading of 284°. Holds this heading for approximately 40 seconds (ch. 13), and then completes the procedure turn by turning left to intercept the localizer course inbound.

(5) Upon intercepting the localizer inbound, continues inbound on the approach course (059°).

(6) Checks the localizer needle (centered) and ADF bearing pointer (relative bearing 0°). If the aircraft flies off track, corrects by turning toward the localizer and ADF needle.

(7) Checks the position of the glidepath needle. It should be above center ("fly up" position), indicating that the actual glidepath is above the procedure turn altitude. Notes that the profile view on the approach chart (fig. 15–12) indicates that the glidepath will be intercepted just prior to passing the LOM inbound. Rechecks to verify that the glidepath *OFF* flag is out of sight. Begins descent on glidepath.

*b.* Upon reaching the LOM inbound, the pilot observes the marker beacon signal, ADF reversal, checks for centered glidepath needle at 1,984 feet indicated altitude, and reports—

(1) Pilot: "Cairns approach control, 64264 outer marker inbound at 41, over."

(2) Approach control: "Roger 264, cleared to land runway 06. Wind: zero four five degrees at 8 kt, altimeter 29.97, over."

(3) Pilot: " 264, Roger, out."

*c.* During this final phase of the approach, the aviator—

(1) Sets up the proper rate of descent on the glidepath for final approach. This rate can be achieved by proper coordination of aircraft controls and power.

(2) Checks the glidepath pointer. Makes glidepath corrections by following the needle (i.e., "fly up" and "fly down"). Corrections are made by varying the rate of descent through the coordinated use of aircraft controls and power.

(3) Checks the deviations of the localizer needle. Only minor corrections should be necessary to keep the needle in the centered position.

(4) Observes the altimeter. The approach chart (fig. 15–12) shows the decision height to be 505 feet. This is the lowest *indicated* altitude to which the aircraft may descend on the glideslope unless the pilot establishes visual contact with the runway environment and can reasonably expect to maintain visual contact throughout the landing.

*Note.* For a localizer only approach, the pilot would descend to the minimum descent altitude (MDA) of 630 feet (fig. 15–12). After passing the final approach fix, he will continue toward the field until the missed approach point is reached. The missed approach point is based on distance from the FAF and is determined by the groundspeed/time relationship shown on the approach chart.

(5) Checks for visual contact with approach lights. At the time definite visual contact is established, the approach can be completed without further reference to localizer and glidepath instruments.

(6) Observes middle marker passage. The marker beacon signal will be medium-pitched alternating dots and dashes.

(7) If visual contact is established, completes the approach.

(8) If visual contact is not established, or if for other reasons the approach cannot be completed, executes the missed approach procedure as published on the approach chart (fig. 15–12). In this event the pilot would report, "Cairns approach control, 64264 missed approach at (time of missed approach) . . . ." This report would be followed by a request for another approach or a clearance to the alternate airfield. If requesting clearance to the alternate airfield, the pilot would give the necessary flight plan data.

### Runway Visual Range (RVR)

*a.* A number of airports are equipped with transmissometers and related equipment for determining the distance that can be seen down the runway from the approach end. If the determination is based on *high-intensity* runway lights, this distance is known as the RVR. RVR values have proven to be accurate and reliable; consequently approach minimums, in many cases, will be based on RVR.

*b.* Figure 15-13 shows a published DH for a straight-in ILS of 207/24. This means that an RVR value of 2,400 feet is authorized as a minimum for beginning the approach. After starting the approach, the pilot must not descend

below the DH published for the approach unless visual contact is established and the landing can be made as described in Approach paragraph. Where the pilot can approach to within 200 feet of the lights, he usually will have little difficulty in establishing visual contact if the reported RVR is equal to or better than the published RVR minimum.

*Note.* Refer to current navigational publications and regulations for additional information on RVR. As improved equipment becomes operational, changes in procedures and regulations will occur.

### Use of the Back Course

The localizer transmitter produces a front and a back course; the front course is usually equipped with all required components for a complete ILS. The back course is frequently used as an additional approach course, but it provides azimuth guidance only.

Altitude on the final approach is controlled with reference to the altimeter, as in VOR and ADF approaches. An approach fix must be established at some suitable location on the back course. Such a fix is necessary to establish minimums for the procedure turn and the descent on final approach. Figure 15–4 illustrates a back course with a fix established by a radial from a nearby VOR station. Since the localizer and the VOR are both VHF facilities with only one receiver available, it is necessary to tune first one facility and then another. Use of dual VHF navigation receivers solves this problem; in most instances, the approach procedure is restricted to aircraft equipped with dual receivers. Notations on the approach chart indicate such restrictions.

| CATEGORY | A | B | C | D |
|---|---|---|---|---|
| S-ILS-5 | | 207/24 200 (200-½) | | |
| S-LOC-5 | | 360/24 353 (400-½) | | |
| S-VOR-5 | | 400/24 393 (400-½) | | |
| CIRCLING | 440-1 433 (500-1) | 460-1 453 (500-1) | 460-1½ 453 (500-1½) | 560-2 553 (600-2) |

*Figure 15–13. Approach minimums based on RVR.*

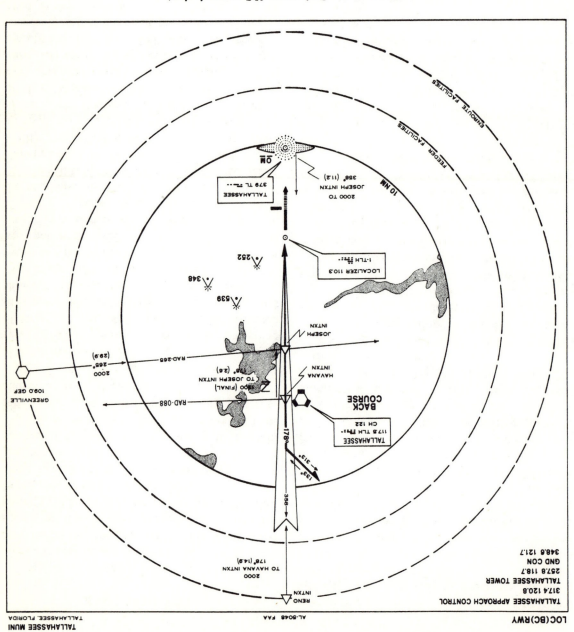

*Figure 15-14. Back course ILS approach chart.*

# 16 RADAR PRINCIPLES AND FLIGHT

## BASIC PRINCIPLES

The manner in which a bat successfully avoids the walls and jutting stalactites of a totally dark cave makes fascinating reading. Investigation has shown that if its mouth is gagged or its hearing impaired the bat can no longer avoid such obstacles. From this, it has been concluded that the bat's uncanny ability to navigate derives from the emission of cries—inaudible to the human ear—which are then reflected from any obstacles in its path. These reflected echoes allow the bat to orient himself with respect to the obstacles and thus maneuver to avoid them.

The fundamental principle of radar is closely akin to that used by the bat. The basis of the system has been known theoretically since the time of Hertz, who in 1888 successfully demonstrated the transfer of electromagnetic energy in space and showed that such energy is capable of reflection. The *transmission* of electromagnetic energy between two points was developed as "radio," but it was not until 1922 that practical use of the *reflection* properties of such energy was conceived. The idea of measuring the elapsed time between the transmission of a radio signal and receipt of its reflected *echo* from a surface originated nearly simultaneously in the United States and England. In the United States, two scientists working with air-to-ground signals noticed that ships moving in the nearby Potomac River distorted the pattern of these signals. In 1925, the same scientists were able to measure the time required for a short burst, or pulse, of radio energy to travel to the ionosphere and return. Following this success, it was realized that the radar principle could be applied to the detection of other objects, including ships and aircraft.

By the beginning of World War II, the Army and Navy had developed equipment appropriate to their respective fields. During and following the war, the rapid advance in theory and technological skill brought improvements and additional applications of the early equipment. By suitable instrumentation it is now possible to measure accurately the distance and direction of a reflecting surface in space—whether it is an aircraft, a ship, a hurricane, or a prominent feature of the terrain —even under conditions of darkness or restricted visibility. For these reasons, radar has become a valuable navigational tool.

There are many different types of radar sets currently in use in the Air Force. For the navigator, they can be divided into two classes: (1) those designed for navigation, search, and weather identification, to which no computer has been attached, and (2) those designed principally for use with high accuracy bombing-navigation computers. With the former, fixing can be accomplished only by using ranges or bearings, or both. With the latter, the position of the aircraft in latitude and longitude is continuously displayed on the navigation computer which is periodically corrected by the radar.

In view of the great variety of radar sets in use, it would be impractical to cover in this manual the specific operating procedures for each. Therefore, the following material is limited to the *general* procedures for using radar as an aid to navigation. For detailed information concerning a specific set, consult the appropriate manual or technical order.

As noted previously, the fundamental principle of radar may be likened to that of relating sound to its echo. Thus, a ship sometimes determines its distance from a cliff at the water's edge by blowing its whistle and timing the interval until the echo is received. Since sound travels at a known speed (approximately 1,100 feet per second in air), the total distance traveled equals the speed times the time interval, and one-half this distance (because the sound must travel to the cliff and *back* before the echo is heard) is the distance to the cliff. Thus, if the time interval is 8 seconds, the distance to the cliff is

$$\frac{1,100 \times 8}{2} = 4,400 \text{ ft} = 0.725 \text{ NM}$$

The same principle applies to radar, which uses the reflected echo of electromagnetic radiation traveling at the speed of light. This speed is approximately 162,000 nautical miles per second; it may also be expressed as 985 feet per microsecond. If the interval between the transmission of the signal and return of the echo is 200 microseconds, the distance to the target is

$$\frac{985 \times 200}{2} = 98,500 \text{ ft} = 16.2 \text{ NM}$$

## TYPICAL RADAR SET

### Components

In addition to the power supply, a radar set contains five major units: the timer, the transmitter, the antenna, the receiver, and the indicator as shown in figure 16-1. Their functions are briefly described as follows:

The *timer,* or modulator, is the heart of the radar system. Its function is to insure that all circuits connected with the system operate in a

It is composed of an electron gun, a focusing magnet, and a set of deflection coils. The function of the electron gun is to produce a thin electron stream, or beam. This beam is so acted upon by the focusing magnet that the focal point of the beam in a properly adjusted CRT is at the face of the tube. The CRT face is coated with a fluorescent compound which glows when struck by the electron beam. The picture seen is thus a small bright spot.

The location of an object by means of these five units of a radar set involves the simultaneous

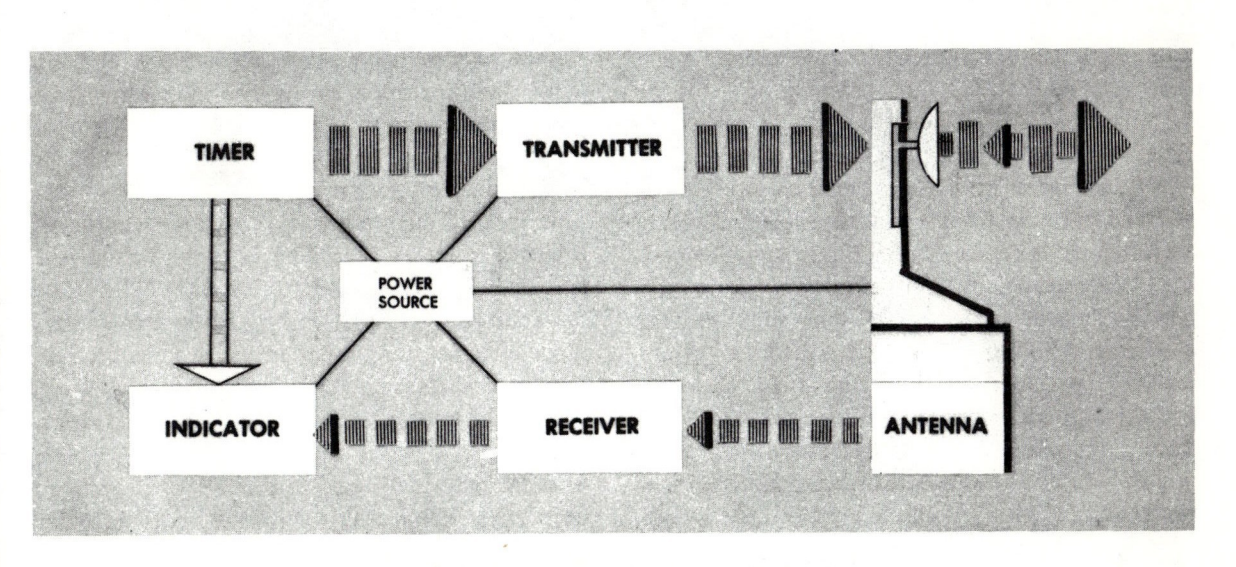

Figure 16-1 *Components of a Typical Radar System*

definite time relationship with each other and that the time interval between transmitted pulses is of the proper duration. The timer "triggers" the other units into operation.

The *transmitter* is similar to the radio transmitters described in the chapter on radio except that it operates in the super high frequency band and operates for very short periods of time.

The *antenna* is a "parabolic dish" which directs the transmitted energy into a narrow wedge or into a conical beam like that of a searchlight. It serves not only for transmission but for reception as well.

The *receiver,* like the transmitter, is similar to a radio receiver. Its main function is to amplify the relatively weak echoes which return to the antenna.

The *indicator* for the radar set is a special vacuum tube called a *cathode ray tube (CRT).* It is similar to the picture tube of a television receiver. The CRT is commonly called the *scope* (from oscilloscope). An electromagnetic CRT, the type used in most radar sets, is illustrated in figure 16-2.

solution of two separate problems. First, the time interval between the transmitted pulse and its echo must be accurately measured and presented as *range* on the indicator. Second, the direction in which each signal is transmitted must be indicated, giving the *bearing* of the object.

### Measurement of Range

The radar cycle may best be explained by discussing first the life of a single pulse as it is affected by each of the components shown in figure 16-1 . The cycle begins when the timer triggers the transmitter. A powerful pulse of radio energy is generated there which is then emitted from the antenna. As soon as the pulse leaves the antenna, the antenna is automatically disconnected from the transmitter and connected to the receiver. If an echo returns, it is amplified by the receiver and sent to the indicator for display.

At the same time that the timer triggers the transmitter, it also sends a trigger signal to the

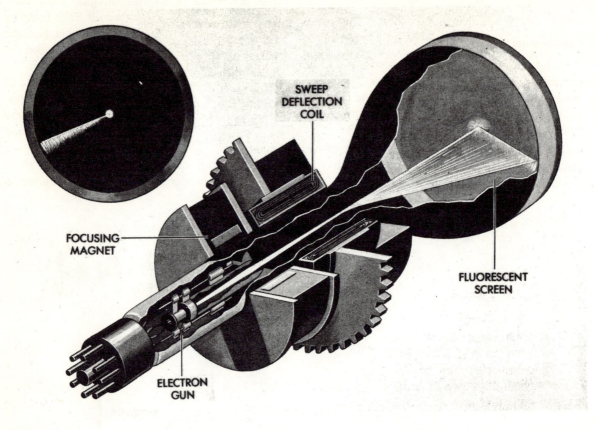

*Figure 16-2  Electromagnetic Cathode Ray Tube (CRT)*

indicator.  Here, a circuit is actuated which causes the current in the deflection coils to rise at a linear (uniform) rate.  The rising current, in turn, causes the spot to be deflected radially outward from the center of the scope.  The spot thus traces a faint line on the scope; this line is called the *sweep*.  If no echo is received, the intensity of the sweep remains uniform throughout its entire length.  However, if an echo is returned, it is so applied to the CRT that it intensifies the spot and brightens momentarily a short segment of the sweep.  Since the sweep is linear and begins with the emission of the transmitted pulse, the point at which the echo brightens the sweep will be an indication of the range to the object causing the echo.

The progressive positions of the pulse in space also indicate the corresponding positions of the electron beam as it sweeps across the face of the CRT.  If the radius of the scope represents 40 miles and the "return" appears at three-quarters of the distance from the scope center to its periphery, the target is represented as being about 30 miles away.

Of interest here is the extremely short time scale which is used.  In the preceding example, the radar

is set for 40-mile range operation.  The sweep circuits will thus operate *only* for an equivalent time interval, so that targets beyond 40 miles will not appear on the scope.  The time equivalent to 40 miles of radar range is only 496 microseconds (0.000496 seconds).  Thus, 496 microseconds (plus an additional period of perhaps 100 microseconds to allow the sweep circuits to recover) after a pulse is transmitted, the radar is ready to transmit the next pulse.  The actual pulse repetition rate in this example is about 800 pulses per second.  The return will therefore appear in virtually the same position along the sweep as each successive pulse is transmitted, even though the aircraft and the target are moving at appreciable speeds.

### Measurement of Bearing

The deflection coils, illustrated in figure 16-2, are mounted so they surround the neck of the CRT. *The direction taken by the sweep corresponds directly to the orientation of the deflection coils.* In the previous discussion of range, it was assumed that the deflection coils were stationary, so that the sweep formed by the deflection current always fell along the same radial. However, if the antenna (which forms the transmitted energy into a narrow beam) is rotated slowly in azimuth and the deflection coils are rotated mechanically in synchronism with the antenna, the sweep trace will point in the same direction as the antenna. Therefore, the relative bearing, as well as the range, of reflecting objects will be shown on the scope.

## Plan Position Indicator (PPI)

The type of cathode ray tube display just discussed is called a *plan position indicator (PPI)*. It presents a map-like picture of the terrain below and around the aircraft. This is the presentation usually available in radar-equipped aircraft. In figure 16-3, the center of the scope represents the aircraft radar antenna. A radar target is shown at its correct bearing and at a distance from the scope center proportional to its range. To facilitate the measurement of range, "false" echoes called *fixed range markers* are generated electrically at regular intervals after each transmitted pulse and introduced at the indicator in the same manner as the actual echoes.

Since these markers always appear at the same position on the sweep for a given range setting, they trace concentric range circles on the scope. On some radar sets, the range mark circles are traced at appropriate intervals, such as every 2 miles when maximum range is 5 or 10 miles, every 5 miles on the 30-mile scale, and every 25 miles on the 100- or 200-mile scales.

To facilitate the measurement of bearing, a compass rose surrounds the scope, providing a convenient bearing reference. On most sets, the bearing of any particular return may be determined by bisecting the return with a cursor (a radial line inscribed on a transparent overlay) which can be rotated manually about an axis coincident with the center of the scope. The corresponding bearing is then read on the compass rose at the periphery of the scope.

In the plan position indicator illustrated in figure 16-3, the range scale is 100 miles; the range marker interval is 25 miles. Thus, the large target shown is at a relative bearing of 060° at a range of 50 miles.

## SCOPE INTERPRETATION

The plan position indicator presents a map-like picture of the terrain below and around the aircraft. Just as map reading skill is largely dependent upon

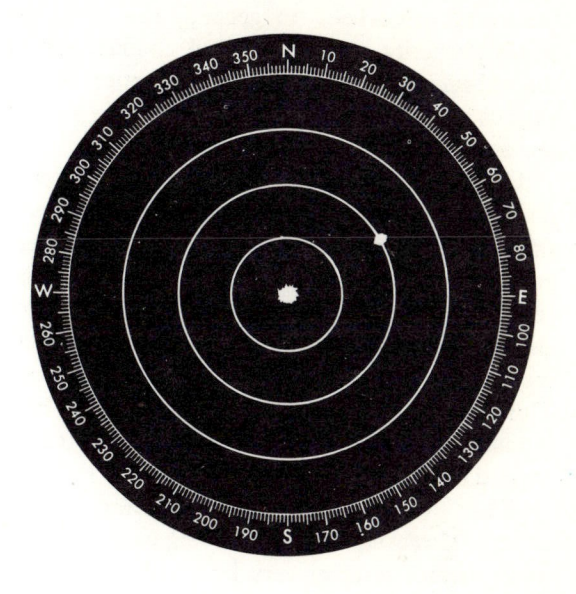

Figure 16-3 Plan Position Indicator (PPI)

the ability to correlate what is seen on the ground with the symbols on the chart; so "radar" skill is largely dependent upon the ability to correlate what is seen on the scope with the chart symbols. Accordingly, a sound knowledge of the factors affecting radar reflection is a prerequisite to intelligent radarscope interpretation. Furthermore, a knowledge of these factors, applied in reverse, enables the navigator to *predict* the probable radarscope appearance of any area.

## Factors Affecting Reflection

There are four primary factors which determine whether a usable portion of the transmitted energy will be reflected from a given object. No one factor, however, determines the amount of reflection; total reflection is based on a combination of all four factors. These factors are:

- Vertical and horizontal size.
- Material of the object.
- Radar range.
- Reflection angle.

VERTICAL AND HORIZONTAL SIZE. Assuming equal width, the *taller* of two structures presents the greater reflecting area. Its radar reflection potential is therefore greater.

Assuming equal height, the *wider* of two structures presents the greater reflecting area. This will cause it to appear as the larger return on the scope.

MATERIAL OF THE OBJECT. It is a basic law of

physics that all substances reflect some electromagnetic energy. All substances also absorb some electromagnetic energy. Whether or not a substance may be considered a good radar reflector depends upon the ratio of the energy reflected to the energy absorbed. In general, this ratio depends primarily on the electrical conductivity of the substance. The reflection properties of the various materials from the best through the poorest are as follows:

*Metals.* Because of their high electrical conductivity, steel, iron, copper, aluminum, and most other metals are excellent reflectors of radio energy. A metal-clad building usually has high radar reflection potential. Those buildings which use steel or some other metal as structural reinforcement follow closely behind metal-clad buildings. This group includes concrete structures built around steel frameworks, commonly called "mixed" structures.

*Water.* Water is an excellent reflector of radio energy. However, an aircraft flying over a smooth water surface receives very little reflected energy from the surface because the angle of reflection is away from the aircraft. If the sea surface is not smooth, considerable energy is reflected resulting in large areas of *sea clutter* similar to *ground clutter*.

*Masonry.* Masonry by itself is a fair reflector. Masonry materials include stone, concrete, and clay products. However, masonry is often mixed with structural steel. These mixed structures have greater radar reflection potential than masonry alone.

*Wood.* In comparison to metal or masonry, wood is a poor reflector. Wooden structures do not normally reflect sufficient radar energy for display unless other important reflecting factors (such as unusual size) are present.

*Dirt, Sand, and Stone.* Although their reflection potential will vary to some extent with the particular chemical composition involved, dirt, sand, and stone are not normally considered good reflecting materials. Also, radar return from flat land areas is not nearly so great as that from land which exhibits irregular surfaces, such as sand dunes, hills, gravel pits, quarries, and similar prominences.

*Glass.* Glass by itself is not a good radar reflector. However, structures using glass walls usually have steel framing and such structures often can be considered among the better reflectors.

The diagram in figure 16-4 shows the relative reflection potential of the basic structural materials. These relationships will hold true, of course, only on the assumption that the other factors affecting reflection—size, range, and reflection angle—are equal.

RADAR RANGE. Range affects radar reflection in two ways. First, the *nearer* of two objects of equal size occupies a greater percentage of the beam width and so has more energy incident upon it. It will therefore appear as the brighter return. Second, as the radar waves travel through space,

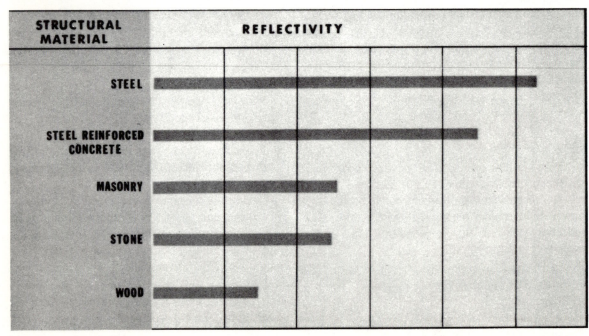

*Figure 16-4  Relative Reflectivity of Structural Materials*

many minute particles of dust and other foreign matter intercept the beam. The resulting absorption decreases the strength of, or attenuates, the signal. The farther the energy travels, the more attenuation occurs. The attenuation effect is compensated for by automatic increase of the power output as the operating range is increased to the next higher increment.

REFLECTION ANGLE. In all cases of true reflection, the angle of reflection is equal to the angle of incidence. Therefore, maximum reflection back to a radar set occurs when the reflecting surfaces are at right angles to the radar beam. For this reason, flat surfaces can give a momentarily intense directional reflection, just as the windshield of an automobile does when it flashes in the sunlight. However, such radar echoes, though very strong, are normally not usable for navigational purposes because the brevity of their existence does not allow sufficient time for identification. Fortunately, the great majority of radar targets are multifaceted rather than smooth, so that at any given instant some part of the target will be at right angles to the beam. The more closely the incident wave approaches a direct right angle with respect to the overall surface, the greater will be the radar reflection.

*Horizontal.* The effect of the horizontal reflection angle is shown in figure 16-5. Because of the particular configuration between the aircraft and reflecting objects, the aircraft at *A* gets no return from building 1 and maximum return from build-

ing 2 and the railroad tracks. At *B,* the aircraft gets no reflection from building 2 and maximum reflection from the railroad tracks. At *C,* maximum reflection is again being obtained from building 2. This entire discussion oversimplifies the problem for purposes of clarity. It should be emphasized again that smooth surfaces such as those shown are rarely encountered, and for normal structures, the amount of reflection would be *minimum* where *zero* reflection has been implied.

*Vertical.* The effect of the vertical reflection angle is not nearly so pronounced as that of the horizontal reflection angle. This is because of the much greater beam width in the vertical dimension as compared to the horizontal when the normal mapping beam is used. The vertical beam width may be on the order of 50 degrees and the horizontal beam width on the order of 1½ degrees. The effect of the vertical reflection angle cannot, however, be neglected. Figure 16-6 shows that the vertical energy pattern contains a main lobe. To obtain maximum reflection, this main lobe should strike the overall reflecting surface at right angles to it. In most buildings, the overall reflecting surface is vertical. Thus, as an aircraft approaches a building or group of buildings, the actual vertical reflection angle becomes more and more removed from the optimum. The final result is that, when the vertical reflection angle approaches 45 degrees, many of the structures scanned by the beam do not reflect enough energy back to the set to appear on the scope (see figure 16-7). The zone in which

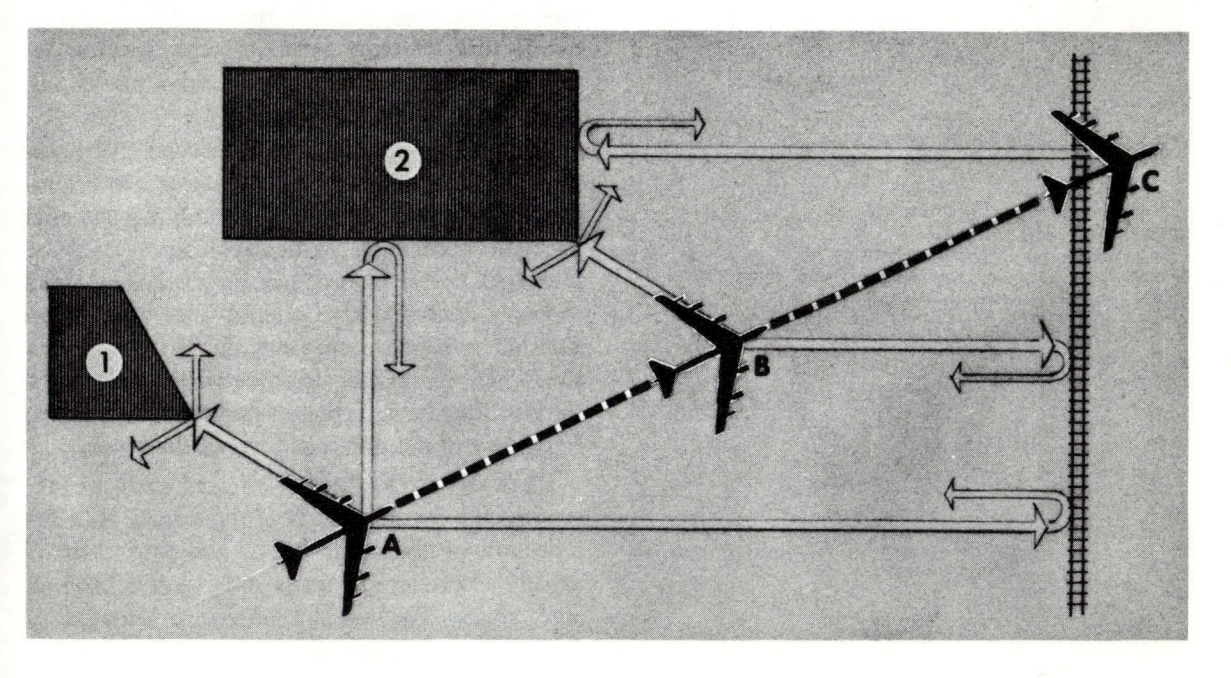

*Figure 16-5 Horizontal Reflection Angle*

this lack of reflection is likely to occur is equal in radius to the altitude of the aircraft. For example, an aircraft flying at extremely high altitudes might display only the returns from the very best types of reflectors—such as steel mills—in the area within eight or ten miles from the aircraft.

Adjustment of the antenna tilt control can compensate to a degree for the above effect. By tilting the antenna downward as the target is approached, more energy is made to strike the target at the optimum reflection angle. However, this reduces the overall coverage of the radar scan because the energy directed out ahead of the aircraft is reduced.

SUMMARY. It is important to re-emphasize that no one factor determines the reflection potential of a given reflector. However, a single factor may prevent sufficient reflection to create a return on the scope. The following general rules will apply:
1. The greatest return potential exists when the radar beam forms a horizontal right angle with the frontal portion of the reflector.
2. Radar return potential is roughly proportional to structure size and the reflective properties of the construction material used.
3. Radar return potential is greatest within the zone of the greatest radiation pattern of the antenna, as illustrated in figure 16-6.
4. Radar return potential decreases as altitude increases because the vertical reflection angle becomes more and more removed from the optimum. (There are many exceptions to this general rule since there are many structures which may present better reflection from roof surfaces than from frontal surfaces.)

5. Radar return potential decreases as range increases because of the greater beam width at long ranges and because of atmospheric attenuation.
6. All of the factors affecting reflection must be considered to determine the radar return potential.

**Typical Scope Returns**

There are certain characteristics peculiar to returns from specific ground features such as rivers, lakes, mountains, railroads, cities, etc. To consolidate the knowledge of the general rules on the factors affecting reflection, this section describes and illustrates how these rules may be applied to specific ground features.

The principal problem in radarscope interpretation is finding the meaning of contrasts in brightness. This comes ahead of the purely navigational problem because a particular feature must be identified before it becomes useful. Figure 16-8 shows a representative ground area and its associated radarscope presentation. In the discussion which follows, it would be well to refer back to this illustration, noting in particular the relative brightness of the different parts of the scope.

RETURNS FROM LAND. All land surfaces present minute irregular parts of the total surface for reflection of the radar beam, and thus there is usually a certain amount of radar return from all land areas. The amount of return varies considerably according to the nature of the land surface. This variance is caused by (1) the difference in reflecting materials of which the land area is composed, and (2) the texture of the land surface. These are the primary factors governing

*Figure 16-6 Radiation Pattern of Antenna*

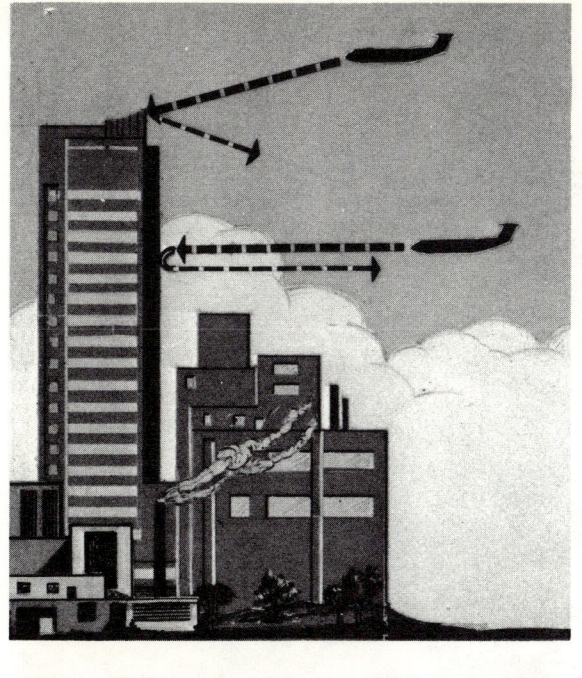

*Figure 16-7 Vertical Reflection Angle Varies With Aircraft Position and Altitude*

the total radar return from specific land areas.

*Flat land.* A certain amount of any surface, however flat in the overall view, is irregular enough to reflect the radar beam. Surfaces which are apparently flat are actually textured and may cause return on the scope. Ordinary soil absorbs some of the radar energy and thus the return that emanates from this type of surface is not strong. Ray *B* in figure 16-9 shows that flat land will reflect most of the energy away from the aircraft.

*Plowed Fields.* Plowed fields and other irregularly textured land areas present more surface to the radar beam than flat land and thus create more return. Returns from plowed fields may be a problem to the radar navigator because such returns are not readily identifiable, are often not persistent, and tend to confuse the overall radarscope interpretation problem. The returns from plowed fields and other irregularly textured land areas are most intense when the radar beam scans the upright furrows or similar features at a right angle. This type of return is exemplified by ray *F* in the illustration.

*Hills and Mountains.* Hills and mountains will normally give more radar return than flat land because the radar beam is more nearly perpendicular to the sides of these features. The typical return is a bright return from the near side of the feature and an area of no return on the far side. This is shown by ray *A*. The area of no return is

called a mountain shadow and exists because the radar beam cannot pentrate the mountain, and its line-of-sight transmission does not allow it to intercept targets behind the mountain. The shadow area will vary in size, depending upon the height of the aircraft with respect to the mountain. As an aircraft approaches a mountain, the shadow area becomes smaller and smaller. Furthermore, the *shape* of the shadow area and the brightness of the return from the peak will vary as the aircraft's position changes.

Recognition of mountain shadow is important because any target in the area behind the mountain cannot be seen on the scope. A striking example of this is illustrated in figure 16-10 . Note that the mountains obscure the area in which the town of Adams lies.

In areas with isolated high peaks or mountain ridges, contour navigation may be possible because the returns from such features assume an almost three-dimensional appearance. This allows specific peaks to be identified.

In more rugged mountainous areas, however, there may be so many mountains with resulting return and shadow areas that contour navigation is almost impossible. Note the complexity of the mountain returns in figure 16-11 . In an area such as this, the radar might still be used for general orientation. For example, the aircraft is flying over the line of demarkation between a mountainous and a level area. This line of demarkation might serve as a *line of position* even though the complexity of the scope picture makes positive position finding impossible.

*Coastlines and Riverbanks.* The contrast between water and land is very sharp, so that the configuration of coasts and lakes are seen with map-like clarity. When the radar beam scans the banks of a river, lake, or larger body of water, there is little or no return from the water surface itself, but there is usually a strong return from the adjoining land. This may be seen from rays *C* and *D* in figure 16-9 . The more rugged the bank or coastline, the more return will be experienced.

A nearby shoreline will be very clearly delineated when an aircraft is out over the adjoining water. However, the shoreline tends to lose its clearcut separation from the water when the aircraft is flying over the land. This is because the coastline presents upright banks to the overwater radar beam and relatively flat land to the overland beam. This also explains why the *far* side of a

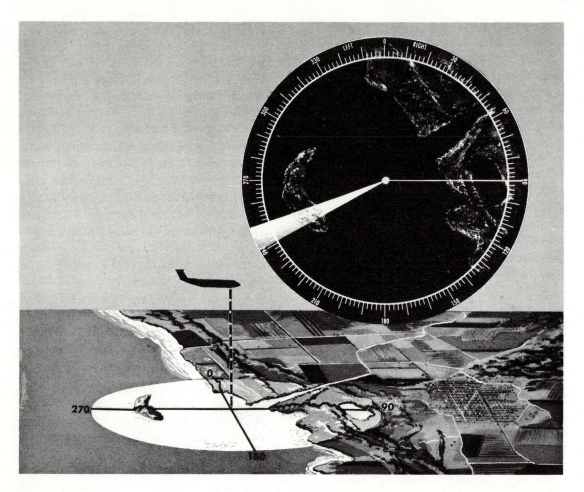

Figure 16-8   Radar Returns

Figure 16-9   Reflections from Various Types of Terrain

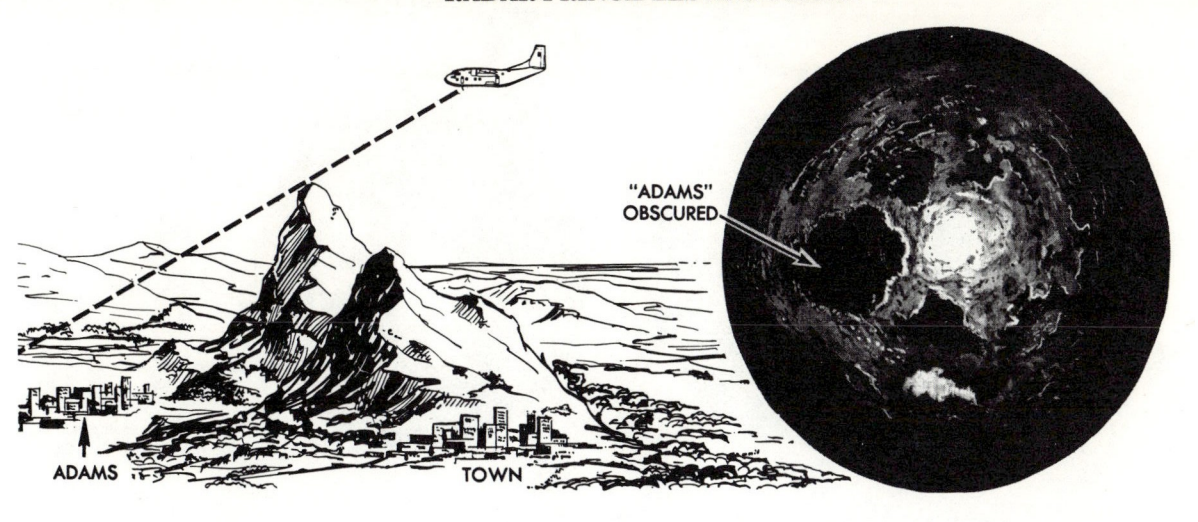

*Figure 16-10  Significant Areas are Sometimes Obscured by Mountains*

lake will appear brighter than the near side. Notice in figure 16-12 that the lake at 320° shows a much stronger return from the far side than from the near side.

Since both mountains and lakes present a "dark" area on the scope, it is sometimes fairly easy to mistake a mountain shadow for a lake.

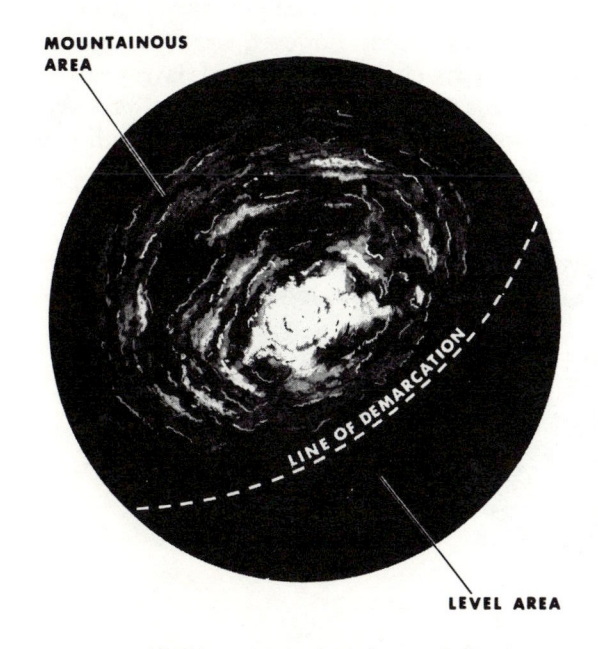

*Figure 16-11  Individual Peaks are Difficult to Identify*

This is particularly true when navigating in mountainous areas which also contain lakes.

The essential difference between returns from mountain areas and lakes is that returns from mountains are bright on the near side and dark on

the far side; returns from lakes are bright on the far side and vague on the near side. Another characteristic of mountain returns is that the dark area changes shape and position quite rapidly as the aircraft moves, returns from lakes change only slightly.

URBAN AREAS.  The overall size and shape of the radar return from any given city can usually be determined with a fair degree of accuracy by referring to a map of the area. However, the brightness of one urban area as compared to another may vary greatly, and this variance can hardly be forecast by reference to the navigation chart. In general, the industrial and commercial centers of cities and towns produce much greater brightness than the outlying residential areas.

ISOLATED STRUCTURES.  Many isolated or small groups of structures create radar returns. The size and brightness of the radar returns these features give are dependent on their construction. Most of these structures are not plotted on the navigation charts; hence, they are of no navigational value. However, some of them give very strong returns— such as large concrete dams, steel bridges, etc.— and if they can be properly identified they give valuable navigational assistance.

GLITTER AND CARDINAL POINT EFFECT.
*Glitter,* Glitter or glint, is the cause of most fluctuating returns that appear on the scope. It occurs when the radar beam scans a relatively small reflector which is able to present sufficient reflection for display on the scope only when the radar beam is at right angles to the structure. Since the aircraft is moving, the conditions necessary for the display of such an object exist only momentarily; hence, the return appears only momentarily on the scope (see figure 16-13).

199

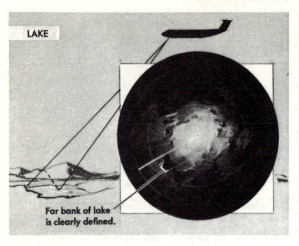

LAKE

Far bank of lake is clearly defined.

MOUNTAIN SHADOWS

Near side of mountain is clearly defined.

*Figure 16-12 Mountains and Lakes*

*Cardinal Point Effect.* This is glitter that occurs when the radar beam scans a large number of structures at right angles. In this case, the return may remain for more than the customary instant. Because most cities and towns are laid out along north-south and east-west axes, cardinal point effect usually occurs when the true bearing of a city is equal to one of the cardinal directions. However, the effect can occur on any axis according to the particular geographical alignment involved. Cardinal point effect is illustrated in figure 16-14.

WEATHER RETURNS. Cloud returns which appear on the scope are called *precipitation or meteorological echoes.* They are of interest for two reasons. First, since the brightness of a given cloud return is an indication of the intensity of the weather within the cloud, intense weather areas can be avoided by directing the pilot through the darker scope areas or by completely circumnavigating the entire cloud return. Second, cloud returns obscure useful natural and cultural features on the ground. They may also be falsely identified as a ground feature, which can lead to gross errors in radar fixing.

Clouds must be reasonably large to create a return on the scope. However, size alone is not the sole determining factor. Cloud layers covering hundreds of square miles will not always create returns. As a matter of fact, clouds covering such

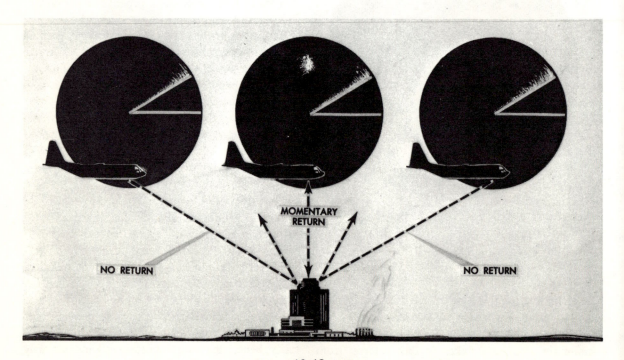

MOMENTARY RETURN

NO RETURN

NO RETURN

*Figure 16-13 Glitter*

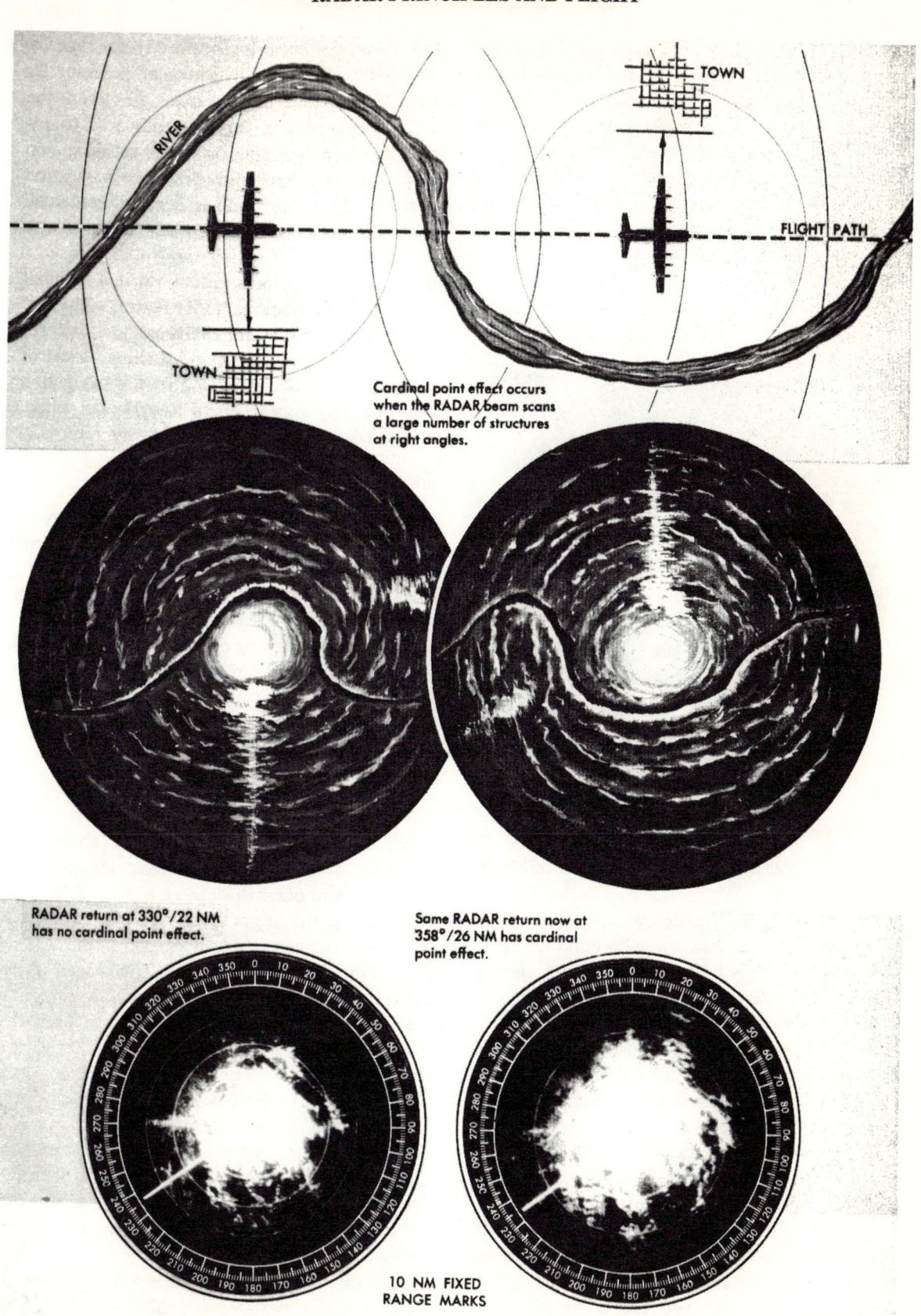

Cardinal point effect occurs when the RADAR beam scans a large number of structures at right angles.

RADAR return at 330°/22 NM has no cardinal point effect.

Same RADAR return now at 358°/26 NM has cardinal point effect.

10 NM FIXED RANGE MARKS

Figure 16-14  Cardinal Point Effect

wide areas rarely show on the scope. The one really important characteristic that causes clouds to create radar returns is the size of the water droplets forming them. Radar waves are reflected from large water droplets falling through the atmosphere as rain or suspended in the clouds by strong vertical air currents. Thunderstorms are characterized by strong vertical air currents; therefore, they give very strong radar returns (see figure 16-15).

*Figure 16-15 Thunderstorm Returns*

Cloud returns may be identified by the following characteristics:
• Brightness varies considerably, but the average brightness is greater than normal ground return.
• Returns generally present a hazy, fuzzy appearance around their edges.
• Terrain features are not present.
• Returns often produce a shadow area similar to mountain shadow because the radar beam does not penetrate clouds with sufficient strength to create echoes beyond them.
• Returns do not fade away as the antenna tilt is raised, whereas ground returns tend to decrease in intensity with an increase in antenna tilt.
• Returns appear in the altitude hole when altitude delay is not used and the distance to the cloud is less than the altitude.

EFFECTS OF SNOW AND ICE. The effects of snow and ice, most prevalent in arctic regions, are similar to the effect of water. If a land area is covered to any great depth with snow, the radar beam will reflect from the snow rather than from the features which lie beneath. The overall effect is to *reduce* the return which would normally come from the snow-blanketed area.

Ice will react in a slightly different manner, depending upon its roughness. If an ice coating on a body of water remains smooth, the return will appear approximately the same as a water return. However, if the ice is formed in irregular patterns, the returns created will be comparable to terrain features of commensurate size. For example, ice ridges or ice mountains would create returns comparable to ground embankments or mountains, respectively. Also, offshore ice floes tend to disguise the true shape of a coastline so that the coastline may appear vastly different in winter as compared to summer.

### Inherent Scope Errors

Another factor which must be considered in radarscope interpretation is the inherent distortion of the radar display. This distortion is present to a greater or lesser degree in every radar set, depending upon its design. It can be minimized by a proper understanding of its causes and by proper use of the radar controls. Inherent scope errors may be laid to three causes: the width of the beam, the length (time duration) of the transmitted pulse, and the diameter of the electron spot.

BEAM WIDTH ERROR. The magnitude of this error may be as great as the width of the radar beam. Suppose, for example, that an isolated structure, 1000 feet wide, is being scanned by a beam one and one-half degrees wide at a range of 20 miles. If the target is an excellent reflector, it will begin to create a return as soon as the leading edge of the beam contacts it, while the center of the beam is still half the beam width away. Similarly, the target continues to create a return until the trailing edge of the beam leaves it. Thus, the target appears too wide by half the beam width on either side, or by a total of one beam width. However, this is the maximum error that could occur. The beam seldom intercepts targets of such excellent reflective characteristics as to induce the maximum error. Also, the beam subtends less linear arc as the range decreases so that beam width error decreases with range.

Actually, beam width error is not overly significant in radar navigation (although it must be taken into account in radar bombing). Since the distortion is essentially symmetrical, it may be nullified by bisecting the return with the bearing cursor when bearing is measured. Beam width

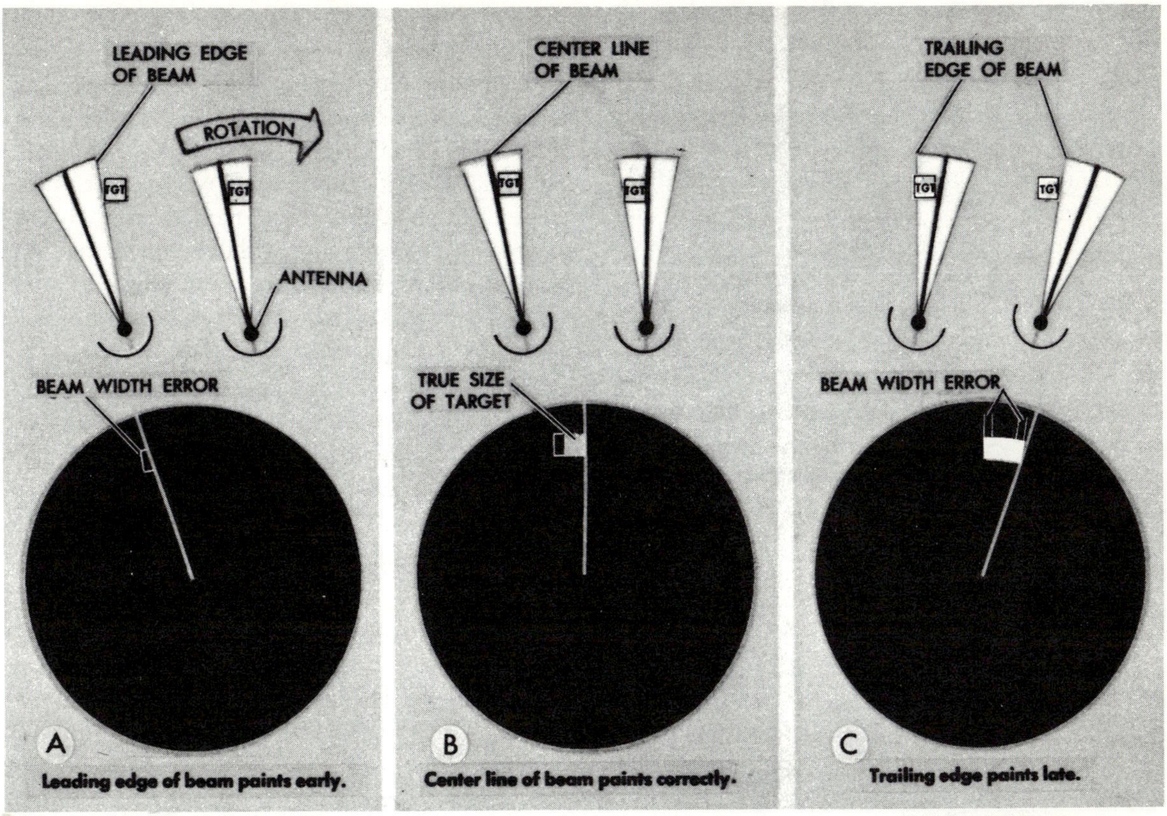

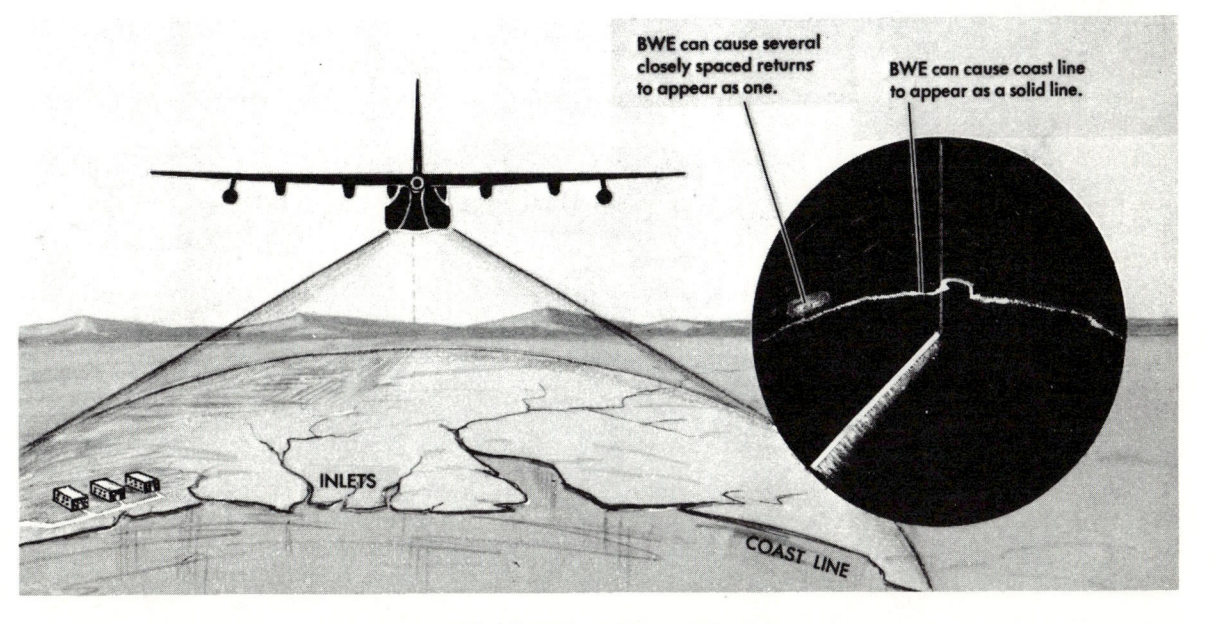

Figure 16-16  Beam Width Error

A  Leading edge of beam paints early.

B  Center line of beam paints correctly.

C  Trailing edge paints late.

Figure 16-17  Effect of Beam Width Error

distortion is also lessened by reduction of the receiver gain control. The principle and effect of beam width error on reflecting targets are illustrated in figures 16-16 and 16-17.

PULSE LENGTH ERROR. Pulse length error is caused by the fact that the radar transmission is not instantaneous but lasts for a brief period of time. Imagine a huge steel plate so large that no radar pulse can penetrate beyond it. Such a reflector has no range dimension, therefore, the return on the scope should have no range dimension. However, if the pulse emitted by the radar transmitter is of one microsecond duration, this pulse will continue to strike the steel plate for one microsecond. It is obvious, then, that the target will reflect energy throughout this period of time, and a return will be displayed on the scope throughout this period. Since time is translated into distance on the scope face (1 microsecond equals 490 feet) and corrected for a radar mile, the target will appear to be approximately 490 feet longer than it actually is. Thus, there is a distortion in the

range direction on the far side of the reflector, and this pulse length error is equal to the range equivalent of one-half the pulse time.

Since pulse length error occurs on the far side of the return as shown in figure 16-18, it may be nullified by reading the range to (and plotting from) the near side of the target when taking radar ranges. It should also be mentioned that the radar set is so designed that the pulse length (and hence the pulse length error) increases as the operating range increases. Therefore, failure to employ the proper range measuring technique will produce greater errors at the longer ranges.

The pulse length also has a primary effect on the *range resolution* of a radar set. Theoretically, two sharply defined targets at the same bearing should be resolved into two separate returns on the scope if their ranges differ by the distance light travels during one-half the pulse length. Thus, with a one microsecond pulse length, two targets would have to be more than 490 feet apart in range

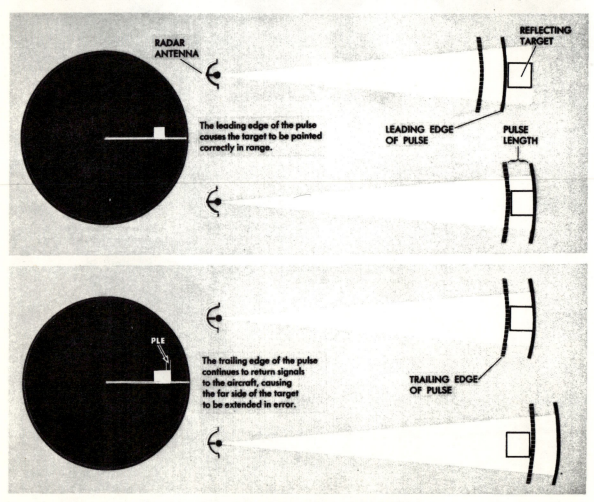

Figure 16-18  Pulse Length Error

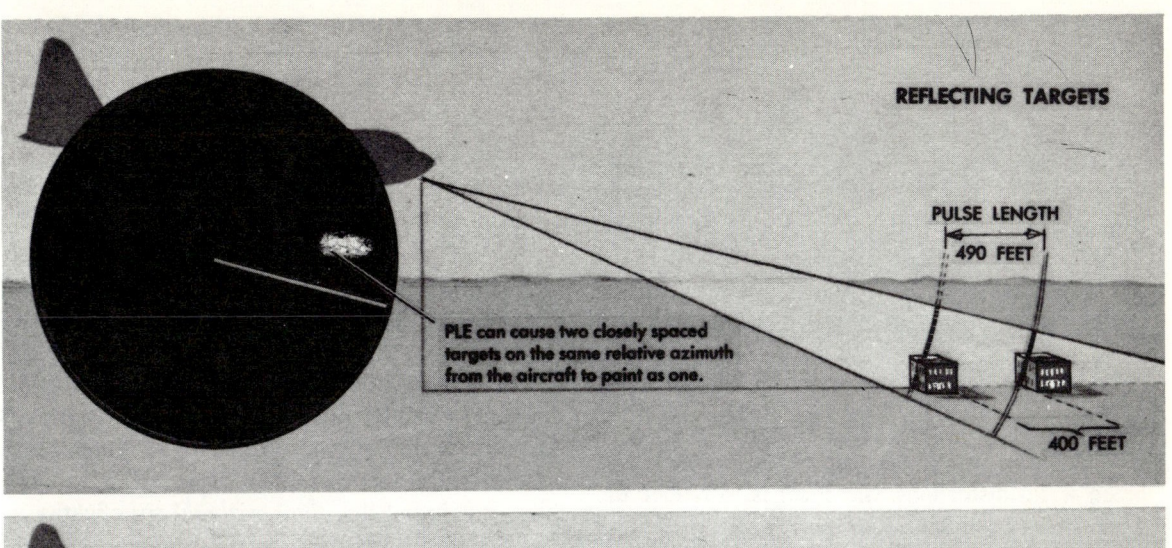

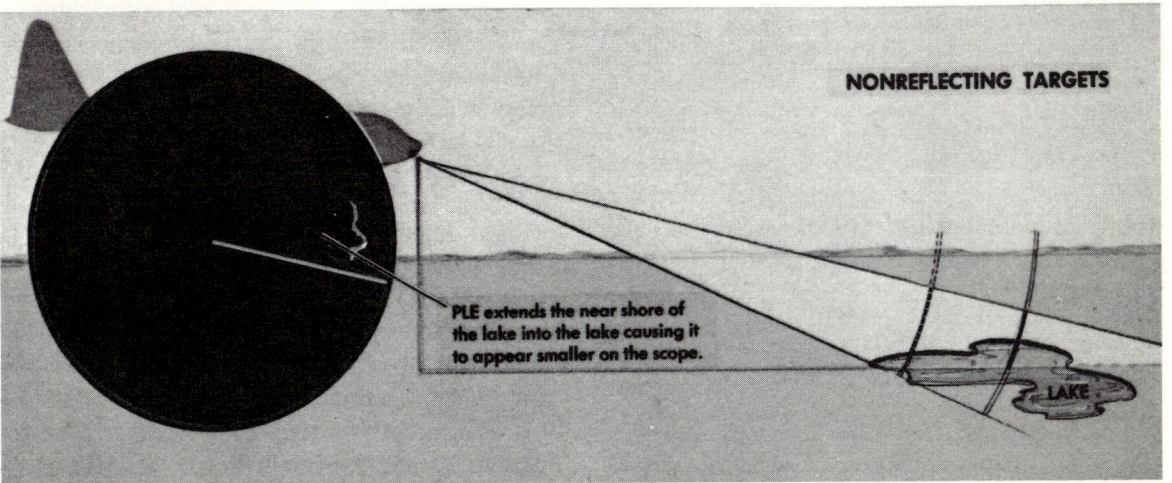

Figure 16-19  Effect of Pulse Length Error

Figure 16-20  Effect of Spot Size Error

to appear as two separate returns on the scope. (See figure 16-19.)

SPOT SIZE ERROR. Spot size error is caused by the fact that the electron beam which displays the returns on the scope has a definite physical diameter. No return which appears on the scope can be smaller than the diameter of the beam. Furthermore, a part of the glow produced when the electron beam strikes the phosphorescent coating of the CRT radiates laterally across the scope. As a result of these two factors, all returns displayed on the scope will appear to be slightly larger in size than they actually are. (See figure 16-20.)

Here again, as in the case of pulse length and beam width distortions, the effect is to reduce the *resolution* of the set. This may make it difficult for the radar navigator to isolate and identify returns.

Spot size distortion may be reduced by using the lowest practicable receiver gain, video gain, and bias settings and by keeping the operating range at a minimum so that the area represented by each spot is kept at a minimum. Further, the operator should check the focus control for optimum setting.

TOTAL DISTORTION. The radar presentation is distorted by three inherent errors: pulse length, beam width, and spot size. The combined effect of these errors is illustrated in figure 16-21. For navigational purposes, the errors are often negligible. However, the radar navigator should realize that they do exist and that optimum radar accuracy demands that they be taken into account.

## NAVIGATIONAL FEATURES OF RADAR

The airborne radar sets used throughout the Air Force vary slightly in the navigational refinements offered. All of the sets in use, however, possess some of the following features.

### Radar Range Selection

The range of any radar system depends upon the transmitted power, the receiver sensitivity, the antenna gain, and the various factors affecting reflection. However, since radar is a line-of-sight instrument, the ultimate limitation on the range of any radar is the curvature of the earth. In practice, it is found that the horizontal stratification of the atmosphere causes refraction of the microwave rays, usually bending them downward so that they tend to follow the curvature of the earth. This results in a theoretical radar range which may be determined by the following formula:

$$D = \sqrt{2h} \times .87$$

when $D$ is nautical miles and $h$ is the altitude in feet.

Though it is possible to receive returns from objects at great distances, it is not always necessary or desirable to have the maximum range portrayed on the scope. Therefore, the operating range of all radar sets can be controlled by the operator. The various sets differ as to the maximum and minimum ranges obtainable. Most sets currently in use may be varied from 3 to 400 miles. The

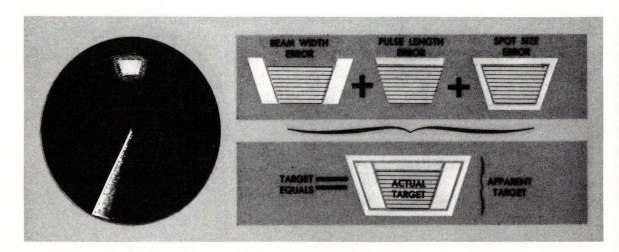

*Figure 16-21 Combined Effect of Inherent Errors*

range selected determines the comparative size of the various returns, and proper use of the range selector control can simplify the scope interpretation problem. For example, decreasing range will show a particular area in much greater detail, while increasing range (altitude permitting) will give a much greater coverage.

**Fixed Range Markers**

Fixed range markers have already been discussed under the heading, Plan Position Indicator. These markers describe concentric circles on the scope at fixed range intervals. They may be used to measure the *slant range* (the direct line-of-sight range) to any return appearing on the scope. It is often necessary to interpolate when using the fixed range markers because the return appears between two markers. On some sets, one of the

several range intervals available—2, 5, 10, etc.— may be arbitrarily selected. On other sets, the range mark interval is automatically determined by the setting of the range selector control.

**Variable Range Marker and Crosshairs**

Most radar sets also provide a range marker which may be moved within certain limits by the radar operator. This variable range marker permits more accurate measurement of range because the marker can be positioned more accurately on the scope. Furthermore, visual interpolation is not necessary when using the variable range marker because it can be placed at the particular return being considered.

On many radar sets, an electronic *azimuth marker* has been added to the variable range marker to facilitate fixing. The intersection of the

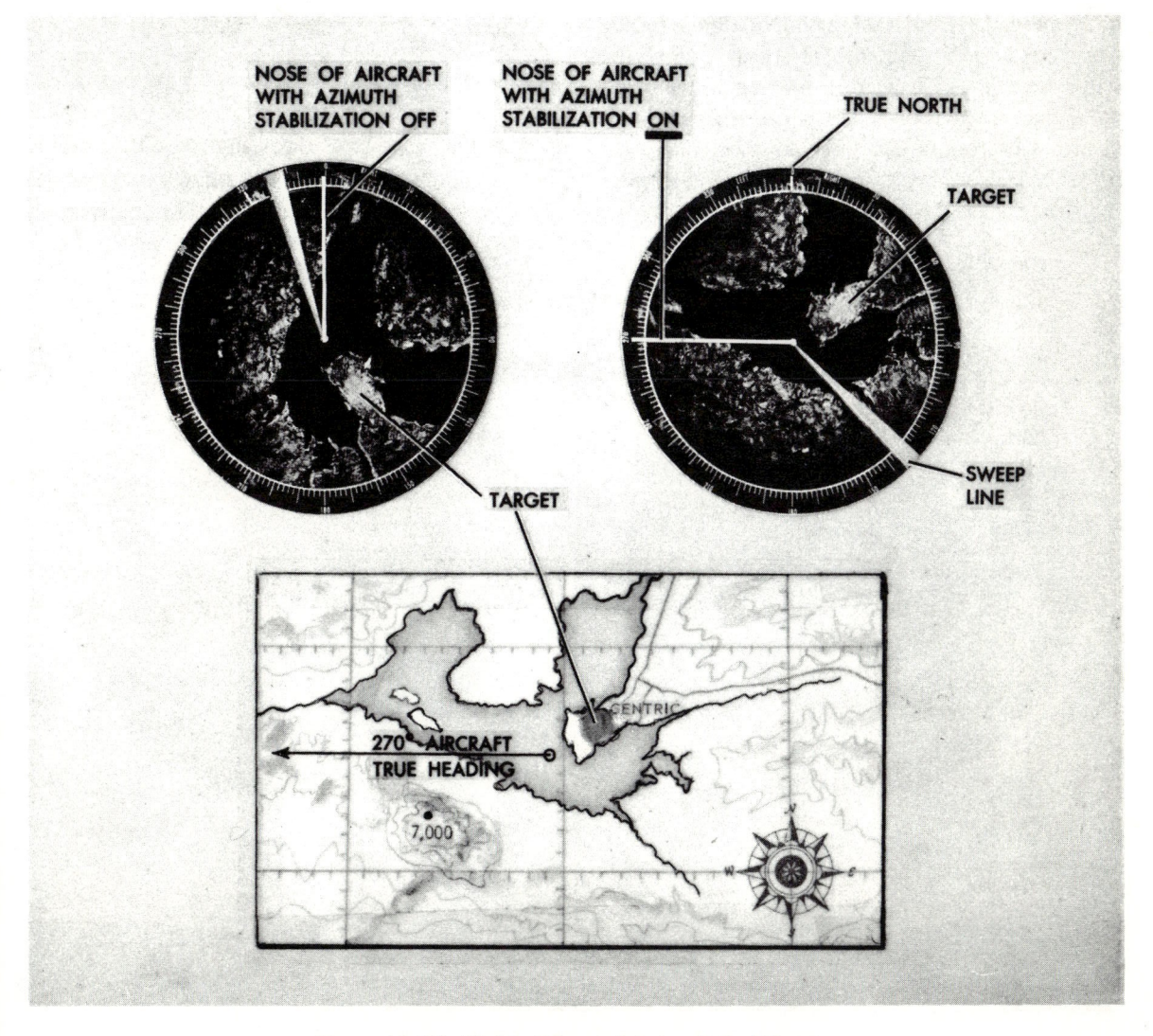

*Figure 16-22   Visible Effects of Azimuth Stabilization*

207

azimuth marker and the variable range marker is defined as the *radar crosshairs*.

### Azimuth Stabilization

To preclude the necessity for converting relative bearings to true bearings, a system has been devised which orients the scope so that true north always appears at the top. This system, known as *azimuth stabilization*, is so connected to the compass system that the sweep assumes the "12 o'clock" position when the antenna is pointed north, if the correct variation is set into the radar. An electronic *heading marker* indicates the true heading of the aircraft, and all bearings taken are *true bearings*. With the azimuth stabilization system off, the top of the scope represents the heading of the aircraft and all bearings taken are *relative* bearings. Figure 16-22 shows the visible effects of azimuth stabilization. The azimuth stabilization should be used at all times except in those regions where the tie-in to the compass system is impractical because of rapid changes in variation (as in polar areas).

BEARING CORRECTION. For optimum accuracy, it may sometimes become necessary to correct the bearings taken on the various targets. This necessity arises whenever: (1) the heading marker reading does not agree with the true heading of the aircraft when azimuth stabilization is used, or (2) the heading marker reading does not agree with 360° when azimuth stabilization is not used.

EXAMPLE. If the TH is 125° and the heading marker reads 120°, all of the returns on the scope will indicate a bearing which is 5° *less* than it should be. Therefore, if a target indicates a bearing of 50°, 5° must be *added* to the bearing before it is plotted. Conversely, should the heading marker read 45° when the TH is 040°, all of the scope returns will indicate a bearing which is 5° *more* than it should be. Therefore, if a target indicates a bearing of 275°, 5° must be *subtracted* from the bearing before it is plotted. The greater the distance of the target from the aircraft, the more important this *heading marker correction* becomes.

### Altitude Delay

It is obvious that the ground directly beneath the aircraft is the closest reflecting object. Therefore, the first return which can appear on the scope will be from this ground point. Since it takes some finite period of time for the radar pulse to travel to the ground and back, it follows that the sweep must travel some finite distance radially from the

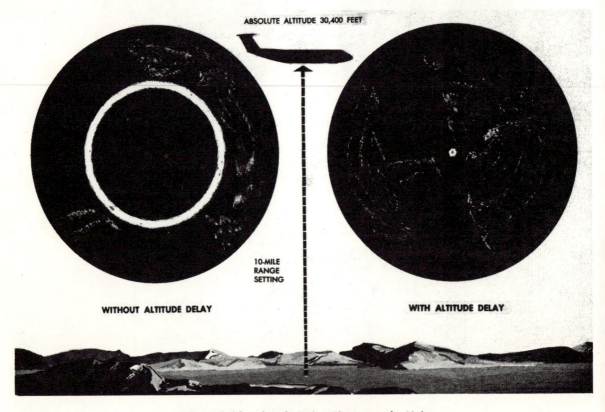

ABSOLUTE ALTITUDE 30,400 FEET

10-MILE RANGE SETTING

WITHOUT ALTITUDE DELAY          WITH ALTITUDE DELAY

*Figure 16-23  Altitude Delay Eliminates the Hole*

center of the scope before it displays the first return. Consequently, a *hole* will appear in the center of the scope within which no ground returns can appear. Since the size of this hole is proportional to altitude, its radius can be used to measure altitude. Thus, if the radius of the *altitude hole* is 12,000 feet, the absolute altitude of the aircraft is 12,000 feet.

Although the altitude hole may be conveniently used to measure altitude, it occupies a large portion of the scope face, especially when the aircraft is flying at a high altitude and using a short range. This may be seen in figure 16-23. In this particular case, the range selector switch is set for a 10-mile range presentation. Without altitude delay, the return shown on the first five miles of the scope consists of the altitude hole, and the return shown on the remaining five miles is a badly distorted presentation of *all of the terrain within ten miles of the point below the aircraft.* To obviate such a condition, many radar sets incorporate an *altitude delay circuit* which permits the removal of the altitude hole. This is accomplished by delaying

the start of the sweep until the radar pulse has had time to travel to the ground point directly below the aircraft and back. Hence, the name altitude *delay* circuit. The altitude delay circuit also minimizes distortion of the type shown in the left-hand scope of figure 16-23. On this scope, it is evident that the area displayed is a greatly compressed version of the actual ground area. The circuit makes it possible for the radarscope to present a *ground picture,* which preserves the actual relationships between the various ground objects.

**Sweep Delay**

Sweep Delay is a feature which delays the start of the sweep until after the radar pulse has had time to travel some distance into space. In this respect, it is very similar to altitude delay. The use of sweep delay enables the radar operator to obtain an enlarged view of areas at extended ranges.

For example, two targets which are 45 miles from the aircraft can only be displayed on the

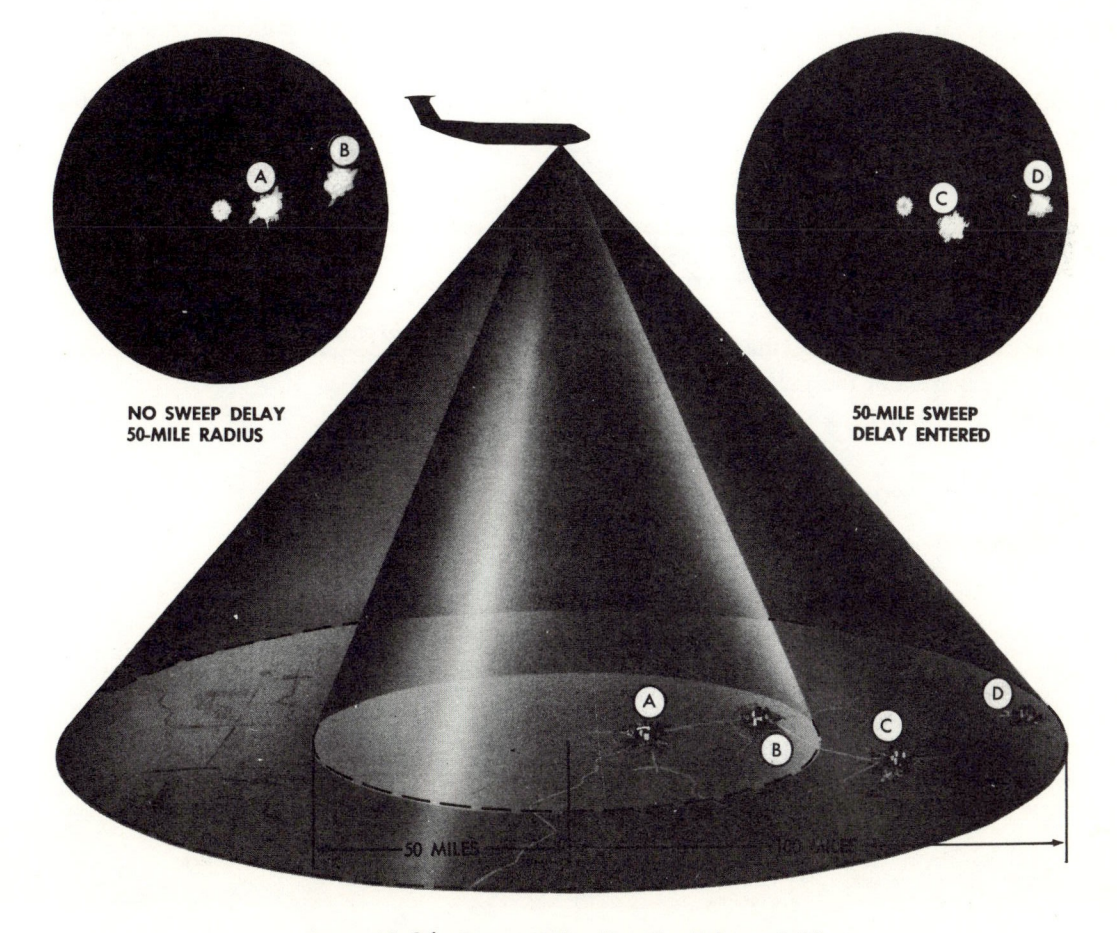

**NO SWEEP DELAY
50-MILE RADIUS**

**50-MILE SWEEP
DELAY ENTERED**

*Figure 16-24  Sweep Delay Provides Telescopic View*

scope if a range scale *greater* than 45 miles is being used. On the 50-mile range scale, the two targets might appear very small and close together. By introducing 40 miles of sweep delay, the display of the two targets will be enlarged *as long as the range displayed on the scope is less than that displayed before sweep delay was introduced.* For instance, if the scope is on a 50-mile range scale, as in the preceding example, introducing 40 miles of sweep delay would have no enlarging effect unless the range being displayed is reduced to some value below 50 miles. The more this range is reduced, the greater will be the enlarging effect. On some sets, the range displayed during sweep delay operation is fixed by the design of the set and cannot be adjusted by the operator. Figure 16-24 shows graphically the effect of sweep delay.

## RADAR FIXING

Each radar set in current use may provide methods of fixing which are peculiar to that set. If would be impossible to describe all of these methods. Therefore, only the basic methods employed with most sets are discussed here.

### Measuring Range

Most radar sets offer a choice between the use of fixed range markers or a variable range marker for the measurement of range.

FIXED RANGE MARKERS. The fixed range markers are separated by specific intervals. These intervals are governed either by a range mark interval selector switch or by the range selector switch. In either case, the range marker interval must first be known before the markers can be

used. Next, the position of the radar return with respect to the range markers is determined. Interpolation will often be necessary; thus, certain limits are imposed on the accuracy of this range measurement.

*Slant Range.* Once a return has been identified, it may be used to fix the position of the aircraft by measuring its bearing and distance from the known geographic point. Of particular significance in any discussion of radar ranging is the subject of slant range versus ground range. Slant range is the straight-line distance between the aircraft and the target, while ground range is the range between the point on the earth's surface directly below the aircraft and the target. Figure 16-25 shows how these two ranges may differ, and how the difference becomes less as the slant range increases. Figure 16-26 is a slant range/ground range table for various altitudes and ranges.

To fix the position of the aircraft, the navigator is interested in his *ground range* from the fixing point. Yet his fixed range markers give him slant range. His problem, then, is to determine the critical range below which he must convert slant range to ground range in order not to introduce significant errors in his fixes. This critical range may be determined by a simple formula:

$$\frac{Absolute\ Altitude - 5,000}{1,000} = Critical\ slant\ range\ (in\ NM)$$

*Example:*

$$\frac{30,000\ (Absolute\ Altitude\ in\ feet) - 5,000}{1,000}$$

$$= 25\ NM$$

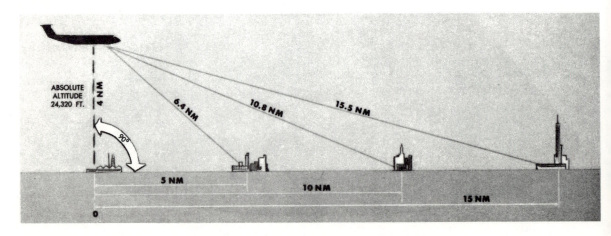

*Figure 16-25   Slant Range Compared to Ground Range*

In this example, if the slant range is *less* than 25 miles, it should be converted to ground range before the fix is plotted. At ranges in excess of 25 miles, slant range would so closely approximate ground range that conversion would be unnecessary.

*Computer Method of Slant Range Conversion.* Slant range may be converted to ground range on the hand-held computer by constructing the right triangle of ground range, slant range, and altitude as follows:

1. Convert the absolute altitude from feet to nautical miles by dividing by 6,080.
2. Set the computer azimuth with the index on a cardinal heading.
3. Using the wind side of the slide, measure down from the grommet the amount of the absolute altitude in nautical miles. This is shown in figure 16-27 where a scale of 10 on the computer is equal

| Absolute Altitude Thousands of feet) | SLANT RANGE | | | | | | | |
|---|---|---|---|---|---|---|---|---|
| | 5NM | 10NM | 15NM | 20NM | 25NM | 30NM | 35NM | 40NM |
| | CORRESPONDING GROUND RANGE (NM) | | | | | | | |
| 13 | 4.5 | 9.8 | 14.9 | 19.9 | 24.9 | 29.9 | 34.9 | 40.0 |
| 14 | 4.5 | 9.8 | 14.8 | 19.9 | 24.9 | 29.9 | 34.9 | 39.9 |
| 15 | 4.4 | 9.7 | 14.8 | 19.9 | 24.9 | 29.9 | 34.9 | 39.9 |
| 16 | 4.2 | 9.6 | 14.8 | 19.8 | 24.9 | 29.9 | 34.9 | 39.9 |
| 17 | 4.1 | 9.6 | 14.7 | 19.8 | 24.8 | 29.9 | 34.9 | 39.9 |
| 18 | 4.0 | 9.5 | 14.7 | 19.8 | 24.8 | 29.9 | 34.9 | 39.9 |
| 19 | 3.9 | 9.5 | 14.7 | 19.8 | 24.8 | 29.8 | 34.9 | 39.9 |
| 20 | 3.7 | 9.4 | 14.6 | 19.7 | 24.8 | 29.8 | 34.8 | 39.9 |
| 21 | 3.6 | 9.4 | 14.6 | 19.7 | 24.8 | 29.8 | 34.8 | 39.9 |
| 22 | 3.5 | 9.3 | 14.6 | 19.7 | 24.7 | 29.8 | 34.8 | 39.8 |
| 23 | 3.3 | 9.3 | 14.5 | 19.7 | 24.7 | 29.8 | 34.8 | 39.8 |
| 24 | 3.0 | 9.2 | 14.5 | 19.6 | 24.7 | 29.7 | 34.8 | 39.8 |
| 25 | 2.8 | 9.1 | 14.4 | 19.6 | 24.7 | 29.7 | 34.8 | 39.8 |
| 26 | 2.5 | 9.0 | 14.4 | 19.5 | 24.6 | 29.7 | 34.7 | 39.8 |
| 27 | 2.2 | 8.9 | 14.3 | 19.5 | 24.6 | 29.7 | 34.7 | 39.8 |
| 28 | 2.0 | 8.9 | 14.3 | 19.5 | 24.6 | 29.7 | 34.7 | 39.7 |
| 29 | 1.5 | 8.8 | 14.2 | 19.4 | 24.5 | 29.6 | 34.7 | 39.7 |
| 30 | .8 | 8.7 | 14.2 | 19.4 | 24.5 | 29.6 | 34.7 | 39.7 |
| 31 | | 8.6 | 14.1 | 19.3 | 24.5 | 29.6 | 34.6 | 39.7 |
| 32 | | 8.5 | 14.1 | 19.3 | 24.4 | 29.5 | 34.6 | 39.6 |
| 33 | | 8.4 | 14.0 | 19.2 | 24.4 | 29.5 | 34.6 | 39.6 |
| 34 | | 8.3 | 13.9 | 19.2 | 24.4 | 29.5 | 34.6 | 39.6 |
| 35 | | 8.2 | 13.9 | 19.1 | 24.3 | 29.4 | 34.5 | 39.6 |
| 36 | | 8.1 | 13.8 | 19.1 | 24.3 | 29.4 | 34.5 | 39.6 |
| 37 | | 7.9 | 13.7 | 19.0 | 24.2 | 29.4 | 34.5 | 39.5 |
| 38 | | 7.8 | 13.6 | 19.0 | 24.2 | 29.3 | 34.4 | 39.5 |
| 39 | | 7.7 | 13.6 | 18.9 | 24.2 | 29.3 | 34.4 | 39.5 |
| 40 | | 7.5 | 13.5 | 18.9 | 24.1 | 29.3 | 34.4 | 39.5 |
| 41 | | 7.4 | 13.4 | 18.8 | 24.1 | 29.2 | 34.4 | 39.4 |

*Figure 16-26 Slant Range/Ground Range Table*

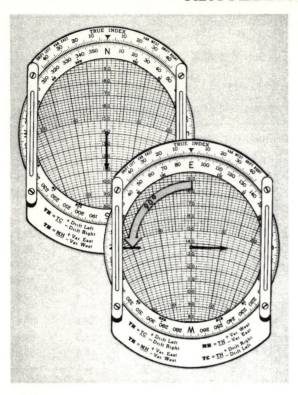

*Figure 16-27 Computer Method of Slant Range Conversion*

to one mile and the absolute altitude is three miles.

4. Rotate the computer 90° to another cardinal heading as in the illustration.

*NOTE: Slant range is represented by a groundspeed line on the computer slide. For instance, if the slant range is 14 nautical miles, use the 140 groundspeed line.*

5. Adjust the slide so that the correct groundspeed line is under the *outer end* of the plotted altitude vector, as in the illustration.

6. Read the ground range under the grommet.

7. Divide by 10 for the actual ground range.

*Variable Range Marker.* The variable range marker or crosshair is used for more precise measurement of position. The marker is placed on a selected radar return, and the position of the aircraft with respect to that return is indicated on a separate dial or dials. The indicated range is the ground range if the radar set is equipped with a computer and the absolute altitude of the aircraft is known and used in the computer. A limitation on the use of the variable range marker is the fact that the maximum range of the marker is far less than the ranges available using the fixed range markers. Fixes obtained using the radar crosshairs are called *precision fixes.*

## Single Range and Bearing Fix

The single range and bearing fix is the method most widely used to obtain radar fixes. This fix is just what the name implies. The bearing of the return is measured on the calibrated azimuth ring or by using the electronic azimuth marker. If azimuth stabilization is off and the top of the scope represents the nose of the aircraft, this bearing is a relative bearing and must be added to the true heading in order to obtain a true bearing. If azimuth stabilization is on and the proper variation is set, the bearing measured is true bearing and may be plotted as such. (The heading marker correction must, of course, be applied to all bearings taken). The reciprocal of the bearing to the radar target becomes a line of position which is crossed with the range measurement to obtain a fix (see figure 16-28 ).

## Multiple Bearing Fix

Sometimes it is not possible to measure range with the radar. When this occurs, accurate fixes are obtained by crossing azimuth bearings from two or more identified targets. The best results are obtained when using three targets 120° apart. If only two targets are used, 90° cuts give the best results. The three-bearing fix, of course, is generally more accurate.

The following is one method of taking a multiple bearing fix:

1. Identify three returns approximately 120° apart.

2. Using the bearing cursor, take the bearing of each return in quick succession.

3. Note the time of the bearings.

4. Plot the reciprocals of the bearings on the chart as shown in figure 16-29 .

The intersection of the bearing lines is the radar fix. The time of the fix is the time noted in step three.

## Multiple Range Fixes

If the scale illumination light is burned out, the cursor cannot be moved, or if the scope presentation cannot be centered on the cursor plate, it would be necessary to position the aircraft by multiple range fixes derived by measuring the range to two or more targets and plotting these ranges on the chart. The intersection of the range circles is the fix. If only two targets are used, the range circles cross at two points. The resulting ambiguity is resolved by using dead reckoning information. If three targets are used, there is no ambiguity.

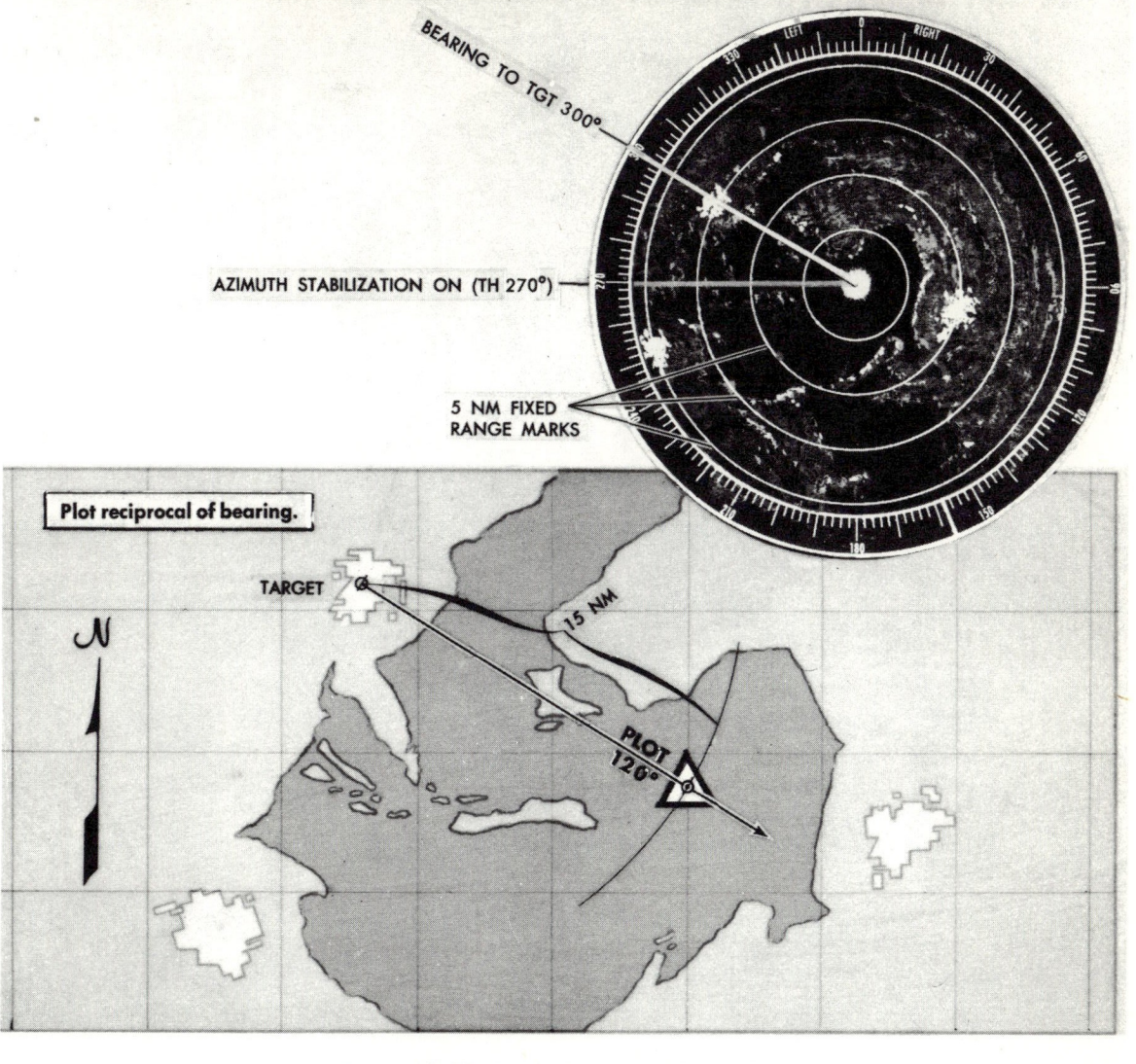

*Figure 16-28 Single Range and Bearing Fix*

The following procedure is used to obtain a multiple range fix:

1. Identify three targets approximately 120° apart.

2. Note the range to each target being used and the time that this observation is made.

3. Set a draftsman's compass to one of the ranges observed.

4. Place the point of the compass on the chart symbol corresponding to the return and draw an arc on the chart.

5. Repeat this for each range observed. The intersection of the arcs (see figure 16-30) is the fix. The time of the fix is the time noted in step two.

## RADAR COMPUTERS

Because of the increasing complexities of navigation and bombing problems in modern high speed aircraft, various computer systems have been devised to assist the navigator in the completion of his mission.

Radar computer systems provide such capabilities as precision position determination, wind runs, and bomb runs. These computers range from the relatively simple to the enormously complex. Only the more basic types are discussed here and only that information relative to navigation is included.

To attain precision, the navigator must observe a particular area or point on the scan presentation.

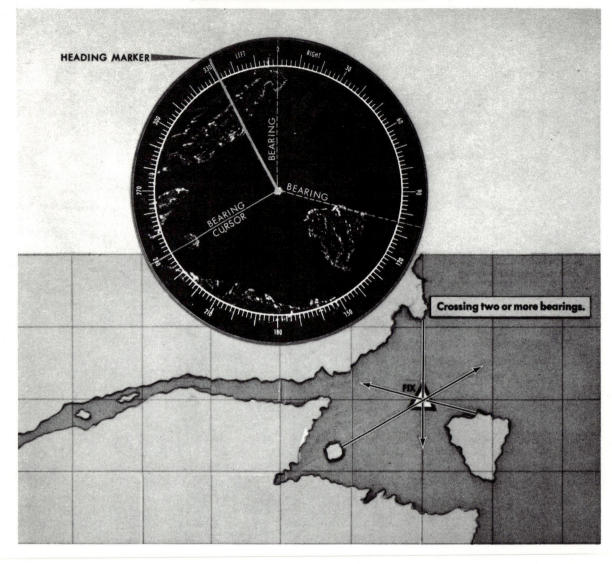

*Figure 16-29 Multiple Bearing Fix*

This capability is provided by the use of sector scan. As illustrated in figure 16-31 , sector scan paints only a portion of the usual full scan presentation. Rather than scanning 360° degrees, the antenna scans only a short arc. It scans in one direction, then reverses itself to scan the same area again. Since the position of the sweep is controlled by the antenna, the sweep paints the area scanned with a clockwise and counterclockwise motion. The resulting presentation is a pie-shaped wedge of radar returns. The vertex of the sector—which may be in the center, on the edge of the radarscope, or beyond the edge of the radar scope—represents the aircraft position. If the heading marker is not in the sectored area, it will not show on the radarscope; however, the fixed range marks will be present.

Other features which may be used with sector scan include the displaced center and the option of having fixed range marks and heading markers or having crosshairs (one variable range marker and an azimuth marker) in which case the heading marker does not appear.

Crosshairs, when used, are placed on the target with a tracking control in such a manner that the variable range marker and the azimuth marker intersect on the desired point as shown in figure 16-32.

**Determining Position**

Regardless of the type, all computers require certain information that must be fed into the system either manually or electronically. In turn, the computer will furnish the navigator with the position of the aircraft at any time. If a wind run

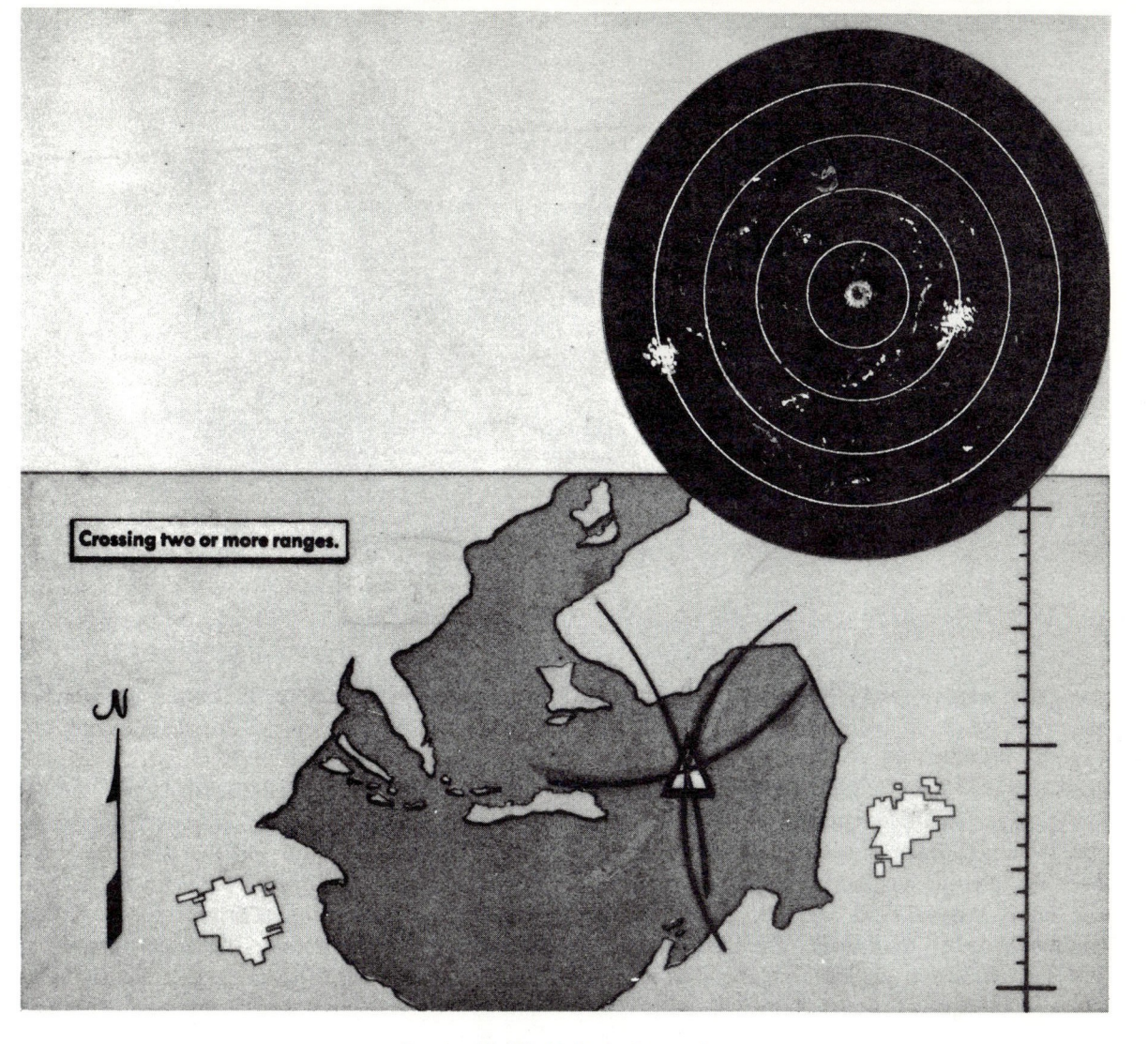

*Figure 16-30 Multiple Range Fix*

has been accomplished, this position will be an accurate DR position; if no wind is set into the system, the position will be an air position.

The computer system accomplishes this in the following manner. First, the navigator sets in a departure point by taking a fix on a known object, or by zeroing his fix dials over a known point. From this point on, the system computes the departure from that point. Normally, the navigator will accomplish a "wind run" at the same time that he is taking a fix. This serves to set into the system the latest known wind, and the computer will continue to use this wind until another is set in. The navigator may also set in any known wind, if a "wind run" is not feasible at departure. Since the navigator manually sets in the departure point and the wind, some system must be incorporated to provide TH and TAS. This is accomplished electrically by the compass and pitot-static systems.

As the aircraft progresses along track, the navigator can determine the DR position at any time by referring to the "fix dials." In addition, the latest known wind may be ascertained by reference to the "wind dials." Both of these dials will furnish information in "rectangular" coordinates or forces. The fix dials will reflect the distance traveled from departure, in components north, south, east or west. Thus, if after one hour, the position shown on the dials is 150 miles north and 120 miles east, then the position in "polar" coordinates becomes 190 miles from the departure point on a true bearing of 040. By reference to the departure point, a DR position in latitude and longitude can be quickly determined.

## Wind Runs

As mentioned earlier, there are radar computers

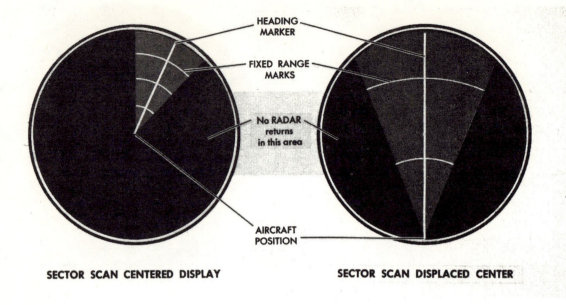

Figure 16-31   Sector Scan Displays

that greatly aid the navigator by furnishing an instantaneous DR position at any time. However, for this DR position to be accurate, frequent wind runs must be made so that an accurate wind is integrated into the system.

With any computer, the wind run is a simple time and distance problem. In the more basic systems the operator notes a particular return, determines the altitude of the return (commonly called a target), and sets the altitude of the aircraft with respect to the target into the set. Next, the crosshairs are placed on the target and the set placed in the "wind run" position. When the crosshairs appear to drift off the target, they are repositioned on the target. This is known as synchronization. When the crosshairs and target are accurately synchronized; i.e., when the crosshairs do not appear to drift off the target, then the wind run is complete and the "wind run" switch is turned off.

At that time, the rectangular coordinates of the wind may be read on the wind run dials. As with the position dials mentioned earlier, these dials read the forces of the wind in components of north, south, east or west. Thus, if the dials read north 30 and east 15, the resultant wind would be from 026 degrees at 34 knots.

Such a wind run is an accurate spot wind and is automatically fed into the computer system. It remains in the system until another wind run is made or a manual setting of the wind is accomplished.

## OTHER RADAR USES

### Airborne Radar Approach

The *airborne radar approach (ARA)* may be described as GCA in reverse. It is used only as an *emergency procedure* in marginal weather conditions if air-to-ground communication is impossible and no radio navigation aids are available. However, it should be practiced often against the time when it may be the only method of effecting a safe landing. Even when the letdown and approach are directed by ground radar, such direction should be monitored by the navigator on the *airborne* radar.

The ARA involves the use of the airborne radar set to guide the pilot to a point on the final approach where he can either complete the landing visually or decide that the landing cannot be safely completed. This point will be defined by the "minimums" for PPI approaches at the base at which the ARA is made.

There are two main phases in the airborne radar approach—the letdown and the approach itself. The letdown is normally accomplished from a nearby VOR or TACAN station in accordance with published procedures. If these are not available, the letdown is made from a point over the runway using a standard procedure for the type of aircraft flown.

After completion of the letdown, the approach itself is begun. Again, the procedures used vary with the capabilities of the radar used. In general,

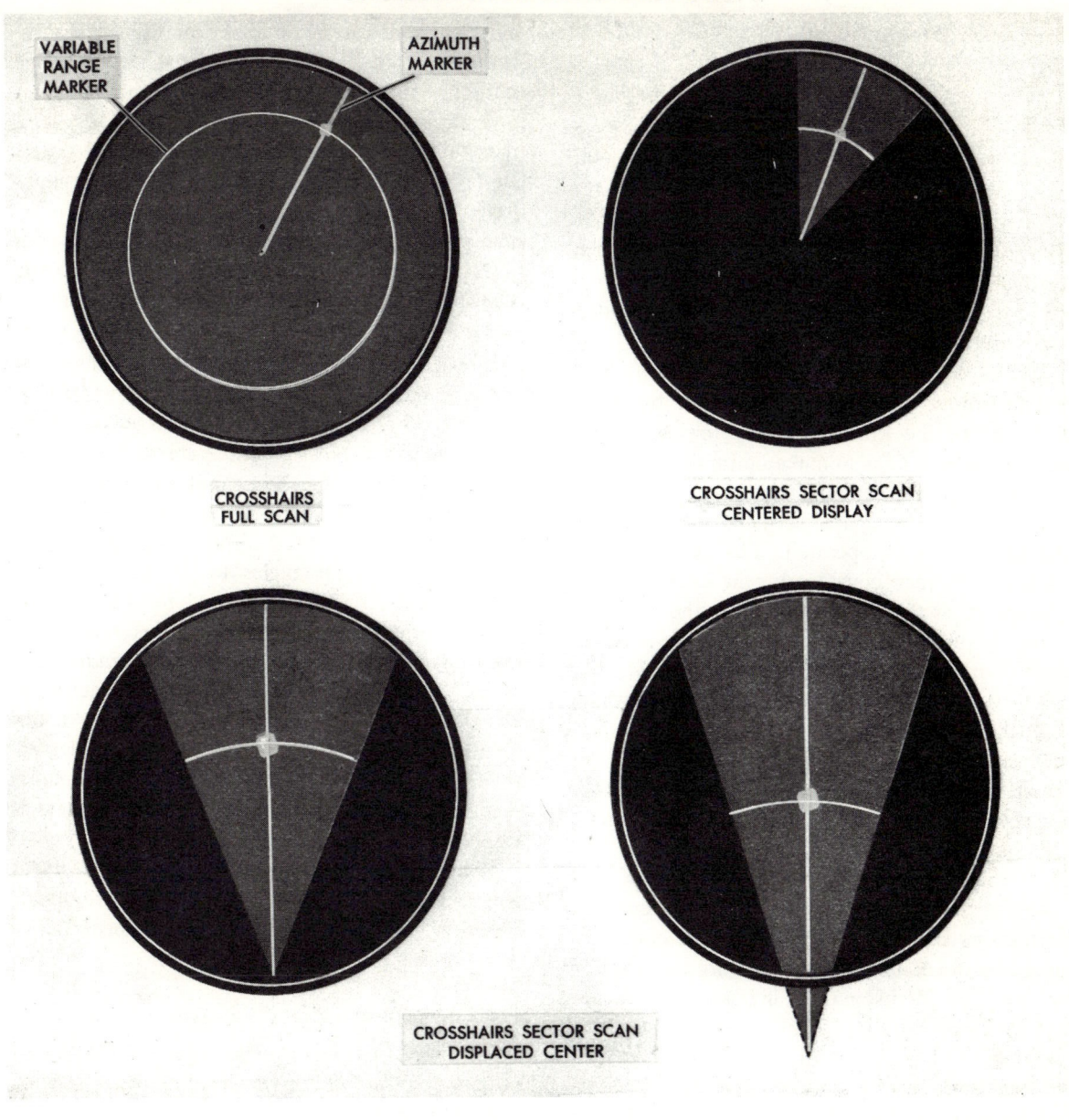

Figure 16-32  *Sector Scans Using Crosshairs*

on radar sets equipped with computers, the crosshairs are used to monitor the distance and direction to the end of the runway; on the other sets, the fixed range markers are used. The specific steps for each type of radar are included in the appropriate aircraft flight handbook.

**Station Keeping**

Station keeping is a technique wherein radar is used to maintain a fixed position relative to one or more aircraft in flight. With the advent of nuclear weapons, the number of aircraft required to bomb a given area effectively has greatly decreased.

Hence, the "tight" bomber formations of World War II are no longer used. Today's bomber formations may require separations on the order of a mile or more.

At distances such as this, radar provides a means of station keeping more accurately. In addition, it is superior to the visual method of formation flying since it is an all weather method.

Positions within the formation are maintained with fixed or variable range markers or with the crosshairs. With radar sets having a highly directional "pencil type" beam, returns appear on any portion of the scope. On other sets, the various aircraft in the formation appear in the altitude hole.

## Navigation Through Weather

The use of radar for weather avoidance has become increasingly important in recent years from the standpoint of both safety and operational flexibility. The severe turbulence, hail, and icing associated with thunderstorms constitute severe hazards to flight. It is mandatory that these thunderstorms be avoided whenever possible. Airborne weather radar, if operated and interpreted properly, can be an invaluable aid in avoiding thunderstorm areas.

Several factors affect the radar returns from thunderstorms, and the operator must be aware of these and the limitations they impose on the weather radar if he is to make optimum use of this tool. Some of these factors are non-meteorological and depend on the characteristics of the radar set and the way it is operated. The same weather "target" can vary considerably in its radarscope appearance as the operator changes the operating characteristics of the set. The operator must insure that he is using the set as designed for weather avoidance. Primary meteorological factors which affect the radar return are the size, shape, number, and phase of the water droplets in the weather "target" and atmospheric absorption characteristics between the radar antenna and the "target." The most important factors are the number and size of the water droplets.

The operator must realize that the predominant weather-induced returns on most radarscopes are caused by precipitation-size water droplets, not by entire clouds. Intense returns indicate the presence of very large droplets. These large droplets are generally associated with the most hazardous phenomena, since strong vertical currents are necessary to maintain these droplets in the cloud. It is possible however to encounter these phenomena in an echo-free area or even in an adjacent cloud-free area, so avoiding areas giving intense returns will not necessarily guarantee safe flight in the vicinity of thunderstorms. Various researchers have empirically determined what they consider to be safe distances for avoiding these intense returns to avoid hazards. These distances vary with altitude and echo characteristics. Although the avoidance procedures recommended by these researchers vary and change somewhat as research continues, they are similar and rarely recommend passage closer to intense echoes than 10 miles at low altitudes. Avoidance by even greater distances is recommended at higher altitudes.

Weather avoidance with radar is mainly of two types: (1) avoidance of isolated thunderstorms and (2) penetration of a line of thunderstorms. The process of avoiding an isolated return is one of first identifying the return and then circumnavigating it at a safe distance. Penetration of a line of thunderstorms presents a somewhat different problem. Since the line may extend for hundreds of miles, circumnavigation is not often practical nor even possible. If the flight must continue and the line penetrated, the main objective is to avoid the more dangerous areas in the line.

An example of frontal penetration using radar is shown in figure 16-33. Upon approaching the line, the navigator determines an area which has weak or no returns and which is large enough to allow avoidance of all intense returns by the recommended distances throughout penetration. He selects this as the penetration point. The pilot directs the aircraft to that point and makes the penetration at right angles to the line so as to remain in the bad weather area for the shortest possible time. *Great care must be taken to avoid the dangerous echoes by the recommended distances.* It should be emphasized that penetration of a line of severe thunderstorms is always a potentially dangerous procedure. It should be attempted only when continuation of the flight is mandatory and the line cannot be circumnavigated.

Some airborne radar sets have special features which are extremely useful in weather avoidance. The user should refer to the operating instructions of these sets for further information.

## Procedure Turns

A procedure turn is one so planned that the aircraft will roll on the proper heading to make good a predetermined track. For example, in military, the bomb run to the target must begin from a fixed point, called the IP (initial point). To roll out on the track from the IP to the target, the turn onto the bomb run must begin at some point *prior* to the IP. This situation is shown in figure 16-34. There are several methods of executing the procedure turn.

COMPUTING THE POINT TO START THE TURN. The point at which a procedure turn should begin is computed on the basis of the amount of turn to be made, the turning *rate*, and the groundspeed of the aircraft. The turn angle and the groundspeed must be determined prior to the turn, through normal navigational procedures. The turn rate is governed by local directives, but it is usually a one-quarter or one-half needlewidth turn. A one-quarter needlewidth turn is an 8-minute,

RADAR PRINCIPLES AND FLIGHT

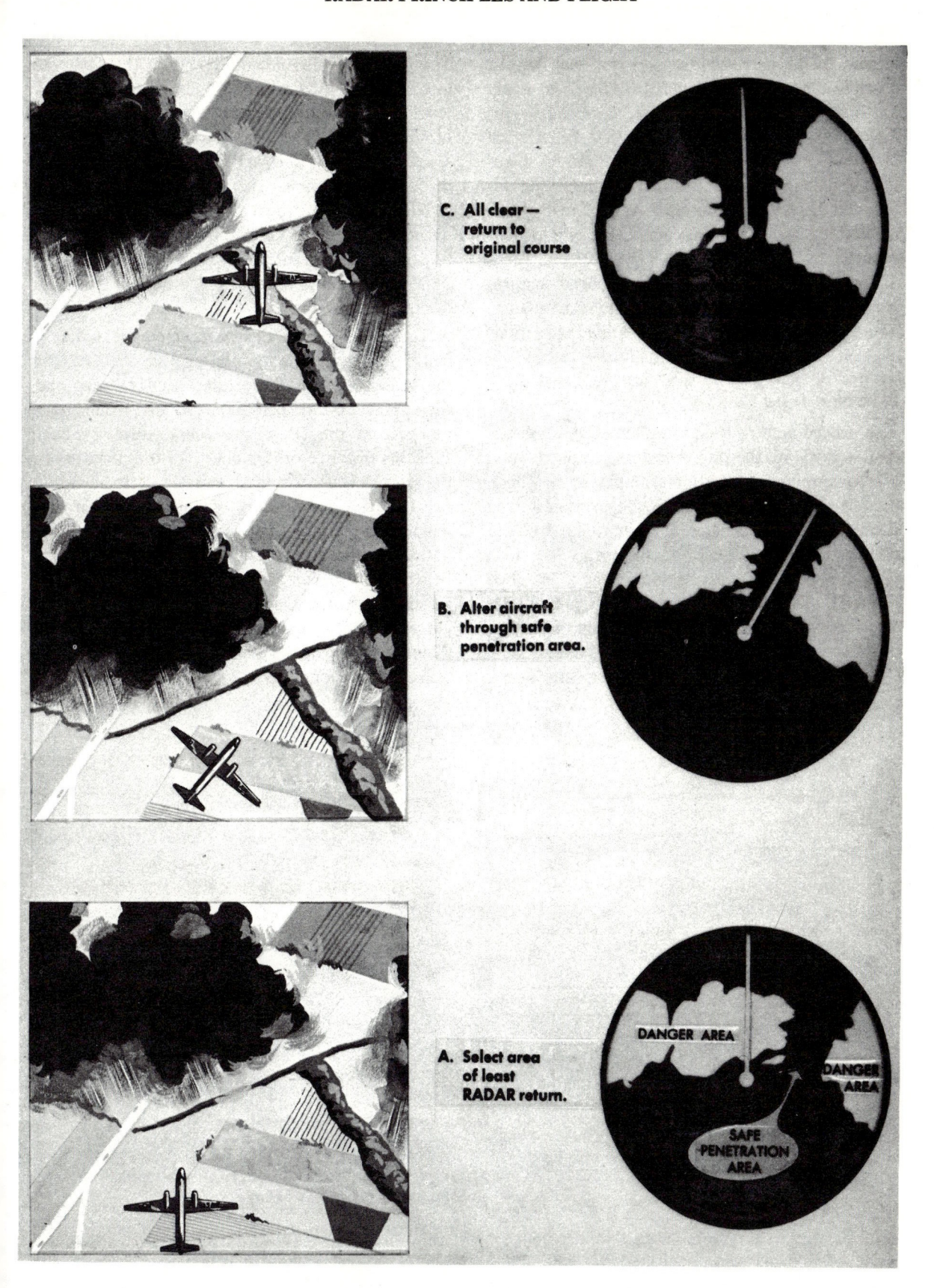

Figure 16-33  Penetration of Thunderstorm Area

219

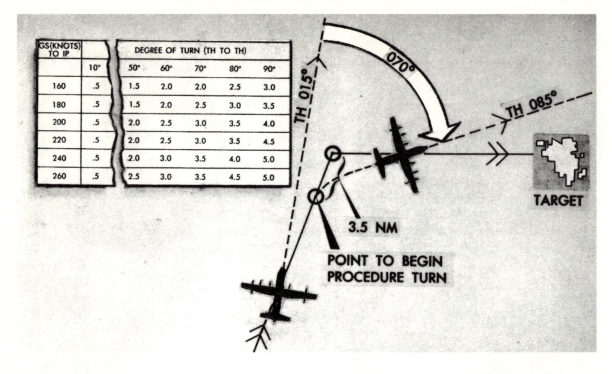

| GS(KNOTS) TO IP | | DEGREE OF TURN (TH TO TH) | | | | | |
|---|---|---|---|---|---|---|---|
| | 10° | 50° | 60° | 70° | 80° | 90° |
| 160 | .5 | 1.5 | 2.0 | 2.0 | 2.5 | 3.0 |
| 180 | .5 | 1.5 | 2.0 | 2.5 | 3.0 | 3.5 |
| 200 | .5 | 2.0 | 2.5 | 3.0 | 3.5 | 4.0 |
| 220 | .5 | 2.0 | 2.5 | 3.0 | 3.5 | 4.5 |
| 240 | .5 | 2.0 | 3.0 | 3.5 | 4.0 | 5.0 |
| 260 | .5 | 2.5 | 3.0 | 3.5 | 4.5 | 5.0 |

*Figure 16-34 Executing a Procedure Turn*

360° turn, and a one-half needlewidth turn is a 4-minute, 360° turn.

To find the turn distance, enter the table in figure 16-34 for the turning rate to be used. At the intersection of the groundspeed column and the degree of turn column, read the ground distance from the IP to the actual turning point Plot this distance on the chart. In the illustration, the aircraft is approaching the IP on a true heading of 015° at a groundspeed of 250 knots. The true heading to the target is 085°. Enter the table with 250 knots and 70° (085° — 015°), the difference between the new heading and the old one. The distance from the IP to start the turn is 3.5 nautical miles. The problem now becomes one of determining when the aircraft is at the turning point. This is found in several ways.

*Fixed Range Marker Method.* The slant range corresponding to the ground range distance can be computed (refer back to Radar Fixing). When the IP appears at the proper slant range on the scope as indicated by the fixed range markers, execute the turn.

*Variable Range Marker Method.* When using a radar set on which there is a variable range marker, execute the turn when the IP intersects the VRM.

*Bearing Method.* If the bearing from the turning point to some easily identifiable ground point is measured, make the turn when the bearing of the

aircraft to this point reaches the measured bearing. For greatest accuracy, the line of bearing from the turning point to the reference point should intersect the track at approximately a right angle.

OFF COURSE METHOD. If the aircraft is not on the desired track to the IP, each of the three methods described above will be inaccurate. For example, if the aircraft is several miles to the left of track and the radar operator waits until the IP is 3.5 nautical miles from the aircraft, the aircraft will roll out on a track above that desired. However, if a line is drawn *on the scope* through the IP and in the direction of the new track, make the turn when the *line*—not the IP—is 3.5 nautical miles from the aircraft.

## IMPROVEMENTS TO BASIC RADAR

Improvements in navigation equipment are continually being made. This section deals with three improvements to the basic radar set.

### Sensitivity Time Constant (STC)

Most radar sets produce a "hot spot" in the center of the radarscope because the high gain setting required to amplify the weak echoes of distant targets over-amplifies the strong echoes of nearby targets. If the receiver gain setting is reduced sufficiently to eliminate the "hot spot," distance returns are weakened or eliminated entirely. The difficulty is most pronounced when radar is used during low-level navigation; to make best use of the radar, the navigator is forced to adjust the receiver gain setting constantly.

STC solves the problem by increasing the gain as the electron beam is deflected from the center to the edge of the radarscope, automatically providing an optimum gain setting for each range displayed. In this manner, the "hot spot" is removed while distant targets are amplified sufficiently.

STC controls vary from one model radar set to another. Refer to the appropriate technical order for operating instructions.

**Terrain Avoidance Radar**

With the increased emphasis on low-level flights, better equipment was needed for flying safety. Terrain avoidance radar (TAR) gives the aircrew all-weather, low-level capability. As mentioned earlier, interpreting mountain shadows on a normal radarscope can be confusing. There is no time for indecision at low levels and at high speeds. TAR increases safety and eliminates confusion by displaying only those vertical obstructions which project above a selected clearance plane.

The two basic types of presentation used with terrain avoidance radar are illustrated in figure 16-35.

PLAN DISPLAY. The plan display is a sector scan presentation that indicates the range and

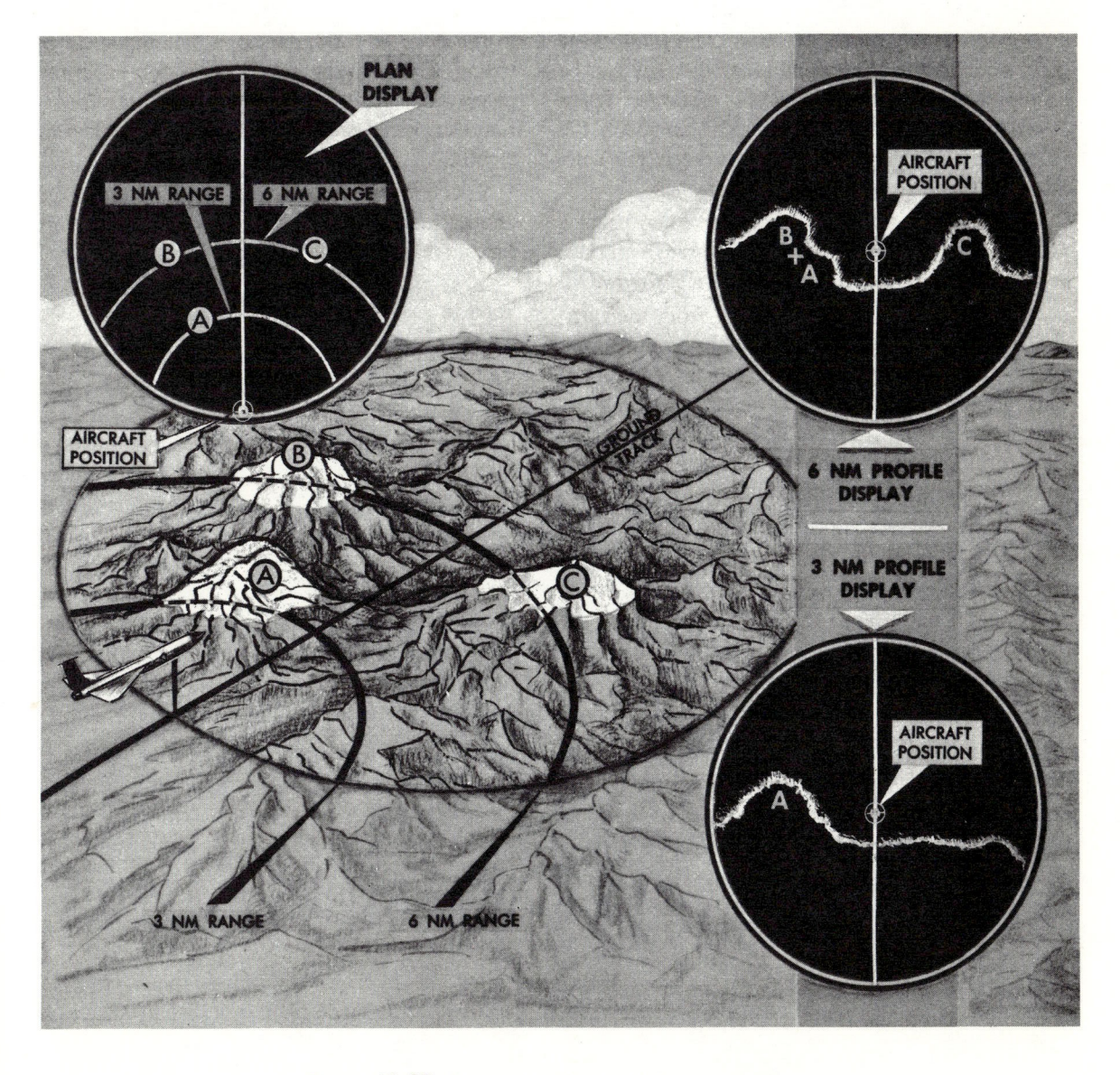

Figure 16-35 Terrain Avoidance Radar Presentations

direction of obstructions projecting above a selected clearance plane. The clearance plane can
be manually set at any level from 3000 feet below
the aircraft up to the level of the aircraft. In
figure 16-35, assume that the clearance plane,
represented by the shaded area, is set 1000 feet
below the aircraft. Only those peaks projecting
above the clearance plane are displayed; all other
returns are inconsequential and are eliminated.
The sector scan presentation limits the returns to
those ahead of the aircraft. The ground track of
the aircraft is represented by the vertical line and
ranges are determined by range marks.

PROFILE DISPLAY. The profile display, normally
received only by the pilot, provides an outline of
the terrain 1500 feet above and below the clearance plane. Elevations of returns are represented
vertically; azimuth is represented horizontally.
This display gives the operator a look "up the
valley." The returns which are seen represent the
highest terrain within the selected range. The
position of the aircraft is represented by an en-
graved aircraft symbol on the indicator overlay.
Figure 16-35 shows both a 3-mile and a 6-mile
presentation.

# RADIO COMMUNICATIONS, FACILITIES AND EQUIPMENT 17

What communications equipment do you need for IFR operations? In uncontrolled airspace, none is required by regulation. What you need is up to your own judgment.

For operations under IFR in controlled airspace, two regulations relate directly to the minimum communications equipment:

a. (FAR 91.33) Your aircraft must be equipped with a two-way radio communications system appropriate to the ground facilities being used.

b. (FAR 91.125) The pilot in command shall have a continuous listening watch maintained on the appropriate frequency and shall report by radio as required.

Information on making radio reports and on the functions and services of Air Traffic Control agencies are considered in Chapter 19.

## Ground Facilities

For civil IFR operations in controlled airspace, the ground facilities available for radio communication include the following compo-

nents of Air Traffic Control:
1. Ground control.
2. Tower control.
3. Departure control.
4. Enroute control.
5. Arrival control.

Communications with these controlling units are normally conducted on VHF frequencies between 108 MHz and 136 MHz. A complete list of VHF frequencies can be found in the *Airman's Information Manual.*

The frequencies with which you should be familiar are listed below. The ILS and VOR frequencies listed may be used as receiving frequencies for communications as well as for navigation.

### AIR NAVIGATION AIDS

108.1–111.9 MHz: ILS localizer with simultaneous radio-telephone channel operating on odd-tenth decimal frequencies (108.1, 108.3 etc.)

108.2–111.8 MHz: VOR's operating on even-tenth decimal frequencies (108.2, 108.4 etc.).

112.0–117.9 MHz: Airway track guidance (VORs)

## COMMUNICATIONS

**118.0–121.4 MHz:** Air Traffic Control Communications
**121.5 MHz:** Emergency (World-Wide)
**121.6–121.95 MHz:** Airport Utility (Ground Control)
**122.0 MHz:** FSS's, Weather, Selected Locations, Private Aircraft and Air Carriers
**122.1 MHz:** Private Aircraft to Flight Service Stations
**122.2, 122.3 MHz:** FSS's, Private Aircraft, Selected Locations
**122.4, 122.5, 122.7 MHz:** Private Aircraft to Towers
**122.6 MHz:** FSS's, Private Aircraft
**122.8, 123.0, 122.85, 122.95 MHz:** Aeronautical Advisory Stations (UNICOM)
**122.9 MHz:** Aeronautical Multicom Stations
**123.05 MHz:** Aeronautical Advisory Stations (UNICOM) Heliports
**123.1–123.55 MHz:** Flight Test and Flying School
**123.6 MHz:** FSS's, Airport Advisory Service
**123.6–128.8 MHz:** Air Traffic Control Communications
**128.85–132.0 MHz:** Aeronautical Enroute Stations (Air Carrier)
**132.05–135.95 MHz:** Air Traffic Control Communications

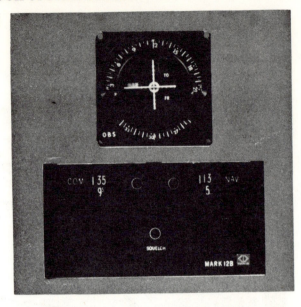

FIGURE 17-1 "One and one-half" system.

## Airborne Equipment

An aircraft equipped to communicate on all available VHF frequencies will have 360 transmitting frequencies (118.0–136.00 MHz with 50 kHz spacing) and 560 receiving frequencies (108.00–136.00 MHz with 50 kHz spacing), covering the entire VHF aircraft radio spectrum. For training, however, you need only a transmitter and receiver suitable for communications with the ground facilities in your training area. Any lightweight equipment which provides the minimum standard frequencies included in the table above, may be sufficient for training but may be inadequate for the instrument rated pilot.

Communications equipment of more recent design reflects the concern of manufacturers for the operational needs of the instrument pilot. Under the best conditions, a pilot flying on instruments must think, decide, and move quickly. Equipment design that contributes to indecision, uncertainty or fatigue creates unnecessary problems when the pilot is flying by reference to instruments.

## "One and One-half" Systems

Figure 17-1 illustrates a dual-purpose radio typical of the "one and one-half" systems. These systems incorporate communications and navigation radios in a single compact unit. This is a welcome change from the older installations that required an extensive cockpit search to locate switches, selectors, and associated indicators and too much time to tune and operate. The "one and one-half" radio enables you to communicate with the necessary ground facilities on the transceiver (combined transmitter/receiver) while simultaneously tuned on the separate "one-half" of the set to a VOR station. This radio has controls that are easily identifiable and crystal-tuned frequencies that can be selected with only a glance or two for tuning.

Operation of the communications section of these radios is simple. The volume switch on the left is the ON/OFF switch for both COMM and NAV sections. The squelch knob is rotated to cut down the background noise generated by the radio tubes. While the set is warming up, set the squelch knob fully clockwise; then turn the control counterclockwise to increase the squelch until the background noise is cut out. Further increase of the squelch setting will decrease receiver sensitivity. If you are tuning in a weak signal, decrease the squelch. Otherwise, once the control is established at a comfortable level, no further adjustment is necessary except to make an occasional check of receiver sensitivity. Transmit/receive frequencies are selected by rotation of the inner and outer knobs on the COMM side of the set.

The "one and one-half" system described has the following communication/navigation frequency coverage:

*Communications*
Transmit/receive 360 channels (118.0–135.95 MHz)

*Navigation*
All VOR and localizer frequencies (108.0–117.9 MHz)

## Expanded NAV/COMM Systems

The advantages of the "one and one-half" system can be expanded with installation of additional radios and centralized control units for rapid selection of receivers and transmitters. The "building block" concept, from which most light aircraft radio design evolves, provides for progressive expansion of your equipment as your training and operational needs increase. For example, if you depart under IFR from an uncontrolled airport and proceed via Victor airway to another uncontrolled airport, your communication/navigation needs may be as little as one VOR frequency and one transmitting frequency for communication with FSS.

When you progress to all-weather IFR flying in and out of unfamiliar terminal areas, your workload can be excessive unless you have sufficient standby equipment for frequent changes of communications channels. Figure 17-2 shows how communications equipment can be grouped for quick reference and operation with minimum distraction from the problem of aircraft control.

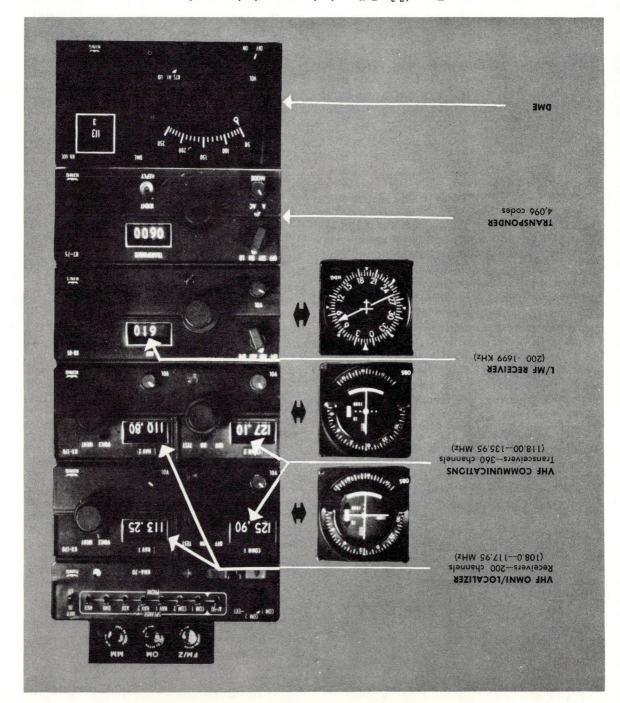

FIGURE 17-2  Full navigation-communications panel.

**DME**

**TRANSPONDER** 4,096 codes

**L/MF RECEIVER** (200 - 1699 KHz)

**VHF COMMUNICATIONS** Transceivers—360 channels (118.00-135.95 MHz)

**VHF OMNI/LOCALIZER** Receivers—200 channels (108.0—117.95 MHz)

## Radiotelephone Procedure

From the time you contact ground control for taxi instructions the effectiveness of your co-ordination with Air Traffic Control will depend upon your competence in communications and your knowledge of traffic procedures under Instrument Flight Rules. Many students have no serious difficulty in learning basic aircraft control and radio navigation, but stumble through even the simplest radio communications. During the initial phase of training in Air Traffic Control procedures and radiotelephone techniques, some students experience difficulty.

Why should talking and listening to a controller pose any problems? The average person takes his speaking and listening habits for granted and has had no occasion to develop specialized skills associated with radio communications. Studies of listening comprehension show that most people listen with low efficiency, even when consciously·attempting to remember what they hear. The poor listener is easily distracted. From habit he tolerates conditions unfavorable to concentration. His mind wanders when he hears anything unexpected or difficult to understand. He is inclined to be more concerned with what he is about to say than with what he should be listening to. When in a confusing situation he is more easily aroused emotionally, and may have trouble comprehending what he hears.

These deficiencies are intensified for the pilot in a busy air traffic environment. In addition to attending to cockpit duties demanding rapid division of attention, quick judgment, concentration, and careful planning, you the pilot must be continuously alert to communications from Air Traffic Control. You should be prepared to listen and to transmit in the brief and unmistakably clear terms vital to orderly control.

You attain proficiency in radiotelephone technique just as you do in developing any other skill. You should first recognize that radio communications under Instrument Flight Rules, though not difficult, require speaking and listening habits different from those you have been accustomed to. Skill in transmitting and listening will come rapidly once you have studied and practiced the basic terminology.

The FAA controller is intensively trained to speak clearly and concisely in an abbreviated terminology understandable to airmen, but somewhat unintelligible to others. He uses standard words and phrases to save time, reduce radio congestion, and lessen the chances of misunderstanding and confusion. However, the most competent controller won't "get through" to you under the best conditions unless you are ready to listen and understand.

Communication is a two-way effort, and the controller expects you to work toward the same level of competency that he strives to achieve. Tape recordings comparing transmissions by professional pilots and inexperienced or inadequately trained general aviation pilots illustrate the need for effective radiotelephone technique. In a typical instance, an airline pilot reported his position in 5 seconds; whereas a private pilot reporting the same fix took 4 minutes to transmit essentially the same information. The difference lay, not in equipment and flight experience, but in communication technique. The novice forgot to tune his radio properly before transmitting, interrupted other transmissions, repeated unnecessary data, forgot other essential information, requested instructions repeatedly, and created the general impression of cockpit disorganization. As encouragement to the novice who is embarrassed and concerned about broadcasting his inexperience, let him remember that every pilot had to make a beginning and was not expected at first to communicate like a veteran airline pilot. But the private pilot who learns and practices standardized words and phrases until they become part of his normal radio vocabulary will be able to communicate effectively even under adverse reception conditions.

## Phonetic Alphabet

It is often necessary in transmitting to identify certain letters and/or groups of letters, or to spell out difficult words, since certain sounds have low intelligibility when mixed with a background of other noises. The standard phonetic alphabet (Fig.17-3) identifies each letter of the alphabet with a word that is easily understood. These words are pronounced to make the message clear when individual letters are transmitted, and are used to spell out words that are hard to understand on the air.

| Phonetic alphabet | | Numerals |
|---|---|---|
| A—Alpha | N—November | 0—Zero |
| B—Bravo | O—Oscar | 1—Wun |
| C—Charlie | P—Papa | 2—Too |
| D—Delta | Q—Quebec*** | 3—Tree |
| E—Echo | R—Romeo | 4—Fo-wer |
| F—Foxtrot | S—Sierra | 5—Fife |
| G—Golf | T—Tango | 6—Six |
| H—Hotel | U—Uniform | 7—Seven |
| I—India | V—Victor | 8—Ait |
| J—Juliett | W—Whiskey | 9—Ni-ner |
| K—Kilo* | X—X-ray | |
| L—Lima** | Y—Yankee | |
| M—Mike | Z—Zulu | |

* Key-lo  ** Lee-mah  ***Keh-beck

FIGURE 17-3 Phonetic alphabet.

## Use of Numbers

Numbers are usually of extreme importance in radio messages and are difficult to hear among other noises. The standard pronunciations in Figure 17-3 have been adopted because they have been found most intelligible.

a. Normally, numbers are transmitted by speaking each number separately. For example, 3284 is spoken as "tree too ait fo-wer."

b. There are certain exceptions to the above rule. Figures indicating hundreds and thousands in round numbers, up to and including 9,000 are spoken in hundreds or thousands as appropriate. 500 is spoken as "fife hundred"; 1,200 as "wun thousand too hundred." Beginning with 10,000, the individual digits in thousands of feet are spoken. For 13,000, say "wun tree thousand", for 14,500, say "wun-fo-wer thousand fife hundred."

c. Aircraft identification numbers are spoken as individual digits/letters. 1234Q is spoken as "wun too tree fo-wer Keh-beck."

d. Time is stated in four digits according to the 24-hour clock. The first two digits indicate the hour; the last two, minutes after the hour (Fig. 17-4), as in "wun niner too zero"; "zero niner fo-wer fife."

e. Field elevations are transmitted with each number spoken separately, as in Figure 17-4.

## Procedural Words and Phrases

The words and phrases in Figure 17-5 should be studied and practiced until they are readily and easily used and clearly enunciated. To pilots and controllers, their meanings are very specific. Careless or incorrect use can cause both delay and confusion.

## Voice Control

Students inexperienced in the use of the microphone are usually surprised at the quality of their own transmissions when they are taped and played back. Words quite clear when

spoken directly to another person can be almost unintelligible over the radio. Effective radio-telephone technique sounds self-conscious and unnatural when you practice it, both because the terminology is new and because you are habitually more concerned with *what* you are saying than in *how* it sounds. Maximum readable radio-telephone transmissions depend on the following factors:

1. *Volume.*—Clarity increases with volume up to a level just short of shouting. Speaking loudly, without extreme effort or noticeably straining the voice, results in maximum intelligibility. To be understood, the spoken sound must be louder at the face of the microphone than the surrounding noises. Open the mouth so the tone will carry to the microphone. A higher-pitched tone is easier to hear than a lower one. A distinct and easily readable side tone in your earphones or speaker is a reliable index of correct volume.

2. *Tempo.*—Effective rate of speech varies with the speaker, the nature of the message, and conditions of transmission and reception. Note the following suggestions for improving your rate of transmission:

| Time | Field elevation |
|---|---|
| 0000—Zero Zero Zero Zero | 10 ft.—Field elevation Wun Zero. |
| 0920—Zero Niner Too Zero | 75 ft.—Field elevation Seven Fife. |
| 1200—Wun Too Zero Zero | 583 ft.—Field elevation Fife Ait Tree. |
| 1645—Wun Six Fower Fife | 600 ft.—Field elevation Six Zero Zero. |
| | 1,250 ft.—Field elevation Wun Too Fife Zero. |
| | 2,500 ft.—Field elevation Too Fife Zero Zero. |

FIGURE 17-4  Expressing time and field elevation.

| Word or phrase | Meanings |
|---|---|
| Acknowledge | Let me know that you have received and understand this message. |
| Affirmative | Yes. |
| Correction | An error has been made in this transmission. The correct version is................ |
| Go ahead | Proceed with your message. |
| How do you hear me? | Self explanatory. |
| I say again | Self explanatory. |
| Negative | That is not correct. |
| Out | This conversation is ended and no response is expected. |
| Over | My transmission is ended and I expect a response from you. |
| Read back | Repeat all of this message back to me exactly as received after I have given "over." |
| Roger | I have received all of your last transmission. (To acknowledge receipt; shall not be used for any other purpose.) |
| Say again | Self explanatory. |
| Speak slower | Self explanatory. |
| Stand by | If used by itself it means, I must pause for a few seconds. If the pause is longer than a few seconds or if "standby" is used to prevent another station from transmitting, it must be followed by the word "out." |
| That is correct | Self explanatory. |
| Verify | Check with the originator. |
| Words twice | As a request—Communication is difficult: Please say every word twice. |

FIGURE 17-5  Radiotelephone words and phrases.

a. Talk slowly enough so that each word and phrase is spoken distinctly, particularly key words and phrases.

b. Talk slowly enough so the listener will have time, not only to hear, but to absorb the meaning.

3. *Pronunciation and Phrasing.*—As you notice the differences in the transmission of various pilots and controllers, you can readily identify those with exceptional skill. They sound natural and unhurried. The words are grouped for easy readability. They pronounce every word clearly and distinctly without apparent effort, without unnecessary words, and without "uh's" and "ah's." They create the impression of competence that any expert conveys after enough study and practice.

## Practice

Many excellent audio training aids are available for practicing radiotelephone procedures. With tapes or records, a microphone, and writing materials, you can develop communications skills under excellent simulated conditions. Practice until you can transmit concisely, hear accurately, and listen critically. Hearing is largely a matter of having an adequate receiver and knowing how to tune it. Critical listening is a more complicated skill. You are ready to listen to a controller when you are thoroughly familiar with your communications equipment and are ready to copy his transmission, evaluate what he says, and if necessary read it back to him without neglecting any other cockpit duties that may demand your attention. Study of Air Traffic Control procedures under Instrument Flight Rules will enable you to "keep ahead" of communications—just as you keep ahead of your basic flying and navigation—by knowing what is ahead of you and attending to details in the proper sequence at the appropriate time.

## Reminders on Use of Equipment

1. Maintain a "readiness" to communicate. With your flight log handy, charts in order, and other necessary materials readily available, you can eliminate fumbling and confusion. You cannot organize an intelligible message, or listen to one, in a disorganized cockpit.

2. Know your radiotelephone equipment and practice tuning it. Check the knobs, switches and selectors *before* you transmit. Monitor the frequency you are using before transmitting. If you hear nothing on a normally busy terminal frequency, for example, check your volume control; you may be interrupting another transmission.

3. Check your clock before transmitting to Flight Service. Other pilots are listening to scheduled weather broadcasts beginning at 15 minutes past the hour.

4. Never subordinate aircraft control to communications. Don't turn your aircraft loose in your haste to transmit.

5. Learn to take notes as you listen. Make written notes of times, altitudes, and other information as you hear it. You have enough to think about in planning ahead without having to waste time thinking back.

# RADIOTELEPHONE PHRASEOLOGY AND TECHNIQUES

## CONTACT PROCEDURE

1. Initiate radio communications with a ground facility by using the following format:

a. Identification of the unit being called.

b. Identification of the aircraft.

c. The type of message to follow, when this will be of assistance.

d. The word 'OVER.'

**Example:**

NEW YORK RADIO, MOONEY THREE ONE ONE ONE ECHO, OVER.

> Note.—When a reply is not received to your initial call to an FSS, check your receiver frequency, make sure it is correct. If you just departed from an outlying airport, check your altitude, you may be too low. If a reply is not received on your second call, listen on the other FSS frequencies to determine if the specialist is busy with another aircraft.

2. Reply to callup from a ground facility by using the following format:

a. Identification of the unit initiating the callup.

b. Identification of the aircraft.

c. The word 'OVER.'

**Example:**

PITTSBURGH TOWER, CESSNA TWO SIX FOUR FIVE ZULU OVER.

> Note.—The word 'OVER' may be omitted if the message obviously requires a reply.

3. Use the same format as for initial callup and reply after communication has been established except, after stating your identification, state the message to be sent or acknowledgment of the message received. The acknowledgment is made with the word 'ROGER' or 'WILCO' and pilots are expected to comply with ATC clearances/instructions when they acknowledge by using either 'ROGER' or 'WILCO.'

**Example:**

APACHE ONE TWO THREE XRAY, ROGER.

4. After contact has been definitely established, it may be continued without further call-up or identification.

5. Pilots operating under provisions of FAR Part 135, ATCO, certificate are urged to prefix their aircraft identification with the phonetic word "Tango" on the initial contact with ATC facilities unless they have been assigned FAA authorized call signs.

**Example:**

TANGO AZTEC 2464 ALFA.

> Note.—The prefix "Tango" may be dropped on subsequent contacts on the same frequency.

## MICROPHONE TECHNIQUE

1. Proper microphone technique is important in radiotelephone communications. Transmissions should be concise and in a normal conversational tone.

> Note.—Identification of Aircraft—Pilots are requested to exercise care that the identification of their aircraft is clearly transmitted in each contact with an ATC facility. Also pilots should be certain that their aircraft are clearly identified in ATC transmissions before taking action on an ATC clearance.

2. When originating a radiotelephone call-up to any air-ground facility, indicate the channel on which reply is expected, if other than normal.

**Example:**

The New York FSS transmits on several VOR frequencies in the area. These VOR's are shown on charts, each having a different name, identification, and frequency. If you are tuned to the Riverhead VOR and wish to call the New York FSS, call "RIVERHEAD RADIO." This tells the New York FSS which frequency you are listening to. If you call "NEW YORK RADIO", you have to tell them what frequency you are listening to—"REPLY ON RIVERHEAD VOR."

3. Keep your contacts as brief as possible. Pilots should not read back altimeter setting, taxi instructions, wind and runway information to towers except for verification or clarification of instructions. Other pilots are waiting to use the channel.

4. Contact the nearest Flight Service Station. Don't continually attempt to see how far your transmitter will reach. If in doubt about the frequency to contact an FSS, transmit on 122.1 MHz and advise them the frequency you are listening on.

5. Avoid calling stations at 15 minutes past the hour, because of interference with scheduled weather broadcasts.

6. When making a position report, pilots should in all cases state the name of the reporting point over which, or in relation to which, they are reporting. The phrase "OVER YOUR STATION" should not be used.

## AIRCRAFT CALL SIGNS

Garbled aircraft identifications in radiotelephone transmissions should never be taken for granted but should always be checked.

1. Use the complete aircraft identification during initial contact with a ground station.

a. Civil itinerant aircraft pilots should state the aircraft type, model or make or the word November followed by the digits and letters of the registration number.

**Example:**

BONANZA ONE TWO THREE FOUR TANGO.

b. Air Taxi or other commercial operators not having FAA authorized call signs should prefix normal identification with the phonetic word "Tango."

2. Air carriers and commuter air carriers having FAA authorized call signs identify themselves by stating the call sign, followed by the trip number spoken as a group.

**Example:**

UNITED TWENTY-FIVE OR COMMUTER SIX-ELEVEN.

3. Military aircraft pilots may use whichever of the following is applicable:

a. The service name or designated prefix followed by the last 5 digits of the serial number.

**Example:**

AIR FORCE FOUR FOUR NINER THREE TWO or MAC FOUR FOUR NINER THREE TWO.

b. The service name or designation followed by the word "RESCUE" and the 5 digits of the serial number.

c. Assigned voice call signs consisting of a selected authorized code word followed by a two-digit flight number.

**Example:**

ANDY TWO ZERO.

d. Assigned double-letter, two-digit flight numbers.

**Example:**

ALFA KILO ONE FIVE.

4. Civilian airborne ambulance flights (aircraft carrying ambulatory or litter patients, organ donors, or

organs for transplant) will be expedited and necessary notification will be made when the pilot requests. When filing flight plans for such flights, add the word "Lifeguard" in the remarks section. In radio communications use the call sign "Lifeguard" followed by the type, digits, and letters of the registration number. Pilots should use discretion in the use of this term. It should be used for those missions of an urgent nature.

**Example:**

LIFEGUARD CESSNA TWO SIX FOUR SIX.

5. Abbreviated aircraft identifications are not to be used except when initiated by the ground station. FAA personnel may abbreviate aircraft call-signs by using the identification prefix and the last three digits or letters of aircraft identificaton after communications have been established and the type aircraft is known. Aircraft with similar sounding identifications, or the identifications of an air carrier and other civil aircraft having an FAA approved call-sign will not be abbreviated.

**Example:**

"TRI PACER SIX TWO YANKEE."

## GROUND STATION CALL SIGNS

Ground station call signs shall comprise the name of the location or airport, followed by the appropriate indication of the type of station:

OAKLAND TOWER (airport traffic control tower);

MIAMI GROUND (ground control position in tower);

DALLAS CLEARANCE DELIVERY (IFR clearance delivery position);

KENNEDY APPROACH (tower radar or nonradar approach control position);

ST. LOUIS DEPARTURE (tower radar departure control position);

WASHINGTON RADIO (FAA Flight Service Station);

NEW YORK CENTER (FAA Air Route Traffic Control center).

## PROCEDURE WORDS AND PHRASES

The following words and phrases should be used where practicable in radiophone communications:

| Word or Phrase | Meaning |
|---|---|
| ACKNOWLEDGE | "Let me know that you have received and understood this message." |
| AFFIRMATIVE | "Yes." |
| CORRECTION | "An error has been made in this transmission. The correct version is . . ." |
| GO AHEAD | "Proceed with your message." |
| HOW DO YOU HEAR ME? | Self-explanatory. |
| I SAY AGAIN | Self-explanatory. |
| NEGATIVE | "No" or "Permission not granted" or "That is not correct." |
| OUT | "This conversation is ended and no response is expected." |
| OVER | "My transmission is ended and I expect a response from you." |
| READ BACK | "Repeat all of this message back to me." |
| ROGER | "I have received all of your last transmission." (To acknowledge receipt, shall not be used for other purposes.) |
| SAY AGAIN | Self-explanatory. |
| SPEAK SLOWER | Self-explanatory. |
| STAND BY | "If used by itself means 'I must pause for a few seconds.' If the pause is longer than a few |

seconds, or if 'STAND BY' is used to prevent another station from transmitting, it must be followed by the ending 'OUT'."

| | |
|---|---|
| THAT IS CORRECT | Self-explanatory. |
| VERIFY | Confirm. |
| WILCO | I have received your message, understand it, and will comply. |
| WORDS TWICE | (a) As a request: "Communication is difficult. Please say every phrase twice." |
| | (b) As information: "Since communication is difficult, every phrase in this message will be spoken twice." |

## TIME

1. The Federal Aviation Administration utilizes Greenwich Mean Time (GMT or "Z") for all operational purposes.

| To Convert From: | To Greenwich Mean Time: |
|---|---|
| Eastern Standard Time | Add 5 hours |
| Eastern Daylight Time | Add 4 hours |
| Central Standard Time | Add 6 hours |
| Central Daylight Time | Add 5 hours |
| Mountain Standard Time | Add 7 hours |
| Mountain Daylight Time | Add 6 hours |
| Pacific Standard Time | Add 8 hours |
| Pacific Daylight Time | Add 7 hours |

2. The 24-hour clock system is used in radiotelephone transmissions. The hour is indicated by the first two figures and the minutes by the last two figures.

**Examples:**

| | |
|---|---|
| 0000 | ZERO ZERO ZERO ZERO |
| 0920 | ZERO NINER TWO ZERO |

3. Time may be stated in minutes only (two figures) in radio telephone communications when no misunderstanding is likely to occur.

4. Current time in use at a station is stated in the nearest quarter minute in order that pilots may use this information for time checks. Fractions of a quarter minute less than eight seconds are stated as the preceding quarter minute; fractions of a quarter minute of eight seconds or more are stated as the succeeding quarter minute.

**Examples:**

| Time | |
|---|---|
| 0929:05 | TIME, ZERO NINER TWO NINER |
| 0929:10 | TIME, ZERO NINER TWO NINER AND ONE-QUARTER |
| 0929:28 | TIME, ZERO NINER TWO NINER AND ONE-HALF |

## FIGURES

1. Figures indicating hundred and thousands in round number, as for ceiling heights, and upper wind levels up to 9900 shall be spoken in accordance with the following examples:

| | |
|---|---|
| 500 | FIVE HUNDRED |
| 1300 | ONE THOUSAND THREE HUNDRED |
| 4500 | FOUR THOUSAND FIVE HUNDRED |
| 9900 | NINER THOUSAND NINER HUNDRED |

2. Numbers above 9900 shall be spoken by separating the digits preceding the word "thousand." Examples:

| | |
|---|---|
| 10000 | ONE ZERO THOUSAND |
| 13000 | ONE THREE THOUSAND |
| 18500 | ONE EIGHT THOUSAND FIVE HUNDRED |
| 27000 | TWO SEVEN THOUSAND |

3. Transmit airway or jet route numbers as follows:

**Examples:**

| | |
|---|---|
| V12 | VICTOR TWELVE |
| J533 | J FIVE THIRTY THREE |

4. All other numbers shall be transmitted by pronouncing each digit.

**Examples:**

| | |
|---|---|
| 10 | ONE ZERO |
| 75 | SEVEN FIVE |
| 583 | FIVE EIGHT THREE |
| 1850 | ONE EIGHT FIVE ZERO |
| 18143 | ONE EIGHT ONE FOUR THREE |
| 26075 | TWO SIX ZERO SEVEN FIVE |

The digit "9" shall be spoken "NINER".

5. When a radio frequency contains a decimal point, the decimal point is spoken as "POINT."

**Examples:**

| | |
|---|---|
| 122.1 | ONE TWO TWO POINT ONE |
| 126.7 | ONE TWO SIX POINT SEVEN |

(ICAO Procedures require the decimal point be spoken as "DECIMAL" and FAA will honor such usage by military aircraft and all other aircraft required to use ICAO Procedures.)

## FLIGHT ALTITUDES

1. Up to but not including 18,000′ MSL—by stating the separate digits of the thousands, plus the hundreds, if appropriate.

**Examples:**

| | |
|---|---|
| 12,000 | ONE TWO THOUSAND |
| 12,500 | ONE TWO THOUSAND FIVE HUNDRED |

2. At and above 18,000′ MSL (FL 180) by stating the words "flight level" followed by the separate digits of the flight level.

**Examples:**

| | |
|---|---|
| 190 | FLIGHT LEVEL ONE NINER ZERO |
| 275 | FLIGHT LEVEL TWO SEVEN FIVE |

## ● DIRECTIONS

The three digits of the magnetic course, bearing, heading or wind direction. All of the above should always be magnetic. The word "true" must be added when it applies.

**Examples:**

| | |
|---|---|
| (magnetic course) 005 | ZERO ZERO FIVE |
| (true course) 050 | ZERO FIVE ZERO TRUE |
| (magnetic bearing) 360 | THREE SIX ZERO |
| (magnetic heading) 100 | ONE ZERO ZERO |
| (wind direction) 220 | TWO TWO ZERO |

## SPEEDS

The separate digits of the speed followed by the word 'knots'. The controller may omit the word "knots" when using speed adjustment procedures, "Reduce/Increase Speed To One Five Zero."

**Examples:**

| | |
|---|---|
| 250 | TWO FIVE ZERO KNOTS |
| 185 | ONE EIGHT FIVE KNOTS |
| 95 | NINER FIVE KNOTS |

## PHONETIC ALPHABET

1. The International Civil Aviation Organization (ICAO) phonetic alphabet is used by FAA personnel when communications conditions are such that the information cannot be readily received without their use. Air traffic control facilities may also request pilots to use phonetic letter equivalents when aircraft with similar sounding identifications are receiving communications on the same frequency.

2. Pilots should use the phonetic alphabet when identifying their aircraft during initial contact with air traffic control facilities. Additionally use the phonetic equivalents for single letters and to spell out groups of letters or difficult words during adverse communications conditions.

| Letter | Morse | Word | Pronunciation |
|---|---|---|---|
| A | ● ▬ | Alfa | (AL-FAH) |
| B | ▬ ● ● ● | Bravo | (BRAH-VOH) |
| C | ▬ ● ▬ ● | Charlie | (CHAR-LEE) (or SHAR LEE) |
| D | ▬ ● ● | Delta | (DELL-TAH) |
| E | ● | Echo | (ECK-OH) |
| F | ● ● ▬ ● | Foxtrot | (FOKS-TROT) |
| G | ▬ ▬ ● | Golf | (GOLF) |
| H | ● ● ● ● | Hotel | (HOH-TEL) |
| I | ● ● | India | (IN-DEE-AH) |
| J | ● ▬ ▬ ▬ | Juliett | (JEW-LEE-ETT) |
| K | ▬ ● ▬ | Kilo | (KEY-LOH) |
| L | ● ▬ ● ● | Lima | (LEE-MAH) |
| M | ▬ ▬ | Mike | (MIKE) |
| N | ▬ ● | November | (NO-VEM-BER) |
| O | ▬ ▬ ▬ | Oscar | (OSS-CAH) |
| P | ● ▬ ▬ ● | Papa | (PAH-PAH) |
| Q | ▬ ▬ ● ▬ | Quebec | (KEH-BECK) |
| R | ● ▬ ● | Romeo | (ROW-ME-OH) |
| S | ● ● ● | Sierra | (SEE-AIR-RAH) |
| T | ▬ | Tango | (TANG-GO) |
| U | ● ● ▬ | Uniform | (YOU-NEE-FORM) (or OO-NEE-FORM) |
| V | ● ● ● ▬ | Victor | (VIK-TAH) |
| W | ● ▬ ▬ | Whiskey | (WISS-KEY) |
| X | ▬ ● ● ▬ | Xray | (ECKS-RAY) |
| Y | ▬ ● ▬ ▬ | Yankee | (YANG-KEY) |
| Z | ▬ ▬ ● ● | Zulu | (ZOO-LOO) |
| 1 | ● ▬ ▬ ▬ ▬ | Wun | |
| 2 | ● ● ▬ ▬ ▬ | Too | |
| 3 | ● ● ● ▬ ▬ | Tree | |
| 4 | ● ● ● ● ▬ | Fow-er | |
| 5 | ● ● ● ● ● | Fife | |
| 6 | ▬ ● ● ● ● | Six | |
| 7 | ▬ ▬ ● ● ● | Sev-en | |
| 8 | ▬ ▬ ▬ ● ● | Ait | |
| 9 | ▬ ▬ ▬ ▬ ● | Nin-er | |
| 0 | ▬ ▬ ▬ ▬ ▬ | Zero | |

## Clearance Shorthand

The shorthand system given here is recommended by the Federal Aviation Administration. Applicants for the Instrument Rating may use *any* shorthand system, in any language, which insures accurate compliance with ATC instructions. No shorthand system is required by regulation and no knowledge of shorthand is required for the written test; however, because of the vital necessity for safe coordination between the pilot and controller, clearance information should be unmistakably clear.

As an instrument pilot, you should make a written record of all ATC clearances and instructions that consist of more than a few words; and any portions that are complex, or about which there is any doubt, should be verified by a repeat back. Safety demands that you receive correctly and do not forget any part of your clearance.

Occasionally ATC will issue a clearance that differs from the original request. In such cases, the pilot must be particularly alert to be sure that he receives and understands the clearance given.

The following symbols and contractions represent words and phrases frequently used in clearances. Most of them are regularly used by ATC personnel. Learn them along with the location identifiers which you will use.

By using this shorthand, omitting the parenthetical words, you will be able, after some practice, to copy long clearances as fast as they are read.

| *Words and Phrases* | *Shorthand* |
|---|---|
| ABOVE | ABV |
| ADVISE | ADV |
| AFTER (PASSING) | < |
| AIRPORT | A |
| (ALTERNATE INSTRUCTIONS) | ( ) |
| ALTITUDE 6,000–17,000 | 60-170 |
| AND | & |
| APPROACH | AP |
| FINAL | F |
| OMNI | O |
| PRECISION | PAR |
| STRAIGHT-IN | SI |
| SURVEILLANCE | ASR |
| APPROACH CONTROL | APC |
| AT (USUALLY OMITTED) | |
| (ATC) ADVISES | CA |
| (ATC) CLEARS OR CLEARED | C |
| (ATC) REQUESTS | CR |
| BEARING | BEAR |
| BEFORE | > |
| BELOW | BLO |
| BOUND | B |
| EASTBOUND, etc. | EB |
| INBOUND | IB |

| *Words and Phrases* | *Shorthand* |
|---|---|
| OUTBOUND | OB |
| CLIMB (TO) | ↑ |
| CONTACT | CT |
| CONTACT DENVER APPROACH CONTROL | DEN |
| CONTACT DENVER CENTER | (DEN |
| COURSE | CRS |
| CROSS | X |
| CROSS CIVIL AIRWAYS | |
| CRUISE | → |
| DELAY INDEFINITE | DLI |
| DEPART | DEP |
| DESCEND (TO) | ↓ |
| DIRECT | DR |
| EACH | ea |
| EXPECT APPROACH CLEARANCE | EAC |
| EXPECT FURTHER CLEARANCE | EFC |
| FAN MARKER | FM |
| FLIGHT PLANNED ROUTE | FPR |
| FOR FURTHER CLEARANCE | FFC |
| FOR FURTHER HEADINGS | FFH |
| HEADING | HDG |
| HOLD (DIRECTION) | H-W |
| IF NOT POSSIBLE | or |
| INTERSECTION | XN |
| (ILS) LOCALIZER | L |
| MAINTAIN OR MAGNETIC | M |
| (MAINTAIN) VFR CONDITIONS ABOVE: | |
| ALL CLOUDS | VFR |
| ALL HAZE | $\frac{VFR}{H}$ |
| ALL DUST | $\frac{VFR}{D}$ |
| ALL SMOKE | $\frac{VFR}{K}$ |
| ALL FOG | $\frac{VFR}{F}$ |
| OMNI (RANGE) | O |
| OUTER COMPASS LOCATOR | LOM |
| OUTER MARKER | OM |
| OVER (Iden.) | OKC-0 |
| RADAR VECTOR | R V |
| RADIAL | RAD |
| REMAIN WELL TO LEFT SIDE | LS |
| REMAIN WELL TO RIGHT SIDE | RS |
| REPORT DEPARTING | RD |

| Words and Phrases | Shorthand |
|---|---|
| REPORT LEAVING | RL |
| REPORT ON COURSE | R-CRS |
| REPORT OVER | RO |
| REPORT PASSING | RP |
| REPORT REACHING | RR |
| REPORT STARTING PROCEDURE TURN | RSPT |
| REQUEST ALTITUDE CHANGES ENROUTE | RACE |
| REVERSE COURSE | RC |
| RUNWAY | RY |
| STANDARD JET PENETRATION | SJP |
| STANDBY | STBY |
| TAKEOFF (DIRECTION) | T→N |
| TOWER | Z |
| TRAFFIC IS | TFC |
| TRACK | TR |
| TURN LEFT AFTER TAKEOFF | LT |

| Words and Phrases | Shorthand |
|---|---|
| TURN RIGHT AFTER TAKEOFF | RT |
| UNTIL | / |
| UNTIL ADVISED (BY) | UA |
| UNTIL FURTHER ADVISE | UFA |
| VICTOR | V |
| WHILE IN CONTROL AREAS | Δ |
| WHILE IN CIVIL AIRWAYS | = |

## EXAMPLE

ATC clears (Iden.) to St. Louis Airport via Victor 16, Victor 9. Maintain one five thousand. Turn left after departure. Proceed direct Dallas Omni. Cross Dallas Omni at 3000. Report leaving 3, 4, and 5000.

C STL-A V16 V9. M 150. LT. DR DAL-OXDAL-O 30. RL 3, 4, 50.

# 18 THE FEDERAL AIRWAYS SYSTEM AND CONTROLLED AIRSPACE

## System Details

Up to this point, your instrument training has been concerned largely with problems within the cockpit. While you have been acquiring proficiency in the use of basic flight instruments and NAV/COM equipment, your instructor has kept a watchful eye on other traffic. He has told you what maneuvers to execute, what radials to fly, when and where to go. Instrument flying would be relatively simple if your instrument training ended with mastery of these basic techniques and with a safety pilot aboard to keep you headed in the right direction, safely separated from other traffic. Your problem now is to learn how to use the facilities, services, and procedures established by the Air Traffic System to provide directional guidance, terrain clearance, and safe separation for aircraft operating under Instrument Flight Rules. The extent of this system and the facilities maintained for airspace users can be appreciated by visualizing the tremendous expansion of aviation since 1903 when only one airplane used the national airspace.

Figure 18-1 shows the low-frequency radio facilities serving the Los Angeles-Oakland area in 1934. The navaids provided only one route between those two terminals via the Los Angeles, Fresno, and Oakland range legs.

In 1964, 30 years later, the system in this area included the VOR facilities and routes shown in Figure 18-2 as well as instrument approach aids, radar coverage, and numerous other facilities and services.

By the 1960's, the Nation's air fleet approached 150,000 aircraft, with as many as 70,000 people aloft at any busy hour, most of them converging on, or departing from, major metropolitan areas. In 1969, the 2,110 navigation aids needed to keep these aircraft moving safely throughout the 50 States were listed as follows (exclusive of nondirectional homing beacons, standard broadcasting stations, precision approach radar systems, and non-Federal or special use aids):

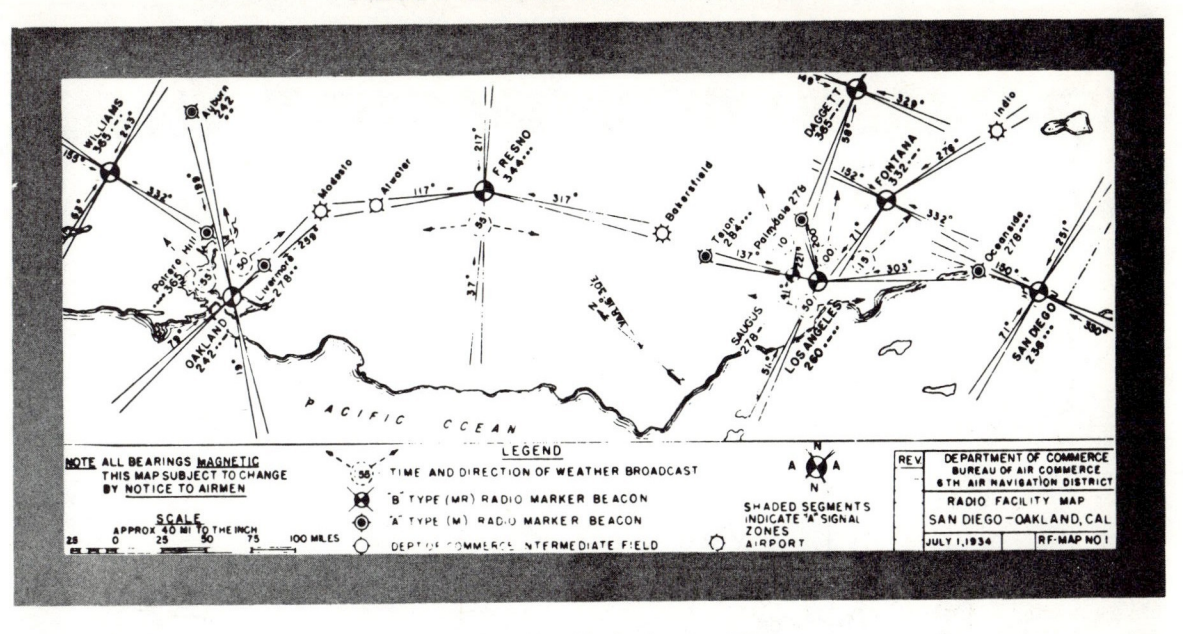

FIGURE 18-1 Navaids during the 1930's.

27 Air Route Traffic Control Centers (includes 2 Center/RAPCONS and 1 Center/Tower).
90 Air Route Surveillance Radar (ARSR).
155 Airport Surveillance Radar (ASR)—(includes 36 military radars which provide service for civil airports).
952 VOR/VORTACS—(includes 27 non-Federal and 42 military which have been incorporated into the "Common System").
279 Instrument Landing Systems (ILS).
256 Approach Light Systems with Sequence Flashers (ALS).
351 Towers and Combined Station/Towers (CS/T—includes 30 non-Federal).

The airways system resembles its automotive counterpart in many ways. Whether you travel by Federal airway or Federal highway, the system must provide a controlling agency, procedural rules, directional guidance, highways between population centers, and a means of access between highways and terminals.

## Federal Airways

The Federal airways network is based upon the electronic aids already discussed. The navigation system has three component parts: the pilot, the airborne receiver, and the ground navigation facility. When the recognized errors which each contributes are considered, a total system accuracy can be determined. When this accuracy is applied, the area in which obstruction clearance should be provided becomes apparent. Inherent in this concept is the premise that the pilot fly, as closely as possible, the prescribed courses and altitudes.

Obstruction clearance criteria are based on the airborne receiver contributing not more than ±4.2° error to the total system error. Pilot performance must assure a tracking accuracy within ±2.5° (quarter scale) needle deflection. Where these standards are not assured by the pilot, the safety and accuracy normally provided by the criteria are impaired. The VOR and VORTAC facilities are the foundation of the system (Fig. 18-3). Connecting these facilities is the network of routes forming the Victor Airway System.

Each Federal Airway is based on a centerline that extends from one navigation aid or intersection to another navigation aid (or through several navigation aids or intersections) specified for that airway. The infinite number of radials transmitted by the VOR permits 360 possible separate airway courses to or from the facility, one for each degree of azimuth. Thus, a given VOR located within approximately 100 miles of several other VORs may be used to establish many different airways. For example, 11 radials at Houston VORTAC define 19 different low-altitude airways (Fig. 18-4).

The enroute airspace structure consists of three strata: Airways (generally 700 feet or 1,200 feet above the surface up to but not including 18,000 MSL), the Jet Route structure (18,000 feet to flight level 450–45,000 feet), and the airspace above FL 450, which is for point-to-point operation.

To the extent possible, these route systems have been aligned in an overlying manner to facilitate transition between each. At certain airports, Standard Terminal Arrival Routes (STARs) have been established for application to arriving IFR aircraft.

Like highways, Victor Airways are designated by number—generally north/south airways are *odd*; east/west airways are *even*. When airways coincide on the same radial, the airway segment shows the numbers of all the airways on it. For example, the Houston 340 radial in the excerpt

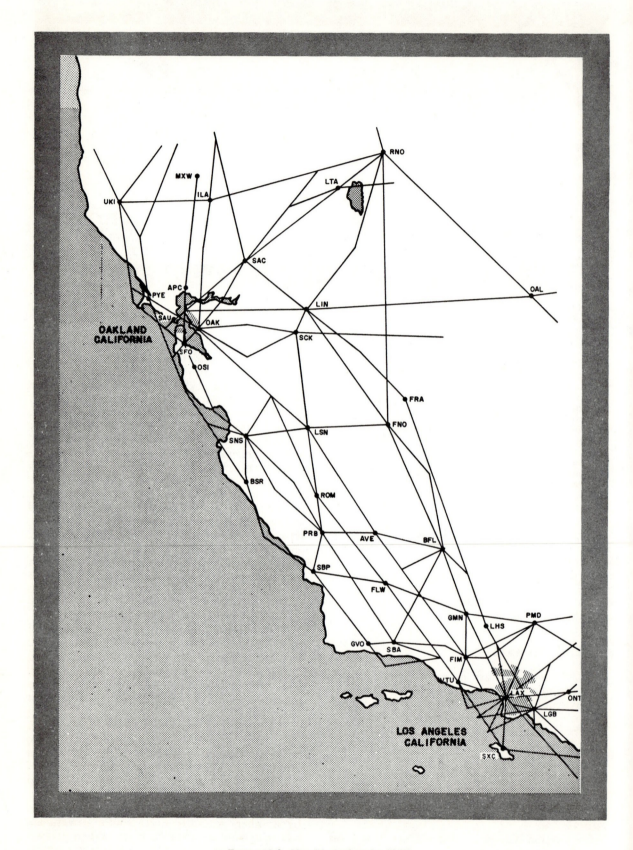

FIGURE 18-2  Navaids during the 1960's.

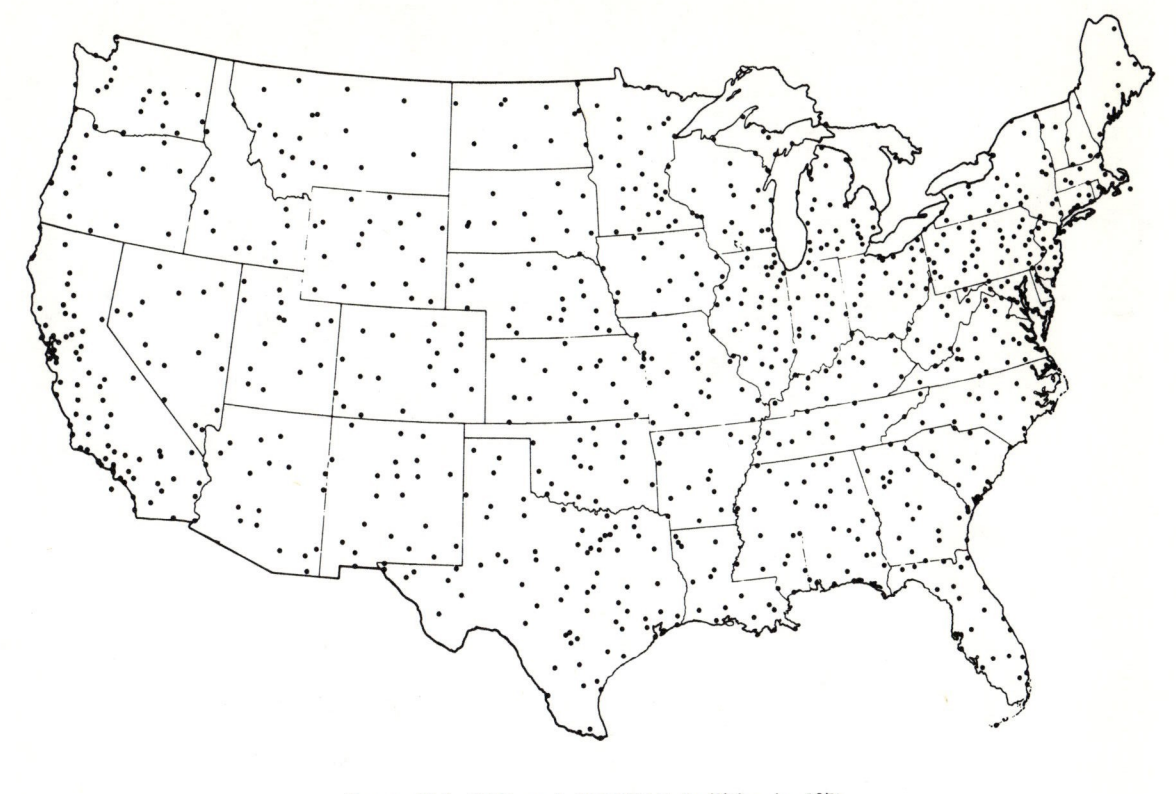

FIGURE 18-3 VOR and VORTAC facilities in 1971.

defines three airways, V–13–15E–477.

*Alternate Victor Airways* are used for lateral separation when traffic conditions require it, or as "one-way" routes for efficient control of traffic flow to and from terminal areas. For example, V–14 serves as a normal arrival route into Oklahoma City from Tulsa, Oklahoma. V–14N, lying to the north of V–14, is an alternate arrival route from Tulsa. V–14S, to the south of V–14, is a normal departure route from Oklahoma City.

*Preferred Routes* have been established between major terminals to guide pilots in planning their routes of flight, to minimize route changes, and to aid in the orderly management of air traffic using Federal airways.

*Terminal Routes (Standard Terminal Arrival Routes [STARs] and Standard Instrument Departures [SIDs])*—These Standard Routes have been established and published as an air traffic control aid in certain complex terminal areas. They help reduce verbiage on clearance delivery and control frequencies and provide the pilot with a description of his terminal routing. SIDs are currently published in the National Ocean Survey SID booklet. STARs are currently published in AIM Part 3. Instructions for pilot use of these coded routes are contained in the AIM. Certain complex terminal areas are covered by Area Charts.

## Controlled Airspace

Traffic separation on the Federal airways, terminal routes, and in the vicinity of airports requires the definition of the airspace in which control is necessary. Such airspace is called controlled airspace, which includes the airspace designated as control zone, transition area, control area, and continental control area, within which some or all aircraft may be subject to air traffic control. Figure 18-5 is a three dimensional portrayal of certain features of the airspace structure.

*Control Zones* extend upward from the earth's surface, terminating at the continental control area. A control zone is normally a circular area of 5 miles radius and may include one or more airports. Extensions are provided where necessary for instrument approach and departure paths.

*Transition Areas* consist of controlled airspace extending upward from 700 feet or more above the surface of the earth when designated in conjunction with an airport for which an approved instrument approach procedure has been prescribed; or from 1,200 feet or more above the surface of the earth when designated in conjunction with airway route structures or segments. Unless otherwise limited, transition areas terminate at the base of the overlying controlled airspace.

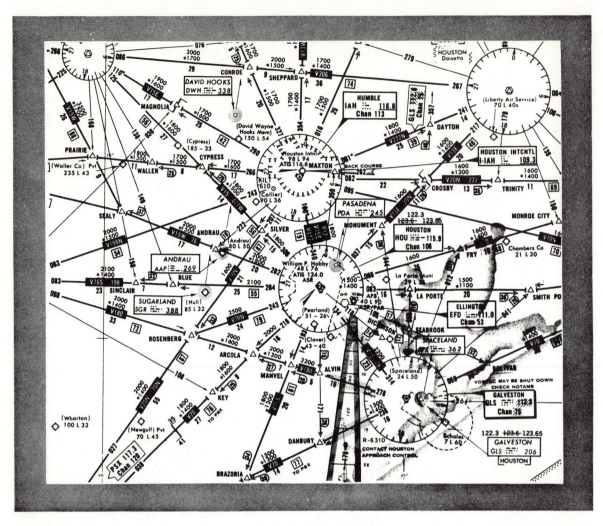

FIGURE 18-4  VOR radials and airways.

*Control Areas* include the following airspace in the Federal Airways structure:

a. The airspace within parallel boundary lines 4 nautical miles on each side of the centerline of the airway, and where the changeover point is more than 51 miles from either of the navigation aids, the airspace between lines diverging at 4.5 degrees from the centerline at the navigation aid most distant from the changeover point (either NAVAID when equidistant from this point) extending until they intersect with the bisector of the angle of the centerlines at the changeover point, and the airspace between the lines connecting these points and the other navigation aid.

b. The airspace that extends upward from 700 feet (unless designated from 1,200 feet or more) above the surface of the earth.

c. Unless otherwise designated, the airspace between a main and associated alternate VOR Federal Airway, with the vertical extent of this area corresponding to the vertical extent of the main airway.

Control areas do not include the airspace of the Continental Control Area. They may or may not include all or part of the airspace within a restricted area.

*Additional Control Areas and Control Area Extensions* are airspace of individually defined dimensions which may be found in Subpart E of Part 71 of the Federal Aviation Regulations.

*The Continental Control Area* consists of the airspace of the 48 conterminous States, part of Alaska, and the District of Columbia at and above 14,500 MSL, but does not include—

a. The airspace less than 1,500 feet above the surface of the earth; or

b. Prohibited and restricted areas, other than restricted area military climb corridors and other special restricted areas designated by regulation.

## Radio Navigation Charts

Radio navigation data are shown on many types of aeronautical charts, including the Sec-

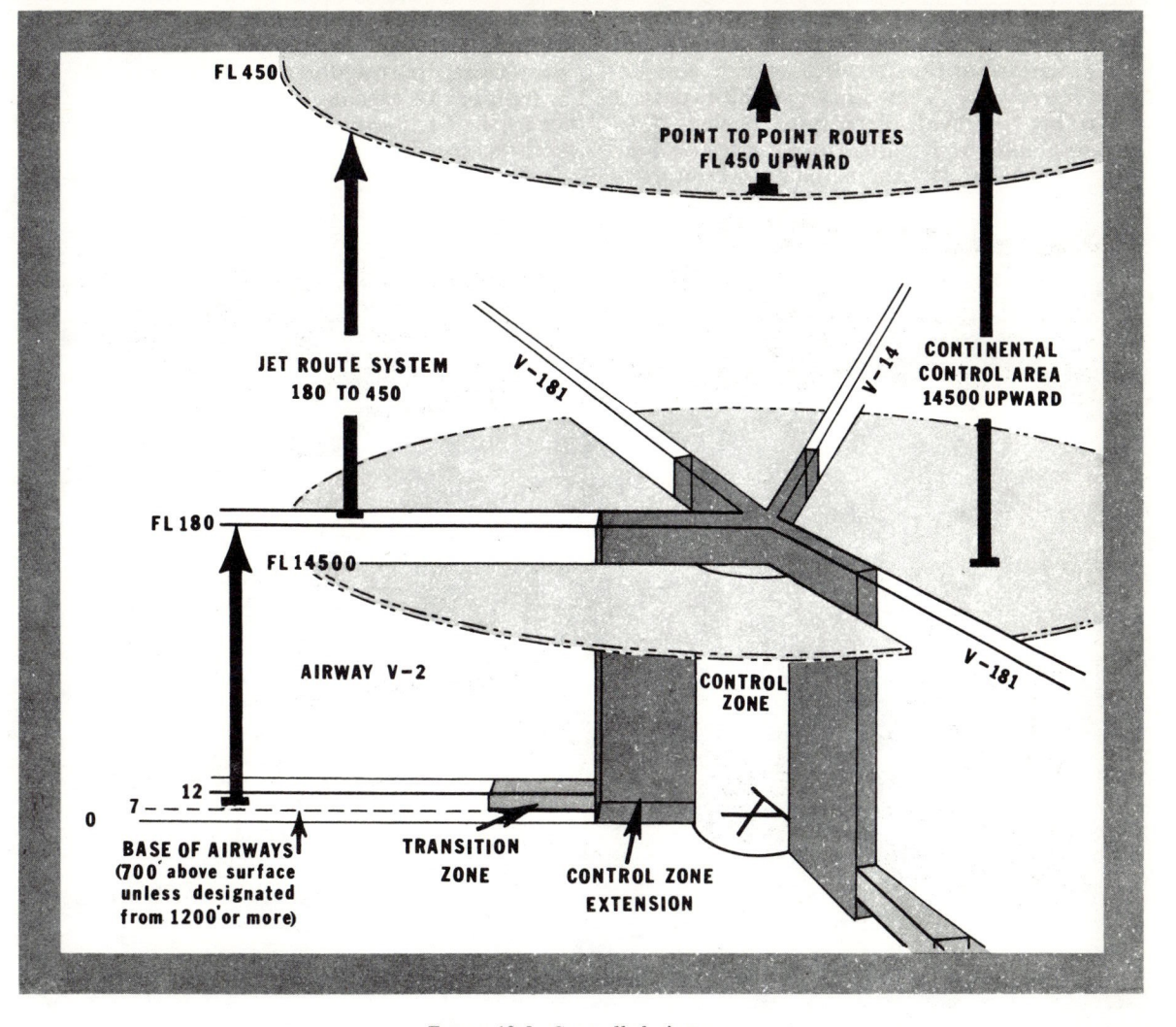

FIGURE 18-5 Controlled airspace.

tional and WAC charts familiar to the VFR pilot. More specialized charts, compiled and printed by the National Ocean Survey, include several types for use by the instrument pilot.

*Standard Instrument Departure (SID) Charts* are designed to expedite clearance delivery and to facilitate transition between takeoff and enroute operations. These charts, published in a bound book, provide departure routing clearance in graphic and textual form. *Enroute Low Altitude Charts* provide aeronautical information for enroute navigation (IFR) in the low altitude stratum. *Area Charts,* which are part of this series, furnish terminal data in a larger scale in congested areas. *Enroute High Altitude Charts* provide aeronautical information for enroute instrument navigation (IFR) in the high altitude stratum. *Instrument Approach Procedure Charts* portray the aeronautical data which is required to execute instrument approaches to airports. Each procedure is designated for use with a specific type navigational aid.

# 19  AIR TRAFFIC CONTROL

There is normally nothing very difficult involved in takeoff, climb, cruise from point to point, and descent solely by reference to instruments. The complications arise when you must execute these maneuvers at precise times, at specified altitudes, over designated routes and geographic positions, and in an orderly sequence with other aircraft. An understanding of the Air Traffic Control system will impress upon you the importance of the training necessary for you to apply the proficiency you have acquired in basic instrument flying and radio navigation techniques.

Federal regulation of civil aviation began with the Air Commerce Act of 1926 and the creation of the Aeronautics Branch in the U.S. Department of Commerce. The Department was concerned with the promotion of air safety, licensing of pilots, development of air navigation facilities, and issuing of flight information. Until the volume of air traffic increased, there was no need for air traffic control, since the likelihood of aircraft colliding in flight was remote.

The need for controlling air traffic was recognized in the 1930's as the aviation industry produced bigger, faster, and safer aircraft, and air transportation became an accepted mode of public travel. A number of large cities, concerned with regulating the increasing air traffic at their airports, built control towers and inaugurated a control service on, and in the immediate vicinity of, the airports. Airline companies, eager to expand and improve their operations, established control centers at Cleveland, Chicago and Newark to provide their pilots with position and estimated time of arrival information during instrument flights between those cities.

In 1936, the Federal Government assumed the responsibility for operation of the centers, employing eight controllers. As aviation has grown, so have the Federal Government functions and the agency charged with the promotion and safety of civil aviation. Today, at the beginning of the 1970's, approximately 23,000 ATC personnel provide direction and assistance to over 100 million flights annually.

The number of active aircraft has increased from 29,000, all flying at relatively slow and uniform speeds, to more than 100,000 aircraft operating at various speeds ranging to more than 1,000 miles per hour. The aerial highways have expanded from a few intercity routes to more than 187,000 miles of very high frequency routes utilizing approximately 1,000 VOR and VORTAC stations. A continued increase in traffic volume is expected during the coming decade.

The difficulties associated with mixed IFR and VFR traffic and with diverse pilot training and varying aircraft capabilities, the trend toward automated electronic equipment, and other aspects of control will have a profound effect on flight operations, under both Visual and Instrument Flight Rules.

As an instrument pilot normally operating under the jurisdiction of Air Traffic Control, your understanding of the present system and its operation will better enable you to make full use of ATC services.

## Structure and Functions of Air Traffic Service

Air Traffic Service of FAA is responsible for three major general functions: developing plans, establishing standards, and implementing systems for control of air traffic. The two specific functions of immediate concern to the instrument pilot are:

1. Providing preflight and inflight service to all pilots.
2. Keeping aircraft safely separated while operating in controlled airspace.

The preflight and inflight services to pilots are the responsibility of the Flight Service Stations (FSS). An extensive teletype and interphone system permits relay of information from many sources. Many of the services provided by tower and flight service personnel are familiar to the VFR pilot. Aircraft separation is the primary responsibility of both Airport Traffic Control Towers and Air Route Traffic Control Centers (ARTCC). Knowledge of the physical setup and services provided by each type of facility enables the instrument pilot to get information and assistance and to communicate with the appropriate controllers with confidence and efficiency.

## Airport Traffic Control Towers

*Jurisdiction.*—The ATC tower is responsible for control of aircraft on and in the immediate vicinity of airports. Terminals handling a large traffic volume employ specialized personnel for operations; they use light signals, radio, and ASDE radar (Airport Surface Detection Equipment) for control of surface traffic. Less congested airports have fewer controllers to handle the workload with less specialization.

Organization of the tower operations falls into the following units:

*Local Control* is concerned mainly with VFR traffic in and around the traffic pattern and with ground traffic. The local controller works with the other IFR controllers to integrate VFR and IFR flights into a smooth, safe traffic flow in and out of the airport.

*Ground Control* directs the movement of aircraft on the airport surface, working closely with other tower positions. The controller relays clearances from ARTCC to departing IFR flights unless a special position is assigned that function.

*Clearance Delivery* is accomplished by a separate controller at most busy terminals where heavy ground traffic and frequent IFR departures require division of the workload.

*Departure Control* originates departure clearances and instructions to provide separation between departing and arriving IFR flights. Although the physical location of this controller varies at different terminals—in the tower at some locations, in separate radar installations at others—he coordinates closely with the Approach Controller and Local Controller.

*Approach (Arrival) Control* formulates and issues approach clearances and instructions to provide separation between arriving IFR aircraft, using radar if available.

*Tower services provide:*

1. Control of aircraft on, and in the vicinity of, the airport.
2. Coordination with pilots and Air Route Traffic Control Centers for IFR clearances.
3. Air traffic advisories to pilots concerning observed, reported, and estimated positions of aircraft that might present a hazard to a particular flight.
4. Flight assistance, including transmission of pilot reports and requests, and weather advisories.

*Approach/Departure Control services provide:*

1. Navigation assistance by radar vector for departing and arriving aircraft.
2. DF assistance to lost aircraft, and cooperation with other facilities in Search and Rescue operations.

## Air Route Traffic Control Centers

*Jurisdiction.*—The primary function of the ARTCC is the direction of all aircraft operating under Instrument Filght Rules in controlled airspace. The Center has no jurisdiction over flights operating (1) under Visual Flight Rules on or off airways, (2) under Instrument Flight Rules outside of controlled airspace, and (3) below the floor of low-altitude Federal Airways.

*Organization.*—The ARTCC facilities are located throughout the continental United States at central points in areas over which they exercise control. Each Center controls IFR traffic within its own area and coordinates with adjacent Centers for the orderly flow of traffic from area to area. Each Center's control area is divided into Sectors, based upon traffic flow patterns and controller workload.

Each Sector Controller, with one or more assistants, is responsible for traffic within his sector. As an IFR flight departs, the affected Sector Controller follows the progress of the flight, maintaining a continuing written record of route, altitude, and time, and monitors the flight with long-range radar equipment when available. Each Sector Controller has a sector discrete frequency for direct communication with IFR flights within his sector. As an IFR flight progresses to adjacent sectors and centers, and finally to the destination terminal facility, the IFR pilot is requested to change to appropriate frequencies.

*ARTCC Services provide:*

1. Control of aircraft operating under Instrument Flight Rules in controlled airspace.

2. Air traffic advisories to aircraft concerning potential hazards to flight, anticipated delays, and any other data of importance to the pilot for the safe conduct of the flight.

3. Navigation assistance by radar vectors for detouring thunderstorms and expediting routing.

4. Transmission of pilot reports and weather advisories to enroute aircraft.

5. Flight assistance to aircraft in distress.

## Flight Service Stations

Flight Service Stations provide preflight and inflight services to pilots. Although this type of facility has no air traffic control authority over either VFR or IFR traffic, it renders extensive assistance to all air traffic. Many of these services are familiar to the VFR pilot.

*FSS Services provide:*

1. Preflight weather briefings for both VFR and IFR flights. Using the latest aviation weather information, the pilot briefer is qualified to interpret and discuss airway weather, weather maps, forecasts, pilot reports, and other significant route information related to the pilot's flight planning.

2. Handling of flight plans. VFR flight plans are filed with the Flight Service Station, which forwards the flight plan data to the FSS at, or nearest, the destination airport. IFR flight plans filed with the Flight Service Station are forwarded to the Center having jurisdiction over the departing flight.

3. Weather reports (scheduled and on request) and weather advisories.

4. Airport data concerning uncontrolled airports, without tower control, information concerning enroute and terminal facilities, and other special information requested by the pilot.

5. Relay of communications between aircraft and Centers for aircraft not able to communicate directly with the Center, including IFR clearances to pilots and position reports to the Center.

6. Emergency and Rescue service to distressed or lost aircraft.

## IFR Control Sequence

To illustrate the typical IFR flight sequence of control, follow an imaginary trip from Oklahoma City, Okla. (OKC), to Washington, D.C. (DCA).

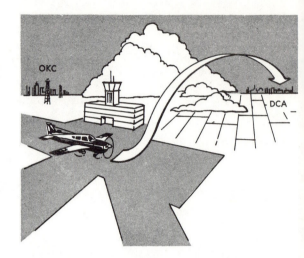

FIGURE 19-1  IFR—Oklahoma City to Washington, D.C.

The pilot of N543K has filed the flight plan, either in person or by telephone, with the OKC Flight Service Station giving the information necessary for the Center to coordinate his flight with other IFR traffic. (At airports where no Flight Service Station is available, the flight plan may be filed by local or long distance telephone to a **Flight Service Station, an ATC Tower, or ARTCC.**)

FIGURE 19-2  Filing the flight plan.

The OKC Flight Service Station then transmits pertinent flight plan data by interphone or teletype to the Fort Worth Air Route Traffic Control Center, which controls the departure point.

FIGURE 19-3 Copying the clearance.

FIGURE 19-4 Cleared for departure.

Upon receipt of the flight plan in the Fort Worth Center, a specialist at the Flight Data Position prepares flight progress strips for designated fixes along the route of flight within the Fort Worth Center's area of jurisdiction. The flight progress strips are held temporarily, with no time estimates or altitude entries made, until the flight is ready for takeoff.

Prior to taxiing, the pilot should call clearance delivery (if available at the airport) or ground control and request IFR clearance. If there is to be a delay, you would normally be so advised at this time. Airports with "Pre-Taxi Clearance Procedures" will probably have your clearance and issue it to you immediately. The clearance would be issued as follows:

**NOVEMBER FIVE FOUR THREE KILO CLEARED TO THE WASHINGTON AIRPORT VIA VICTOR FOURTEEN SOUTH, TULSA, FLIGHT PLAN ROUTE. CROSS SHAWNEE INTERSECTION AT FIVE THOUSAND, MAINTAIN SEVEN THOUSAND. DEPARTURE CONTROL FREQUENCY WILL BE ONE TWO FOUR POINT SEVEN.**

The pilot should copy this clearance and read it back substantially the same as issued. This clearance was originated in the Fort Worth Center and the clearance delivery controller is merely relaying it to you. However, he may add certain restrictions required by the Oklahoma City departure controller.

After receipt and acknowledgement of the IFR clearance, the pilot should contact ground control and request taxi clearance to the runway in use.

After engine run-up, the pilot changes to tower frequency (118.3), advises that he is ready for takeoff, and receives the following clearance:

**NOVEMBER FIVE FOUR THREE KILO LEFT TURN AFTER TAKEOFF HEADING ZERO NINE ZERO, CLEARED FOR TAKEOFF.**

Upon departure, the local controller notifies the Oklahoma City departure controller and the Fort Worth Center that N543K has departed.

When approximately ½ mile from end of runway, the local controller will advise:

**NOVEMBER FIVE FOUR THREE KILO CONTACT DEPARTURE CONTROL.**

The departure controller will radar identify your aircraft and issue further instructions to provide separation from other aircraft in the terminal area and establish you on your intended route.

**NOVEMBER FIVE FOUR THREE KILO, OKLAHOMA CITY DEPARTURE CONTROL, RADAR CONTACT, FLY HEADING ZERO NINE ZERO TO INTERCEPT VICTOR FOURTEEN SOUTH.**

The phrase "TO INTERCEPT VICTOR FOURTEEN SOUTH" indicates that the pilot should resume normal navigation upon intercepting the airway without further clearance. In the absence of instructions "TO INTERCEPT" the pilot should maintain the last assigned heading even though it will take him through the assigned airway. It is possible that the departure controller is using this means to provide separation from another aircraft. Normally, the controller would advise "NOVEMBER FIVE FOUR THREE KILO MAINTAIN HEADING ZERO NINE ZERO FOR VECTORS EAST OF VICTOR FOURTEEN SOUTH TO PASS TRAFFIC AHEAD," however, time does not always permit this advisory to be issued.

As N543K approaches Shawnee intersection, the pilot is advised to contact Fort Worth Center on a specified frequency. Since you have been radar identified, your correct initial call would be:

**FORT WORTH CENTER THIS IS NOVEMBER FIVE FOUR THREE KILO LEAVING FIVE THOUSAND.**

Fort Worth Center replies:

**NOVEMBER FIVE FOUR THREE KILO, FORT WORTH CENTER, RADAR CONTACT, THREE MILES EAST OF SHAWNEE, REPORT LEAVING SIX THOUSAND.**

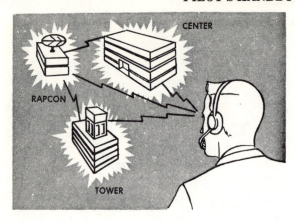

FIGURE 19-5 Enroute control.

Since N543K is operating in a radar environment, and has been advised "radar contact," his only reports would be those specifically requested by the controller; however, each time the aircraft is changed to a different control facility or to a different sector within a facility, the pilot should report his altitude on the initial call.

When advised "RADAR SERVICE TERMINATED" or "RADAR CONTACT LOST," the pilot should resume normal position reporting procedures.

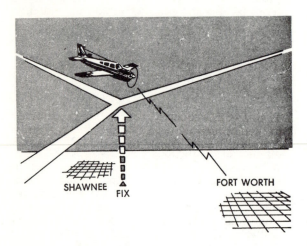

FIGURE 19-6 Position report.

As the flight progresses along airways, the flight data information is passed along from sector to sector within each Center, and from Center to Center along the route. Each controller receives advance notification of this flight and determines if there is any conflict with traffic within his sector. He is responsible for issuing control instructions which will provide standard separation between all aircraft under his control. This may be accomplished by route changes, altitude changes, requiring holding or providing radar vectors off course.

As the flight nears the Washington Terminal area, the Center controller initiates coordination

with Washington approach control. As you enter the terminal area, the Center controller will effect what is known as a "Radar Hand-off" to Washington approach control. This is a method whereby one controller points out a target to another controller and thereby transfers control without an interruption in radar service. In the absence of a "Radar Handoff" it would be necessary for the approach controller to use one of several other means to identify your aircraft, such as identifying turns, position reports, or transponder.

When both controllers are satisfied with the identification of N543K, the pilot will be advised to contact Washington approach control on a specified frequency.

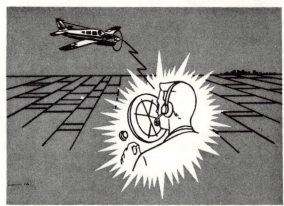

FIGURE 19-7 Positive identification by approach control.

Prior to this frequency change, the pilot should have determined that ATIS (Automatic Terminal Information Service) is available at Washington National Airport and he should have tuned in the proper frequency and listened to the information. If you have received the ATIS broadcast, you should advise the approach controller. The ATIS provides all the necessary approach, runway, NOTAM, and other pertinent information for landing at Washington National. The controller need provide only necessary control instructions. Your initial call to Washington approach control would be:

WASHINGTON APPROACH CONTROL, NOVEMBER FIVE FOUR THREE KILO SEVEN THOUSAND, INFORMATION BRAVO.

The approach controller's action now depends on the conditions and traffic at the airport. If no approach delay were expected, his reply to N543K would be:

NOVEMBER FIVE FOUR THREE KILO, WASHINGTON APPROACH CONTROL, RADAR CONTACT, FIVE MILES WEST OF HERNDON, FLY HEADING ONE TWO ZERO FOR VECTORS TO FINAL APPROACH COURSE, OVER.

(NOTE: ATIS information BRAVO advised that ILS approach to runway 36 would be expected.)

FIGURE 19-8 ILS approach.

The approach controller will now provide radar navigation to position your aircraft so as to intercept the final approach course prior to the outer marker. You will be advised to descend in sufficient time to reach the proper glide path interception altitude. Your clearance for an ILS approach will be issued when the controller has provided proper spacing between N543K and other aircraft under his control and when N543K is on a course which will intercept the ILS course, outside the outer marker, on not more than a 30° angle. This clearance will be issued as follows:

NOVEMBER FIVE FOUR THREE KILO, THREE MILES FROM OUTER MARKER, CLEARED FOR ILS APPROACH, CONTACT TOWER ON ONE ONE EIGHT POINT FIVE OVER THE OUTER MARKER.

(NOTE: Pilot should not turn on the ILS course unless this clearance is received.)

N543K will now turn on the ILS when he intercepts it and will complete his approach to a landing or to the "MISSED APPROACH POINT." A procedure turn should not be made unless ATC is advised.

If the traffic conditions at the airport required a delay, the approach controller would have issued instructions to N543K to proceed to a holding point to hold until his turn for approach arrived. This clearance would have been:

NOVEMBER FIVE FOUR THREE KILO, WASHINGTON APPROACH CONTROL, RADAR CONTACT, FIVE MILES WEST OF HERNDON, MAINTAIN SEVEN THOUSAND, PROCEED DIRECT TO HERNDON VORTAC, HOLD WEST ON VICTOR FOUR. EXPECT APPROACH CLEARANCE AT 1135.

At the appropriate time, N543K would have been instructed to depart Herndon on a specific heading and would have been vectored to final approach as in the preceding example.

After the approach and landing is completed, the local controller will advise the pilot where to turn off the runway and when to contact ground control.

In summary, the foregoing picture of an IFR flight illustrates the need for adherence to carefully established procedures for safe, orderly, and expeditious traffic flow. The pilot and controller must know what to expect of each other, as well as knowing the details of their own special authority and responsibility. The controller, for example, must be aware of the performance capabilities of the various aircraft operating in

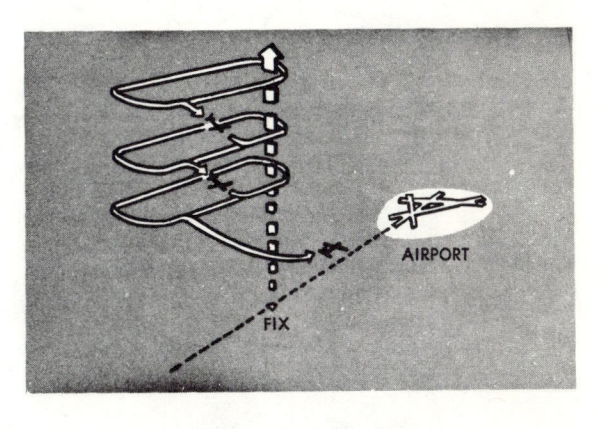

FIGURE 19-10 Shuttle.

the system for he must issue clearances and instructions with which the aircraft are capable of complying. Likewise, the instrument pilot should understand the overall traffic control problem and the standardized IFR control procedures so that he will not be caught unawares by an unexpected request or change in clearance. In fact, the experienced pilot usually knows beforehand what the controller will say and when he will say it. During periods of heavy traffic conditions, when clearances sometimes may seem complicated or unduly restrictive to the pilot, the utmost cooperation between the pilot and controller is especially required.

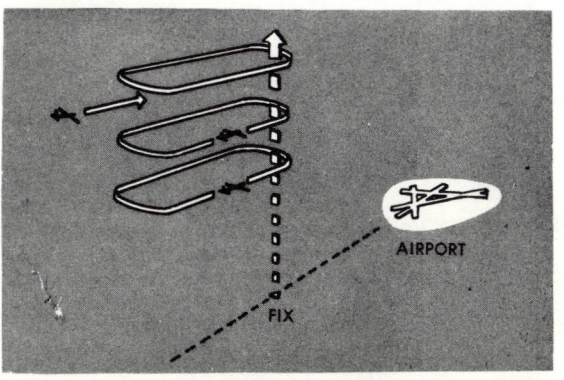

FIGURE 19-9 Stacking.

FIGURE 19-11 Cleared to land.

# 20 ATC OPERATIONS AND PROCEDURES

## Using the System

Once you understand the overall operation of the traffic control system, the many procedural details can be put into the appropriate sequence as you learn them. The problem now is to review the regulations and procedures involved in using the system under Instrument Flight Rules. Pilot responsibilities relating to air traffic control are stated in the Federal Aviation Regulations, Part 91, General Operating and Flight Rules.

More detailed procedures can be found under "Air Traffic Control" in the *Airman's Information Manual*. The instrument pilot should purchase copies of the Regulations and the *Airman's Information Manual* and maintain them with current revisions included in the subscription prices.

## IFR Flight Plan

As specified in FAR 91, if you plan to operate an aircraft in controlled airspace under Instrument Flight Rules, you must file an IFR flight plan and receive an appropriate ATC clearance.

If your flight is to be in controlled airspace, your initial contact with Air Traffic Service involves your flight plan. The information to be entered on an IFR flight plan is listed in FAR 91, in the *Airman's Information Manual*, and on the Federal Aviation Agency Flight Plan, Form 7233-1 (Figure 20-1), which is available at Flight Service Stations, flight planning rooms in airport terminal buildings, and at other convenient locations. Following is an explanation of the data you need for a proposed IFR flight in the order of the 18 numbered blocks on the form.

Block 1. An IFR Flight Plan must be filed for flight in controlled airspace under Instrument Flight Rules, regardless of weather conditions. Filing an IFR Flight Plan means that—

a. You are requesting Air Traffic Control to provide separation from other IFR traffic in controlled airspace.

b. You possess a current and valid instrument rating.

c. Your aircraft is equipped for flight as prescribed in FAR 91 General Operating and Flight Rules.

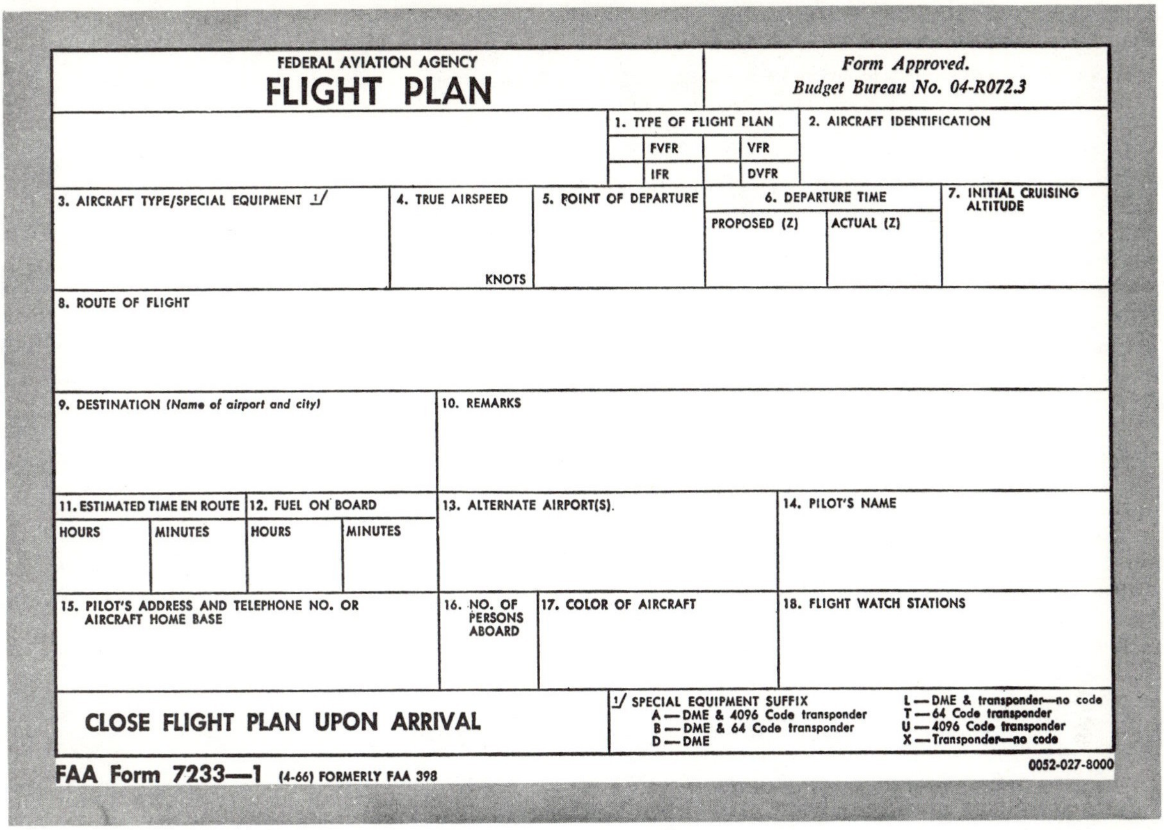

FEDERAL AVIATION AGENCY
FLIGHT PLAN

*Form Approved.*
*Budget Bureau No. 04-R072.3*

1. TYPE OF FLIGHT PLAN: FVFR, VFR, IFR, DVFR
2. AIRCRAFT IDENTIFICATION
3. AIRCRAFT TYPE/SPECIAL EQUIPMENT 1/
4. TRUE AIRSPEED — KNOTS
5. POINT OF DEPARTURE
6. DEPARTURE TIME — PROPOSED (Z) — ACTUAL (Z)
7. INITIAL CRUISING ALTITUDE
8. ROUTE OF FLIGHT
9. DESTINATION *(Name of airport and city)*
10. REMARKS
11. ESTIMATED TIME EN ROUTE — HOURS — MINUTES
12. FUEL ON BOARD — HOURS — MINUTES
13. ALTERNATE AIRPORT(S).
14. PILOT'S NAME
15. PILOT'S ADDRESS AND TELEPHONE NO. OR AIRCRAFT HOME BASE
16. NO. OF PERSONS ABOARD
17. COLOR OF AIRCRAFT
18. FLIGHT WATCH STATIONS

CLOSE FLIGHT PLAN UPON ARRIVAL

1/ SPECIAL EQUIPMENT SUFFIX
A — DME & 4096 Code transponder
B — DME & 64 Code transponder
D — DME
L — DME & transponder—no code
T — 64 Code transponder
U — 4096 Code transponder
X — Transponder—no code

FAA Form 7233—1 (4-66) FORMERLY FAA 398
0052-027-8000

FIGURE 20-1 Flight Plan form.

d. You will conform to all provisions of the Instrument Flight Rules.

Many airline and general aviation pilots file IFR on all flights in controlled airspace, regardless of weather conditions. The newly rated instrument pilot should also file IFR for practice even though VFR weather conditions exist. Air Traffic Control will provide welcome assistance to controlled aircraft entering congested areas, and the practice and experience gained in flight planning and coordination with ATC should be part of transition training for flying under instrument weather conditions.

Block 2. The aircraft identification is the full serial number of the aircraft, e.g., N5432Z.

Block 3. Aircraft type data includes specific make and basic model, e.g., Cessna 182, Beech Baron, etc. This information identifies the performance characteristics of the aircraft for the controller. DME and transponder capability must be included in this block.

Block 4. The estimated true airspeed is the computed value at your requested cruising altitude, in knots.

Block 5. Point of Departure is the airport of departure. If you file IFR in flight, the point of departure is the radio navigation fix at which you request ATC to assume control.

Block 6. The departure time proposed should be at least 30 minutes after the flight plan is filed with Flight Service to allow the Center sufficient time to clear your route and formulate your clearance. The time entered is "Z" time (Greenwich Mean Time) based on the 24-hour clock. This identifies the time regardless of differences in local time at various ATC locations. To illustrate, a proposed 9:00 A.M. departure at Denver, Colo. (on Mountain Standard Time), is entered as 1600 on the flight plan. Note the hourly differences in the time belts:

| Zulu | Z minus 5 | Z minus 6 | Z minus 7 | Z minus 8 |
|---|---|---|---|---|
| 1600 = | 1100EST = | 1000CST = | 0900MST = | 0800PST |

**If departure is from a controlled airport, the tower relays your takeoff time to Departure Control. If departure is from an airport without a control tower, you report the actual departure time by radio to Flight Service or as directed in your clearance.**

**Block 7. Initial cruising altitude is your requested altitude, but is not necessarily the altitude that ATC will assign. During your flight planning, you select altitudes on the basis of such factors as aircraft performance, favorable winds, precipitation, turbulence, minimum IFR altitudes for the route, etc. Without assuming an altitude, you cannot make the computations necessary for your flight plan. However, ATC assigns altitudes on the basis of your proposed departure time, your requested altitude, and the position of known traffic.**

**Block 8. Route of Flight should be described in accordance with detailed procedures found in the *Airman's Information Manual*. The route**

requested may not be the route assigned for reasons stated earlier.

Block 9. Destination (Airport and City), e.g., Wiley Post, Oklahoma City: Will Rogers, Oklahoma City; Tinker, Oklahoma City. Where a number of airports serve a metropolitan area, Approach Control must know the specific destination airport filed. The Controller must be informed well in advance of your arrival at the terminal area in order to vector you from an enroute position to the appropriate approach route in a complex system of arrival routings.

Block 10. Remarks. In this block enter your aircraft limitations as to radio frequencies, both for communications and radio navigation. This information is of utmost importance to Air Traffic Control in clearing your route, conducting communications, and controlling your arrival. If your communications equipment is limited, ATC will not request that you use frequencies normally utilized by better-equipped aircraft. Regulations require this entry so that there will be no confusion between pilot and controller as to navigation and communication capability. Desired altitude changes may also be listed in this block.

Block 11. Estimated Time Enroute is the estimated elapsed time from takeoff until over the point of first intended landing. Does this mean from takeoff to touchdown at the destination airport? If so, the complications involved in an accurate computed estimate would tax the patience of an expert. Assume that "over the point of first intended landing" means the final approach fix from which you intend to approach and land. Your estimated enroute time, then, is the sum of the following times:

a. Climb from takeoff to the initial cruising altitude requested.

b. Enroute time based upon the estimated true airspeed at altitudes requested in blocks 7 and 10 of your flight plan.

c. Time from the enroute navigation aid to the final approach fix from which you intend to land.

After a few practice IFR cross-country flights, during which ATC may request changes in altitude and routing, you may question the value of time-consuming pre-flight estimates. Why bother with careful estimates if you have to revise them in the air? Apart from the fact that a competent pilot constantly revises almost everything that he does in the air, careful flight planning is your guide to a safe arrival in the event of emergencies, which will be discussed later.

Block 12. FAR 91 requires that your fuel on board be based on the estimate of elapsed time from takeoff to the first intended point of landing, thence to the alternate airport, if required, with a reserve of 45 minutes at normal cruising speed. An accurate estimate of the fuel required involves fuel consumption based upon the following time estimates:

a. Estimated enroute time in block 11 of your flight plan.

b. Estimated time from the final approach fix through the execution of a missed approach, thence to the alternate airport, if required.

c. 45 minutes reserve at normal cruise fuel consumption.

Block 13. FAR 91 contains the requirement for listing an alternate airport. Under certain specified forecast ceiling, visibility, and time period conditions, an alternate is not required.

Blocks 14–17. These entries are self-explanatory.

Block 18. Flight Watch Stations are not listed on an IFR Flight Plan.

Your Flight Plan completed, Flight Service relays to the Center the data you have entered in blocks 1 through 11. The other information has no bearing on control of your flight unless—

a. You execute a missed approach at your destination and request ATC clearance to your alternate, or

b. An emergency occurs during your flight that makes the additional data of importance in search and rescue operations. The completed Flight Plan form is retained at the Flight Service Station for later reference if necessary.

## Filing in Flight

An IFR flight plan may be filed in the air under various conditions, including:

a. An IFR flight outside of controlled airspace prior to proceeding into IFR conditions in controlled airspace.

b. A flight on a VFR flight plan expecting IFR weather conditions enroute in controlled airspace.

c. A flight on a VFR flight plan enroute to a destination having high air traffic volume. Even in VFR weather conditions, more efficient handling is provided a flight on an IFR clearance. However, acceptance of an IFR clearance does not relieve the pilot of his responsibility to maintain separation from other traffic when operating in VFR conditions.

d. A flight departing under VFR conditions from an airport where no means of communication with Flight Service is available on the ground.

In any of these situations, the flight plan may be filed with the nearest Flight Service Station or directly with the Center. If the pilot files with Flight Service, he submits the information normally entered during pre-flight filing, except for "point of departure," together with his present position and altitude. The Center will then clear him from his present position or from a specified enroute navigation fix. If the pilot files direct with the Center, he reports his present position and altitude, and submits only the flight plan information normally relayed by Flight Service to the Center.

## Clearances

*Definition.*—An Air Traffic Clearance is "an authorization by air traffic control, for the purpose of preventing collision between known IFR traffic, for an aircraft to proceed under specified traffic conditions within controlled airspace."

As the definition implies, an ATC clearance can be very simple or quite complicated, depending on traffic conditions. Your departure clearance will contain the following items, as required:

1. Aircraft identification.
2. Clearance limit.
3. Route of flight.
4. Altitude data.
5. Departure procedure.
6. Holding instructions.
7. Any special information.
8. Instructions for contacting the controlling facility, if direct communication is being maintained between pilot and controllers.

*Examples:* A flight filed for a short distance at relatively low altitude in an area of low traffic density might receive a clearance as follows:

SKYBIRD 1234D CLEARED TO THE DOEVILLE AIRPORT, DIRECT, CRUISE THREE THOUSAND.

The term "cruise" in such a clearance means that the pilot is cleared to the destination airport, to climb to 3,000 and descend at his discretion—that is, without further ATC clearance.

If a flight plan is filed, and the clearance received, through Flight Service by telephone, ATC would specify appropriate instructions as follows:

SKYBIRD 1234D CLEARED TO THE SKYLINE AIRPORT VIA THE CROSSVILLE 055 RADIAL VICTOR 18, MAINTAIN FIVE THOUSAND, CLEARANCE VOID IF NOT OFF BY 1330.

Under more complex traffic conditions, the clearance is more involved:

SKYBIRD 1234D CLEARED TO THE TULSA AIRPORT VIA VICTOR 14, TURN RIGHT AFTER DEPARTURE, PROCEED DIRECT TO THE OKLAHOMA CITY VORTAC. MAINTAIN THREE THOUSAND TO THE OKLAHOMA CITY VORTAC. HOLD WEST ON THE OKLAHOMA CITY 277 RADIAL. CLIMB TO FIVE THOUSAND IN THE HOLDING PATTERN BEFORE PROCEEDING ON COURSE. MAINTAIN FIVE THOUSAND UNTIL CROSSING THE PONCA CITY 167 RADIAL. CLIMB TO AND MAINTAIN SEVEN THOUSAND. DEPARTURE CONTROL FREQUENCY WILL BE 121.1.

None of the foregoing clearances are especially difficult to copy, understand, and comply with—assuming that you—

a. Have properly tuned your radio.
b. Are concentrating on what you hear.
c. Can copy fast enough to keep up with the clearance delivery.
d. Are familiar with the area.

Suppose, on the other hand, that you are awaiting departure clearance at a busy metropolitan terminal (your first IFR departure from this airport). On an average date, the tower at this airport controls departures at a rate of one every 2 minutes to maintain the required traffic

flow. Sequenced behind you are a number of aircraft ready for departure, including jet transports burning 25 gallons of fuel per minute while they wait at idle power.

You request and receive the following clearance, called a Standard Instrument Departure (SID) (Fig 20-2):

SKYBIRD 1234D CLEARED TO WORCHESTER MUNICIPAL AIRPORT, BAYVILLE EIGHT DEPARTURE, WATERBURY TRANSITION BDR, FLIGHT PLANNED ROUTE, MAINTAIN NINE THOUSAND, CONTACT NEW YORK CENTER 125.1 WHEN AIRBORNE.

This clearance can be copied readily in shorthand as the following:

WORA/BV-8BDR/FPR/M90  (NY 125.1)

The information contained in this clearance for a Standard Instrument Departure is an abbreviation of Air Traffic instructions too complicated and extensive for you to follow and copy, regardless of your proficiency in using clearance shorthand. Study of the route specified in the clearance shows the importance of the Standard Instrument Departure. At common operating speeds in modern light aircraft, the clearance allows no time for extensive reference to your departure chart. You will be too busy flying your aircraft, navigating, and communicating with ATC to familiarize yourself with the clearance data after you have accepted it. You must know the locations of the specified navigation facilities, together with the route and point-to-point times, *before* accepting the clearance.

The Standard Instrument Departure, which you have available during pre-flight planning, enables you to study and understand the details of your departure before filing your IFR flight plan. It permits you to set up your communications and navigation equipment and to be ready for departure before requesting IFR clearance from the tower. The SID eliminates unnecessarily long delays in clearance delivery that would result in inconvenience and expense to airspace users as well as revisions in flight planning for pilots awaiting departure.

Regardless of the nature of your clearance, it is imperative that you are prepared to understand it, and having accepted it, comply with ATC instructions to the letter. It is your privilege to request a clearance different from that issued by ATC if you consider another course of action more practicable or if your aircraft equipment limitations or other considerations make acceptance of the clearance inadvisable. Though regulations do not require that you accept a clearance, they are very specific as to your *privileges and responsibilities.*

1. **Responsibility and authority of the pilot in command.**

(a) The pilot in command of an aircraft is directly responsible for, and is the final authority as to, the operation of that aircraft.

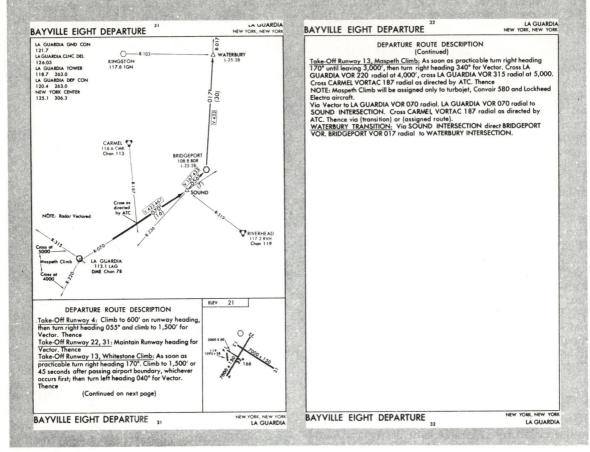

FIGURE 20-2 Standard Instrument Departure (SID).

(b) In an emergency requiring immediate action, the pilot in command may deviate from a rule to the extent required to meet that emergency.

(c) Each pilot in command who deviates from a rule shall, upon the request of the Administrator, send a written report of that deviation to the Administrator.

2. Compliance with ATC clearances and instructions.

(a) When an ATC clearance has been obtained, no pilot in command may deviate from that clearance, except in an emergency, unless he obtains an amended clearance. However, except in positive controlled airspace, this paragraph does not prohibit him from cancelling an IFR flight plan if he is operating in VFR weather conditions.

(b) Except in an emergency, no person may, in an area in which air traffic control is exercised, operate a aircraft contrary to an ATC instruction.

(c) Each pilot in command who deviates, in an emergency, from an ATC clearance or instruction shall notify ATC of that deviation as soon as possible.

(d) Each pilot in command who (though not deviating from a rule of this subpart) is given priority by ATC in an emergency, shall, if requested by ATC, submit a detailed report of that emergency within 48 hours to the chief of that ATC facility.

An ATC clearance presupposes that you are equipped and will comply with the applicable Regulations (Part 91), as follows:

1. Instruments and equipment IFR flight. The following instruments and equipment are required: . . . Two-way radio communications system and navigational equipment appropriate to the ground facilities to be used. . . .

2. IFR radio communications. The pilot in command of each aircraft operated under IFR in controlled airspace shall have a continuous watch maintained on the appropriate frequency. . . .

3. Course to be flown. Unless otherwise authorized by ATC, no person may operate an aircraft within controlled airspace, under IFR, except as follows:

(a) On a Federal Airway, along the centerline of that airway.

(b) On any other route, along the direct course between the navigational aids or fixes defining that route.

However, this section does not prohibit maneuvering the aircraft to pass well clear of other air traffic or the maneuvering of the aircraft, in VFR conditions, to clear the intended flight path both before and during climb or descent.

# Departure Procedures

*Standard Instrument Departures (SIDs).*—To simplify air traffic control clearances and relay and delivery procedures, Standard Instrument Departures have been established for the most frequently used departure routes in areas of high traffic activity. You will normally use a SID (Fig. 20-2) where such departures are available since this is advantageous to both users and Air Traffic Control. The following points are important to remember if you file IFR out of terminal areas where SIDs are in use:

1. SIDs are published in booklet form every eight weeks with every other issue of the National Ocean Survey Approach and Enroute Charts. Three booklets are issued, one for the eastern half of the United States, one for the western half, and one for Alaska. The descriptions are both graphic and textual. AIM, Part 1, describes SID procedures, and Part 3 lists currently published SIDs plus those newly approved, revised, or cancelled.

2. ATC will not issue a SID clearance unless requested by the pilot, except for military and air carrier pilots and certain other exceptions as noted in the AIM "Notices to Airmen."

3. Your request for a SID means that you are familiar with it and have it in the cockpit for reference. It is your responsibility to accept or refuse the clearance issued.

4. If you accept a SID in your clearance, you must comply with it, just as you comply with other ATC instructions.

*Preferred Routes.*—In the major terminal and enroute environments, preferred routes have been established to guide pilots in planning their routes of flight, to minimize route changes, and to aid in the orderly management of air traffic using the Federal Airways. You file via SID and preferred route for the same reasons that for long automobile trips you drive via expressway and interstate superhighway. The route is quicker, easier, and safer.

*Radar-Controlled Departures.*—On your IFR departures from airports in congested areas, you will normally receive navigational guidance from Departure Control by radar vector. When your departure is to be vectored immediately following takeoff, you will be advised before takeoff of the initial heading to be flown and the frequency on which you will contact Departure Control. This information is vital in the event that you experience (complete) loss of two-way radio communications during departure. This possibility will be considered later in this chapter under "Emergencies."

The radar departure is normally simple. Following takeoff, you contact Departure Control on the assigned frequency upon release from Tower Control. Awaiting your call-up, Departure Control verifies contact, tells you briefly the purpose of the vector (airway, point, or route to which you will be vectored), and gives headings, altitude, and climb instructions, and other information to move you quickly and safely out of the terminal area. You listen to instructions and fly basic instrument maneuvers (climbs, level-offs, turns to predetermined headings, and straight-and-level flight) until the controller tells you your position with respect to the route given in your clearance, whom to contact next, and to "resume normal navigation."

Departure Control will vector you either to a navigation facility or an enroute position appropriate to your departure clearance, or you will be transferred to another controller with further radar surveillance capabilities. It is just like having your instructor along to tell you what to do and when to do it. The procedure is so easy, in fact, that inexperienced pilots are often inclined to depend entirely on radar for navigational guidance, unconcerned about the consequences of loss of radar contact and indifferent to common-sense precautions associated with flight planning.

A radar-controlled departure does NOT relieve you of your responsibilities as pilot-in-command of your aircraft. You should be prepared before takeoff to conduct your own navigation according to your ATC clearance, with navigation receivers checked and properly tuned. While under radar control, you should monitor your instruments to ensure that you are continuously oriented to the route specified in your clearance and you should record the time over designated check points.

*Departures from Uncontrolled Airports.*—Occasionally, you will depart from airports which have neither a Tower nor Flight Service Station. Under these circumstances, it is desirable that you telephone your flight plan to the nearest FAA facility at least 30 minutes prior to your estimated departure time. If weather conditions permit, you could depart VFR and request IFR clearance as soon as radio contact is established with an FAA facility. If weather conditions made it undesirable to attempt to maintain VFR, you could again telephone the facility which took your flight plan and request clearance by telephone. In this case, the controller would probably issue a short range clearance pending establishment of radio contact and might also restrict your departure time to a certain period. For example:

CLEARANCE VOID IF NOT OFF BY 0900.

This would authorize you to depart within the allotted time period and proceed in accordance with your clearance. In the absence of any specific departure instructions, the pilot would be expected to proceed on course via the most direct route.

# Enroute Procedures

Normal procedures enroute will vary according to your proposed route, the traffic environment, and the ATC facilities controlling your flight. Some IFR flights are under radar surveillance and control from departure to arrival; others rely entirely on pilot navigation. Flights proceeding from controlled to uncontrolled airspace are outside ATC jurisdiction as soon as the aircraft is outside of controlled airspace.

Where ATC has no jurisdiction, it does not issue an IFR clearance. It has no control over the flight; nor does the pilot have any assurance of separation from other traffic.

With the increasing use of the national airspace, the amount of uncontrolled airspace is diminishing, and the average pilot will normally file IFR via airways and under ATC control. For IFR flying in uncontrolled airspace, there are few regulations and procedures to comply with. The advantages are also few, and the hazards can be many. For rules governing altitudes and course to be flown in uncontrolled airspace, see FAR 91.

*Enroute Separation.*—For enroute control, Departure Control normally requests that you contact Air Route Traffic Control (the appropriate Center) on a specified frequency as you approach the limit of terminal radar jurisdiction. At this point Departure Control, in coordination with the Center, has provided you with standard separation from other aircraft on IFR clearances. Separation from other IFR aircraft is provided thus:

a. Vertically by assignment of different altitudes.

b. Longitudinally by controlling time separation between aircraft on the same course.

c. Laterally by assignment of different flight paths.

d. By radar—including all of the above.

ATC does NOT provide separation for an aircraft operating—

a. Outside controlled airspace.

b. On an IFR clearance:

(1) With "VFR conditions on top" authorized instead of a specific assigned altitude.

(2) Specifying climb or descent in "VFR conditions."

(3) At any time in VFR conditions, since uncontrolled VFR flights may be operating in the same airspace.

Of more importance to you are the reporting procedures by which you convey information to the Center Controller. With the data you transmit, he follows the progress of your flight by entering time/position/altitude information on a flight progress strip, relating your flight to the progress strips of other aircraft. The accuracy of your reports can affect the progress and safety of every other aircraft operating in the area on an IFR flight plan because ATC must correlate your reports with all the others to provide separation and expedite aircraft movements.

*Reporting Requirements.*—Federal Aviation Regulations require pilots to maintain a listening watch on the appropriate frequency and unless operating under radar control, to furnish position reports over certain reporting points. Any unforecast weather conditions or other information related to the safety of flight must also be reported.

Position reports are required by all flights regardless of altitude, including those operating in accordance with a "VFR conditions-on-top" clearance, over each designated compulsory reporting point (shown as solid triangles) along the route being flown. Along direct routes, reports are required of all flights over each reporting point used to define the route of flight. Reports over an "on request" reporting point (shown as open triangles) are made only when requested by ATC.

*Position Reports.*—The use of standardized reporting procedures generally makes for faster and more effective communication. A standardized communication with ATC normally complies with the Radiotelephone Contact Procedure comprising call-up, reply, message, and acknowledgement or ending. Use this procedure when making position reports to a Flight Service Station for relay to the center controlling your flight. Your IFR position reports should include the following items:

1. Identification.

2. Position.

3. Time. Over a VOR, the time reported should be the time at which the first complete reversal of the TO/FROM indicator is noted. Over a nondirectional radio beacon, the time reported should be the time at which the ADF needles make a complete reversal, or indicates that you have passed the facility.

4. Altitude.

5. Type of flight plan, if your report is made to a Flight Service Station. This item is not required if your report is made direct to an ATC center or approach controller, both of whom already know that you are on an IFR flight plan.

6. Estimated time of arrival (ETA) over next reporting point.

7. The name only of the next succeeding (required) reporting point along the route of flight.

8. Remarks, when required.

During your early communications training, your reports will be clear and concise if you learn and adhere to the procedures outlined in the *Airman's Information Manual.* Often these procedures are abbreviated by both pilots and controllers when clarity and positive identification are not compromised. With increasing experience in ATC communications, you will readily learn to reduce verbiage when high radio congestion makes it advisable.

Enroute position reports are submitted normally to the ARTCC Controllers via direct controller-to-pilot communications channels, using the appropriate ARTCC frequencies listed on the enroute chart. Unless you indicate the limitation of your communications equipment, under "Remarks" on your IFR flight plan, ATC will expect direct pilot-to-center communications enroute, advising you of the frequency to be used and when a frequency change is required.

Failure to provide ATC with frequency information on your flight plan contributes to radio congestion until ATC can assign you a frequency suitable to your equipment limitations.

In order to reduce congestion, pilots reporting direct to an ARTCC follow special voice procedures when making the initial call-up. Whenever an initial center contact is to be followed by a position report, the name of the reporting point should be included in the call-up. This alerts the controller that such information is forthcoming.

When a full position report is not required (for example, when a pilot has been instructed to contact a Center at a specified time or point other than at a compulsory reporting point), the pilot includes in his call-up an estimate for the next reporting point, his actual altitude and, if appropriate, the altitude to which climb or descent is being made. The controller simply acknowledges the contact and adds control instructions if required. Examples of such reports are given in the *Airman's Information Manual*.

*Malfunction Reports.*—Pilots of aircraft operated in controlled airspace under IFR are required to report immediately to ATC any of the following malfunctions of equipment occurring in flight:

(1) Loss of VOR, TACAN, or ADF receiver capability.

(2) Complete or partial loss of ILS receiver capability.

(3) Impairment of air/ground communications capability.

In each such report, pilots are expected to include aircraft identification, equipment affected, and degree to which IFR operational capability in the ATC system is impaired. The nature and extent of assistance desired from ATC must also be stated.

*Additional reports.*—The *Airman's Information Manual* specifies the following reports to ATC, without request by the controller:

1. The time and altitude/flight level reaching a holding fix or point to which cleared.

2. When vacating any previously assigned altitude/flight level for a newly assigned altitude/flight level.

3. When leaving any assigned holding fix or point.

4. When leaving final approach fix inbound on final approach.

5. When an approach has been missed. (Request clearance for specific action, i.e., to alternate airport, another approach, etc.)

6. A corrected estimate any time it becomes apparent that a previously submitted estimate to a reporting point will be in error in excess of 3 minutes.

7. That an altitude change will be made if operating on a clearance specifying "VFR conditions-on-top."

## Arrival Procedures

ATC arrival procedures and your cockpit workload are affected by weather conditions, traffic density, aircraft equipment, and radar availability.

*Standard Terminal Arrival Routes (STARs).*—These routes have been established to simplify clearance delivery procedures to arriving aircraft at certain areas having high density traffic. STARs serve a purpose parallel to that of SIDs for departing traffic. Procedures are published in Part 1 of AIM and textual descriptions of current STARs are given in Part 3. Booklets such as those published for SIDs are not available at present.

*Uncontrolled Airports (No Tower).*—On a flight in an uncongested area into an airport with no tower, the controlling ARTCC advises you to contact the Flight Service Station at or near the destination airport for airport advisory information. This includes the current local altimeter setting, wind direction and velocity, runway information, and known traffic. Further reports are relayed by Flight Service to the Center until you have canceled your IFR flight plan.

*Airports With Tower.*—Your workload can be much greater on arrival at a terminal area where a combination of low ceiling and visibility, heavy traffic, and type of available approach aids requires considerable delay in the issuance of approach clearances. At an airport equipped with ILS facilities, the local controllers can normally handle an arrival every 2 minutes. At the same airport, with the ILS system unavailable and the VORTAC located 8 miles from the field, the arrival interval may be increased to 10 minutes.

Whatever the reasons for delaying instrument approaches—including arrival intervals, traffic density, deteriorating weather, missed approaches, etc.—holding may be necessary. The order of priority of issuance of approach clearances is normally established on a first-come-first-served basis. The first aircraft estimated over the fixes from which approaches are begun will be the first to receive an approach clearance, followed by the aircraft in the order of their estimated or actual times of arrival over the several fixes.

*Holding.*—"Holding" is maneuvering an aircraft along a predetermined flight path within prescribed airspace limits with respect to a geographic fix. The fix may be identified visually (without reference to instruments) as a specified location, or by reference to instruments as a radio facility or intersection of courses. VORs, radio beacons, and airway intersections are used as holding points.

The diagram in Figure 20-3 illustrates control procedures used when a number of aircraft are stacked at an approach fix (Outer Compass

Locator) on an ILS front course, with additional aircraft holding in the stack at an outer fix (VOR). Successive arriving aircraft are cleared to the approach fix until the highest altitude/ flight level to be assigned is occupied, and thereafter to the outer fix at an appropriate altitude/ flight level above the highest level occupied at the approach fix. This permits aircraft subsequently cleared to the approach fix to proceed in descending flight. The illustration shows an interval of 2 minutes between successive approaches. The #1 and #2 aircraft have already passed the Outer Locator (LOM) on final approach, and the #3 aircraft has been cleared for approach and to depart the LOM 2 minutes after the #2 aircraft reported leaving the LOM inbound on final approach.

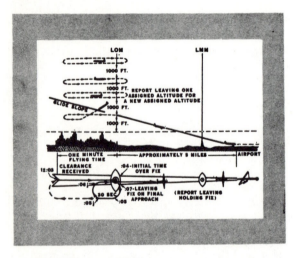

Figure 20-3 ATC procedures—timed approaches.

As Figure 20-3 shows, to fly a holding pattern involves a combination of simple basic maneuvers—two turns and two legs in straight-and-level flight or descending when cleared by ATC. Although these maneuvers are far less difficult than, for example, absolute-rate climbs or descent to predetermined headings and altitudes (see Chapter V, "Attitude Instrument Flying"), holding procedures are a common source of confusion and apprehension among instrument pilot trainees.

There are many reasons for this apprehension, among them the idea that holding implies uncertainty, delay, procedural complications, and generally an increased workload at a time when you are already busy reviewing the details of your instrument approach. Another reason involves the normal psychological pressure attending approach to your destination, when you become increasingly conscious of the fact that your margin of error is narrowing. The closer you get to touchdown, the more decisions you must make, and the decisions must be quick, positive, and accurate as you have fewer chances to correct the inaccuracies. Like any other flight problem, holding complications become routine

after sufficient study of the procedures in their normal sequence.

*Standard Holding Pattern (No Wind).*—At or below 14,000 feet, the standard holding pattern (Fig. 20-4) is a racetrack pattern requiring ap-

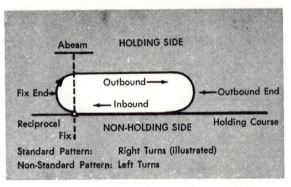

FIGURE 20-4 Standard holding pattern—no wind.

proximately 4 minutes to execute. The aircraft follows the specified course inbound to the holding fix, turns 180° to the right, flies a parallel straight course outbound for one minute, turns 180° to the right, and flies the inbound course to the fix.

*Nonstandard Holding Pattern.*—A nonstandard holding pattern is one in which fix end and outbound turns are made to the left. Your ATC clearance will always specify left turns when a nonstandard pattern is to be flown.

*Standard Holding Pattern With Wind.*—In compliance with the holding pattern procedures given in the *Airman's Information Manual,* the symmetrical racetrack pattern cannot be tracked when a wind exists. Pilots are expected to—

a. Execute all turns during entry to and while in the holding pattern at 3° per second, or a 30° bank angle, or a 25° bank angle if a flight director system is used; whichever requires the least bank angle.

b. Compensate for the effect of a known wind except when turning.

Figure 20-5 illustrates the holding track followed with a left crosswind. Further details of compensation for wind effect are given in the *Airman's Information Manual* and explained in this manual in Chapter in Chapter 5, "Wind and Its Effect."

The effect of wind is thus counteracted by correcting for drift on the inbound and outbound legs, and by applying time allowances to be discussed under "Time Factors."

*Holding Instructions.*—If you arrive at your clearance limit before receiving clearance beyond the fix, ATC expects you to maintain the last assigned altitude and execute a standard holding pattern on the course on which you approached the fix, until further clearance is received. Normally, when no delay is anticipated, ATC will issue further clearance as soon as practicable and, in any event, prior to your

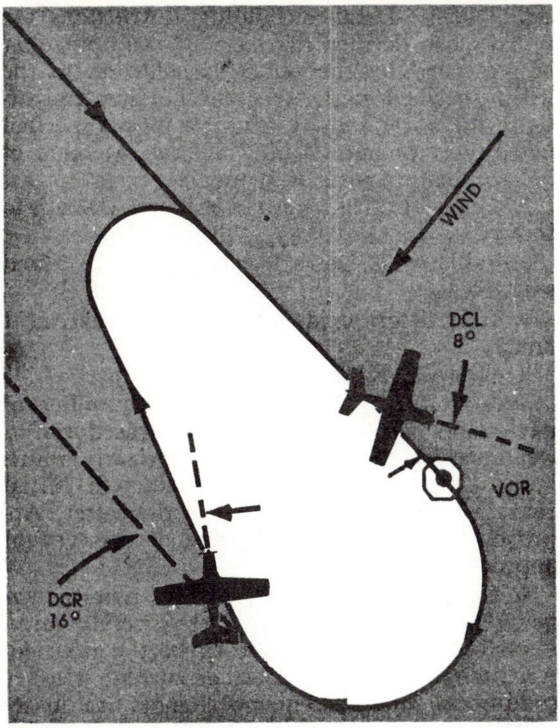

FIGURE 20-5 Drift correction in holding pattern.

arrival at the clearance limit. If delay is anticipated, ATC will issue holding instructions at least 5 minutes before your estimated arrival at the fix. General Holding Instructions specify the following:

1. Direction of holding from the fix, using magnetic directions and referring to one of eight general points of the compass (north, northeast, east, etc.).

2. Name of the holding fix.

3. Radial, course, or airway on which holding is to be accomplished.

4. Direction of holding pattern, if other than right turns. Detailed holding instructions are issued when deemed necessary by the Controller or requested by the pilot. In addition to the items specified under general holding instructions, the outbound leg length is always specifed in minutes, or miles DME, and the direction of holding pattern turns is included.

5. Outbound leg length in miles, if DME is to be used.

Suitable ATC instructions will also be issued whenever—

1. It is determined that delay will exceed 1 hour.

2. A revised EAC or EFC is necessary.

3. In a terminal area having a number of navigation aids and approach procedures, a clearance limit may not indicate clearly which approach procedures will be used. On initial contact, or as soon as possible thereafter, Approach Control will advise you of the type of approach you may anticipate.

4. Ceiling and/or visibility is reported as being at or below the highest "circling minimums" established for the airport concerned. ATC will transmit a report of current weather conditions, and subsequent changes, as necessary.

5. Aircraft are holding while awaiting approach clearance, and pilots thereof advise that reported weather conditions are below minimums applicable to their operation. In this event, ATC will issue suitable instructions to aircraft which desire either to continue holding while awaiting weather improvement or proceed to another airport.

*Standard Entry Procedures.*—The entry procedures given in the *Airman's Information Manual* evolved from extensive experimentation under a wide range of operational conditions. These standardized procedures should be followed to ensure that you remain within the boundaries of the prescribed holding airspace.

*Maximum airspeed* for propeller-driven aircraft holding below 14,000 feet is 175 knots Indicated Airspeed (for procedures appropriate to higher altitudes and high-speed aircraft, refer to the *Airman's Information Manual*). You are expected to reduce airspeed to 175 knots, or less if practicable, within approximately 3 minutes prior to ETA at the holding fix. The purpose of the speed limitation is to prevent overshooting the holding airspace limits, especially at locations where adjacent holding patterns are close together. The *exact* time at which you reduce speed is not important, so long as you arrive *at the fix* at your preselected holding speed within 3 minutes of your submitted ETA. If it takes more than 3 minutes for you to complete a speed reduction and ready yourself for identification of the fix, adjustment of navigation and communications equipment, entry to the pattern, and reporting, make the necessary time allowance.

Technique will vary with pilot experience, cockpit workload, aircraft performance, equipment used, and other factors. With crystal-tuned dual VOR receivers, dual transceivers, distance measuring equipment, Integrated Flight System (Flight Director), autopilot, and a copilot to share the workload, holding is far less of a problem than it would be for an inexperienced solo pilot holding with a single obsolescent VOR receiver without crystal tuning ("coffee grinder" receiver).

*Entry.*—Aircraft Heading on arrival at the fix determines the direction of entry turn. Fig. 20 6 . Entry procedures are oriented to a line at a 70° angle with the inbound course. This applies to both standard and nonstandard patterns.

*Parallel procedure* (Fig. 20-6 area 1)
Parallel holding course, turn left, and return to the holding fix or intercept inbound course.

*Tear drop procedure* (Fig.20-6, area 2)
Proceed on outbound heading at a 30° angle

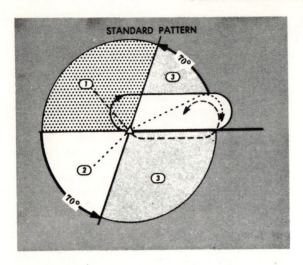

FIGURE 20-6 Holding pattern entry.

(or less) with the holding course; turn right to intercept the holding course.

*Direct entry procedure* (Fig. 20-6, area 3)
Turn right and fly the pattern.

*Turns.*—Make all turns during entry and holding at whichever of the following requires the lesser degree of bank:

1. 3° per second.
2. 30° bank angle.
3. 25° bank angle provided a flight director system is used.

Holding at speeds below 175 knots, 3° per second will require the lesser degree of bank.

*Time factors.*—

*Entry time* reported to ATC (see "Additional Reports" in this chapter) is the initial time of arrival over the fix.

*Initial outbound leg* is flown for 1 minute at or below 14,000 feet MSL. Timing for subse-quent outbound legs should be adjusted as necessary to achieve proper inbound leg time.

*Outbound timing* begins *over* or *abeam* the fix, whichever occurs later. If the abeam position cannot be determined, start timing when turn to outbound is completed (Fig 20-7).

No other time adjustments are necessary unless ATC specifies a time to leave the holding fix, as on a timed approach (see Fig. 20-3). Any necessary reduction in holding time will depend upon the time required to complete a circuit of the pattern. If, for example, a complete circuit requires 4½ minutes, and the fix departure time is 12 minutes after passage over the fix, fly two more patterns (9 minutes) and shorten the outbound leg on the last circuit to provide for a total of 1 minute except for the turns, which at 3° per second will take 2 minutes. Precise timing requires rapid cross-check and planning, in addition to the attention devoted to basic attitude control, tracking, and ATC communications.

*Expect Approach Clearance times* (EAC) and *Expect Further Clearance times* (EFC) require no time adjustment since the purpose for issuance of these times is to provide for possible loss of two-way radio communications. You will normally receive further clearance prior to your EAC or EFC. If you fail to receive it, request it.

*Time leaving* the holding fix must be known to ATC before succeeding aircraft can be cleared to the airspace you have vacated (see "Additional Reports" in this chapter). Leave the holding fix—

1. When ATC issues either further clearance enroute or approach clearance.

2. As prescribed in FAR 91 (for IFR operations; two-way radio communications failure and responsibility and authority of the pilot in command) or

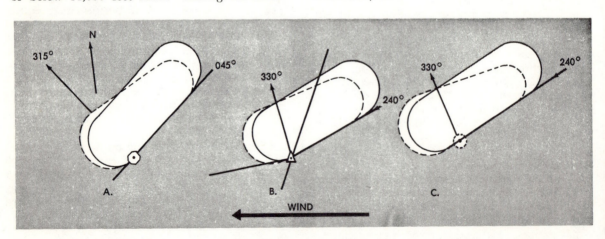

| *VOR* | *AIRWAY INTERSECTION* | *COMPASS LOCATOR* |
|---|---|---|
| Outbound timing starts when To/ From indicator reverses. | Outbound timing starts at completion of outbound turn, since 330° magnetic bearing cannot be determined. | Outbound timing starts when ADF rel. bearing is 90° minus drift correction angle. |

FIGURE 20-7 Holding—outbound timing.

3. After you have canceled your IFR flight plan, if you are holding in VFR conditions.

The diagram in Figure 20-8 shows the application of the foregoing procedures. Assume approach to the holding fix on the following ATC clearance, estimating Fox 1200:

SKYROD 234D HOLD EAST OF FOX INTERSECTION ON VICTOR 140. MAINTAIN SEVEN THOUSAND. EXPECT FURTHER CLEARANCE AT 1220.

Steps 1–10 show you overheading the fix inbound at 1203, with 17 minutes remaining to hold. If the first complete holding circuit takes 4 minutes, 30 seconds, you should therefore expect further clearance (1220) on the fourth circuit approximately 1 minute from the fix, inbound.

*DME Holding.*—The same entry and holding procedures apply to DME holding except distances (nautical miles) are used instead of time values. The outbound course of the DME holding pattern is called the outbound leg of the pattern. The length of the outbound leg will be specified by the controller and the end of this leg is determined by the DME odometer reading.

## Approaches

*Instrument approaches to civil airports.*—Unless otherwise authorized, each person operating an aircraft shall, when an instrument letdown to an airport is necessary, use a standard instrument approach procedure prescribed for that airport. Instrument approach procedures are depicted on National Ocean Survey Approach and Landing Charts. Newly established,

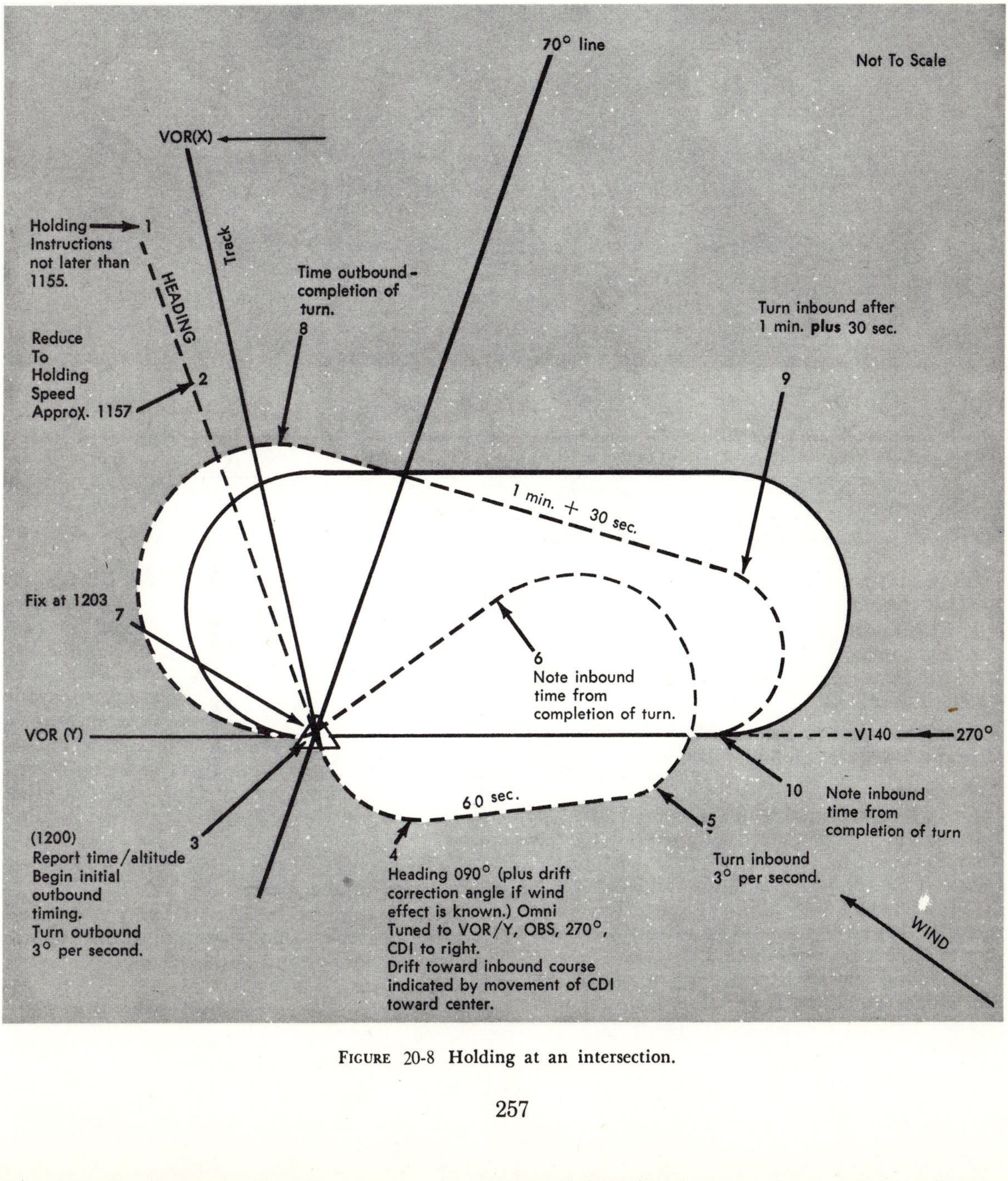

FIGURE 20-8 Holding at an intersection.

revised, or canceled procedures are listed in the "Special" section of the *Airman's Information Manual*.

ATC approach procedures depend upon the facilities available at the terminal area, the type of instrument approach executed, and existing weather conditions. The ATC facilities, navigation aids, and associated frequencies appropriate to each standard instrument approach are given on the AL Chart. Individual charts are pub-

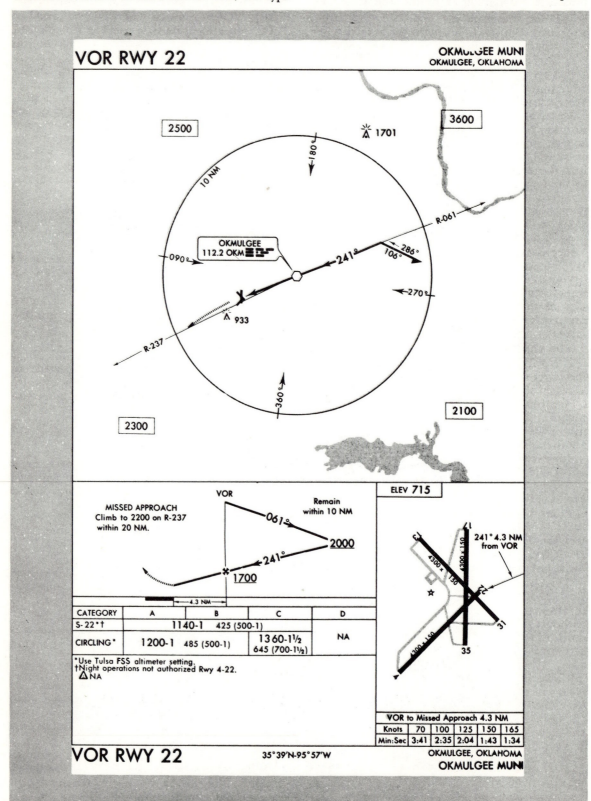

FIGURE 20-9  Instrument approach procedure chart—minimum ATC facilities.

lished for the standard approach procedures associated with the following types of facility:

1. Nondirectional homing beacon (ADF).
2. Very high frequency omnirange (VOR).
3. Very high frequency omnirange with distance measuring equipment (VORTAC).
4. Instrument landing system (ILS).

Radar approach procedures may be associated with any of the standard procedures listed above and do not require separate AL Charts. While executing a radar controlled approach, you monitor the standard approach chart appropriate to the approach specified by the controller and comply with the FAA radar ceiling and visibility minimums published as a separate section along with the standard instrument approach procedures.

*Approach to an Airport Without Tower and Without AAS.*—On an approach to an airport having minimum ATC facilities, it is of the utmost importance that you understand the details of the AL Chart to be used. Figure 20-9 prescribes the instrument approach for such an airport.

The chart shows the following information:

a. No ATC service of any kind is available at the airport. This can be determined by the absence of frequency information in the upper left corner of the chart plan view and the notation "NA" below the minimums box, which means the airport is *not* authorized for use as an alternate. No airport can be used as an alternate unless weather service is available through FSS, UNICOM, or a suitable commercial source.

b. The VOR is remotely controlled by Tulsa Flight Service Station.

From this information, you can anticipate the ATC· procedures associated with your approach to Okmulgee. If you are arriving on a "cruise" clearance, ATC will not issue further clearance for approach and landing.

The term "cruise" removes the altitude restriction usually contained in a clearance and authorizes you to continue to your destination without further clearance. However, you are required to comply with communications and reporting procedures as on any other IFR clearance. If an approach clearance is required, ARTCC will authorize you to execute your choice of standard instrument approaches (more than one may be published for the airport) with the phrase "CLEARED FOR APPROACH" and the communications frequency change required, if any. Inbound from the procedure turn, you will have no contact with ATC. Accordingly, you must close your IFR flight plan before landing, if in VFR conditions, or by telephone after landing.

Unless the approach is begun from a holding pattern on the primary navigation facility (OKM VOR) or unless otherwise authorized by ATC, you are expected to execute the complete instrument approach procedure shown on the chart.

*Approach to an Airport With AAS Only.*—Figure 20-10 shows the approach procedure at an airport where a Flight Service Station is located. When direct communication between pilot and controller is no longer required, the enroute controller will clear you for an instrument approach and advise you to contact the FSS for airport advisory information. During this approach you will have the benefit of information concerning current weather, known traffic, and any other conditions affecting your flight.

The AIM lists McAlester as an airport with an FSS from which you can anticipate the procedures noted above.

*Approach to an Airport With Tower but No Radar.*—Where an AL Chart indicates an Approach Control without radar at your destination airport, ARTCC will clear you to an approach/outer fix with the following information and instructions:

1. Name of the fix.
2. Altitude to be maintained.
3. Holding information and expected approach clearance time, if appropriate.
4. Instructions regarding further communications, including:
a. facility to be contacted.
b. time and place of contact.
c. frequency/ies to be used.

At the time of your first radio contact, or, in any event, prior to issuance of your approach clearance, approach control will inform you of the following:

1. Altimeter setting.
2. Wind direction and velocity.
3. Runway information.
4. If the landing runway is to be other than that aligned with the direction of the instrument approach, instructions to circle to the runway in use; and
5. Other information, as appropriate.

Other information, noted above, includes current weather affecting your approach and landing, as follows:

1. When ceiling and/or visibility is reported through official weather reports as being at or below the highest "circling minimums" established for the airport, approach control will report the current weather and subsequent changes.

2. When the official weather report indicates that weather conditions are below the minimums published for the approach being executed, or to be executed, approach control will—

a. Issue the weather report.

b. Advise you that the weather conditions are below the published minimums and request your intentions.

c. Issue approach clearance, landing clearance, or other clearance and/or instructions, as appropriate, in accordance with your stated intentions and the traffic situation.

Although the instrument approach to a controlled airport is conducted with direct pilot-to-

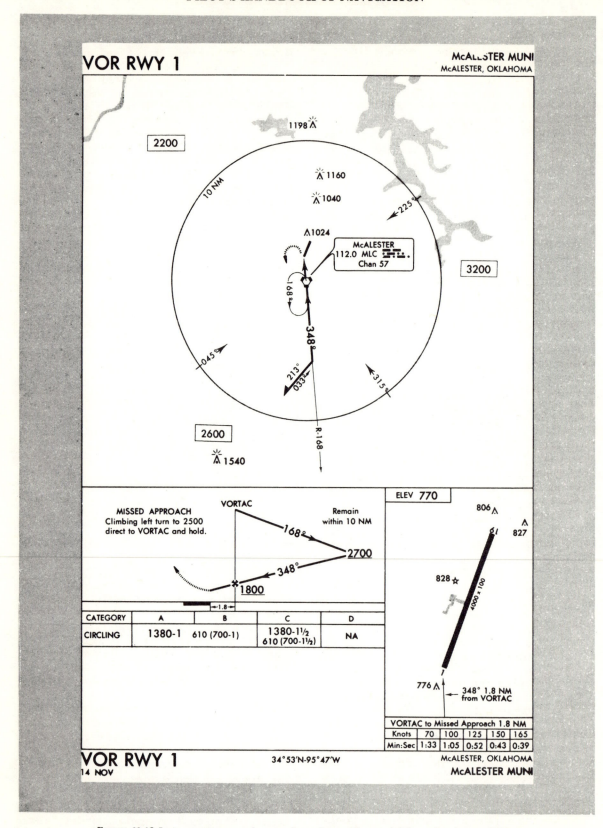

FIGURE 20-10 Instrument approach procedure chart—Airport Advisory Service available.

controller communications, without radar you are "on your own" in navigating and in maintaining separation from other aircraft unknown to the local controllers. Preflight study of the latest information about your destination airport and understanding of the details of the standard instrument approach procedures is essential for the execution of your approach.

*Approach to a Controlled Airport With Radar Available—Use of radar in instrument approach procedures.* When radar is approved at certain locations for ATC purposes, it may be used not only for surveillance and precision radar approaches, as applicable, but also may be used in conjunction with instrument approach procedures predicated on other types of radio navigational aids. Radar vectors may be authorized to provide course guidance through the segments of an approach procedure to the final approach fix or position. Upon reaching the final approach fix or position, the pilot will either complete his instrument approach in accordance with the procedure approved for the facility, or will continue a surveillance or precision radar approach to a landing.

FAA radar units operate continuously at the locations shown in the *Airman's Information Manual* and are available to all pilots.

Terminal area radar is authorized to control air traffic within the airspace encompassing the routes from outer fixes to the airport, missed approach courses, and the vector areas necessary to position, space, and control departing and arriving aircraft. Coordination between enroute radar control, approach control, and pilot is illustrated by reference to the excerpt from the Dallas-Ft. Worth Area Chart (Fig. 20-11) and the AL Chart for the ILS approach to runway 13 at Love Field (Fig. 20-12).

Assume that you are inbound to Dallas (Love Field) on an IFR flight plan filed via Ardmore (see area chart) Victor 15 Dallas. ATC has cleared you via V15W to Denton Intersection. You are conducting your own navigation, in direct contact with Ft. Worth ARTCC radar.

The available radar listed in the *Airman's Information Manual* for Love Field includes Radar Approach Control Services (IFR Arrival Control, IFR Departure Control, and Radar Traffic Information Service), and Surveillance Radar Approaches. From initial contact with radar to final authorized landing minimums, the instructions of the radar controller are mandatory except when—

1. Visual contact is established on final approach at or before descent to the authorized landing minimums, or
2. At pilot's discretion if it appears desirable to discontinue the approach.

Thus on this approach you will have maximum ATC assistance. Following radar identification of your flight between Ardmore and Denton, the Center instructs you to contact Dallas Approach Control. From Denton inbound, you will be vectored to the final approach serving the runway in use at Love Field unless radar contact is lost, complete loss of communications occurs, or you request nonradar routing. The dotted line on the area chart shows a typical radar routing to the Dallas outer compass locator from Denton, via Argyle and Lewisville Intersections and the Dallas ILS front course. The route from an outer fix to intercep-

tion of the final approach course will be, in any case, the most direct route consistent with minimum radar vectoring altitudes, radar coverage, and other traffic.

Insofar as practicable, radar-vectored routes overlie navigation courses established by other navigation aids. This provides backup in the event of radar failure and reduces the amount of radar navigation service required. Instructions from Approach Control will include the following:

1. Notification whenever—
a. Radar identification is established or lost; or
b. Radar service being provided for your flight is discontinued.
2. Purpose of the vector.
3. Procedures to be followed in the event radio communications are lost while under radar vector.
4. If radar contact is lost, an appropriate clearance via navigational aids for an instrument approach, or an alternative clearance if necessary.
5. Prior to your arrival at the approach gate (a point not less than 5 miles from the approach end of the runway) —
a. Position of your aircraft with respect to the final approach fix.
b. The vector to intercept the final approach course, if required.
c. Clearance for approach.
d. Instructions to do one of the following—
(1) Report over the approach fix to the tower on the local control frequency.
(2) Establish radio communications with the tower on the local control frequency and report over the approach fix, or
(3) Establish radio communications on the final approach frequency if a surveillance or precision approach will be executed.

*Surveillance Approach Radar.*—Initial radar contact for either a surveillance or precision approach is made with airport surveillance radar. The pilot must comply promptly with all control instructions when conducting either type procedure.

You can determine the radar approach facilities (Surveillance and/or Precision) available at a specific airport by referring to the Airport/Facility Directory of AIM, Low Altitude Enroute Charts, and Instrument Approach Procedure Charts. Surveillance and precision radar minimums are listed in the pages titled, "FAA RADAR CEILING AND VISIBILITY MINIMUMS," in the first part of the Instrument Approach Chart booklet.

On a surveillance approach, the controller vectors you to a point where you can begin a descent to the airport or to a specific runway. Course guidance, and after passing the final approach fix, distance information are issued each mile from the runway/airport down to the last mile. If requested by the pilot, recommended altitudes may be issued each mile from

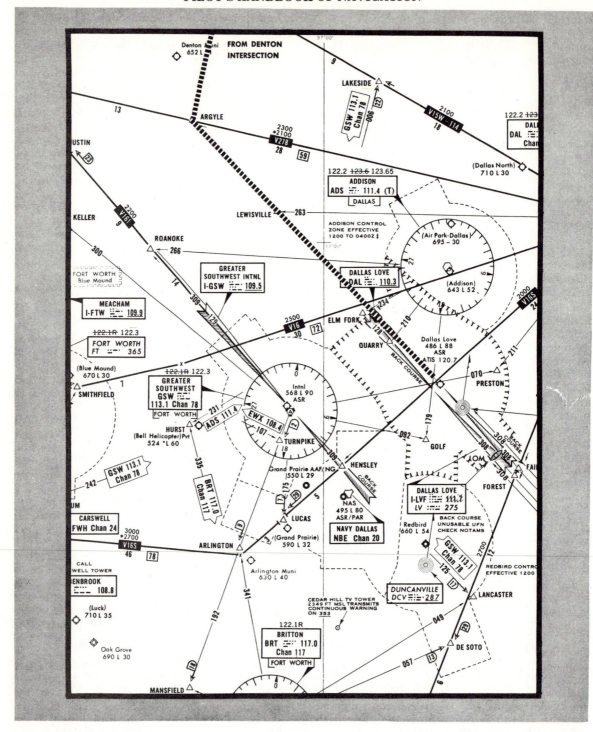

FIGURE 20-11 Terminal area ATC control.

the runway/airport down to the last mile, where the altitude is at or above the minimum descent altitude (MDA). The recommended altitudes on final approach decrease 300 feet each mile (approximate 3° descent slope). The pilot should adjust his rate of descent to achieve a rate consistent with recommended altitudes. If the MDA is reached before the missed approach point (MAP), the pilot should maintain this altitude to the MAP. The controller will advise the pilot when he reaches the MAP or one mile from the runway/airport whichever is greater, and if at this point the airport, runway or runway environment is not in sight, a missed approach should be commenced. If, on final, communication is lost for more than 15 seconds, the pilot should take over visually; if unable, he should execute the missed approach procedure.

*Precision Approach Radar.*—A PAR serves the same purpose as an Instrument Landing System (ILS) except guidance information is presented to the pilot through aural rather than visual means. If a PAR is available, it is normally aligned with an ILS.

The precision approach starts when the aircraft is within range of the precision radar and contact has been established with the PAR controller. Normally, this occurs approximately eight miles from touchdown, a point to which the pilot is vectored by surveillance radar or is positioned by a non-radar instrument approach procedure. Prior to glide path interception, the final approach airspeed and configuration should be established. When the controller advises the aircraft is intercepting the glide path, power and attitude should be adjusted to maintain a pre-determined rate of descent and airspeed. The rate of descent will vary with aircraft approach speeds and airport glide path angles, but will be near 500 feet per minute.

Prior to intercepting the glide path, the pilot will be advised of communications failure/missed approach procedures and told not to acknowledge further transmissions. The controller will give elevation information as, "slightly/well above" or "slightly/well below glide path" and course information as "slightly/well right" or "slightly/well left of course." Extreme accuracy in maintaining and correcting headings and rate of descent is essential. The controller will assume the last assigned heading is being maintained and will base further corrections on this assumption. Elevation and azimuth guidance is continued to the landing threshold of the runway, at which point the pilot will be instructed to take over visually. If the runway or runway environment is not in sight at the decision height (DH), a missed approach must be executed. If commuication is lost for more than 5 seconds on final, the pilot should take over visually; if unable, he should execute the missed approach procedure.

NOTE: Facilities for PAR approaches have been discontinued at a number of civil airports. Those remaining can be determined by referring to AIM.

*No-Gyro Approach Under Radar Control.*—If you should experience failure of your directional gyro or for other reasons need more positive radar guidance, ATC will provide a no-gyro vector or approach on request. Prior to commencing such an approach, you will be advised as to the type of approach (surveillance or precision approach and runway number) and the manner in which turn instructions will be issued. All turns are executed at standard rate, except on final approach; then at half standard rate. The controller tells you when to start and stop turns, recommends altitude information and otherwise provides guidance and information essential for the completion of your approach. You can execute this approach in an emergency with an operating communications receiver and primary flight instruments.

*Compliance With Published Standard Instrument Approach Procedures.*—Compliance with the approach procedures shown on the AL Charts provides necessary navigation guidance information for alignment with the final approach courses, as well as obstruction clearance. Under certain conditions, execution of the complete published procedure is not permissible.

In the case of a radar initial approach to a final approach fix or position, or a timed approach from a holding fix, or where the procedure specifies "NoPT" or "FINAL," no pilot may make a procedure turn unless, when he receives his final approach clearance, he so advises ATC. Execution of the procedure turn is not required in the following instances:

1. When the final approach can be executed from an established holding point on and aligned with the final approach course or from a final approach fix specified in the procedures.

2. When the symbol "NoPT" appears on the approach course on the Plan View of the AL Chart.

3. When your approach clearance specifies "Cleared for straight-in (type of approach) approach."

4. When a contact approach has been requested by the pilot and approved by ATC. A "contact" approach is defined as an approach wherein an aircraft on an IFR flight plan, operating clear of clouds with at least 1-mile flight visibility and having received an air traffic control authorization, may deviate from the prescribed instrument approach procedure and proceed to the airport of destination by visual reference to the surface. Approval of your request for a contact approach does not constitute cancellation of your IFR flight plan, and the controller must issue alternative procedures in the event that conditions less than those specified for a contact approach are encountered following approval.

5. At any time you can complete the approach in VFR conditions and cancel your IFR flight plan. Unless you cancel, or unless otherwise authorized by ATC, you are required to comply with the prescribed instrument approach procedure, regardless of weather conditions.

*Circling and Low-Visibility Approaches.*—There are various methods for executing a circling approach, depending on aircraft performance, ceiling and visibility conditions, wind direction and velocity, final approach course alignment, distance from the final approach fix to the runway, and ATC instructions. During the circling approach, you maintain visual contact with the field and fly no lower than the published circling minimums until landing is assured. It is essential, then, that you understand maneuvering procedures and select one that will keep you oriented to the landing runway and clear of obstructions. Preflight study of the AL charts for your destination and alternate airports will indicate which procedure is most suitable, particularly at uncontrolled

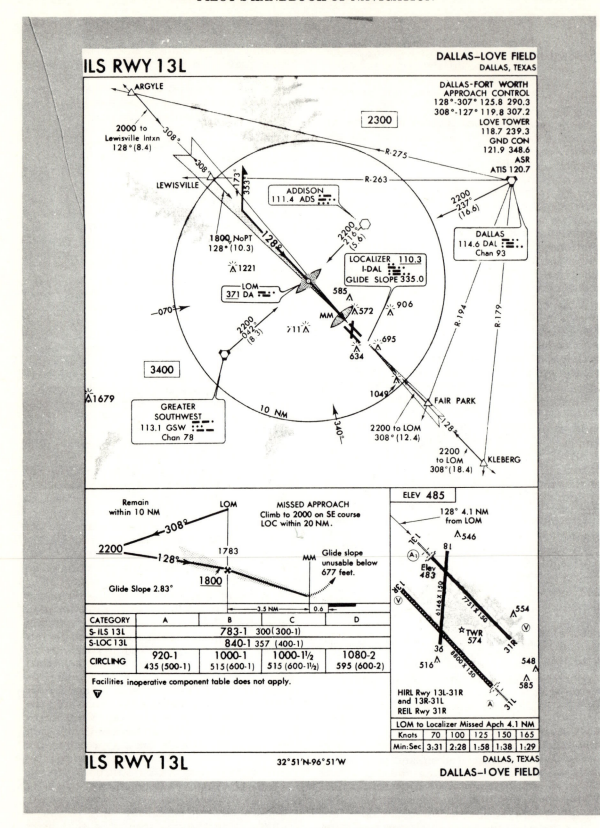

FIGURE 20-12 Instrument approach procedure chart—maximum ATC facilities.

airports where no ATC guidance is provided. Methods for completing the circling approach are shown in Figure 20-13.

Approach "A" can be made when the runway is sighted in time for a turn to a downwind leg, with the runway kept in sight throughout the approach. Approaches "B" and "C" can be made when the runway is sighted late (night approach to a field with minimum lighting, for example). On either of these low-visibility approaches, you execute the turns both by outside references and by reference to instruments until completion of the turn for alignment with the landing runway. These are applications of procedure turns that you have already learned.

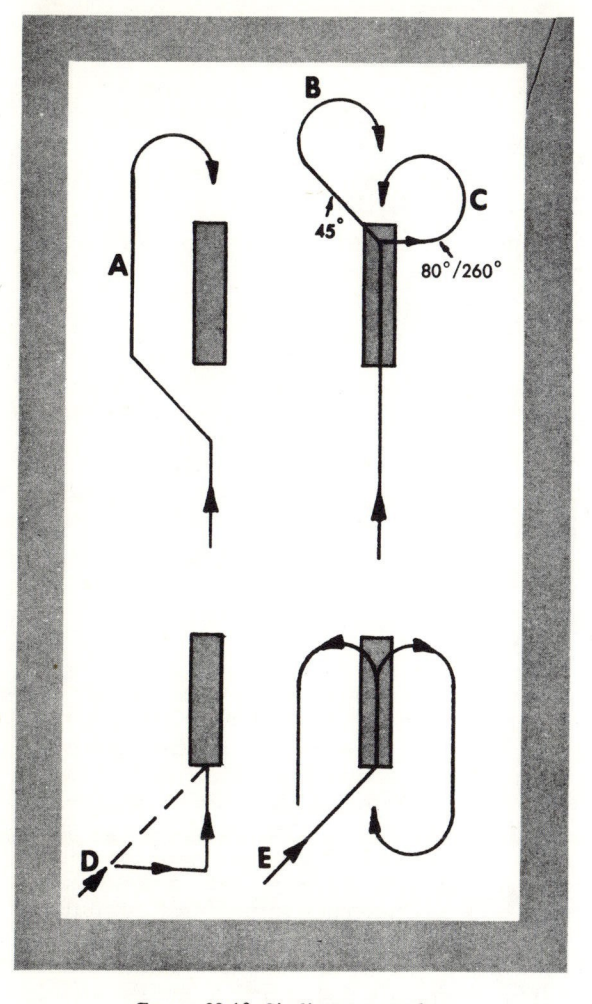

FIGURE 20-13 Circling approaches.

## Missed Approaches

A missed approach procedure is formulated for each published instrument approach. The missed approach is normally made on a course that most nearly approximates a continuation of the final approach course. A missed approach will be initiated at the point where the aircraft has descended to authorized landing minimums at a specified distance from the facility if visual contact is not established, or if the landing has not been accomplished, or when directed by Air Traffic Control. The procedure is shown on the AL Chart in narrative and pictorial form. Since the execution of a missed approach occurs when your cockpit workload is at a maximum, the procedure should be studied and mastered before beginning the approach.

Applicable landing minimums are listed on the AL Chart under *circling* or "S" (straight-in). Straight-in minimums apply if landing is to be made on the runway aligned with the final approach course. Circling minimums apply when it is necessary to circle the airport or maneuver for landing, or when no straight-in minimums are specified on the AL Chart.

The following subjects are a partial review of those regulations published in FAR Part 91 which prescribes takeoff and landing minimums under IFR.

*Landing minimums.* Unless otherwise authorized by the Administrator, no person operating an aircraft (except a military aircraft of the United States) may land that aircraft using a standard instrument approach procedure prescribed in Part 97 of this chapter unless the visibility is at or above the landing minimum prescribed in that Part for the procedure used. If the landing minimum in a standard instrument approach procedure prescribed in Part 97 is stated in terms of ceiling and visibility, the visibility minimum applies. However, the ceiling minimum shall be added to the field elevation and that value observed as the MDA or DH, as appropriate to the procedure being executed.

*Descent below MDA or DH.* No person may operate an aircraft below the prescribed minimum descent altitude or continue an approach below the decision height unless—

(1) The aircraft is in a position from which a normal approach to the runway of intended landing can be made; and

(2) The approach threshold of that runway, or approach lights or other markings identifiable with the approach end of that runway, are clearly visible to the pilot.

If, upon arrival at the missed approach point or decision height, or at any time thereafter, any of the above requirements are not met, the pilot shall immediately execute the appropriate missed approach procedure.

*Inoperative or unusable components and visual aids.* The basic ground components of an ILS are the localizer, glide slope, outer marker, and middle marker. The approach lights are visual aids normally associated with the ILS. In addition, if an ILS approach procedure in Part 97 of this chapter prescribes a visibility minimum of 1,800 feet or 2,000 feet RVR, high intensity runway lights, touchdown zone lights, centerline lighting and marking and RVR are aids associated with the ILS for those minimums. Compass locator or precision radar may be substituted for the outer or middle marker. Surveillance radar may be substituted for the outer marker. Unless otherwise specified by the Administrator, if a ground component, visual aid, or RVR is inoperative, or unusable, or not utilized, the straight-in minimums prescribed in any approach procedure in Part 97 are raised in accordance with the following tables. If the related airborne equipment for a ground component is

inoperative or not utilized, the increased minimums applicable to the related ground component shall be used. If more than one component or aid is inoperative, or unusable, or not
utilized, each minimum is raised to the highest
minimum required by any one of the components or aids which is inoperative, or unusable, or not utilized.

*Missed Approach from ILS Front Course.*—A
missed approach is reported and executed in the
following instances:

1. If, at the Decision Height (DH), the runway approach threshold, approach lights or other markings identifiable with the approach end
of the runway, are not clearly visible to the
pilot.

2. When directed by ATC.

3. If a landing is not accomplished.

*Missed Approach While Under Radar Con-
trol.*—Missed approach instructions will be issued by the controller before the final approach
and the missed approach point, as required.
Except when the controller directs otherwise
prior to the final approach, you will report and
execute a missed approach—

1. When communication on final approach is
lost for more than 5 seconds during a PAR
approach, or for more than 15 seconds during an
ASR approach.

2. When directed by the controller.

3. When the runway approach threshold, approach lights or other markings identifiable with
the approach end of the runway are not clearly visible at the Minimum Descent Altitude
(MDA) 1 mile from the end of the runway on
an ASR approach, or at the DH on a PAR
approach.

4. If a landing is not accomplished.

*Missed Approach—ILS Back Course, VOR
and ADF Approaches.*—For these, the missed
approach procedures are related to the location
of the final approach fix and are initiated in the
following instances:

1. If, at the MDA and the missed approach
point, the runway approach threshold, approach
lights, or other markings identifiable with the
approach end of the runway, are not clearly
visible to the pilot.

2. When directed by ATC.

3. If a landing is not accomplished.

When the final approach fix is not located on
the field, the missed approach procedure specifies the distance from the facility to the missed
approach point. The "Aerodrome Data" on the
AL Chart shows the time from the facility to
missed approach at various ground speeds,
which you must determine from airspeed, wind,
and distance values. At this time, you report
and execute a missed approach if you do not
have applicable minimums.

## Landing

The official weather report applicable to published minimums is the weather report, RVR
reading and/or runway visibility report, as appropriate. Whenever the reported weather is
below the published minimums for a particular
approach, the controller will so advise you and
request your intentions. Clearance to land will
be based solely on your stated intentions and the
traffic situations. If you request landing, the
controller will qualify the landing clearances as
follows:

CLEARED TO LAND IF YOU HAVE
LANDING MINIMUMS

The decision to land under such circumstances is
thus your own responsibility.

## Canceling IFR Flight Plan

You may cancel an IFR flight plan at any
time you are in VFR weather conditions unless
you are flying in a positive control area. If a
destination airport has a functional tower, the
flight plan is closed automatically upon landing.
At airports without a functional tower, close
your flight plan through the Flight Service Station by radio or by telephone after arrival.

## Emergencies

Many inflight emergency procedures are special procedures established to meet situations
that have been foreseen and for which an
immmediate solution is available as a standard
procedure. Because of the number of variable
factors involved, it is impossible to prescribe
ATC procedures covering every possible inflight
emergency. You are expected to know thoroughly the emergency procedures formulated
to prevent emergencies from developing into
accidents.

The emergency for which a published solution is available is just another procedure if you
are properly prepared for it. FAR Part 91
prescribes procedures to follow in the event of
communications failure.

## Loss of Communications While
## Under Radar Control

If two-way radio communications are lost
while you are under radar control, communications will be transmitted on all suitable air/-
ground radio frequencies as well as on the voice
feature of all available radio navigation or approach aids requesting you to acknowledge by
executing suitable turns. If the turns are observed as directed, ATC so advises you, and
radar control continues. If the radar controller

observes no acknowledging turns, he assumes loss of receiver as well as transmitter capability. If, at this time, you are being vectored off the route specified in your last ATC clearance delivered prior to issuance of vectors, ATC expects you to proceed to such route by the most direct course practicable and then proceed in accordance with two-way radio failure procedures. Pilots of aircraft equipped with coded radar beacon transponders may alert ATC of their radio failure by adjusting their transponders to reply on Mode A/3, Code 7600.

## Emergency Radar Flight Patterns

Radar controllers are on the alert for emergency radar flight patterns indicating two-way communication failure. The patterns designed to alert radar systems are as shown in Figure 20-14.

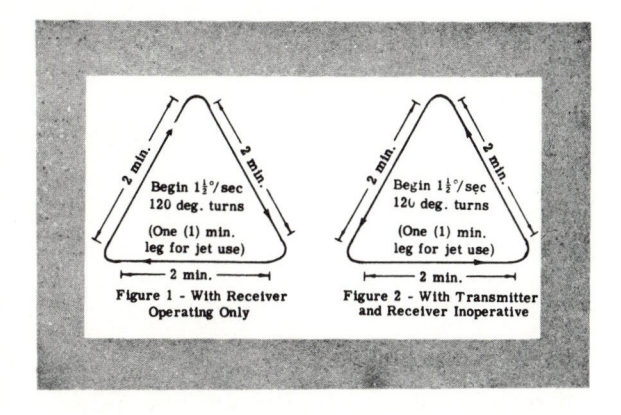

FIGURE 20-14 Emergency radar flight patterns.

## VHF/UHF DF Steer

VHF/UHF equipment is installed at many locations equipped with airport surveillance radar to assist in:

1. Locating aircraft;

2. Obtaining an accurate bearing for the purpose of;

a. Vectoring an aircraft to an airport by the most direct route, when required.

b. Locating lost aircraft or aircraft in distress which are within communications range but outside the area of radar coverage. This information will furnish the controller with the proper heading required to bring aircraft within the antenna pattern of the surveillance radar equipment.

c. Obtaining and coordinating cross bearings to establish a fix in localities where the VHF/UHF communications range of two or more VHF/UHF DF installations overlap.

VHF/DF capability is indicated in the Airport/Facility Directory of AIM.

## VHF/DF Homing Procedure

Request for an emergency DF steer is normally made on 121.5 MHz. If you cannot transmit on this frequency, request DF service on any standard frequency from any FAA tower or Flight Service Station. Such requests will be relayed without delay to the nearest VHF/DF facility in the DF network. Practice steers may also be obtained, using 121.5 MHz to establish initial contact. If an emergency exists, make the fact clear in your initial callup.

The following is a sample DF homing procedure:

1. (pilot)—"Boston Homer, this is Skyrod 1234 Delta, request homing (or "request emergency homing"), over."

2. (VHF/DF station)—"Skyrod 1234 Delta, this is Boston Homer, transmit for homing on ——— MHz, over."

3. (pilot)—"Boston Homer, this is Skyrod 1234 Delta," (depress VHF tone button for two periods of 10-seconds each. If your aircraft is not equipped with a tone button, transmit a voice signal—steady ah-h-h—for the same periods at a volume as constant as possible), "Skyrod 1234 Delta, over."

4. (VHF/DF station)—"Skyrod 1234 Delta, this is Boston Homer, course with zero wind 060, over."

5. (pilot)—"Boston Homer, this is Skyrod 1234 Delta, course 060, out."

The course given is the magnetic direction you must steer to reach the DF station assuming zero wind. Bearing information (true bearing of your aircraft from the station) is available on request. During an emergency, the DF station will monitor your position and verify or revise course information as necessary. Remember that the accuracy of a VHF/DF steer is subject to line-of-sight limitations as to height and distance, as well as fluctuations in the volume of the tone you transmit.

# 21 FLIGHT PLANNING

## Sources of Flight Planning Information

In addition to the Enroute Charts, Area Charts, and Approach and Landing Charts previously discussed in other chapters, the FAA publishes the *Airman's Information Manual* for flight planning in the National Airspace System. The manual is issued in four Parts, with each revised on a periodic schedule. The revision cycle of each Part is based upon the relative stability of the different types of information presented.

Because AIM is published for all airspace users, much of the information applies to flight operations of no more than casual interest to the instrument flying student familiar only with typical light plane equipment. For example, TACAN, CONSOLAN, and Jet Advisory Service data relating to navigation and radar equipment are of concern to pilots of military and large transport-type aircraft. Specific other information may be important to you only under occasional or rare circumstances. In this category are "Search and Rescue Procedures," "Visual Emergency Signals," "ADIZ Procedures," and "VHF Direction Finding Procedures and Facilities."

The sections of Part 1 entitled "Basic Flight Manual and ATC Procedures," applicable to VFR as well as IFR operations, contain basic operational information especially helpful to inexperienced pilots and those "rusty" on current procedures. Descriptions, diagrams, and general discussions about airport lighting and marking, altimeter settings, radar services, approach lighting systems, and navigation facilities help to clarify these important elements of aeronautical knowledge. Finally, of more immediate importance for the planning and conduct of instrument flights are such listings as "Weather Reporting Stations," "Flight Service Stations," and "VOR Receiver Check Points," found in other parts of AIM. Review of the contents of AIM will help you to determine which sections are useful for frequent or occasional reference.

As you become familiar with this publication, you will be able to plan your IFR flights quickly and easily.

Preferred routes are designed to provide for the systematic flow of air traffic in the major terminal and enroute flight environments. The *Airman's Information Manual* lists the preferred routes from the major terminals alphabetically, to other major terminals in the airspace system. Although this routing is not mandatory, filing via preferred routes has obvious advantages. Flight planning is simplified because of the probability that the route filed will be approved by ATC. Further, the clearance for a flight via SID and preferred route is in brief form, eliminating the complicated route descriptions familiar to instrument pilots.

Since the preferred routes provide for the most efficient flow of traffic in and out of termi-

nal areas, route changes and traffic delays are minimized. For example, when planning a flight from New Orleans, La., to Washington, D.C., you find a number of possible routes shown on the four Low-Altitude Charts to be used. A check of the preferred routes listed will show this entry (Fig. 21-1).

**NEW ORLEANS METRO AREA**

| | |
|---|---|
| Dallas | 0000–2359 GMT Walker V114N AEX V114 GSW 081 Fitch (L-17, 13) |
| Dulles | V20 SBV V39 GVE V39E Brandy (L-17, 18, 20, 22) |
| Kennedy | V20 V213 ENO V44 Beachwood (L-17, 18, 20, 22, 24) |
| Newark | V20 V213 ENO V29 V433 RockyHill (L-17, 18, 20, 22, 24) |
| Washington | V20 SBV V39 GVE V16 Ironsides (L-17, 18, 20, 22) |

FIGURE 21-1

The preferred route is V20 SBV V39 GVE V16 Ironsides. Ironsides is a fix within the Washington terminal area. The Enroute Low-Altitude Charts pertinent to the flight are in parentheses. By thorough study of the navigation and communications data appropriate to your selected route, you avoid uncertainty and confusion during IFR departures from busy terminal areas.

The "Notices to Airmen" list alphabetically by State, operation hazards and restrictions, navigation facility limitations, amendments to instrument flight procedures, and other current airspace data that should be checked during preflight planning, for either VFR or IFR operations.

The "Airport/Facility Directory" tabulates the facilities and services available at major airports which have terminal NAVAIDS and communications facilities. This listing, by State and city, includes runway data, radio frequencies, radar services, and terminal information essential to flight planning. Reference to this section provides a check as to availability of instrument landing facilities at your destination and alternate airport.

If your preliminary check of weather indicates marginal conditions, the facilities and services available at your destination will determine whether or not you can complete your flight. For complete instrument approach and landing data, refer to the "Instrument Approach and Landing Charts."

On the outside flap of each Enroute Low Altitude Chart are listed the Air Route Traffic Control Center/Remote Frequencies used by

ATC for direct controller-to-pilot enroute control of IFR flights in the area covered by that chart. Figure 21-2 is an excerpt from the outside flap of charts L–13 and L–14.

As your flight progresses from sector to sector, you will be advised of the frequency to be used and when the frequency change is required. If your communications equipment is limited, the limitations should be specified in your flight plan. The Sector Discrete Frequencies listed in Figure 21-2 for Albuquerque, Atlanta, Fort Worth, and Memphis Centers range from 119.3 to 135.6 MHz for control in the low altitude structure. Accordingly, if you have 90 channel capability (118.0–126.9 MHz), enter "VHF T/R 90 ch." under "Remarks" on your flight plan. This entry tells ATC that your flight enroute can be controlled directly by the centers

1. A landing area condition which precludes safe operation of aircraft.

2. An unscheduled change in, or irregular operation of, any component of the National Airspace System which precludes the use of a facility for normal aircraft operations.

3. Any scheduled and published change to components of the National Airspace System that is rescheduled or modified.

4. New or modified instrument procedures or changes in operating minimums.

The example below shows a NOTAM given at the end of a Baltimore sequence report:

$$\rightarrow BAL\ 4/2\ QAPES$$

Decoded, the NOTAM says that the transmitting station is Baltimore, the NOTAM is the

---

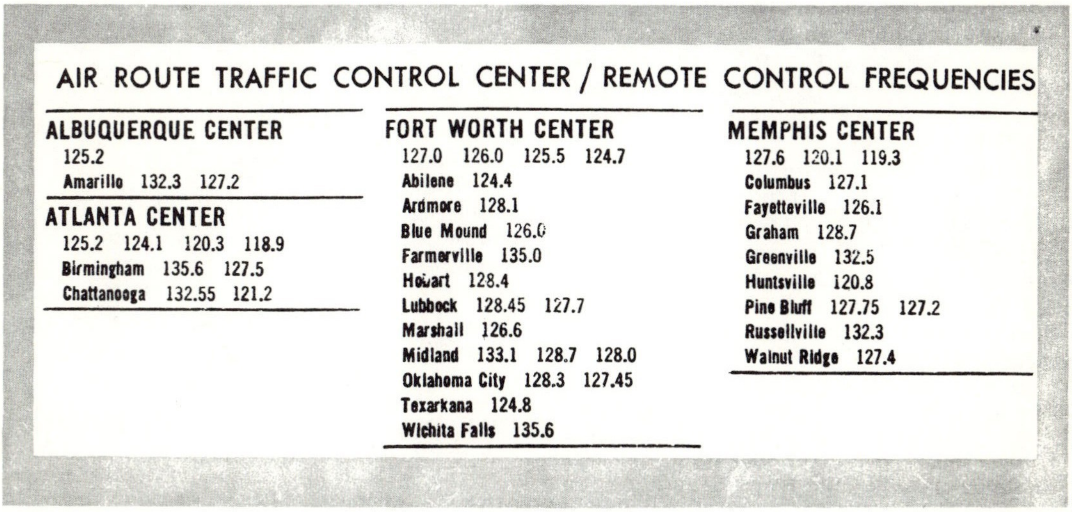

# AIR ROUTE TRAFFIC CONTROL CENTER / REMOTE CONTROL FREQUENCIES

**ALBUQUERQUE CENTER**
125.2
Amarillo  132.3  127.2

**ATLANTA CENTER**
125.2  124.1  120.3  118.9
Birmingham  135.6  127.5
Chattanooga  132.55  121.2

**FORT WORTH CENTER**
127.0  126.0  125.5  124.7
Abilene  124.4
Ardmore  128.1
Blue Mound  126.0
Farmerville  135.0
Hobart  128.4
Lubbock  128.45  127.7
Marshall  126.6
Midland  133.1  128.7  128.0
Oklahoma City  128.3  127.45
Texarkana  124.8
Wichita Falls  135.6

**MEMPHIS CENTER**
127.6  120.1  119.3
Columbus  127.1
Fayetteville  126.1
Graham  128.7
Greenville  132.5
Huntsville  120.8
Pine Bluff  127.75  127.2
Russellville  132.3
Walnut Ridge  127.4

FIGURE 21-2

on Sector Discrete Frequencies below 127.0 MHz.

If your equipment includes only the standard FSS VHF frequencies, enter this limitation on your flight plan so that the ARTCC will communicate to you through Flight Service. This information not only facilitates enroute communications; it also affects the separation standards (radar vs nonradar) applied to your flight. The use of radar separation minimums requires instantaneous interference-free controller-to-pilot communications.

## NOTAMS

In addition to the basic flight information found in the publications already discussed, you will be interested in the current information concerning the National Airspace System as it is distributed for rapid dissemination through the Flight Service Stations.

NOTAMS are transmitted by teletype when the basic flight publications do not contain current information concerning:

second issued since the beginning of the fourth month, and the Baltimore VOR range and associated voice facilities are out of service.

The Hourly Weather Sequence Reports at a particular Flight Service Station apply only to the area within approximately 400 miles of the station. NOTAM data beyond this radius can be obtained from the FSS by request.

## Flight Planning Problem

The flight planning exercise presented here is not intended as an inflexible procedure for the instrument pilot to follow. Many factors affect the sequence of your preflight activity and the content of your flight log entries. For example, the preflight time and inflight attention to fuel consumption are minor problems for a 1-hour IFR flight in an aircraft with 5 hours of fuel aboard, as compared with fuel management problems in a jet aircraft on a maximum range IFR flight to a destination where the weather is just above minimums. In the first instance, your fuel problem involves only routine inflight

269

checks on fuel consumption for subsequent planning purposes and the usual management of the aircraft fuel system. In the second case, there is little or no margin for error, either in preflight computations or enroute record.

The low-time instrument pilot can benefit from experience in the use of all the facilities and sources of information available to him. Through experience you will find a reliable basis for adapting planning methods to operational needs. The professional corporation pilot familiar with his proposed route and terminal facilities may be concerned primarily with weather and loading data. Other professional pilots limit their preflight computations to those necessary for the IFR flight plan (estimated true airspeed, estimated time enroute, and fuel on board). Then they prepare a tentative flight log suitable for quick and continuous inflight computations and revisions. Regardless of their experience and differences in planning methods, they overlook no detail that might invite unnecessary inflight problems.

The following problem involves preparation for an IFR training flight from Oklahoma City to Dallas (Love Field) via Okmulgee, then return to Oklahoma City. Instrument approaches are planned at Okmulgee and Love Field. The problem begins with the preliminary check with the weather forecaster and ends with the completed flight plan.

Although the 10 steps and the flight log may need adapting to your particular operational needs, they include the aeronautical knowledge applicable to an IFR flight in controlled airspace. Since the FAA written tests for the Instrument Rating are based on comparable problems, the exercise serves as a guide to detailed preparation for the tests. Each step discussed in the preflight action is related to the appropriate Federal Aviation Regulation and sources of information related to the planning procedure.

Aircraft equipment, performance and operating data, weight and balance information, and weather reports and forecasts are included. The Enroute Low Altitude cross referenced to pages 277, 278 and 280 applies to this exercise. For completion of this exercise and further practice, you will need, in addition to the information supplied in this handbook, the following "tools of the trade":

1. *Airman's Information Manual,* including Part 2, *Airport Directory.*
2. *Aviation Weather*
3. A navigation computer.

### AIRCRAFT DATA.—FASTFLIGHT N6864S

The aircraft is a 5- to 7-place twin-engine "Fastflight" typical of various light twins currently in use. It is appropriately equipped for instrument flight and has the following radio installations:

One L/MF receiver.
One automatic direction finder (ADF).

Dual VHF receivers (both with frequency range 108.0–126.9 MHz) with omni heads of the three-element type (course selector, course deviation indicator, and TO-FROM indicator) and instrument landing equipment (ILS-localizer and glide slope receivers).
One marker beacon receiver.
One 360 channel transceiver (range 118.0–135.9 MHz).
One 90 channel transmitter (range 118.0–126.9 MHz).

EMPTY WEIGHT.—As equipped, 4,940 lb.
MAXIMUM ALLOWABLE GROSS WEIGHT.—7,000 lb.
OIL CAPACITY.—10 gallons (5 gal./engine).
USABLE FUEL.—
Main, 156 gal.
Aux., 67 gal. (38.5 gal. in each of 2 aux.).
FUEL CONSUMPTION.—40 gal./hr. (20 gal./hr./engine).
BAGGAGE COMPARTMENT.—See Weight and Balance information.
CALIBRATED AIRSPEEDS.—
Climb .................... 145 knots.
Cruise .................... 165 knots.
Approach .................... 110 knots.
Stall .................... 85 knots.
ALTITUDES.—All flight altitudes will be in terms of MEAN SEA LEVEL (MSL) unless otherwise specified.
RADIO CALL.—Fastflight 6864S.
DE-ICING EQUIPMENT.—Aircraft is equipped with propeller anti-icing, and wing, and vertical and horizontal stabilizer de-icers.
COMPASS CORRECTION CARD—

| FOR (MH) | 0 | 30 | 60 | 90 | 120 | 150 | 180 | 210 | 240 | 270 | 300 | 330 |
|---|---|---|---|---|---|---|---|---|---|---|---|---|
| STEER (CH) | 1 | 30 | 58 | 90 | 122 | 153 | 179 | 210 | 240 | 272 | 300 | 330 |

## Preflight Action

As you proceed with the preflight steps, refer to the flight logs and the sources listed within the brackets. Step 1, for example, refers you to (a) Federal Aviation Regulations, which specify what checks are *required* and (b) *Aviation Weather,* which explains the *meaning* of the weather information.

*Step 1. Preliminary weather check.*—Regardless of how good the current weather looks in your departure area, you may save planning time by making use of forecast services well in advance of your expected departure time. Usually, a detailed forecast is accurate for about 6 hours in advance. Beyond 24 hours, only general weather outlooks are possible. It is therefore important to know the types of forecasts available, their filing times, and their valid times. Familiarization with weather service facilities as you plan practice IFR flights in VFR weather conditions is excellent preparation for all-weather flying capability. [See FAR 91, *Aviation Weather.*

# FLIGHT LOG

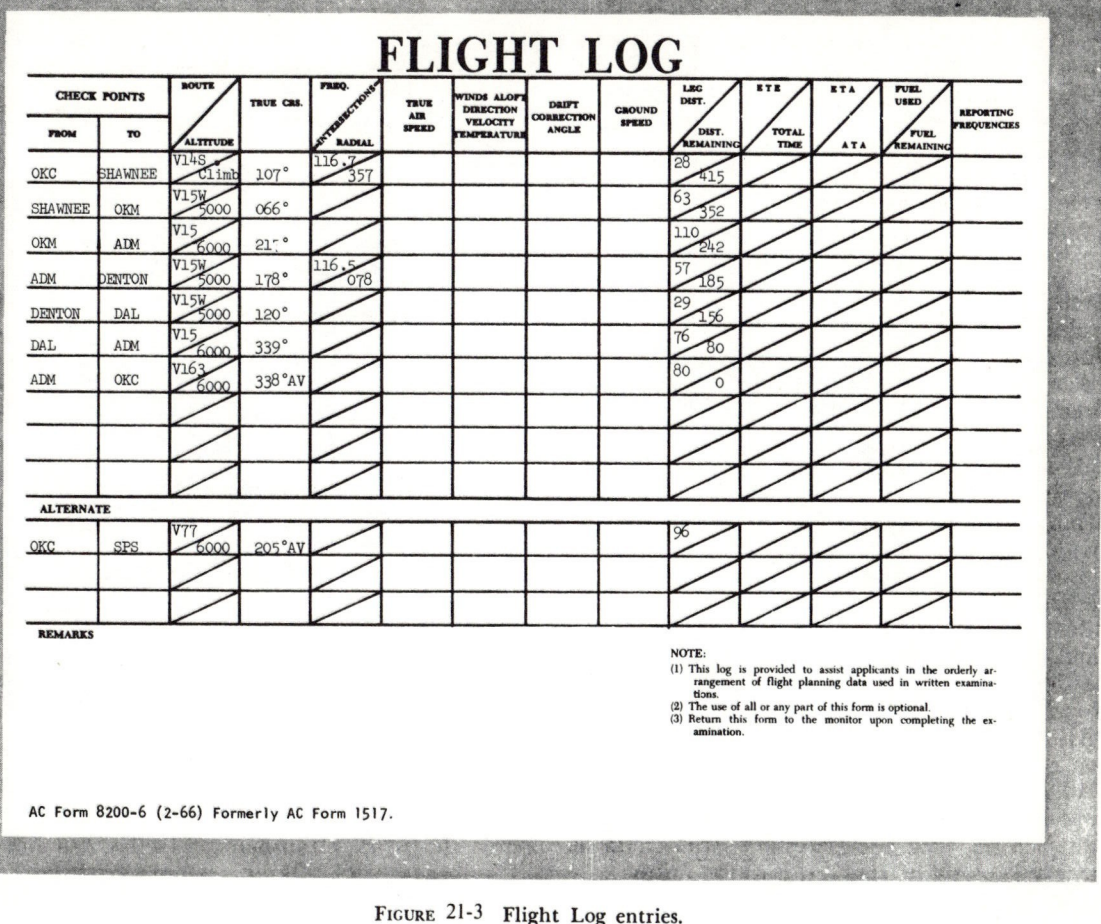

| CHECK POINTS | | ROUTE / ALTITUDE | TRUE CRS. | FREQ. / INTERSECTIONS / RADIAL | TRUE AIR SPEED | WINDS ALOFT DIRECTION VELOCITY TEMPERATURE | DRIFT CORRECTION ANGLE | GROUND SPEED | LEG DIST. / DIST. REMAINING | ETE / TOTAL TIME | ETA / ATA | FUEL USED / FUEL REMAINING | REPORTING FREQUENCIES |
|---|---|---|---|---|---|---|---|---|---|---|---|---|---|
| FROM | TO | | | | | | | | | | | | |
| OKC | SHAWNEE | V14S Climb | 107° | 116.7 / 357 | | | | | 28 / 415 | | | | |
| SHAWNEE | OKM | V15W 5000 | 066° | | | | | | 63 / 352 | | | | |
| OKM | ADM | V15 6000 | 21.° | | | | | | 110 / 242 | | | | |
| ADM | DENTON | V15W 5000 | 178° | 116.5 / 078 | | | | | 57 / 185 | | | | |
| DENTON | DAL | V15W 5000 | 120° | | | | | | 29 / 156 | | | | |
| DAL | ADM | V15 6000 | 339° | | | | | | 76 / 80 | | | | |
| ADM | OKC | V16S 6000 | 338°AV | | | | | | 80 / 0 | | | | |
| | | | | | | | | | | | | | |
| | | | | | | | | | | | | | |
| | | | | | | | | | | | | | |
| ALTERNATE | | | | | | | | | | | | | |
| OKC | SPS | V77 6000 | 205°AV | | | | | | 96 | | | | |
| | | | | | | | | | | | | | |
| | | | | | | | | | | | | | |
| REMARKS | | | | | | | | | | | | | |

NOTE:
(1) This log is provided to assist applicants in the orderly arrangement of flight planning data used in written examinations.
(2) The use of all or any part of this form is optional.
(3) Return this form to the monitor upon completing the examination.

AC Form 8200-6 (2-66) Formerly AC Form 1517.

FIGURE 21-3 Flight Log entries.

*Step 2. Tentative route(s) to destination and alternate(s).*—If possible, choose preferred routes and select more than one alternate if weather is marginal. Start preparation of flight log (Fig. 21-3). [FAR 91.83 (a) (5); Enroute Chart (s): *Airman's Information Manual;* Low Altitude Area Chart; Winds Aloft Forecasts.]

a. *Proposed Altitude.*—This will depend on factors such as aircraft performance and equipment, winds aloft, freezing level, turbulence, cloud tops, and minimum instrument altitudes. [FAR 91.119; FAR 91.121 (a) and (b).]

b. *Route and Check Points.*—The Dallas-Ft. Worth Area Chart (Fig. 21-1) shows normal arrival route (1) from Denton Intersection direct Argyle and Lewisville Intersections; and (2) from Denton Intersection V15W to Dallas VORTAC, then V16S to Dallas Love Airport. Either route may be used for flight planning. [FAR 91.123; Fig. 21-1; Enroute LA Chart segment inside back cover.]

c. *True or Magnetic Courses.*—The courses entered on the sample flight log are true courses. Some pilots reference their computations to true direction; others to magnetic. If you start with magnetic direction in your computations, winds aloft data must be corrected for variation; if you

use winds aloft as reported or forecast (true direction/knots), magnetic courses shown on the chart must be corrected for variation.

d. *Distances.*—The first leg (28 nmi) entered on the log is measured from the departure airport (Will Rogers World International) to the first enroute fix. Note that the flight log shows DAL VORTAC as the destination fix for the second leg of the flight. You may log the 16 miles between the enroute facility and Love Field, and an additional 16 miles for return to the VORTAC following the instrument approach. The Dallas-Ft. Worth Area Chart shows DAL VORTAC as the route terminating fix. You can terminate your flight log at the VORTAC or at one of the compass locators serving Love Field. In either case, your flight plan will supply ATC with the necessary route data in the event of radio communications failure. In actual practice, the routes and mileages shown on the log are adequate. You can expect to be handed off to Approach Control before arrival at Denton and to be vectored to Love Field by a more direct route than planned on your log.

e. *Communications/navigation frequencies.*—You may wish to list the VOR frequencies with

271

# FLIGHT LOG

| CHECK POINTS FROM | CHECK POINTS TO | ROUTE / ALTITUDE | TRUE CRS. | FREQ. INTERSECTIONS RADIAL | TRUE AIR SPEED | WINDS ALOFT DIRECTION VELOCITY TEMPERATURE | DRIFT CORRECTION ANGLE | GROUND SPEED | LEG DIST. / DIST. REMAINING | ETE / TOTAL TIME | ETA / ATA | FUEL USED / FUEL REMAINING | REPORTING FREQUENCIES Total Fuel: 196 gals. |
|---|---|---|---|---|---|---|---|---|---|---|---|---|---|
| OKC | SHAWNEE | V14S Climb | 107° | 116.7 / 357 | 150 | 2430 | 8°R | 169 | 28 / 415 | 9.9 / 9.9 | 1640 | 6.6 / 189.4 | |
| SHAWNEE | OKM | V15W 5000 | 066° | | 177 | 2625/4 | 2°L | 201 | 63 / 352 | 18.8 / 28.7 | | 12.6 / 176.8 | |
| OKM | ADM | V15 6000 | 217° | | 180 | 2626/2 | 6°R | 160 | 110 / 242 | 41.2 / 1:09.9 | | 27.5 / 149.3 | |
| ADM | DENTON | V15W 5000 | 178° | 116.5 / 078 | 178 | 2730/8 | 10°R | 177 | 57 / 185 | 19.3 / 1:29.2 | | 12.9 / 136.4 | |
| DENTON | DAL | V15W 5000 | 120° | | 178 | 2730/8 | 5°R | 204 | 29 / 156 | 8.5 / 1:36.7 | | 5.7 / 130.7 | |
| DAL | ADM | V15 6000 | 339° | | 182 | 2730/6 | 9°L | 168 | 76 / 80 | 27.1 / 2:03.8 | | 18.1 / 112.6 | |
| ADM | OKC | V163 6000 | 338°AV | | 180 | 2626/2 | 8°L | 173 | 80 / 0 | 27.7 / 2:31.5 | | 18.5 / 94.1 | |
| ALTERNATE | | | | | | | | | | | | | |
| OKC | SPS | V77 6000 | 205°AV | | 180 | 2626/2 | 7°R | 164 | 96 / | 35.2 / | | 23.5 / | |

REMARKS

NOTE:
(1) This log is provided to assist applicants in the orderly arrangement of flight planning data used in written examinations.
(2) The use of all or any part of this form is optional.
(3) Return this form to the monitor upon completing the examination.

AC Form 8200-6 (2-66) Formerly AC Form 1517.

FIGURE 21-4  Completed Flight Log.

the check points. In this case the Ardmore and Bridgeport frequencies and radials used to establish those intersections are listed under "Reporting Frequencies." Enter the ARTCC Sector Discrete Frequency for Ft. Worth Center if your communications equipment is limited. With the 360-channel transceiver assumed for the flight, use this column to note frequencies as ATC assigns them enroute. It is a good idea to record Center Sector Discrete Frequencies for each Center involved in your flight. [FAR 91.125.]

*Step 3.* Low-Altitude Area Charts (if published for terminals on your flight). The dashed lines on the enroute chart segment show that area charts are published for Dallas, but not for Oklahoma City, Okmulgee, or Wichita Falls. Your ATC clearances are not likely to include unexpected fixes, routes, and procedures if you know the following:

a. Departure and arrival routes.

b. Standard Instrument Departures (SIDs), if available.

c. Communications and navigation frequencies.

d. Normal route starting and/or terminating fixes.

[Enroute Chart (s); Low-Altitude Area Charts;

*Airman's Information Manual.*]

*Step 4. Current Instrument Approach Procedure Charts for*—

a. Airport of departure (for the possibility of an emergency return after takeoff).

b. Destination (s).

c. Alternate airport (s).

[FAR 91.83; 91.116; AL Charts appropriate to your aircraft equipment. ILS chart for Love Field and VOR chart for Okmulgee are shown in Chapter XI, "ATC Operations and Procedures."]

*Step 5. Current information on facilities and procedures related to your flight, including*—

a. Notices to Airmen.

b. Special Notices.

c. Services available at destination (s) and alternate (s).

d. Airport conditions, including lighting, obstructions, and other notations under "remarks." [Current *Airman's Information Manual.*]

*Step 6. Contact Weather Bureau or Flight Service Station for briefing.* Departure, enroute, arrival, and alternate weather items of significance are presented to the pilot.

[FAR 91.5; *Airman's Information Manual; Aviation Weather;* weather data in this chap-

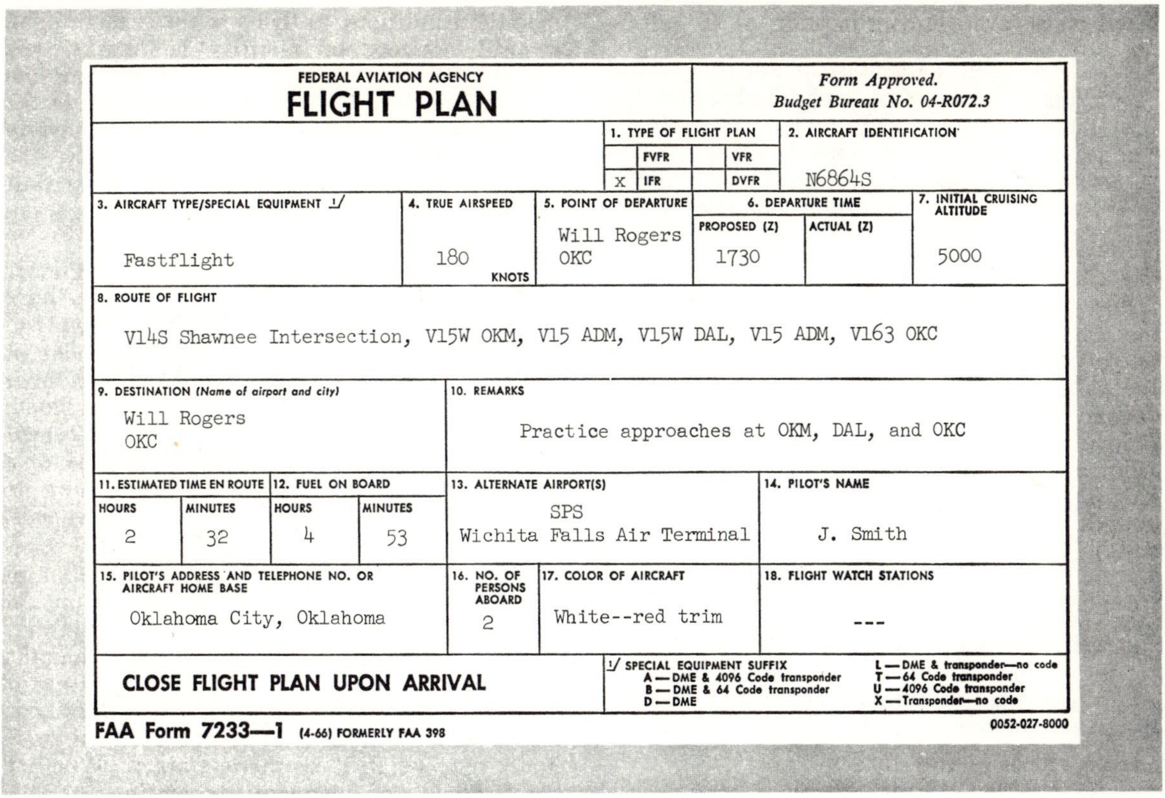

| FEDERAL AVIATION AGENCY **FLIGHT PLAN** | | | Form Approved. Budget Bureau No. 04-R072.3 | | |
|---|---|---|---|---|---|
| | 1. TYPE OF FLIGHT PLAN | | 2. AIRCRAFT IDENTIFICATION | | |
| | | FVFR | VFR | | |
| | X IFR | | DVFR | N6864S | |

| 3. AIRCRAFT TYPE/SPECIAL EQUIPMENT 1/ | 4. TRUE AIRSPEED | 5. POINT OF DEPARTURE | 6. DEPARTURE TIME | | 7. INITIAL CRUISING ALTITUDE |
|---|---|---|---|---|---|
| Fastflight | 180 KNOTS | Will Rogers OKC | PROPOSED (Z) 1730 | ACTUAL (Z) | 5000 |

**8. ROUTE OF FLIGHT**

V14S Shawnee Intersection, V15W OKM, V15 ADM, V15W DAL, V15 ADM, V163 OKC

| 9. DESTINATION (Name of airport and city) Will Rogers OKC | 10. REMARKS Practice approaches at OKM, DAL, and OKC |
|---|---|

| 11. ESTIMATED TIME EN ROUTE | | 12. FUEL ON BOARD | | 13. ALTERNATE AIRPORT(S) | 14. PILOT'S NAME |
|---|---|---|---|---|---|
| HOURS | MINUTES | HOURS | MINUTES | SPS Wichita Falls Air Terminal | J. Smith |
| 2 | 32 | 4 | 53 | | |

| 15. PILOT'S ADDRESS AND TELEPHONE NO. OR AIRCRAFT HOME BASE Oklahoma City, Oklahoma | 16. NO. OF PERSONS ABOARD 2 | 17. COLOR OF AIRCRAFT White--red trim | 18. FLIGHT WATCH STATIONS --- |
|---|---|---|---|

| **CLOSE FLIGHT PLAN UPON ARRIVAL** | 1/ SPECIAL EQUIPMENT SUFFIX A — DME & 4096 Code transponder B — DME & 64 Code transponder D — DME | L — DME & transponder—no code T — 64 Code transponder U — 4096 Code transponder X — Transponder—no code |
|---|---|---|

FAA Form 7233—1 (4-66) FORMERLY FAA 398                    0052-027-8000

FIGURE 21-5 Flight Plan entries.

ter.]

*Step 7. Complete the flight log* (Fig. 21-3). *to include:—*

a. TAS, wind data, groundspeed.

b. Estimated time enroute.

c. Estimated time between reporting points and check points.

d. Fuel required.

On the completed flight log (Fig. 21-4) note that the enroute times are computed to the tenth of a minute. In actual practice, the figures are rounded off. Note also that the groundspeed from OKC to Shawnee is computed from an assumed average climb speed of 150 TAS and from the OKC forecast wind for 3,000 feet MSL. The other true airspeed values are based on 165 knots CAS and on the temperatures given in the winds aloft data for the flight planned altitudes. The only estimated arrival time logged is at Shawnee; other ETAs are added enroute.

The fuel *required* is computed as follows, based on an average consumption of 40 gph.:—

| | | Gallons |
|---|---|---|
| a. | Enroute fuel (OKC-OKM-DAL-OKC) | 101.5 |
| b. | Fuel to alternate | 23.5 |
| c. | 45-minute reserve | 30.0 |
| d. | 30-minute additional total allowance for approaches at OKM, DAL, and OKC | 20.0 |
| | Total required | 175.0 |

[FAR 91.23; Aircraft Flight Handbook data in this chapter; weather data for wind forecasts.]

*Step 8. Compute weight and balance.* To the empty weight and oil data given, enter weights for pilot, instructor, and fuel. Assume 196 gallons total fuel load. The moments are derived from the weight and balance chart and the minimum and maximum moments (shown under "limits" in the following table) are interpolations between 6,500 and 6,550 pounds. [FAR 91.31; Aircraft Flight Handbook data.]

| | Pounds | Moment/1000 |
|---|---|---|
| Empty weight | 4,940 | 854.8 |
| Oil | 75 | 15.0 |
| Pilot and Instructor (standard weight = 170 lbs.) | 340 | 32.0 |
| Fuel (156 gal. main tank) | 936 | 175.0 |
| Fuel (20 gal. each aux.) | 240 | 43.0 |
| Totals | 6,531 | 1,119.8 |
| Limits.—Minimum 1093 (1092.96) Maximum 1139 (1138.96) | | |

*Step 9. Check teletype NOTAMS for the latest information concerning—*

a. Status of navigation and communications facilities to be used enroute.

b. Status of terminal facilities at destination and alternate.

c. Any other information pertinent to your flight [FAR 91.5].

*Step 10. Complete the flight plan* (FAA Form 7233–1) and file with FSS at least 30 minutes before estimated departure time (Fig. 21-5).

## Instrument Weather Flying

Your first flight under instrument conditions—like your first solo, first night flight, and first simulated instrument flight under the hood—will probably involve some normal apprehensions.

Notwithstanding your temperament, quality of your training, and the thoroughness of your flight planning, the decision to fly under instrument conditions is a commitment for which you *alone* are responsible. What affects your decision to go ahead with an IFR flight or to wait out a weather change for VFR conditions? Your instructor will probably elaborate on the following considerations affecting your judgment.

*Flying Experience.*—The more experience, the better—both VFR and IFR. Night flying promotes both instrument proficiency and confidence. Progressing from night flying under clear, moonlit conditions to flying without moonlight, natural horizon, or familiar landmarks, you learn to trust your instruments with a minimum dependence upon what you can see outside the aircraft. The more VFR experience you have in terminal areas with high traffic activity, the more capable you can become in dividing your attention between aircraft control, navigation, communications, and other cockpit duties.

The greater your total experience, the greater the number of unexpected situations you have behind you and the fewer surprises you can expect ahead. If you have had the benefit of instrument instruction under instrument weather conditions as well as under the hood, you may have noted that weather flying seemed routine to the instructor. This is the mark of a professional; the unusual is routine because he expects it and is ready for it before it happens.

Recency of experience is an equally important consideration. You may not act as pilot-in-command of an aircraft under IFR or in weather conditions less than VFR minimums unless within the previous 6 calendar months you have had at least 6 hours of instrument flight under actual or simulated instrument flight conditions, of which not more than 3 hours may be in a synthetic trainer. Whether the 6 hours are adequate preparation for *you* is another question.

*Airborne Equipment and Ground Facilities.*—Here again, regulations specify *minimum* equipment for filing an IFR flight plan. It is your own responsibility to decide on the adequacy of your aircraft and NAV/COM equipment for the conditions expected. A single-engine, well-equipped aircraft in excellent condition and flown by a competent pilot is obviously safer under instrument conditions than a twin, superbly equipped and in perfect condition, in the hands of a reckless or ill-prepared pilot. Whether your aircraft is single-engine or multiengine, its performance limitations, accessories, and general condition are directly related to the weather, route, altitude, and ground facilities pertinent to your flight, as well as to the cockpit workload you can expect.

*Weather Conditions.*—Departure, enroute, arrival, and alternate weather items that should be checked out before determining whether conditions are within your capabilities. Turbulence, icing, and ceiling/visibility at your destination and alternate should be evaluated in terms of their effect on your aircraft, route and altitude, and the cockpit workload that you can safely handle.

IFR flying is a team effort, using a network of facilities manned by many personnel. Your instrument rating puts you on this team, making air safety your responsibility also.

Airspace users' operations and needs are varied. Because of the nature of some oper-tions, restrictions must be placed upon others for safety reasons. The complexity or density of aircraft movements in other airspace areas may result in additional aircraft and pilot requirements for operation within such airspace. It is of the utmost importance that pilots be familiar with the operational requirements for the various airspace segments.

## UNCONTROLLED AIRSPACE

### GENERAL

Uncontrolled airspace is that portion of the airspace that has not been designated as continental control area, control area, control zone, terminal control area, or transition area and within which ATC has neither the authority nor the responsibility for exercising control over air traffic.

### VFR REQUIREMENTS

Rules governing VFR flight have been adopted to assist the pilot in meeting his responsibility to see and avoid other aircraft. Minimum weather conditions and distance from clouds required for VFR flight are con-tained in these rules. (FAR 91.105.) See figure 22-1.

### IFR REQUIREMENTS

Federal Aviation Regulations specifies the pilot and aircraft equipment requirements for IFR flight.

1. The appropriate altitude/flight level for IFR flight in controlled airspace is shown in figure 22-2.

2. Pilots are reminded that in addition to the altitude/flight level indicated in the table, FAR 91.119 includes a requirement to remain at least 1,000 feet (2,000 feet in designated mountainous terrain) above the highest obstacle within a horizontal distance of 5 statute miles from the course to be flown.

## CONTROLLED AIRSPACE

### GENERAL

1. Safety, users' needs, and volume of flight operations are some of the factors considered in the designation of controlled airspace. When so designated, the airspace is supported by ground/air communications, navigation aids, and air traffic services.

2. Controlled airspace consists of those areas desig-nated as Continental Control Area, Control Area, Control Zones, Terminal Control Areas and Transition Areas, within which some or all aircraft may be subject to Air Traffic Control.

### CONTINENTAL CONTROL AREA

The continental control area consists of the airspace of the 48 contiguous States, the District of Columbia and Alaska, excluding the Alaska peninsula west of Longitude 160°00'00''W, at and above 14,500 feet MSL, but does not include:

1. The airspace less than 1,500 feet above the surface of the earth; or

2. Prohibited and restricted areas, other than the restricted areas listed in FAR Part 71 Subpart D.

| ALTITUDE | UNCONTROLLED AIRSPACE | | CONTROLLED AIRSPACE | |
|---|---|---|---|---|
| | Flight Visibility | Distance From Clouds | ** Flight Visibility | ** Distance From Clouds |
| 1200' or less above the surface, regardless of MSL Altitude | *1 statute mile | Clear of clouds | 3 statute miles | 500' below 1000' above 2000' horizontal |
| More than 1200' above the surface, but less than 10,000' MSL | 1 statute mile | 500' below 1000' above 2000' horizontal | 3 statute miles | 500' below 1000' above 2000' horizontal |
| More than 1200' above the surface and at or above 10,000' MSL | 5 statute miles | 1000' below 1000' above 1 statute mile horizontal | 5 statute miles | 1000' below 1000' above 1 statute mile horizontal |

 * Helicopters may operate with less than 1 mile visibility, outside controlled airspace at 1200 feet or less above the surface, provided they are operated at a speed that allows the pilot adequate opportunity to see any air traffic or obstructions in time to avoid collisions.

 ** In addition, when operating within a control zone beneath a ceiling, the ceiling must not be less than 1000'. If the pilot intends to land or takeoff or enter a traffic pattern within a control zone, the ground visibility must be at least 3 miles at that airport. If ground visibility is not reported at the airport, 3 miles flight visibility is required. (FAR 91.105)

Figure 22-1 —MINIMUM VISIBILITY AND DISTANCE FROM CLOUDS—VFR

## CONTROLLED AND UNCONTROLLED AIRSPACE
## VFR ALTITUDES AND FLIGHT LEVELS

| If your magnetic course (ground track) is | More than 3000' above the surface but below 18,000' MSL fly | Above 18,000' MSL to FL 290 (except within Positive Control Area, FAR 71.193) fly | Above FL 290 (except within Positive Control Area, FAR 71.193) fly 4000' intervals |
|---|---|---|---|
| 0° to 179° | Odd thousands, MSL, plus 500' (3500, 5500, 7500, etc) | Odd Flight Levels plus 500' (FL 195, 215, 235, etc) | Beginning at FL 300 (FL 300, 340, 380, etc) |
| 180° to 359° | Even thousands, MSL, plus 500' (4500, 6500, 8500, etc) | Even Flight Levels plus 500' (FL 185, FL 205, 225, etc) | Beginning at FL 320 (FL 320, 360, 400, etc) |

## UNCONTROLLED AIRSPACE – IFR ALTITUDES AND FLIGHT LEVELS

| If your magnetic course (ground track) is | Below 18,000' MSL, fly | At or above 18,000' MSL but below FL 290, fly | At or above FL 290, fly 4000' intervals |
|---|---|---|---|
| 0° to 179° | Odd thousands, MSL, (3000, 5000, 7000, etc) | Odd Flight Levels, FL 190, 210, 230, etc) | Beginning at FL 290, (FL 290, 330, 370, etc) |
| 180° to 359° | Even thousands, MSL, (2000, 4000, 6000, etc) | Even Flight Levels (FL 180, 200, 220, etc) | Beginning at FL 310, (FL 310, 350, 390, etc) |

Figure 22-2 —ALTITUDES AND FLIGHT LEVELS

## CONTROL AREAS

Control areas consist of the airspace designated as Colored Federal airways, VOR Federal airways, Additional Control Areas, and Control Area Extensions, but do not include the Continental Control Area. Unless otherwise designated, control areas also include the airspace between a segment of a main VOR airway and its associated alternate segments. The vertical extent of the various categories of airspace contained in control area is defined in FAR Part 71.

## POSITIVE CONTROL AREA

Positive control area is airspace so designated in Part 71.193 of the Federal Aviation Regulations. This area includes specified airspace within the conterminous United States from 18,000 feet to and including FL600, excluding Santa Barbara Island, Farallon Island, and that portion south of latitude 25°04'N. In Alaska, it includes the airspace over the State of Alaska from FL240 to, and including FL600, excluding the Alaskan Peninsula west of longitude 160°00'W. Rules for operating in positive control area are found in FARs 91.97 and 91.24.

## TRANSITION AREAS

1. Controlled airspace extending upward from 700 feet or more above the surface when designated in conjunction with an airport for which an instrument approach procedure has been prescribed; or from 1,200 feet or more above the surface when designated in conjunction with airway route structures or segments. Unless specified otherwise, transition areas terminate at the base of overlying controlled airspace.

2. Transition areas are designated to contain IFR operations in controlled airspace during portions of the terminal operation and while transitioning between the terminal and en route environment.

## CONTROL ZONES

1. Controlled airspace which extends upward from the surface and terminates at the base of the continental control area. Control zones that do not underlie the continental control area have no upper limit. A control zone may include one or more airports and is normally a circular area within a radius of 5 statute miles and any extensions necessary to include instrument departure and arrival paths.

2. Control zones are depicted on charts (for example— on the sectional charts the zone is outlined by a broken blue line) and if a control zone is effective only during certain hours of the day, this fact will also be noted on the charts. A typical control zone is depicted in figure 22-3. (See SPECIAL VFR CLEARANCES in Chapter 23)

## TERMINAL CONTROL AREA

A Terminal Control Area (TCA) consists of controlled airspace extending upward from the surface or higher to specified altitudes, within which *all aircraft* are subject to operating rules and pilot and equipment requirements specified in Part 91 of the FAR's. TCA's are described in Part 71 of the FAR's. Each such location is designated as a Group I or Group II terminal control area, and includes at least one primary airport around which the TCA is located. (See FAR 71.12)

1. Group I terminal control areas represent some of the busiest locations in terms of aircraft operations and passengers carried, and it is necessary for safety reasons to have stricter requirements for operation within Group I TCA's. (See Far 91.70(c) and FAR 91.90)

2. Group II terminal control areas represent less busy locations, and though safety dictates some pilot and equipment requirements, they are not as stringent as those for Group I locations. (See FAR 91.70(c) and FAR 91.90)

3. As terminal control areas are activated they are depicted in graphic form in AIM Part 4 and charted by the National Ocean Survey in their VFR Terminal Area Charts series.

The following areas have been designated as Group I Terminal Control Areas and are depicted on the TCA charts

| | |
|---|---|
| Atlanta | Miami |
| Boston | New York |
| **Chicago** | San Francisco |
| Dallas | Washington, D.C. |
| Los Angeles | |

# AIR SPACE

The following areas have been designated as Group II Terminal Control Areas and are depicted on VFR TCA charts.

| | |
|---|---|
| Cleveland | Minneapolis |
| Denver | New Orleans |
| Detroit | Philadelphia |
| Houston | Pittsburgh |
| Kansas City | Seattle |
| Las Vegas | St. Louis |

## IFR ALTITUDES/FLIGHT LEVELS

Pilots operating IFR within controlled airspace will fly at an altitude/flight level assigned by ATC. See figure 22-1. When operating IFR within controlled airspace with an altitude assignment of "VFR-ON-TOP", an altitude/flight level appropriate for *VFR flight* is to be flown. See figure 22-1. (FAR 91.121) VFR-ON-TOP is not permitted in certain airspace such as positive control airspace, active Intensive Student Jet Training Areas, certain Restricted Areas, etc. Consequently, IFR flights operating VFR-ON-TOP will avoid such airspace.

## VFR REQUIREMENTS

Minimum flight visibility and distance from clouds have been prescribed for VFR operation in controlled airspace. See figure 22-1. In addition, appropriate altitudes/flight levels for VFR flight in controlled, as well as in uncontrolled airspace have been prescribed in FAR 91.109. See figure 22-2. The ever increasing speeds of aircraft results in increasing closure rates for opposite direction aircraft. This means that there is less time for pilots to see each other and react to avoid each other. By adhering to the altitude/flight level appropriate for the direction of flight, a "built-in" vertical separation is available for the pilots.

## SPECIAL USE AIRSPACE

### GENERAL

Special use airspace consists of that airspace wherein activities must be confined because of their nature, or wherein limitations are imposed upon aircraft operations that are not a part of those activities, or both. These areas are depicted on aeronautical charts.

### PROHIBITED AREA

Prohibited areas contain airspace of defined dimensions identified by an area on the surface of the earth within which the flight of aircraft is prohibited. Such areas are established for security or other reasons associated with the national welfare. These areas are published in the Federal Register and depicted on aeronautical charts.

### RESTRICTED AREA

Restricted areas contain airspace identified by an area on the surface of the earth within which the flight of aircraft, while not wholly prohibited, is subject to restrictions. Activities within these areas must be confined because of their nature or limitations imposed upon aircraft operations that are not a part of those activities or both. Restricted areas denote the existence of unusual often invisible, hazards to aircraft such as artillery firing, aerial gunnery, or guided missiles. Penetration of restricted areas without authorization from the using or controlling agency may be extremely hazardous to the aircraft and its occupants. Restricted areas are published in the Federal Register and constitute Part 73 of the Federal Aviation Regulations. (See OPERATION IN RESTRICTED AIRSPACE in Chapter 23)

### WARNING AREA

Warning areas are airspace which may contain hazards to nonparticipating aircraft in international airspace. Warning areas are established beyond the 3 mile limit. Though the activities conducted within warning areas may be as hazardous as those in Restricted areas, Warning areas cannot be legally designated because they are over international waters.

Penetration of Warning areas during periods of activity may be hazardous to the aircraft and its occupants. Official descriptions of Warning areas may be obtained on request to the FAA, Washington, D.C.

### INTENSIVE STUDENT JET TRAINING AREA (ISJTA)

Intensive student jet training area is that airspace which contains the intensive training activities of military student jet pilots and in which restrictions are imposed on IFR flight. Information on these training areas may be obtained from any FSS within 200 miles of the area.

NOTE: The FAA is currently in the process of converting these areas to Military Operations Areas.

### MILITARY OPERATIONS AREAS (MOA)

Military Operations Areas consist of airspace of defined vertical and lateral limits established for the purpose of separating certain military training activities from IFR traffic. Whenever an MOA is being used, nonparticipating IFR traffic may be cleared through an MOA if IFR separation can be provided by ATC. Otherwise, ATC will reroute or restrict nonparticipating IFR traffic.

Pilots operating under VFR should exercise extreme caution while flying within an MOA when military activity is being conducted. Information regarding activity in MOA's may be obtained from any FSS within 200 miles of the area.

These areas will be depicted on Sectional, VFR Terminal and Low Altitude En Route Charts.

### ALERT AREA

Alert areas are depicted on aeronautical charts to inform nonparticipating pilots of areas that may contain a high volume of pilot training or an unusual type of aerial activity. Pilots should be particularly alert when flying in these areas. All activity with an Alert Area shall be conducted in accordance with Federal Aviation Regulations, without waiver, and pilots of participating aircraft as well as pilots transiting the area shall be equally responsible for collision avoidance. Information concerning these areas may be obtained upon request to the FAA, Washington, D.C.

NOTE: The FAA plans to convert Alert Areas to Military Operations Areas.

### VFR LOW ALTITUDE TRAINING ROUTES

VFR Low Altitude Training Routes are developed in accordance with mutually acceptable procedures established by the FAA and the military services for use by the military services in conducting VFR low altitude navigation and tactical training as well as flight testing. Flights are conducted on established routes when the forecasts and weather conditions are equal to or better than a ceiling of 3000 feet and a visibility of five miles. The routes are flown in one direction only by aircraft whose speeds exceed 250 Kts. Flights will be conducted at or below 1500 feet AGL. Unless current information indicates otherwise, it should be assumed that routes are active on a continuous basis. The latest information may be obtained from the FAA FSS located nearest the route of flight. The flights are conducted in a see-and-avoid environment and pose no restrictions to other aircraft. However, because of the high speeds under which the routes are flown, *the most effective means of avoiding potential conflict is to fly above the 1500 feet AGL limit established for the routes.*

The Department of Defense Flight Information Publication, Area Planning Military Training Routes U.S., AP/1B, contains a narrative description of these routes and includes charts depicting both the VFR Low Altitude Routes and the All Weather Low Altitude Training Routes (Olive Branch). This publication is available to the general public by single copy ($1.75) or annual subscription ($20.70) from the National Ocean Survey, Distribution Division, C-44 Riverdale, Maryland 20840. This DOD FLIP, including charts, is available for pilot briefing at every FAA Flight Service Stations and at many airports.

## ALL WEATHER LOW ALTITUDE TRAINING ROUTES (OLIVE BRANCH ROUTES)

Olive Branch Routes are somewhat similar to VFR Low Level Routes but may extend to higher altitudes and are conducted under both VFR and IFR weather conditions. Further information on Olive Branch Routes is available in the AIM Part 4 and from the same sources listed above for Low Altitude High Speed Training Routes (VFR).

## OTHER AIRSPACE AREAS
### AIRPORT TRAFFIC AREAS

1. Unless otherwise specifically designated (FAR Part 93), that airspace within a horizontal radius of five statute miles from the geographical center of any air-port at which a control tower is operating, extending from the surface up to, but not including, an altitude of 3,000 feet above the elevation of the airport.

2. The rules prescribed for airport traffic areas are established in FAR 91.70, 91.85 and 91.87. They require, in effect, that unless a pilot is landing or taking off from an airport within the airport traffic area, he must avoid the area unless otherwise authorized by ATC. If operating to, from or on the airport served by the control tower, he must also establish and maintain radio communications with the tower. Maximum indicated airspeeds are prescribed. The areas *are not* depicted on charts. See figure 22-3.

### AIRPORT ADVISORY AREA

1. The area within five statute miles of an airport where a control tower is not operating but where a Flight Service Station is located. At such locations,

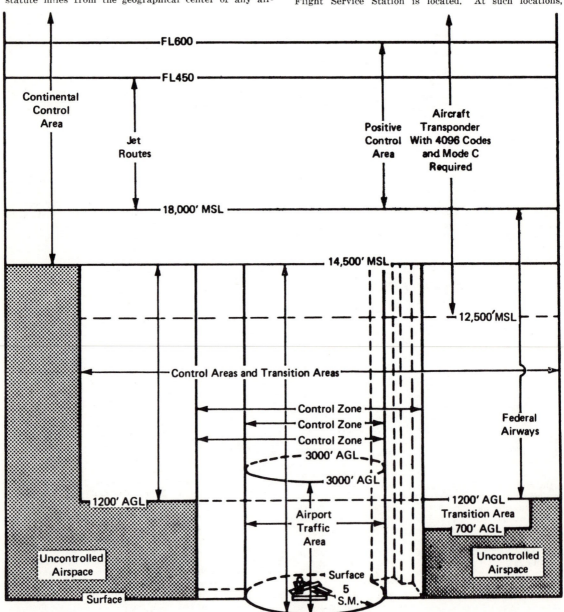

**Figure 22-3 —General Dimensions of Control Zones, Airport Traffic Areas, and the Vertical Extent of Airspace Segments.** Refer to FARs for specific dimensions, exceptions, geographical areas covered, exclusions, specific transponder/equipment requirements, and flight operations. Arrows ending near but not touching reference lines mean "up to/down to" but not including the referenced altitude.

the FSS provides advisory service to arriving and departing aircraft. (See AIRPORT ADVISORIES AT NON TOWER AIRPORTS in Chapter *23*)

**2.** It is not mandatory that pilots participate in the airport advisory service program, but it is strongly recommended that they do.

## TEMPORARY FLIGHT RESTRICTIONS

**1.** Temporary flight restrictions may be put into effect in the vicinity of any incident or event which by its nature may generate such a high degree of public interest that the likelihood of a hazardous congestion of air traffic exists. FAR 91.91, as amended 1 March, 1971, prohibits the operation of nonessential aircraft in airspace that has been designated in a NOTAM as an area within which temporary flight restrictions apply. The revised rule will continue to be implemented in the case of disasters of substantial magnitude. It will also be implemented as necessary in the case of demonstrations, riots, and other civil disturbances, as well as major sporting events, parades, pageants, and similar functions which are likely to attract large crowds and encouraging viewing from the air.

**2.** NOTAM's implementing temporary flight restrictions will contain a description of the area in which the restrictions apply. Normally the area will include the airspace below 2,000 feet above the surface within 5 miles of the site of the incident. However, the exact dimensions will be included in the NOTAM.

**3.** Pilots are not to operate aircraft within such an area described in the NOTAM unless they are one of the following: (1) That aircraft is participating in disaster relief activities and is being operated under the direction of the agency responsible for relief activities; (2) They are operating to or from an airport within the area and such operation will not hamper or endanger relief activities; (3) Their operation is authorized under an IFR ATC clearance; (4) Flight around the area is impracticable because of weather or other considerations and advance notice is given to the Air Traffic facility specified in the NOTAM, and enroute flight through the area will not hamper or endanger relief activities; or (5) They are carrying accredited news representatives or persons on official business concerning the incident, and the flight is conducted in accordance with **FAR 91.79** and a flight plan is filed with the Air Traffic facility specified in the NOTAM.

# PREFERRED IFR ROUTES

A system of preferred routes has been established to guide pilots in planning their route of flight, to minimize route changes during the operational phase of flight, and to aid in the efficient orderly management of the air traffic using federal airways. The preferred IFR routes which follow are designed to serve the needs of airspace users and to provide for a systematic flow of air traffic in the major terminal and en route flight environments. Cooperation by all pilots in filing preferred routes will result in fewer traffic delays and will better provide for efficient departure, en route and arrival air traffic service.

The following lists contain preferred IFR routes for the low altitude stratum and the high altitude stratum. The high altitude list is in two sections; the first section showing terminal to terminal routes and the second section showing single direction route segments. Also, on some high altitude routes low altitude airways are included as transition routes.

The following will explain the terms/abbreviations used in the listing:

1. Preferred routes beginning/ending with an airway number indicate that the airway essentially overlies the airport and flight are normally cleared directly on the airway.

2. Preferred IFR routes beginning/ending with a fix indicate that aircraft may be routed to/from these fixes via a Standard Instrument Departure (SID) route, radar vectors (RV), or a Standard Terminal Arrival Route (STAR).

3. Preferred IFR routes for major terminals selected are listed alphabetically under the name of the departure airport. Where several airports are in proximity they are listed under the principal airport and

categorized as a metropolitan area; e.g., New York Metro Area.

4. Preferred IFR routes used in one direction only for selected segments, irrespective of point of departure or destination, are listed numerically showing the segment fixes and the direction and times effective.

5. Where more than one route is listed the routes have equal priority for use.

6. Official location identifiers are used in the route description for VOR/VORTAC navaids.

7. Intersection names are spelled out.

8. Navaid radial and distance fixes (e.g., ARD201113) have been used in the route description in an expediency and intersection names will be assigned as soon as routine processing can be accomplished. Navaid radial (no distance stated) may be used to describe a route to intercept a specified airway (e.g., MIV MIV101 V39); another navaid radial (e.g., UIM UIM255 GSW081); or an intersection (e.g., GSW081 Fitch).

9. Where two navaids, an intersection and a navaid, a navaid and a navaid radial and distance point, or any navigable combination of these route descriptions follow in succession, the route is direct.

10. The effective times for the routes are in GMT. Pilots planning flight between the terminals or route segments listed should file for the appropriate preferred IFR route.

11. (90-170 incl) altitude flight level assignment in hundred of feet.

12. The notations "pressurized" and "unpressurized" for certain low altitude preferred routes to Kennedy Airport indicate the preferred route based on aircraft performance.

## EXAMPLES
## LOW ALTITUDE

| Terminals | Route | Effective Times (GMT) |
|---|---|---|
| **ALBANY** | | |
| Boston | (60–170 incl) V2 GDM V431 | 1000–0300 |
| Kennedy | (130–170 incl) V489 V157 Empire | 1000–0300 |
| | or | |
| | (90–120 incl) V91 V487 CMK | 1000–0300 |
| La Guardia | (90–170 incl) V91 V487 V123 | 1000–0300 |
| Newark | (80–170 incl) ALB205 SAX034 Monroe | 1000–0300 |
| Philadelphia | (70–170 incl) V449 LHY V149 Mazie | 1100–0300 |

| Terminals | Route | Effective Times (GMT) |
|---|---|---|
| Chicago O'Hare | (60–170 incl) V97 Nello V51W HCH V51 CGT V7 Niles | 1200–0300 |
| Cincinnati | (80–170 incl) V97 LEX V57 FLM | 1200–0300 |
| **ATLANTA METRO AREA** | | |
| Chicago Midway | (60–170 incl) V97 Nello V51W HCH V51 CGT | 1200–0300 |

## AREA NAVIGATION ROUTES

Area navigation systems permit navigation via a selected course to a predefined point without having to fly directly toward or away from a navigational aid. Typical of such systems are doppler radar, inertial and course line computers.

Introduction of the area navigation capability into the National Airspace System provides a means of overcoming many of the constraints of the VOR system. Elimination of the requirement to fly along radials that lead directly to or from the ground station makes it possible to design routes and procedures that better facilitate the movement of traffic.

Guidelines for the implementation of area navigation within the National Airspace System were issued in the form of Advisory Circular 90–45 on 18 August 1969. Advisory Circular 90–45 contains standards and methods for obtaining approval for use of area navigation equipment in IFR operations. It also contains air traffic control procedures applicable to approved area navigation equipment.

Since the VOR system forms the basis for the existing airway and air traffic control systems, area navigation route definitions and air traffic control clearances will be in terms of, and referenced to, VOR ground stations. However, latitude/longitude descriptions of fixes forming area navigation routes are provided to facilitate the use of airborne systems that are not dependent on VOR input signals.

Area navigation routes are numbered in a manner similar to VOR airways and jet routes. The suffix "R" identifies the route as an area navigation route. When filing an IFR flight plan containing an area navigation route, the route number may be filed in the same manner as a VOR airway number. The filed route should clearly define the intended route during transition from the VOR structure to the area navigation structure.

Area navigation route numbering will be as follows:

(1) Series 700 (low) and 800, 900 (high) routes—Routes designated in Parts 71 and 75. Routes expected

to be in frequent use by more than one user and published on appropriate U.S. Government Charts. MEAs will be published in FAR 95.

(2) Series 500 (low) and 600 (high) routes—Routes published in AIM. Routes planned for more than one user but their expected usage does not justify Government charting. MEA's when not published in FAR 95 and AIM, will be provided by ATC.

(3) Series 300 (low) and 400 (high) routes—Routes established primarily for one user. Not carried on Government charts. MEAs will be periodically published in the listing of non-95 routes.

The routes contained herein are those of the 500 and 600 Series.

Each area navigation route listed is based on a centerline which extends between fixes, described by reference to VOR/DME, called waypoints, specified for that route. Each waypoint is listed by geographical name, followed by the VOR/DME fix description in the format: VOR STATION RADIAL/DME MILEAGE (example POM 033/010). Also included are the track to be flown (Mag track), the VOR changeover point (COP, with "MP" indicating Midpoint) the distance (DIST) between waypoints, and the latitude/longitude description.

## EXAMPLES

| WAYPOINT NAME | VOR/DME DESCRIPTION | GEOGRAPHICAL COORDINATES |
|---|---|---|

### J614R

#### Denver, Colorado, to Glenwood Springs, Colorado

| | | |
|---|---|---|
| Denver VORTAC .......... | DEN 000/000 | 39°51'38.8'',104°45'07.8'' |
| MAG TRACK: 244* | | DIST: 36 |
| COP: NOT REQUIRED | | |
| Camp ................ | DEN 244.3/35.5 | 39°43'40.0'',105°30'00.0'' |
| MAG TRACK: 245* | | DIST: 13 |
| COP: DEN RLG | MAG TRACK: 242 | DIST: 13 |
| Dillon ................ | RLG 124.8/29.4 | 39°38'00.0'',106°01'30.0'' |
| MAG TRACK: 242* | | DIST: 62 |
| COP: NOT REQUIRED | | |
| Glenwood Spgs ........ | RLG 213.8/55.4 | 39°22'45.0'',107°19'25.0'' |

\* One direction airway

### J616R

#### Glenwood Springs, Colorado, to Denver, Colorado

| WAYPOINT NAME | DESCRIPTION VOR/DME | GEOGRAPHICAL COORDINATES |
|---|---|---|
| lenwood Spgs ........ | RLG 213.8/55.4 | 39°22'45.0'',107°19'25.0'' |
| MAG TRACK: 100* | | DIST: 23 |
| COP: NOT REQUIRED | | |
| Jerome ............... | RLG 188.9/50.5 | 39°13'35.0'',106°51'45.0'' |
| MAG TRACK: 071* | | DIST: 56 |
| COP: NOT REQUIRED | | |
| Webster ............... | RLG 124.7/55.4 | 39°18'20.0'',105°39'25.0'' |
| MAG TRACK: 037* | | DIST: 12 |
| COP: RLG/DEN | MAG TRACK: 038 | DIST: 42 |
| Denver VORTAC .......... | DEN 000/000 | 39°51'38.8'',104°45'07.8'' |

\* One direction airway

### V513R

#### Arcadia, Washington, to Cove, Washington

| | | |
|---|---|---|
| Arcadia ................ | SEA 212.4/27.2 | 47°10'13.0',122°50'54.0'' |
| MAG TRACK: 017/197 | | DIST: 22 |
| COP: NOT REQUIRED | | |
| Cove ................... | SEA 255.3/8.3 | 47°27'11.0'',122°30'40.0'' |

## TERMINAL CONTROL AREA (TCA)

Terminal Control Area (TCA) configurations are tailored to the operational needs of the individual area. The specific airspace assignments for each TCA location are described in FAR Part 71 and are depicted on VFR terminal area charts.

FAR 91.24 and 91.90 set forth the equipment and operating requirements for flight within TCAs. Regardless of weather conditions, an ATC authorization is required prior to operating within a Group I or Group II TCA. Other FAR 91 operating rules and equipment requirements include:

### 1. Group I TCAs

a. A 4096 code transponder with Mode C automatic altitude reporting equipment, except for helicopters operating at or below 1,000 feet under a letter of agreement. (ATC may authorize deviations from the transponder and altitude reporting equipment requirements. Request for deviation should be submitted to the controlling ATC facility at least four hours before the proposed operation.)

b. A private pilot certificate or better if landing or taking off from an airport within the TCA.

### 2. Group II TCAs

a. A 4096 code transponder, except for helicopters operating at or below 1,000 feet under a letter of agreement, or for IFR flights operating to or from an airport outside of but in close proximity to the TCA when the commonly used transition, approach, or departure procedures to such airport require flight within the TCA. (ATC may authorize deviations from the transponder requirements. Requests for deviation should be submitted to the controlling ATC facility at least four hours before the proposed operation.)

### 3. Group I and II TCAs

a. A two-way radio capable of communicating with ATC on appropriate frequencies.

b. A VOR or TACAN receiver, except for helicopters.

c. Unless otherwise authorized by ATC, large turbine-powered aircraft (over 12,500 lbs.) must operate at or above the floor of the TCA while operating to or from a primary airport. (A clearance for a visual approach is not authorization for an aircraft to operate below the designated floors of the TCA.)

Additionally, there is a 200 knot speed limit for aircraft operating beneath the floors of a TCA or within a VFR corridor through a TCA.

### 2. FLIGHT PROCEDURES.

a. IFR Flights. Aircraft operating within the TCA shall be operated in accordance with current IFR procedures.

b. VFR Flights.

(1) Arriving aircraft should contact ATC on specified frequencies and in relation to geographical fixes shown on VFR terminal area charts. Although arriving aircraft may be operating beneath the floor of the TCA on initial contact, communications should be established with ATC in relation to the points indicated for sequencing and spacing purposes.

(2) Departing aircraft are requested to advise the ground controller the intended altitude and route of flight to depart the TCA.

(3) Aircraft not landing/departing the primary airport (i.e., the airport for which TCA is designated) may obtain an ATC clearance to transit the TCA when traffic conditions permit, provided the requirements of FAR 91 are met. Notwithstanding this, VFR transiting traffic is encouraged to the extent possible to use the VFR corridors (if they exist) through the TCA, or to operate beneath or above the TCA.

**3. ATC PROCEDURES.** All aircraft will be controlled and separated by ATC while operating within a Group I or II TCA. VFR aircraft will be separated in accordance with Stage III radar service procedures. (See Part I—Services Available to Pilots.) IFR aircraft will receive normal IFR separation. Although radar separation will be the primary separation standard used, approved visual separation and other nonradar procedures will be applied as required or deemed appropriate. Traffic information on observed but unidentified radar targets will be provided on a workload permitting basis to aircraft operating outside of the TCA. Initially, this additional service will be provided within TCAs on a traffic permitting basis because of the likelihood of unintentional violations.

In the event of a radar outage, separation and sequencing of VFR aircraft will be suspended as this service is dependent on radar. The pilot will be advised that the service is not available and issued wind, runway information and the time or place to contact the tower. Traffic information will be provided on a workload permitting basis.

Note.—Assignment of radar headings and/or altitudes are based on the provision that a pilot operating in accordance with visual flight rules is expected to advise ATC if compliance with an assigned route, radar heading or altitude will cause the pilot to violate such rules.

# AIR TRAFFIC CONTROL — SERVICES AVAILABLE TO PILOTS 23

## AIR ROUTE TRAFFIC CONTROL CENTERS

Centers are established primarily to provide air traffic service to aircraft operating on IFR flight plans within controlled airspace, and principally during the enroute phase of flight.

## CONTROL TOWERS

Towers have been established to provide for a safe, orderly and expeditious flow of traffic on and in the vicinity of an airport. When the responsibility has been so delegated, towers also provide for the separation of IFR aircraft in the terminal areas (Approach Control).

## FLIGHT SERVICE STATIONS

Flight Service Stations are the Air Traffic Service facilities within the National Airspace System which have the prime responsibility for preflight pilot briefing, en route communications with VFR flights, assisting lost VFR aircraft, originating NOTAMS, broadcasting aviation weather information, accepting and closing flight plans, monitoring radio NAVAIDS, participating with search and rescue units in locating missing VFR aircraft, and operating the national weather teletypewriter systems. In addition, at selected locations, FSSs take weather observations, issue airport advisories, administer airman written examinations, and advise Customs and Immigration of transborder flight.

## PILOT VISITS

Pilots are encouraged to visit air traffic facilities—Towers, Centers and Flight Service Stations and participate in "Operation Raincheck". Operation Raincheck is conducted at these facilities and is designed to familiarize pilots with the ATC system, its functions, responsibilities and benefits. On rare occasions, facilities may not be able to approve a visit because of workload or other reasons. It is therefore requested that pilots contact the facility prior to the visit—give the number of persons in the group, the time and date of the proposed visit, the primary interest of the group. With this information avaialble, the facility can prepare an itinerary and have someone available to guide the group through the facility.

## VFR ADVISORY SERVICE

1. VFR advisory service is provided by numerous nonradar *Approach Control* facilities to those pilots intending to land at an airport served by an approach control tower. This service includes: wind, runway, traffic and NOTAM information, unless this information is contained in the ATIS broadcast and the pilot indicates he has received the ATIS information.

2. Such information will be furnished upon initial contact with concerned approach control facility. The pilot will be requested to change to the *tower* frequency at a predetermined time or point, to receive further landing information.

3. Where available, use of this procedure will not hinder the operation of VFR flights by requiring excessive spacing between aircraft or devious routing. Radio contact points will be based on time or distance rather than on landmarks.

4. Compliance with this procedure is not mandatory but pilot participation is encouraged.

## AIRPORT ADVISORY PRACTICES AT NONTOWER AIRPORTS

There is no substitute for alertness while in the vicinity of an airport. An airport may have a flight service station, UNICOM operator, or no facility at all. Pilots should predetermine what, if any, service is available at a particular airport. Combining an aural/visual alertness and complying with the following recommended practices will enhance safety of flight into and out of uncontrolled airports.

1. **Recommended Traffic Advisory Practices**—As standard operating practice all inbound traffic should continuously monitor the appropriate field facility frequency from 15 miles to landing. Departure aircraft should monitor the appropriate frequency either prior to or when ready to taxi.

a. **Inbound Aircraft**

| Airport | Frequency | Broadcast Position, Altitude, Intentions | Broadcast Position |
|---|---|---|---|
| Part-Time Tower (when closed) | Tower Local Control | 5 Miles | Downwind, Base, Final |
| Part-Time Tower (closed) but Full-Time FSS | Tower Local Control | *15 Miles | *5 Miles |
| Part-Time FSS (closed) | 123.6 | 5 Miles | Downwind, Base, Final |
| Full-Time or Part-Time FSS (open) | 123.6 | *15 Miles | *5 Miles |
| UNICOM | 122.8 | *5 Miles | |
| UNICOM (if unable establish contact) | 122.8 | 5 Miles | Downwind, Base, Final |
| No Facility on Airport | 122.9 | 5 Miles | Downwind, Base, Final |

b. **Outbound Aircraft**

| Airport | Frequency | Broadcast Position And Intentions |
|---|---|---|
| Part-Time Tower (closed) | Tower Local Control | When ready to taxi; and before taking runway for takeoff |
| Part-Time Tower (closed) but Full-Time FSS | Tower Local Control | *When ready to taxi; and before taking runway for takeoff |
| Part-Time FSS (closed) | 123.6 | When ready to taxi; and before taking runway for takeoff |
| Full-Time or Part-Time FSS (open) | 123.6 | *When ready to taxi; and before taking runway for takeoff |
| UNICOM | 122.8 | *When ready to taxi; and before taking runway for takeoff |
| UNICOM (if unable establish contact) | 122.8 | When ready to taxi; and before taking runway for takeoff |
| No facility on airport | 122.9 | When ready to taxi; and before taking runway for takeoff |

*Contact appropriate facility first; e.g., "Zanesville Radio, this is Cessna 12345, Over," before announcing arrival/departure intentions. Except for scheduled air carriers and other civil operators having authorized company call signs, departure aircraft should state the aircraft type, identification number, type of flight planned, i.e., VFR or IFR, and the planned destination.

NOTE.—FSS at part-time non-FAA tower locations do not have tower local control frequency. Use 123.6.

c. **Information Furnished by FSS or UNICOM**

(1) FSSs provide airport advisory service at airports where there is no control tower or when the tower is not in operation (part-time tower location). Advisories provide: wind direction and velocity, favored or designated runway, altimeter setting, known traffic (caution: all aircraft in the airport vicinity may not

be communicating with the FSS), notices to airmen, airport taxi routes, airport traffic patterns, and instrument approach procedures. These elements are varied so as to best serve the current traffic situation. Some airport managers have specified that under certain wind or other conditions, designated runways are to be used. Pilots using other than the favored or designated runways should advise the FSS immediately.

NOTE.—Airport Advisory Service is offered to enhance safety: TRAFFIC CONTROL IS NOT EXERCISED.

### 2. Recommended Phraseologies
#### a. Departures
Example:

**Aircraft:** JOHNSON RADIO, COMANCHE SIX ONE THREE EIGHT, ON TERMINAL BUILDING RAMP, READY TO TAXI, VFR TO DULUTH, OVER

**FSS:** COMANCHE SIX ONE THREE EIGHT, JOHNSON RADIO, ROGER, WIND THREE TWO ZERO DEGREES AT TWO FIVE, FAVORING RUNWAY THREE ONE ALTIMETER THREE ZERO ZERO ONE, CESSNA ONE-SEVENTY ON DOWNWIND LEG MAKING TOUCH AND GO LANDINGS ON RUNWAY THREE ONE

NOTE.—The takeoff time should be reported to the FSS as soon as practicable. If the aircraft has limited equipment and it is necessary to use the navigational feature of the radio aid immediately after takeoff, advise the FSS of this before shifting frequency to the range. In such cases, advisories will be transmitted over 123.6 or the tower local control frequency, as appropriate, and the aid frequency.

#### b. Arrivals
Example:

**Aircraft:** JOHNSON RADIO, TRIPACER ONE SIX EIGHT NINER, OVER KEY WEST, TWO THOUSAND, LANDING GRAND FORKS, OVER

**FSS:** TRIPACER ONE SIX EIGHT NINER, JOHNSON RADIO, OVER KEY WEST AT TWO THOUSAND, WIND ONE FIVE ZERO DEGREES AT FOUR, DESIGNATED RUNWAY FIVE, DC–3 TAKING OFF RUNWAY FIVE, BONANZA ON DOWNWIND LEG RUNWAY FIVE MAKING TOUCH AND GO LANDINGS, COMANCHE DEPARTED RUNWAY ONE SEVEN AT ONE SIX PROCEEDING EASTBOUND, OVER

NOTE.—Pilots should guard 123.6 or the tower local control frequency, as appropriate, until clear of the runway after landing and report leaving the runway to the FSS.

#### c. Transmissions (blind broadcasts) When Not Communicating With an FSS or UNICOM Operator
##### (1) Inbound
Example:

THIS IS APACHE TWO TWO FIVE ZEBRA, FIVE MILES EAST OF STRAWN AIRPORT, TWO THOUSAND DESCENDING TO ENTER DOWNWIND FOR RUNWAY ONE SEVEN STRAWN

##### (2) Outbound
Example:

THIS IS QUEENAIRE SEVEN ONE FIVE FIVE BRAVO, TAXIING ONTO RUNWAY TWO SIX AT STRAWN AIRPORT FOR TAKEOFF

## AERONAUTICAL ADVISORY STATIONS (UNICOM)— 122.8, 123.0, 123.05, 122.85, 122.95 MHz

1. 122.8 MHz is assigned to airports *not* served by a control tower or FSS. Its use is limited to the necessities of safe and expeditious operation of private aircraft pertaining to runway and wind conditions, types of fuel available, weather, and dispatching. Secondarily, com-

munications may be transmitted concerning ground transportation, food and lodging during transit.

2. 123.0 MHz is assigned to airports *served* by a control tower or FSS. Communications on this frequency are identical to those permitted on 122.8 MHz, with the exception of information such as runway and wind conditions, weather, etc., which would be furnished by the tower, or FSS at nontower locations.

3. 123.05 MHz is assigned to stations at landing areas that are used exclusively as heliports. Communications on this frequency are identical to those permitted on 122.8 MHz with the exception of information on runways, wind conditions, weather, etc., if the landing area is served by a control tower or FSS.

4. 122.85 or 122.95 MHz may be assigned to stations at landing areas not open to the public. Communications on these frequencies are identical to those permitted on 122.8 MHz with the exception of information on runways, wind conditions, weather, etc., if the landing area is served by a control tower or FSS.

THIS SERVICE SHALL NOT BE USED FOR AIR TRAFFIC CONTROL PURPOSES, except for the verbatim relay of ATC information limited to the following:

a. Revision of proposed departure time.

b. Takeoff, arrival, or flight plan cancellation time.

c. ATC clearances *provided* arrangements are made between the ATC facility and UNICOM licensee to handle such messages.

## AERONAUTICAL MULTICOM SERVICE 122.9 MHz

1. A mobile service used to provide communications essential to conduct of activities being performed by or directed from private aircraft.

Example:

Ground/air communications pertaining to agriculture, ranching, conservation activities, forest fire fighting, aerial advertising and parachute jumping. THIS SERVICE SHALL NOT BE USED FOR AIR TRAFFIC CONTROL PURPOSES, except for the verbatim relay of ATC information limited to the following:

a. Revision of proposed departure time.

b. Takeoff, arrival, or flight plan cancellation time.

c. ATC clearances *provided* arrangements are made between the ATC facility and UNICOM licensee to handle such messages.

## AUTOMATIC TERMINAL INFORMATION SERVICE (ATIS)

Automatic Terminal Information Service (ATIS) is the continuous broadcast of recorded noncontrol information in selected high activity terminal areas. Its purpose is to improve controller effectiveness and to relieve frequency congestion by automating the repetitive transmission of essential but routine information.

Information to include the time of the latest weather sequence, ceiling, visibility (sky conditions/ceilings below 5000 feet and visibility less than 5 miles will be broadcast; if conditions are at or better than 5000 and 5, sky condition/ceiling and visibility may be omitted) obstructions to visibility, temperature, wind direction (magnetic) and velocity, altimeter, other pertinent remarks, instrument approach and runways in use is continuously broadcast on the voice feature of a TVOR/VOR/VORTAC located on or near the airport, or in a discrete UHF/VHF frequency. Where VFR arrival aircraft are expected to make initial contact with approach control, this fact and the appropriate frequencies may be broadcast on ATIS. Pilots of aircraft arriving or departing the terminal area can receive the continuous ATIS broadcasts at times when cockpit duties are least pressing and listen to as many repeats as desired. ATIS broadcasts shall be updated upon the receipt of any official weather, regardless of content change and reported values. A new recording will also be made when there is a change in other pertinent data such as runway change, instrument approach in use, etc.

Sample Broadcast:

DULLES INTERNATIONAL INFORMATION SIERRA. 1300 GREENWICH WEATHER. MEASURED CEILING THREE THOUSAND OVERCAST. VISIBILITY THREE, SMOKE. TEMPERATURE SIX EIGHT. WIND THREE FIVE ZERO AT EIGHT. ALTIMETER TWO NINER NINER TWO. ILS RUNWAY ONE RIGHT APPROACH IN USE. LANDING RUNWAY ONE RIGHT AND LEFT. DEPARTURE RUNWAY THREE ZERO. ARMEL VORTAC OUT OF SERVICE. ADVISE YOU HAVE SIERRA.

1. Pilots should listen to ATIS broadcasts whenever ATIS is in operation.

2. Pilots should notify controllers that they have received the ATIS broacast by repeating the alphabetical code word appended to the broadcast.
EXAMPLES: "INFORMATION SIERRA RECEIVED."

3. When the pilot acknowledges that he has received the ATIS broadcast, controllers may omit those items contained on the broadcast if they are current. Rapidly changing conditions will be issued by Air Traffic Control and the ATIS will contain words as follows:
"LATEST CEILING / VISIBILITY / ALTIMETER / (OTHER CONDITIONS) WILL BE ISSUED BY APPROACH CONTROL/TOWER."

The absence of a sky condition/ceiling and/or visibility on ATIS indicates a sky condition/ceiling of 5000 feet or above and visibility of 5 miles or more. A remark may be made on the broadcast, "The weather is better than 5000 and 5," or the existing weather may be broadcast.

4. Controllers will issue pertinent information to pilots who do not acknowledge receipt of a broadcast or who acknowledge receipt of a broadcast which is not current.

5. To serve frequency-limited aircraft, Flight Service Stations (FSS) are equipped to transmit on the omnirange frequency at most en route VORs used as ATIS voice outlets. Such communication interrupts the ATIS broadcast. Pilots of aircraft equipped to receive on other FSS frequencies are encouraged to do so in order that these override transmissions may be kept to an absolute minimum.

Pilots are urged to cooperate in the ATIS program as it relieves frequency congestion on approach control, ground control, and local control frequencies.

Part 3 of the Airman's Information Manual indicates airport for which ATIS is provided.

6. Some pilots use the phrase "Have Numbers" in communications with the control tower. Use of this phrase means that the pilot has received wind and runway information ONLY and the tower does not have to repeat this information. It does not indicate receipt of the ATIS broadcast and should never be used for this purpose.

## RADAR TRAFFIC INFORMATION SERVICE

1. A service provided by radar air traffic control facilities. Pilots receiving this service are advised of any radar target observed on the radar display which may be in such proximity to the position of their aircraft or its intended route of flight that it warrants their attention. This service is not intended to relieve the pilot of his responsibility for continual vigilance to see and avoid other aircraft.

**a. Purpose of the Service**—The issuance of traffic information as observed on a radar display is based on the principle of assisting and advising a pilot that a particular radar target's position and track indicates it may intersect or pass in such proximity to his intended flight path that it warrants his attention. This is to alert the pilot to the traffic so that he can be on the lookout for it and thereby be in a better position to take appropriate action should the need arise.

Pilots are reminded that the surveillance radar used by ATC does not provide altitude information unless the aircraft is equipped with Mode C and the Radar Facility is capable of displaying altitude information.

**b. Provision of the Service**—Many factors, such as limitations of the radar, volume of traffic, controller workload and communications frequency congestion, could prevent the controller from providing this service. The controller possesses complete discretion for determining whether he is able to provide or continue to provide this service in a specific case. His reason against providing or continuing to provide the service in a particular case is not subject to question nor need it be communicated to the pilot. In other words, the provision of this service is entirely dependent upon whether the controller believes he is in a position to provide it. Traffic information is routinely provided to all aircraft operating on IFR Flight Plans except when the pilot advises he does not desire the service, or the pilot is operating within positive controlled airspace. Traffic information may be provided to flights not operating on IFR Flight Plans when requested by pilots of such flight.

NOTE.—Participation by VFR pilots in formal programs implemented at certain terminal locations constitutes pilot request. This also applies to participating pilots at those locations where arriving VFR flights are encouraged to make their first contact with the tower on the approach control frequency.

**c. Issuance of Traffic Information**—Traffic information will include the following concerning a target which may constitute traffic for an aircraft that is:

(1) Radar identified:

(a) Azimuth from the aircraft in terms of the twelve hour clock;

(b) Distance from the aircraft in nautical miles;

(c) Direction in which the target is proceeding; and

(d) Type of aircraft and altitude if known.

**Example:**

Traffic 10 o'clock, 3 miles, west-bound (type aircraft and altitude, if known, of the observed traffic). The pilot may, upon receipt of traffic information, request a vector (heading) to avoid such traffic. The vector will be provided to the extent possible as determined by the controller provided the aircraft to be vectored is within the airspace under the jurisdiction of the controller.

(2) Not radar identified:

(a) Distance and direction with respect to a fix;

(b) Direction in which the target is proceeding; and

(c) Type of aircraft and altitude if known.

**Example:**

Traffic 8 miles south of the airport northeast-bound, (type aircraft and altitude if known).

(d) The examples depicted in figures 23-1 and 23-2 point out the possible error in the position of this traffic when it is necessary for a pilot to apply drift correction to maintain this track. This error could also occur in the event a change in course is made at the time radar traffic information is issued.

**Figure 23-1**

In figure 23-1 traffic information would be issued to the pilot of aircraft "A" as 12 o'clock. The actual position of the traffic as seen by the pilot of aircraft "A" would be one o'clock. Traffic information issued to aircraft "B" would also be given as 12 o'clock, but in this case, the pilot of "B" would see his traffic at 11 o'clock.

**Figure 23-2**

In figure 23-2 traffic information would be issued to the pilot of aircraft "C" as two o'clock. The actual position of the traffic as seen by the pilot of aircraft "C" would be three o'clock. Traffic information issued to aircraft "D" would be at an 11 o'clock position. Since it is not necessary for the pilot of aircraft "D" to apply wind correction (crab) to make good his track, the actual position of the traffic issued would be correct. Since the radar controller can only observe aircraft track (course) on his radar display, he must issue traffic advisories accordingly, and pilots should give due consideration to this fact when looking for reported traffic.

## RADAR SAFETY ADVISORY

ATC will issue a radar safety advisory to pilots of radar identified aircraft if the controller observes the aircraft to be at an altitude which, in the controller's judgment, places the aircraft in unsafe proximity to terrain, obstructions or other aircraft.

### 1. Terrain/Obstruction Advisory.

The controller will immediately issue an advisory to the pilot of an aircraft under his control if he observes the aircraft at an altitude which, in the controller's judgment, places the aircraft in unsafe proximity to terrain/obstructions. The primary method of detecting unsafe proximity is through Mode C.

#### Example:

APACHE THREE THREE PAPA, LOW ALTITUDE ALERT, ADVISE YOU CLIMB TO (ALTITUDE) IMMEDIATELY.

### 2. Aircraft Conflict Advisory.

The controller will immediately issue an advisory to the pilot of an aircraft under his control if he observes an aircraft that is not under his control at an altitude which, in the controller's judgment, places both aircraft in unsafe proximity to each other. With the alert the controller will offer the pilot an alternate courses(s) of action when feasible. Any alternate course(s) of action the controller may recommend to the pilot will be predicated only on other traffic under his control.

#### Example:

AMERICAN THREE, CONFLICT ALERT, ADVISE YOU TURN RIGHT/LEFT HEADING (DEGREES) AND/OR CLIMB/DESCEND TO (ALTITUDE) IMMEDIATELY.

The provision of this service is contingent upon the capability of the controller to observe an unsafe condition(s). An awareness of situation(s) involving unsafe proximity to terrain, obstructions and uncontrolled aircraft cannot be mandated, but it can be expected on a reasonable, though intermittent basis. Once the advisory is issued, it is solely the pilot's prerogative to determine what course of action, if any, he will take. This procedure is intended for use in time critical situations where aircraft safety is in question. Non-critical situations should be handled via the normal traffic advisory procedures.

## RADAR ASSISTANCE TO VFR AIRCRAFT

1. Radar equipped FAA Air Traffic Control facilities provide radar assistance and navigation service (vectors) to VFR aircraft provided the aircraft can communicate with the facility, are within radar coverage, and can be radar identified.

2. Pilots should clearly understand that authorization to proceed in accordance with such radar navigational assistance does not constitute authorization for the pilot to violate Federal Aviation Regulations. In effect, assistance provided is on the basis that navigational guidance information issued is advisory in nature and the job of flying the aircraft safely, remains with the pilot.

3. In many cases, the controller will be unable to determine if flight into instrument conditions will result from his instructions. To avoid possible hazards resulting from being vectored into IFR conditions, pilots should keep the controller advised of the weather conditions in which he is operating and along the course ahead.

4. Radar navigation assistance (vectors) may be initiated by the controller when one of the following conditions exist:

   a. The controller suggests the vector and the pilot concurs.

   b. A special program has been established and vectoring service has been advertised.

   c. In the controller's judgment the vector is necessary for air safety.

5. Radar navigation assistance (vectors) and other radar derived information may be provided in response to pilot requests. Many factors, such as limitations of radar, volume of traffic, communications frequency, congestion, and controller workload could prevent the controller from providing it. The controller has complete discretion for determining if he is able to provide the service in a particular case. His decison not to provide the service in a particular case is not subject to question.

## TERMINAL RADAR PROGRAMS FOR VFR AIRCRAFT

### 1. STAGE I SERVICE (Radar Advisory Service for VFR Aircraft)

   a. In addition to the use of radar for the control of IFR aircraft, Stage I facilities provide traffic information and limited vectoring to VFR aircraft on a workload permitting basis.

   b. Vectoring service may be provided when requested by the pilot or with pilot concurrence when suggested by ATC.

   c. Pilots of arriving aircraft should contact approach control on the publicized frequency (sectional aeronautical chart/AIM), give their position, altitude, radar beacon code (if transponder equipped), destination, and request traffic information.

   d. Approach control will issue wind and runway, except when the pilot states 'HAVE NUMBERS' or this information is contained in the ATIS broadcast and the pilot indicates he has received the ATIS information. Traffic information is provided on a workload permitting basis. Approach control will specify the time or place at which the pilot is to contact the tower on local control frequency for further landing information. Upon being told to contact the tower, radar service is automatically terminated.

**2. Stage II Service (Radar Advisory and Sequencing for VFR Aircraft).**

a. This service has been implemented at certain terminal locations (locations listed in Part 3, AIM). The purpose of the service is to adjust the flow of arriving VFR and IFR aircraft into the traffic pattern in a safe and orderly manner and to provide radar traffic information to departing VFR aircraft. Pilot participation is urged but it is not mandatory.

b. Pilots of arriving VFR aircraft should initiate radio contact (frequencies listed in Part 3, AIM) with approach control when approximately 25 miles from the airport at which Stage II services are being provided. On initial contact by VFR aircraft, approach control will assume that Stage II service is requested. Approach control will provide the pilot with wind and runway (except when the pilot states "Have Numbers" or that he has received the ATIS information), routings, etc., as necessary for proper sequencing with other participating VFR and IFR traffic en route to the airport. Traffic information will be provided on a workload permitting basis. If an arriving aircraft does not want the service, the pilot should state NEGATIVE STAGE II, or make a similar comment, on initial contact with approach control.

c. After radar contact is established, the pilot may navigate on his own into the traffic pattern or, depending on traffic conditions, he may be directed to fly specific headings to position the flight behind a preceding aircraft in the approach sequence. When a flight is positioned behind the preceding aircraft and the pilot reports having that aircraft in sight, he will be directed to follow it. If other "non-participating" or "local" aircraft are in the traffic pattern, the tower will issue a landing sequence.

d. Standard radar separation will be provided between IFR aircraft until such time as the aircraft is sequenced and the pilot sees the traffic he is to follow. Standard radar separation between VFR or between VFR and IFR aircraft will not be provided.

e. Pilots of departing VFR aircraft are encouraged to request radar traffic information by notifying ground control on initial contact with their request and proposed direction of flight.

**Example:**

"XRAY GROUND CONTROL, N18 AT HANGAR 6, READY TO TAXI, VFR SOUTHBOUND, HAVE INFORMATION BRAVO AND REQUEST RADAR TRAFFIC INFORMATION."

Following takeoff, the tower will advise when to contact departure control.

f. Pilots of aircraft transiting the area and in radar contact/communication with approach control will receive traffic information on a controller workload permitting basis. Pilots of such aircraft should give their position, altitude, radar beacon code (if transponder equipped), destination, and/or route of flight.

**3. Stage III Service (Radar Sequencing and Separation Service for VFR Aircraft).**

a. This service has been implemented at certain terminal locations. The service is advertised in Parts 3 and 4 of the AIM. The purpose of this service is to provide separation between all participating VFR air-

craft and all IFR aircraft operating within the airspace defined as the Terminal Radar Service Area (TRSA). Pilot participation is urged but it is not mandatory.

b. If any aircraft does not want the service, the pilot should state NEGATIVE STAGE III, or make a similar comment, on initial contact with approach control or ground control, as appropriate.

c. TRSA charts and a further description of the Services Provided, Flight Procedures, and ATC Procedures are contained in Part 4 of the AIM.

d. The TRSA is contained in a radar environment and control is predicated thereon; however, this does not preclude application of nonradar separation when required or deemed appropriate. The type of separation used will depend on prevailing conditions (traffic volume, type of aircraft, ceiling and visibility, etc.).

e. Visual separation is used when prevailing conditions permit and it will be applied as follows:

(1) When a VFR flight is positioned behind the preceding aircraft and the pilot reports having that aircraft in sight, he will be directed to follow it.

(2) When IFR flights are being sequenced with other traffic and the pilot reports the aircraft he is to follow in sight, the pilot may be directed to follow it and will be cleared for a "visual approach."

(3) If other "non-participating" or "local" aircraft are in the traffic pattern, the tower will issue a landing sequence.

(4) Departing VFR aircraft may be asked if they can visually follow a preceding departure out of the TRSA. If the pilot concurs, he will be directed to follow it until leaving the TRSA.

f. Until visual separation is obtained, standard vertical or radar separation will be provided.

(1) 1000 feet vertical separation may be used between IFR aircraft.

(2) 500 feet vertical separation may be used between VFR aircraft, or between a VFR and an IFR aircraft.

(3) Radar separation varies depending on size of aircraft and aircraft distance from the radar antenna. The minimum separation used will be $1\frac{1}{2}$ miles for most VFR aircraft under 12,500 pounds GWT. If being separated from larger aircraft, the minimum is increased appropriately.

g. Pilots operating VFR in a TRSA—

(1) Must maintain an altitude when assigned by ATC.

(2) When not assigned an altitude should coordinate with ATC prior to any altitude change.

h. Within the TRSA, traffic information on observed but unidentified targets will, to the extent possible, be provided all IFR and participating VFR aircraft. At the request of the pilot, he will be vectored to avoid the observed traffic, insofar as possible, provided the aircraft to be vectored is within the airspace under the jurisdiction of the controller.

i. Departing aircraft should inform ATC of their intended destinaton and/or route of flight and proposed cruising altitude.

**4. PILOTS RESPONSIBILITY:** THESE PROGRAMS ARE NOT TO BE INTERPRETED AS RELIEVING PILOTS OF THEIR RESPONSIBILITIES TO SEE AND AVOID OTHER TRAFFIC OPERATING IN BASIC VFR

WEATHER CONDITIONS, TO MAINTAIN APPRO-PRIATE TERRAIN AND OBSTRUCTION CLEAR-ANCE, OR TO REMAIN IN WEATHER CONDITIONS EQUAL TO OR BETTER THAN THE MINIMA RE-QUIRED BY FAR 91.105. WHENEVER COMPLIANCE WITH AN ASSIGNED ROUTE, HEADING AND/OR ALTITUDE IS LIKELY TO COMPROMISE SAID PILOT RESPONSIBILITY RESPECTING TERRAIN AND OBSTRUCTION CLEARANCE AND WEATHER MINIMA, APPROACH CONTROL SHOULD BE SO ADVISED AND A REVISED CLEARANCE OR IN-STRUCTION OBTAINED.

## TERMINAL CONTROL AREA OPERATION

**1. Operating Rules and Equipment Requirements.** Regardless of weather conditions, ATC authorization is required prior to operating within a TCA. Pilots should not request such authorization unless the requirements of FAR 91.24 and 91.90 are met. Included among these requirements are:

### Group I TCAs

a. A two-way radio capable of communicating with ATC on appropriate frequencies.

b. A VOR or TACAN receiver, except for helicopters.

c. A 4096 code transponder with Mode C automatic altitude reporting equipment, except for helicopters operating at or below 1,000 feet AGL under a Letter of Agreement. (ATC may authorize a deviation from the altitude reporting equipment requirement immediately; however, request for a deviation from the 4096 trans-ponder equipment requirement must be submitted to the controlling ATC facility at least 4 hours before the pro-posed operation.)

d. A private pilot certificate or better in order to land or takeoff from an airport within the TCA.

e. Unless otherwise authorized by ATC, each person operating a large turbine engine powered airplane to or from a primary airport shall operate at or above the designated floors while within the lateral limits of the terminal control area.

f. No person may operate an aircraft in the airspace underlying a terminal control area, at an indicated air-speed of more than 200 knots (230 m.p.h.) (FAR 91.70).

### Group II TCAs

a. A two-way radio capable of communicating with ATC on appropriate frequencies.

b. A VOR or TACAN receiver, except for helicopters.

c. A 4096 code transponder, except for helicopters operating at or below 1,000 feet under a letter of agree-ment, or for IFR flights operating to or from an airport outside of but in close proximity to the TCA when the commonly used transition, approach, or departure pro-cedures to such airport require flight within the TCA. (ATC may authorize deviations from the transponder requirements. Requests for deviation should be sub-mitted to the controlling ATC facility at least four hours before the proposed operation.)

d. Unless otherwise authorized by ATC, large tur-bine-powered aircraft must operate at or above the floor of the TCA while operating to or from the primary air-port.

e. No person may operate an aircraft in the airspace underlying a Terminal Control Area, at an indicated airspeed of more than 200 knots (230 m.p.h.) (FAR 91.70).

A synopsis of these requirements is also found in AIM Part 4 and on National Ocean Survey VFR Terminal Area Charts (See Terminal Control Area in AIM Part 1 Chapter 2 for additional information).

**2. Flight Procedures.**

a. *IFR Flights.* Aircraft operating within the TCA shall be operated in accordance with current IFR pro-cedures. A clearance for a visual approach is not authorization for an aircraft to operate below the desig-nated floors of the TCA.

b. *VFR Flights.*

(1) Arriving VFR flights should contact ATC on the appropriate frequency and in relation to geographi-cal fixes shown on local charts. Although a pilot may be operating beneath the floor of the TCA on initial contact, communications with ATC should be established in relation to the points indicated for spacing and sequencing purposes.

(2) Departing VFR aircraft should advise the ground controller of the intended altitude and route of flight to depart the TCA.

(3) Aircraft not landing/departing the primary airport may obtain ATC clearance to transit the TCA when traffic conditions permit and provided the require-ments of FAR 91.90 are met. Such VFR transiting aircraft are encouraged, to the extent possible, to transit through VFR corridors or above or below the TCA.

**3. ATC Clearances and Separations.** While operating within a TCA, pilots are provided the service and sepa-ration as in the Stage III, Terminal Radar Programs For VFR Aircraft in this chapter. In the event of a radar outage, separation and sequencing of VFR aircraft will be suspended as this service is dependent on radar. The pilot will be advised that the service is not avail-able and issued wind, runway information and the time or place to contact the tower. Traffic information will be provided on a workload permitting basis.

**4.** Assignment of radar headings and/or altitudes are based on the provision that a pilot operating in accord-ance with visual flight rules is expected to advise ATC if compliance with an assigned route, radar heading or altitude will cause the pilot to violate such rules.

## RADAR SERVICE FOR VFR AIRCRAFT IN DIFFICULTY

**1.** Radar equipped FAA Air Traffic Control facilities provide radar assistance and navigation service (vec-tors) to VFR aircraft in difficulty provided the aircraft can communicate with the facility, are within radar coverage, and can be radar identified. Pilots should clearly understand that authorization to proceed in accordance with such radar navigational assistance does not constitute authorization for the pilot to violate Federal Aviation Regulations. In effect, assistance pro-vided is on the basis that navigational guidance infor-mation issued is advisory in nature and the job of flying the aircraft safely, remains with the pilot.

**2.** Experience has shown that many pilots who are not qualified for instrument flight cannot maintain con-trol of their aircraft when clouds or other reduced visibility conditions are encountered. In many cases, the controller will be unable to determine if flight into instrument conditions will result from his instructions. To avoid possible hazards resulting from being vectored into IFR conditions, a pilot in difficulty should keep the controller advised of the weather conditions in which he is operating and along the course ahead; and should observe the following:

a. If an alternative course of action is available which will permit flight in VFR weather conditions, noninstrument rated pilots should choose the alternative rather than requesting a vector or approach into IFR weather conditions; or,

**b.** If no alternative course of action is available, the noninstrument rated pilot should so advise the controller and 'declare an emergency.'

**c.** If the pilot is instrument rated and the aircraft is instrument equipped, the pilot should so indicate by filing an IFR flight plan. Assistance will be provided on the basis that the flight can operate safely in IFR weather conditions.

**3.** Some 'DO's' and 'DONT's:'

**a.** DO let ATC know of your difficulty immediately. DON'T wait until the situation becomes an emergency.

**b.** DO give as much information as possible on initial contact with ATC-nature of difficulty, position (in relation to a navaid if possible), altitude, radar beacon code (if transponder equipped), weather conditions, if instrument rated or not, destination, service requested.

**c.** DON'T change radio frequency without informing the controller.

**d.** DO adhere to ATC instructions or information, or if not possible, DO advise ATC immediately that you cannot comply.

## TRANSPONDER OPERATION

### 1. GENERAL

**a.** Air Traffic Control Radar Beacon System (ATC-RBS) is similar to and compatible with military coded radar beacon equipment (Mark X (SIF)). Civil Mode A is identical to military Mode 3.

**b.** Civil and military transponders except Basic Mark X, should be adjusted to the "on" or normal operating position as late as practicable prior to takeoff and to "off" or "standby" as soon as practicable after completing landing roll unless the change to "standby" has been accomplished previously at the request of ATC.

**c.** If entering a U.S. domestic control area from outside the U.S., the pilot should advise on first radio contact with a U.S. radar air traffic control facility that such equipment is available by adding "transponder" or "no code transponder" as appropriate, to the aircraft identification.

**d.** It should be noted by all users of the ATC Transponders that the coverage they can expect is limited to "line of sight." Low altitude or aircraft antenna shielding by the aircraft itself may result in reduced range. Range can be improved by climbing to a higher altitude. It may be possible to minimize antenna shielding by locating the antenna where dead spots are only noticed during abnormal flight attitudes.

**e.** Aircraft are equipped with two basic types of airborne transponders having different select code (4096 or 64) capability on Mode A/3. The 64 select code transponder transmits only the two front digits of the 4096 code scale; e.g., 11 of 1100. 21 of 2100. 31 of 3100, etc. For ATC to utilize one or a combination of the 4096 discrete codes including the 64 basic select codes, FOUR DIGIT CODE DESIGNATION will be used, e.g., code 2100 will be expressed as TWO ONE ZERO ZERO.

**f.** Pilots of a 64 select code transponder equipped aircraft should disregard the last two numerals of the numbered code issued by ATC, e.g., if assigned code 2100 set in numerals 21.

**g.** Pilots with 4096 code capable transponders should be particularly sure to abide by the provisions of subparagraph b above. Additionally, due to the operational characteristics of the rapidly expanding automated air traffic control system, THE LAST TWO DIGITS OF THE SELECTED TRANSPONDER CODE SHOULD ALWAYS READ '00' UNLESS SPECIFICALLY REQUESTED BY ATC TO BE OTHERWISE.

**h.** Some transponders are equipped with a Mode C automatic altitude reporting capability. This system converts aircraft altitude in 100 foot increments, to coded digital information which is transmitted together with MODE C framing pulses to the interrogating radar facility. The manner in which transponder panels are designed differs, therefore, a pilot should be thoroughly familiar with the operation of his transponder so that ATC may realize its full capabilities.

**i.** Adjust transponder to reply on the Mode A/3 code specified by ATC and, if equipped, to reply on Mode C with altitude reporting *capability activated* unless deactivation is directed by ATC or unless the installed aircraft equipment has not been tested and calibrated as required by FAR 91.36. If deactivation is required and your transponder is so designed, turn off the altitude reporting switch and continue to transmit MODE C framing pulses. If this capability does not exist, turn off MODE C.

**j.** Pilots of aircraft with operating Mode C altitude reporting transponders should report exact altitude/flight level to the nearest hundred foot increment when establishing initial contact with an air traffic control facility. Exact altitude/flight level reports on initial contact provide air traffic control with information that is required prior to using Mode C altitude information for separation purposes. This will significantly reduce altitude verification requests.

**k.** The transponder shall be operated only as specified by ATC. Activate the "IDENT" feature only upon request of the ATC controller.

**l.** Under no circumstances should a pilot of a civil aircraft operate the transponder on Code 0000. This code is reserved for military interceptor operations.

**m.** Pilots operating within restricted/warning areas should adjust their transponders to code 4000 unless otherwise advised by ATC.

**n.** When making routine code changes, pilots should avoid inadvertent selection of codes 3100, 7600 or 7700 thereby causing momentary false alarms at automated ground facilities. For example when switching from code 2700 to code 7200, switch first to 2200 then 7200, NOT to 7700 and then 7200. This procedure applies to all discrete codes in the 3100, 7600 or 7700 series (i.e., 3100–3177, 7600–7677, 7700–7777) which will trigger special indicators in automated ATC facilities.

**o.** Specific details concerning requirements, exceptions and ATC authorized deviations for transponder and Mode C operation above 12,500' and below 18,000' MSL are found in FAR 91.24. In general, the FAR requires aircraft to be equipped with Mode A/3 (4096 codes) and Mode C altitude reporting capability when operating in controlled airspace of the 48 contiguous States and the District of Columbia above 12,500 MSL, excluding airspace at and below 2500' AGL. Pilots should insure that their aircraft transponder is operating on an appropriate or ATC assigned VFR/IFR code and Mode C when operating in such airspace. If in doubt about the operational status of either feature of your transponder while airborne, contact the nearest ATC facility or Flight Service Station and they will advise you what facility you should contact for determining the status of your equipment. In-flight requests for "immediate" deviation may be approved by controllers only when the flight will continue IFR or when weather conditions prevent VFR descent and continued VFR flight in airspace not affected by the FAR. All other requests for deviation should be made by contacting the nearest Flight Service/Air Traffic facility in person or by telephone. The nearest ARTC Center will normally be the controlling agency and is responsible for coordinating requests involving deviations in other ARTCC areas.

NOTE: Positive Control Area (PCA) and Terminal Control Area (TCA) deviation requests are handled as they have been in the past.

p. Pilots should be aware that proper application of these procedures will provide both VFR and IFR aircraft with a higher degree of safety in the environment where high-speed closure rates are possible. Transponders substantially increase the capability of radar to see an aircraft and the Mode C feature enables the controller to quickly determine where potential traffic conflicts may exist. Even VFR pilots who are not in contact with ATC will be afforded greater protection from IFR aircraft and VFR aircraft which are receiving traffic advisories. Nevertheless, pilots should never relax their visual scanning vigilance for other aircraft.

## 2. INSTRUMENT FLIGHT RULES (IFR) FLIGHT PLAN

a. If the pilot cancels his IFR flight plan prior to reaching the terminal area of destination, the transponder should be adjusted according to the instructions below for VFR flight.

b. The transponder shall be operated only as specified by ATC. Activate the "IDENT" feature only upon request of the ATC controller.

## 3. VISUAL FLIGHT RULES (VFR)

a. Unless otherwise instructed by an Air Traffic Control Facility adjust Transponder to reply on Mode 3/A code 1200 regardless of altitude.

b. Adjust transponder to reply on Mode C, with altitude reporting *capability activated* if the aircraft is so equipped, unless deactivation is directed by ATC or unless the installed equipment has not been tested and calibrated as required by FAR 91.36. If deactivation is required and your transponder is so designed, turn off the altitude reporting switch and continue to transmit MODE C framing pulses. If this capability does not exist, turn off MODE C.

## 4. EMERGENCY OPERATION

a. When an emergency occurs, the pilot of an aircraft equipped with a coded radar beacon transponder, who desires to alert a ground radar facility to an emergency condition and who cannot establish communications without delay with an air traffic control facility may adjust the transponder to reply on Mode A/3, Code 7700.

b. Pilots should understand that they may not be within a radar coverage area and that, even if they are certain radar facilities are not yet equipped to automatically recognize Code 7700 as an emergency signal. Therefore, they should establish radio communications with an air traffic control facility as soon as possible.

## 5. RADIO FAILURE

a. Should the pilot of an aircraft equipped with a coded radar beacon transponder experience a loss of two-way radio capability the pilot should:

(1) Adjust his transponder to reply on Mode A/3, code 7700 for a period of 1 minute,

(2) then change to code 7600 and remain on 7600 for a period of 15 minutes or the remainder of the flight, whichever occurs first.

(3) repeat steps 1 and 2, as practicable.

b. Pilots should understand that they may not be in an area of radar coverage. Also many radar facilities are not presently equipped to automatically display code 7600 and will interrogate 7600 only when the aircraft is under direct radar control at the time of radio failure. However, replying on code 7700 first increases the probability of early detection of a radio failure condition.

## 6. RADAR BEACON PHRASEOLOGY

Air traffic controllers, both civil and military, will use the following phraseology when referring to operation of the Air Traffic Control Radar Beacon System (ATCRBS) Mark XIFF (SIF). Instructions by air traffic control refer only to Mode A/3 or Mode C operations and do not affect the operation of the transponder on other Modes.

SQUAWK (number)—Operate radar beacon transponder on designated code in Mode A/3.

IDENT—Engage the "IDENT" feature (military I/P of the transponder.

SQUAWK (number) AND IDENT—Operate transponder on specified code in Mode A/3 and engage the "IDENT" (military I/P) feature.

SQUAWK STANDBY—Switch transponder to standby position.

SQUAWK LOW/NORMAL—Operate transponder on low or normal sensitivity as specified. Transponder is operated in "NORMAL" position unless ATC specified "LOW" ("ON" is used instead of "NORMAL" as a master control label on some types of transponders.)

SQUAWK ALTITUDE—Activate MODE C with automatic altitude reporting.

STOP ALTITUDE SQUAWK—Turn off altitude reporting switch and continue transmitting MODE C framing pulses. If your equipment does not have this capability, turn off MODE C.

STOP SQUAWK (mode in use)—Switch off specified mode. (Use for military aircraft when the controller is unaware if a military service requires the aircraft to continue operating on another MODE.)

STOP SQUAWK—Switch off transponder.

SQUAWK MAYDAY—Operate transponder in the emergency position. (Mode A Code 7700 for civil transponder. Mode 3 Code 7700 and emergency feature for military transponder.)

SQUAWK VFR—Operate transponder on code 1200 regardless of altitude.

## GENERAL

1. A pilot in any emergency phase (uncertainty, alert, or distress) should do three things to obtain assistance:

*a. If equipped with a radar beacon transponder (civil) or IFF/SIF (military), and if unable to establish voice communications with an air traffic control facility, switch to Mode 3/A and Code 7700. Military transponder should also be placed in the Emergency position. If crash is imminent and equipped with a Locator Beacon, actuate the emergency signal.*

*b. Contact controlling agency and give nature of distress and pilot's intentions.* If unable to contact controlling agencies, attempt to contact any agency on assigned frequency or any of the following frequencies (transmit and receive):

| Frequency | Emission | Effective Range in Nautical Miles | Guarded By |
|---|---|---|---|
| 121.5 MHz | Voice | Generally limited to radio line-of-sight | All military towers, most civil towers, VHF direction finding stns, radar facilities. Flight Service Stns. Ocean Station Vessels. |
| 243.0 MHz | Voice | Generally limited to radio-line-of-sight | All military towers, most civil towers, UHF direction finding stns, radar facilities. Flight Service Stns. Ocean Station Vessels. |
| 2182 kHz | Voice | Generally less than 300 miles for average aircraft installations | Some ships and boats, Coast Guard stations, most commercial coast stations. |
| 500 kHz | CW | Generally less than 100 miles for average aircraft installations | Most large ships, most Coast Guard radio stations, most commercial coast stations. |
| 8364 kHz | CW | Up to several thousand miles, depending upon propagation conditions. Subject to "skip." | U.S.N. Direction Finding Stations, Ocean Station Vessels, most Coast Guard radio stations and some FAA International Flight Service Stations (IFSS). |

**Transmit as much of the following as possible:**

(1) MAYDAY, MAYDAY, MAYDAY (if distress), or PAN, PAN, PAN (if uncertainty or alert). If CW transmission, use SOS (distress) or XXX (uncertainty or alert).

(2) Aircraft identification repeated three times.

(3) Type of aircraft.

(4) Position or estimated position (stating which).

(5) Heading (true or magnetic) (stating which).

(6) True airspeed or estimated true airspeed (stating which).

(7) Altitude.

(8) Fuel remaining in hours and minutes.

(9) Nature of distress.

(10) Pilot's intentions (bailout, ditch, crash landing, etc.).

(11) Assistance desired (fix, steer, bearing, escort, etc.).

(12) Two 10-second dashes with mike button (voice) or key (CW) followed by aircraft identification (once) and OVER (voice) or K (CW).

NOTE.—ARTCC emergency frequency capability normally does not extend to radar coverage limits. If the ARTCC does not respond to transmission on emergency frequency 121.5 MHz or 243.0 MHz pilots should initiate a call to the nearest Flight Service Station or Airport Traffic Control Tower.

**c. Comply with information and clearance received.** Accept the communications control offered to you by the ground radio station, silence interfering radio stations, and do not shift frequency or shift to another ground station unless absolutely necessary.

2. Pilots of IFR flights experiencing two-way radio failure are expected to adhere to the procedures prescribed under "RADIO COMMUNICATIONS FAILURE" (FAR Part 91.127). However, if a pilot should become lost, or if an emergency or other factors prevent the pilot from being able to comply with prescribed procedures, such pilot, as well as pilots of VFR flight with an emergency and a radio failure, may attempt to alert civil or military radar systems as follows:"

**a. If receiver only is operating:** Fly triangular pattern to the RIGHT (see diagram). Hold each heading for two minutes (1 minute for jet use). Complete a minimum of two such patterns before resuming original course and then repeat pattern at 20-minute intervals. When flying triangular patterns to the right the emergency frequencies (121.5 or 243.0 MHz) should be guarded. If pattern is observed by radar controllers, instructions will be given.

**b. If transmitter and receiver are both inoperative:** Fly a triangular pattern to the LEFT (see diagram), in the same manner as described above. If patterns are observed by radar controllers, an escort will be dispatched, if possible. If distressed aircraft is flying in IFR conditions or if at night, landing lights, navigational lights, etc., should be turned on to aid the interceptor. If intercepted, follow rescue aircraft.

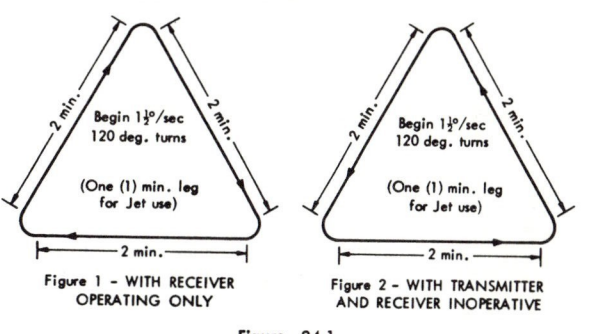

Figure 1 - WITH RECEIVER OPERATING ONLY

Figure 2 - WITH TRANSMITTER AND RECEIVER INOPERATIVE

Figure 24-1

c. The pilot should remember that he has three means of declaring an emergency:

(1) Emergency SQUAWK (Code 7700) from transponders

(2) Sending emergency message

(3) Flying triangular pattern

d. Some ground stations have *four* electronic means of assisting:

(1) Receipt of emergency message

(2) Radar detection of triangular pattern.

(3) DF bearings; and

(4) Detection of transponder emergency SQUAWK (Code 7700).

**e. Pilots should remember the FOUR C's**

**(1) Confess** your predicament to any ground radio station. Do not wait too long. Give SAR a chance!

**(2) Communicate** with your ground link and pass as much of the distress message on first transmission as possible. We need information for best SAR action!

**(3) Climb** if possible for better radar and DF detection. If flying at low altitude, the chance for establishing radio contact is improved by climbing, also chances of alerting radar systems are sometimes improved by climbing or descending.

Note.—Unauthorized climb or descent under IFR conditions within controlled airspace is not permitted except in emergency. Any variation in altitude, in connection with flying radar patterns, will be unknown to Air Traffic Control except at radar locations having height finding capabilities. Air Traffic Control will operate on the assumption that the provisions of FAR 91.127 are being followed by the pilot.

**(4) Comply**—especially *Comply*—with advice and instructions received, if you really want help. Assist the ground "communications control" station to control communications on the distress frequency on which you are working (as that is the distress frequency for your case). Tell interfering stations to maintain silence until you call. Cooperate!

**3.** For bailout, set radio for continuous emission. For ditching or crash landing, radio should, if it is considered that there is no additional risk of fire and if circumstances permit, be set for continuous transmission.

**4.** When a pilot is in doubt of his position, or feels apprehensive for his safety, he should not hesitate to request assistance. Search and Rescue facilities, including Radar, Radio, and DF stations, are ready and willing to help. There is no penalty for using them. Delay has caused accidents and cost lives. *Safety is not a luxury.* Take Action!

## VHF/UHF DIRECTION FINDING INSTRUMENT APPROACH PROCEDURE

**1.** Direction finding (DF) equipment has long been used to locate lost aircraft, to guide aircraft to areas of good weather or to airports; and now at most DF equipped airports, DF instrument approaches may be given to aircraft in emergency.

**2.** Experience has shown that a majority of actual emergencies requiring DF assistance involve pilots with a minimum of flight experience, particularly IFR experience. With this in mind, a DF approach procedure provides for maximum flight stability in the approach by utilizing small degrees of turn, and descents when the aircraft is in a wings level attitude. The DF specialist will give the pilot headings to fly and tell the pilot when to begin descent.

**3.** DF instrument approach procedures are for emergency use, this type of approach will not be given in IFR weather conditions unless the pilot has declared an emergency.

**4.** To become familiar with the procedures and other benefits of DF, pilots are urged to request practice guidance and approaches in VFR weather conditions. DF specialists welcome the practice and, workload permitting, will honor such requests.

## TWO-WAY RADIO COMMUNICATIONS FAILURE

**1.** Whether two-way communications failure constitutes an emergency depends on the circumstances, and in any event is a determination made by the pilot. FAR 91.3 authorizes a pilot to deviate from any rule to the extent required to meet an emergency. A pilot experiencing two-way communications failure should (unless emergency authority is exercised) comply with FAR 91.127 as follows:

**a. General.** Unless otherwise authorized by ATC, each pilot who has two-way radio communications failure when operating under IFR shall comply with the following:

**b. VFR conditions.** If the failure occurs in VFR conditions, or if VFR conditions are encountered after the failure, each pilot shall continue the flight under VFR and land as soon as practicable. This procedure also applies when two-way radio failure occurs while operating in Positive Control Airspace (PCA). A descent out of PCA, under VFR, should be accomplished as soon as practicable.

**c. IFR conditions.** If the failure occurs in IFR conditions, or if paragraph (b) of this section cannot be complied with, each pilot shall continue the flight according to the following:

**(1) Route.**

(i) By the route assigned in the last ATC clearance received;

(ii) If being radar vectored, by the direct route from the point of radio failure to the fix, route, or airway specified in the vector clearance;

(iii) In the absence of an assigned route, by the route that ATC has advised may be expected in a further clearance; or

(iv) In the absence of an assigned route or a route that ATC has advised may be expected in a further clearance, by the route filed in the flight plan.

**(2) Altitude.** At the HIGHEST of the following altitudes or flight levels FOR THE ROUTE SEGMENT BEING FLOWN:

(i) The altitude or flight level assigned in the last ATC clearance received except that the altitude in (ii) shall apply for only the segment/s of the route where the minimum altitude is higher than the ATC assigned altitude.

(ii) The minimum altitude (converted, if appropriate, to minimum flight level as prescribed in § 91.81 (c)) for IFR operations; or

(iii) The altitude or flight level ATC has advised may be expected in a further clearance except that the altitude in (ii) shall apply for only the segment/s of the route where the minimum altitude is higher than the expected altitude.

(iv) The intent of the rule is that a pilot who has experienced two-way radio failure should, during any segment of his route, fly at the appropriate altitude specified in the rule for that *particular segment*. The appropriate altitude is whichever of the three is *highest* in each given phase of flight: (1) the altitude or flight level last assigned; (2) the MEA; or (3) the altitude or flight level the pilot has been advised to expect in a further clearance.

**Example 1.**

A pilot sustaining two-way radio failure had an assigned altitude of 7,000 feet, and while enroute come to a route segment for which the MEA was 9,000 feet. He would climb to 9,000 feet at the time or place where it became necessary to comply with the 9,000 feet MEA. (See FAR 91.119(b)) If later, while still proceeding to his destination, the MEA dropped from 9,000 feet to 5,000 feet, he would descend to *7,000 feet* (the last assigned altitude), because that altitude is *higher* than the MEA.

**Example 2.**

The MEA between A and B—5,000 feet.
The MEA between B and C—5,000 feet.
The MEA between C and D—11,000 feet.
The MEA between D and E—7,000 feet.

A pilot had been cleared via **A, B, C, D,** to **E.** His assigned altitude was 6,000 feet and he was told to expect a clearance to 8,000 feet at **B.** Prior to receiving the higher altitude assignment, he experienced two-way radio failure. The pilot would maintain 6,000 to **B,** then climb to 8,000 feet (the altitude he was advised to expect). He would maintain 8,000 feet, then climb to 11,000 at **C,** or prior to **C** if necessary to comply with an MCA at **c.** (FAR 91.119(b)). Upon reaching **D,** the pilot would descend to *8,000 feet* (even though the MEA was 7,000 feet), as 8,000 was the highest of the altitude situations stated in the rule, FAR 91.127.

**(3) Leave holding fix.** If holding instructions have been received, leave the holding fix at the expect-further-clearance time received, or, if an expected approach clearance time has been received, leave the holding fix in order to arrive over the fix from which the approach begins as close as possible to the expected approach clearance time.

If holding instructions have not been received and the aircraft is ahead of its ETA, the pilot is expected to hold at the fix from which the approach begins. If more than one approach fix is available, it is pilot choice and ATC protects airspace at all of them. Descent for approach begins at the ETA shown in the flight plan, as amended with ATC.

**(4) Descent for approach.** Begin descent from the en route altitude or flight level upon reaching the fix from which the approach begins, but not before—

　　**(i) The** expect-approach-clearance time (if received) ; or

　　**(ii)** If no expect-approach-clearance time has been received, at the estimated time of arrival, shown on the flight plan, as amended with ATC.

**2.** In the event of two-way radio communications failure, ATC service will be provided on the basis that the pilot is operating in accordance with FAR 91.127.

**3.** In addition to monitoring the NAVAID voice feature, the pilot should attempt to reestablish communications by attempting contact :

　**a.** on the previously assigned frequency, or

　**b.** with an FSS or ARINC.

If communications are established with an FSS or ARINC, the pilot should advise of the aircraft's position, altitude, last assigned frequency and then request further clearance from the controlling facility. The preceding does not preclude the use of 121.5 MHz. There is no priority on which action should be attempted first. If the capability exists, do all at the same time.

　　Note.—ARINC is a commercial communications corporation which designs, constructs, operates, leases or otherwise engages in radio activities serving the aviation community. ARINC has the capability of relaying information to/from ATC facilities throughout the country.

**4.** Should the pilot of an aircraft equipped with a coded radar beacon transponder experience a loss of two-way radio capability he should :

　**(a)** adjust his transponder to reply on Mode A/3, Code 7700 for a period of 1 minute,

　**(b)** then change to Code 7600 and remain on 7600 for a period of 15 minutes or the remainder of the flight, whichever occurs first.

　**(c)** repeat steps a and b, as practicable.

The pilot should understand that he may not be in an area of radar coverage. Many radar facilities are also not presently equipped to automatically display Code 7600 and will interrogate 7600 only when the aircraft is under direct radar control at the time of radio failure. However, replying on Code 7700 first increases the probability of early detection of a radio failure condition.

**5.** It is virtually impossible to provide regulations and procedures applicable to all possible situations associated with two-way radio communications failure. During two-way radio communications failure when confronted by a situation not covered in the regulation, pilots are expected to exercise good judgment in whatever action they elect to take. Should the situation so dictate, they should not be reluctant to use the emergency action contained in FAR 91.3(b).

**6. In VFR Conditions.** The primary objective of this provision in FAR 91.127 is to preclude extended IFR operations in the air traffic control system in VFR weather conditions. Pilots should recognize that operation under these conditions may unnecessarily as well as adversely affect other users of the airspace, since ATC may be required to reroute or delay other users in order to protect the failure aircraft. However, it is not intended that the requirement to "land as soon as practicable" be construed to mean "as soon as possible." The pilot retains his prerogative of exercising his best judgment and is not required to land at an unauthorized airport, at an airport unsuitable for the type of aircraft flown, or to land only minutes short of his destination.

**7. Holding.** If holding is necessary at the radio fix to be used for the approach at the destination airport, holding and descent to the initial approach altitude shall be accomplished in a holding pattern in accordance with the pattern depicted on the approach chart. If no holding pattern is depicted, holding and descent will be accomplished in a holding pattern on the side of the final approach course on which the procedure turn is prescribed.

## EMERGENCY LOCATOR TRANSMITTERS

Emergency Locator Transmitters of various types which are independently powered and of incalculable value in an emergency have been developed as a means of locating downed aircraft and their occupants. These electronic, battery operated transmitters are not a fire hazard. They are designed to emit a distinctive downward swept audio tone for homing purposes on 121.5 MHz and/or 243 MHz, preferably on both emergency frequencies. The power source should be capable of providing power for continuous operation from 24 to 48 hours or more at a very wide range of ambient temperatures and can expedite search and rescue operations as well as facilitate accident investigation and analysis.

This equipment is required for most general aviation and small private aircraft. The pilot and other occupants could survive the crash impact only to die of exposure before they are located. These transmitters are made by several electronic manufacturers for civil aviation use.

Once the transmitter has been activated and the signal detected, it will be a simple matter for the search aircraft with homing equipment to locate the scene. Search patterns have been developed that enable search aircraft, equipped with but one receiver, to locate the transmitter site.

Some models of locator transmitter may be of value for transmitting a signal for inflight emergencies. If an emergency situation occurs in conjunction with a radio failure, the pilot could actuate his transmitter signal to supplement other actions taken to declare the emergency. DF stations hearing the signal would take bearings on the aircraft in distress and notify search and rescue authorities.

Caution should be exercised to prevent the inadvertent actuation of locator transmitters in the air or while they are being handled on the ground. Operational testing of transmitters should be carried out only in shielded areas under controlled conditions. False signals on the distress frequencies can interfere with actual distress transmissions as well as decrease the degree of urgency that should be attached to such signals.

"Aircraft operational testing is authorized on 121.5 MHz as follows:

a. Tests should be no longer than three audio sweeps.

b. If the antenna is removable, a dummy load should be substituted during test procedures.

c. Tests shall be conducted *only* within the time period made up of the first five minutes after every hour. Emergency tests outside of this time have to be coordinated with the nearest FSS or Control Tower. Airborne ELT tests are not authorized."

## SEARCH AND RESCUE

### 1. GENERAL

a. Search and Rescue is a life-saving service provided through the combined efforts of the FAA. Air Force, Coast Guard, State Aeronautic Commissions or other similar state agencies who are assisted by other organizations such as the Civil Air Patrol, Sheriffs Air Patrol, State Police, etc. It provides search, survival aid, and rescue of personnel of missing or crashed aircraft.

b. Prior to departure on every flight, local or otherwise, someone at the departure point should be advised of your destination and the route of flight if other than direct. Search efforts are often wasted and rescue is often delayed because of pilots who thoughtlessly take off without telling anyone where they are going.

c. All you need to remember to obtain this valuable protection is:

(1) File a Flight Plan with an FAA Flight Service Station in person or by telephone or radio.

(2) Close your flight plan with the appropriate authority immediately upon landing.

(3) If you land at a location other than the intended destination, report the landing to the nearest FAA Flight Service Station.

(4) If you land en route and are delayed more than 30 min. (15 min. for jets), report this information to the nearest FSS.

(5) Remember that if you fail to report within one-half hour after your ETA, a search will be started to locate you.

d. If a crashed aircraft is observed:

(1) Determine if crash is marked with yellow cross; if so, crash has already been reported and identified.

(2) Determine, if possible, type and number of aircraft and whether there is evidence of survivors.

(3) Fix, as accurately as possible, exact location of crash.

(4) If circumstances permit, orbit scene to guide in other assisting units or until relieved by another aircraft.

(5) Transmit information to nearest FAA or other appropriate radio facility.

(6) Immediately after landing, make a complete report to nearest FAA, Air Force, or Coast Guard installation. Report may be made by long distance collect telephone.

e. To assist survival and rescue in the event of a crash landing the following advice is given:

(1) For flight over uninhabited land areas, it is wise to take suitable survival equipment depending on type of climate and terrain.

(2) If forced landing occurs at sea, chances for survival are governed by degree of crew proficiency in emergency procedures and by effectiveness of water survival equipment.

(3) If it becomes necessary to ditch, distressed aircraft should make every effort to ditch near a surface vessel. If time permits, the position of the nearest vessel can be obtained from a Coast Guard Rescue Coordination Center through the FAA facility.

(4) The rapidity of rescue on land or water will depend on how accurately your position may be determined. If flight plan has been followed and your position is on course, rescue will be expedited.

(5) Unless you have good reason to believe that you will not be located by search aircraft, it is better to remain near your aircraft and prepare means for signalling whenever aircraft approach your position.

f. Search and Rescue facilities include:

(1) Rescue Coordination Centers;

(2) Search and Rescue aircraft;

(3) Rescue vessels;

(4) Pararescue and ground rescue teams;

(5) Emergency radio fixing.

### 2. CLOSE YOUR FLIGHT PLAN

The control tower does not automatically close VFR flight plans since many of the landing aircraft are not operating on flight plans. It remains the responsibility of a pilot to close his own flight plan. This will prevent a needless search.

### 3. NATIONAL SEARCH AND RESCUE PLAN

Under the National Search and Rescue Plan, the U.S. Coast Guard is responsible for coordination of search and rescue for the Maritime Region, and the U.S. Air Force is responsible for coordination of search and rescue for the Inland Region. In order to carry out this responsibility the Air Force and the Coast Guard have established Rescue Coordination Centers to direct search and rescue activities within their regions. This service is available to all persons and property in distress, both civilian and military. Normally, for aircraft incidents, information will be passed to the Rescue Coordination Centers through the appropriate Air Route Traffic Control Center or Flight Service Station.

## Coast Guard Rescue Coordination Center

### (area code listed)

| | |
|---|---|
| Boston, Mass. | Long Beach, Calif. |
| 617–223–3645 | 213–590–2225 |
| New York, N.Y. | San Francisco, Calif. |
| 212–264–4800 | 415–556–5500 |
| Portsmouth, Va. | Seattle, Wash. |
| 804–393–9611 | 206–624–2219 |
| Miami, Florida | Juneau, Alaska |
| 305–350–5611 | 907–586–7351 |
| New Orleans, La. | Honolulu, Hawaii |
| 504–527–6225 | 808–546–7109 |
| Cleveland, Ohio | Kodiak, Alaska |
| 216–522–3984 | 907–487–5888 |
| St. Louis, Mo. | San Juan, Puerto Rico |
| 314–622–4614 | 809–722–2943 |

Coast Guard Rescue Coordination Centers are served by major radio stations which guard 500 kHz (CW) and 2182 kHz (voice). In addition, San Francisco and Honolulu guard 8364 kHz and 247 Coast Guard units along the sea coasts of the United States and shores of the Great Lakes guard 2182 kHz (voice). All of these facilities are available for reporting distress or potential distress. THE CALL "NCU" (CW) or "COAST GUARD" (VOICE) ALERTS ALL COAST GUARD RADIO STATIONS WITHIN RANGE.

# EMERGENCY PROCEDURES

## Air Force Rescue Coordination Center
### (area code listed)
## Inland Region

Scott Air Force Base, IL_____All contiguous U.S.
      Commercial          618–256–4815
      WATS               800–851–3051
      AUTOVON         638–4815
      FTS        618–234–4815 or 4810

Detachment 1
Elmendorf AFB, AL
  907–227–2131

## EMERGENCY PROCEDURES
## GROUND-AIR VISUAL CODE FOR USE BY SURVIVORS

| Message | Symbol | Message | Symbol | Message | Symbol |
|---|---|---|---|---|---|
| REQUIRE DOCTOR SERIOUS INJURIES | I | REQUIRE SIGNAL LAMP WITH BATTERY, AND RADIO | I | REQUIRE FUEL AND OIL | L |
| REQUIRE MEDICAL SUPPLIES | II | INDICATE DIRECTION TO PROCEED | K | ALL WELL | LL |
| UNABLE TO PROCEED | X | AM PROCEEDING IN THIS DIRECTION | ↑ | NO | N |
| REQUIRE FOOD AND WATER | F | WILL ATTEMPT TAKE-OFF | ▷ | YES | Y |
| REQUIRE FIREARMS AND AMMUNITION | ▼ | AIRCRAFT SERIOUSLY DAMAGED | L7 | NOT UNDERSTOOD | JL |
| REQUIRE MAP AND COMPASS | □ | PROBABLY SAFE TO LAND HERE | △ | REQUIRE MECHANIC | W |

IF IN DOUBT, USE INTERNATIONAL SYMBOL **S O S**

### INSTRUCTIONS
1. Lay out symbols by using strips of fabric or parachutes, pieces of wood, stones, or any available material.
2. Provide as much color contrast as possible between material used for symbols and background against which symbols are exposed.
3. Symbols should be at least 10 feet high or larger. Care should be taken to lay out symbols exactly as shown.
4. In addition to using symbols, every effort is to be made to attract attention by means of radio, flares, smoke, or other available means.
5. On snow covered ground, signals can be made by dragging, shoveling or tramping. Depressed areas forming symbols will appear black from the air.
6. Pilot should acknowledge message by rocking wings from side to side.

### GROUND-AIR VISUAL CODE FOR USE BY GROUND SEARCH PARTIES

| NO. | MESSAGE | CODE SYMBOL |
|---|---|---|
| 1 | Operation completed. | LLL |
| 2 | We have found all personnel. | LL |
| 3 | We have found only some personnel. | ⧻ |
| 4 | We are not able to continue. Returning to base. | X X |
| 5 | Have divided into two groups. Each proceeding in direction indicated. | ⚡ |
| 6 | Information received that aircraft is in this direction. | ⇢→ |
| 7 | Nothing found. Will continue search. | NN |

"Note: These visual signals have been accepted for international use and appear in Annex 12 to the Convention on International Civil Aviation."

Figure 24-2

## EMERGENCY PROCEDURES

# VISUAL EMERGENCY SIGNALS

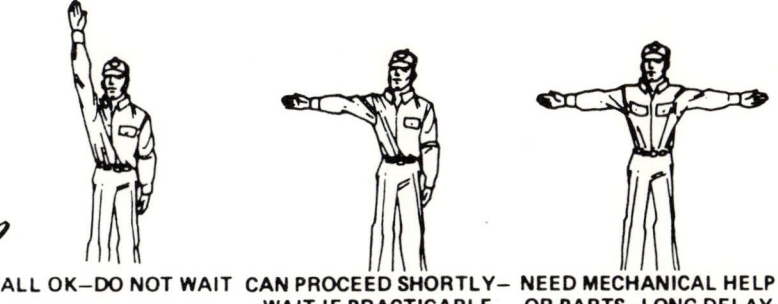

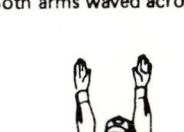

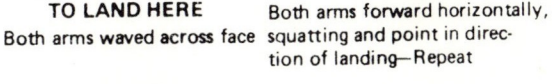

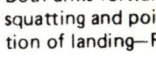

| | | | |
|---|---|---|---|
| **NEED MEDICAL ASSISTANCE—URGENT** | **ALL OK—DO NOT WAIT** | **CAN PROCEED SHORTLY— WAIT IF PRACTICABLE** | **NEED MECHANICAL HELP OR PARTS—LONG DELAY** |
| Used only when life is at stake | Wave one arm overhead | One arm horizontal | Both arms horizontal |

| | | | |
|---|---|---|---|
| **USE DROP MESSAGE** | **OUR RECEIVER IS OPERATING** | **DO NOT ATTEMPT TO LAND HERE** | **LAND HERE** |
| Make throwing motion | Cup hands over ears | Both arms waved across face | Both arms forward horizontally, squatting and point in direction of landing—Repeat |

| | | | |
|---|---|---|---|
| **NEGATIVE (NO)** | **AFFIRMATIVE (YES)** | **PICK US UP— PLANE ABANDONED** | **AFFIRMATIVE (YES)** |
| White cloth waved horizontally | White cloth waved vertically | Both arms vertical | Dip nose of plane several times |

**NEGATIVE (NO)**

Fishtail plane

---

### HOW TO USE THEM

If you are forced down and are able to attract the attention of the pilot of a rescue airplane, the body signals illustrated on this page can be used to transmit messages to him as he circles over your location. Stand in the open when you make the signals. Be sure that the background, as seen from the air, is not confusing. Go through the motions slowly and repeat each signal until you are positive that the pilot understands you.

---

Figure 24-3

297

# TABLE OF CONTENTS

## — APPENDIX —

# PILOT/CONTROLLER ROLES AND RESPONSIBILITIES

The roles and responsibilities of the pilot and controller for effective participation in the ATC system are contained in several documents. Pilot responsibilities are in the Federal Aviation Regulations (FARs) and the air traffic controller's are in the Air Traffic Control Handbook (7110.65). Additional and supplemental information for pilots can be found in the Airman's Information Manual (AIM) and Advisory Circulars. In addition to the AIM and Advisory Circulars, agency directives provide added and supplemental information for the controller.

The pilot in command of an aircraft is directly responsible for, and is the final authority as to the safe operation of that aircraft. In an emergency requiring immediate action, the pilot in command may deviate from any rule in the General Subpart A and Flight Rules Subpart B in accordance with FAR 91.3.

The air traffic controller is responsible to give first priority to the separation of aircraft and to the issuance of radar safety advisories, second priority to other services that are required, but do not involve separation of aircraft and third priority to additional services to the extent possible.

In order to maintain a safe and efficient air traffic system, it is necessary that each party fulfill his responsibilities to the fullest.

The responsibilities of the pilot and the controller intentionally overlap in many areas providing a degree of redundancy. Should one or the other fail in any manner, this overlapping responsibility is expected to compensate, in many cases, for failures that may affect safety.

The following, while not intended to be all inclusive, is a brief listing of pilot and controller responsibilities for some commonly used procedures or phases of flight. More detailed explanations are contained in portions of AIM, Part 1, the appropriate Federal Aviation Regulations, Advisory Circulars and similar publications. The availability of these documents is contained in AIM, Part 1. The information provided is an overview of the principles involved and is not meant as an interpretation of the rules nor is it intended to extend or diminish extant responsibilities.

---

## AIR TRAFFIC CLEARANCE

### Pilot
- Acknowledge receipt and understanding of an ATC clearance.
- Request clarification or amendment, as appropriate, any time a clearance is not fully understood, or considered unacceptable from a safety standpoint.
- Comply with an air traffic clearance upon receipt except as necessary to cope with an emergency. If deviation is necessary, advise ATC as soon as possible and obtain an amended clearance.

### Controller
- Issues appropriate clearances for the operation being or to be conducted in accordance with established criteria.
- Assigns altitudes in IFR clearances that are at or above the minimum IFR altitudes in controlled airspace.

## INSTRUMENT APPROACH

### Pilot
- Follow the procedure as shown on the instrument approach chart unless otherwise instructed by ATC.
- Upon receipt of an approach clearance while on an unpublished route or being radar vectored:
  a. Comply with the minimum altitude for IFR and;
  b. Maintain last assigned altitude until established on a segment of a published route or IAP, at which time published altitudes apply.

### Controller
- Issues an IFR approach clearance only after the aircraft is established on a segment of published route or IAP, or assigns an appropriate altitude for the aircraft to maintain until so established.

## MISSED APPROACH

### Pilot
- Execute a missed approach when one of the following conditions exist:
  a. Arrival at the missed approach point (MAP) or the decision height (DH) and visual reference to the runway environment is insufficient to complete the landing.
  b. Determined that a safe landing is not possible.
  c. Instructed to do so by ATC.

- Comply with the missed approach instructions for the instrument approach procedure being executed unless other *missed approach* instructions are specified by ATC.
- If executing a missed approach prior to reaching the MAP or DH, advise ATC and fly the instrument procedure to the MAP at an altitude at or above the MDA or DH before executing a turning maneuver.
- Radar vectors issued by ATC when informed that a missed approach is being executed supersedes the previous missed approach procedure.
- If making a missed approach from a radar approach, execute the missed approach procedure previously given or climb to the altitude and fly the heading specified by the controller.

### Controller
- Issues an approved alternate missed approach procedure if it is desired that the pilot execute a procedure other than as depicted on the instrument approach chart.
- May vector a radar identified aircraft executing a missed approach when operationally advantageous to the pilot or the controller.

## RADAR SAFETY ADVISORY

### Pilot
- Initiate appropriate action if a radar safety advisory is received from ATC.
- Be aware that this service is not always available and that many factors affect the ability of the controller to observe, on radar, a situation in which unsafe proximity to terrain, obstructions or another aircraft may be developing.
- This service is not a substitute for pilot adherence to safe operating practices.

### Controller
- Issues a radar safety advisory if he observes a radar identified aircraft under his control to be at an altitude believed to place the aircraft in unsafe proximity to terrain, obstruction or another aircraft. Types of radar safety advisories are:
  a. Terrain/Obstruction Advisory.

     Immediately issued to an aircraft under his control that is observed to be at an altitude believed to place the aircraft in unsafe proximity to terrain/obstructions.

b. Aircraft Conflict Advisory.

Immediately issued to an aircraft under his control when an aircraft not under his control is observed at an altitude believed to place the aircraft in unsafe proximity to each other. With the alert, he offers the pilot an alternative, if feasible.

– Discontinues further advisories if informed by the pilot that he is taking action to correct the situation or that he has the other aircraft in sight.

## RADAR VECTORS

### Pilot

– Comply with headings and altitudes assigned to you by the controller.

– Question any assigned heading or altitude believed to be incorrect.

– If operating VFR and compliance with any radar vector or altitude would cause a violation of any FAR, advise ATC and obtain a revised clearance or instruction.

### Controller

– Vectors aircraft in controlled airspace:

a. For separation

b. For noise abatement

c. To obtain an operational advantage for the pilot or controller.

– Vectors aircraft in controlled and uncontrolled airspace when requested by the pilot.

– Vectors IFR aircraft at or above minimum vectoring altitudes.

– Is permitted to vector VFR aircraft below minimum vectoring altitudes.

## SEE AND AVOID

### Pilot

– When meteorological conditions permit, regardless of type of flight plan, whether or not under control of a radar facility, the pilot is responsible to see and avoid other traffic, terrain or obstacles.

### Controller

– Provides radar traffic information to radar identified aircraft operating outside positive control airspace on a workload permitting basis.

– Issues a radar safety advisory to radar-identified aircraft under his control if the aircraft is observed to be at an altitude which is believed to place the aircraft in an unsafe proximity to terrain, obstruction or other aircraft.

## SPEED ADJUSTMENTS

### Pilot

– Advise ATC any time cruising airspeed varies plus or minus 10 knots from that given in the flight plan.

– Comply with speed adjustments from ATC unless:

a. The minimum or maximum safe airspeed for any particular operation is greater or less than the requested airspeed. In such cases, advise ATC.

b. The requested speed adjustment will, in your opinion, cause you to exceed 250 knots (288 mph) or 200 knots (230 mph) indicated airspeed, as specified in FAR 91.70. In such cases, advise ATC.

– When complying with speed adjustment assignment, maintain an indicated airspeed within plus or minus 10 knots of the requested airspeed.

### Controller

– Assigns aircraft to speed adjustments when necessary but not as a substitute for good vectoring technique.

– Adheres to the restrictions of 7110.65 as to when speed adjustment procedures may be applied.

– Avoids speed adjustments requiring alternate decreases and increases.

– Assigns speed adjustments to a specified IAS or to increase or decrease speed utilizing increments of 10 knots or multiples thereof.

– Advises pilots to resume normal speed when speed adjustments are no longer required.

– Gives due consideration to aircraft capabilities to reduce speed while descending.

## TRAFFIC ADVISORIES (Traffic Information)

### Pilot

– Acknowledge receipt of traffic advisories.

– Inform controller if traffic in sight.

– Advise ATC if a vector to avoid traffic is desired.

– Do not expect to receive radar traffic advisories on all traffic. Some aircraft may not appear on the radar display. Be aware that the controller may be occupied with higher priority duties and unable to issue traffic information for a variety of reasons.

– Advise controller if service not desired.

### Controller

– Issues radar traffic to the maximum extent consistent with higher priority duties except in positive controlled airspace.

– Provides vectors to assist aircraft to avoid observed traffic when requested by the pilot.

– Issues traffic information to aircraft in the airport traffic area for sequencing purposes.

## VISUAL APPROACH

### Pilot

– If a visual approach is not desired, advise ATC.

– Comply with controller's instructions for a vector to the airport traffic pattern or to a visual position behind a preceding aircraft.

– After being cleared for a visual approach, proceed to the airport in a normal manner or follow designated traffic, as appropriate, remaining in VMC at all times.

– Acceptance of a visual approach clearance is pilot acknowledgement that he will insure a safe landing interval behind a preceding aircraft, if so cleared, and that he accepts responsibility for his own wake turbulence separation.

– Advise ATC immediately if you are unable to continue following a designated aircraft or encounter less than basic VFR weather conditions.

– Be aware that radar service is automatically terminated without advising the pilot when the aircraft is instructed to contact the tower.

### Controller

– Does not clear an aircraft for visual approach unless it is at the minimum vectoring altitude (MVA) or reports indicate that descent to MVA can be made in VMC.

– Provides radar separation until the pilot accepts a visual approach clearance.

– Continues flight following and traffic information until the aircraft is instructed to contact the tower.

– Issues visual approach clearance when the pilot reports sighting the airport or a preceding aircraft which is to be followed.

## VISUAL SEPARATION

### Pilot

– If instructed by ATC to follow another aircraft for the purpose of maintaining visual separation, notify the controller if you do not see it, are unable to maintain visual contact with it, or for any other reason you cannot accept the responsibility for your separation under these conditions.

– Acceptance of both traffic information and instructions to follow another aircraft is pilot acknowledgement that he sees the other aircraft and will avoid it.

– The pilot also accepts responsibility for wake turbulence separation under these conditions.

### Controller

– Applies visual separation only within a terminal area when a controller has both aircraft in sight or by instructing a pilot who sees the other aircraft to maintain visual separation from it.

# PILOT/CONTROLLER GLOSSARY

This glossary was compiled to promote a common understanding of the terms used in the Air Traffic Control system. It includes those terms which are intended for pilot/controller communications. These are printed in **_bold italics_**. The definitions are primarily defined in an operational sense applicable to both users and operators of the National Airspace System. Use of the glossary will preclude many misunderstandings concerning the system's design, function and purpose.

Because of the international nature of flying, terms used in the "Lexicon" published by the International Civil Aviation Organization (ICAO) are included when they differ from FAA definitions. These terms are _italicized_. For the readers convenience, there are also cross references to related terms in other parts of the glossary and to other documents such as the Federal Aviation Regulations (FARs) and the Airman's Information Manual (AIM).

This glossary will be revised as necessary to maintain a common understanding of the system.

_**ABEAM**_ — An aircraft is "abeam" a fix, point or object when that fix, point or object is approximately 90 degrees to the right or left of the aircraft track. Abeam indicates a general position rather than a precise point.

_**ABORT**_ — To terminate a preplanned aircraft maneuver; e.g., an aborted takeoff.

_**ACKNOWLEDGE**_ — Let me know that you have received and understand my message.

**ACROBATIC FLIGHT** — An intentional maneuver involving an abrupt change in an aircraft's attitude, an abnormal attitude, or abnormal acceleration, not necessary for normal flight. (Refer to FAR Part 91)
_ICAO – ACROBATIC FLIGHT_ – Manoeuvres intentionally performed by an aircraft involving an abrupt change in its attitude, an abnormal attitude, or an abnormal variation in speed.

**ADDITIONAL SERVICES --** Advisory information provided by ATC which includes but is not limited to the following:
1. Traffic advisories.
2. Vectors, when requested by the pilot, to assist aircraft receiving traffic advisories to avoid observed traffic.
3. Altitude deviation information of 300 feet or more from an assigned altitude as observed on a verified (reading correctly) automatic altitude readout (Mode C).
4. Advisories that traffic is no longer a factor.
5. Weather and chaff information.
6. Weather assistance.
7. Bird activity information.
8. Holding pattern surveillance.

Additional services are provided to the extent possible contingent only upon the controller's capability to fit it into the performance of higher priority duties and on the basis of limitations of the radar, volume of traffic, frequency congestion and controller workload. The controller has complete discretion for determining if he is able to provide or continue to provide a service in a particular case. The controller's reason not to provide or continue to provide a service in a particular case is not subject to question by the pilot and need not be made known to him. (See Duty Priorities, Traffic Advisories) (Refer to AIM Part 1)

**ADMINISTRATOR** — The Federal Aviation Administrator or any person to whom he has delegated his authority in the matter concerned.

_**ADVISE INTENTIONS**_ — Tell me what you plan to do.

**ADVISORY** — Advice and information provided to assist pilots in the safe conduct of flight and aircraft movement. (See Advisory Service)

**ADVISORY SERVICE** — Advice and information provided by a facility to assist pilots in the safe conduct of flight and aircraft movement. (See Airport Advisory Service, Traffic Advisories, Deviation Advisories, Radar Safety Advisories, Additional Services, Radar Advisory, En Route Flight Advisory Service) (Refer to AIM Part 1)

**AERIAL REFUELING / INFLIGHT REFUELING** — A procedure used by the military to transfer fuel from one aircraft to another during flight. (Refer to AIM Part 4)

**AERODROME** — A defined area on land or water (including any buildings, installations and equipment) intended to be used either wholly or in part for the arrival, departure and movement of aircraft.

**AERONAUTICAL CHART** — A map used in air navigation containing all or part of the following: Topographic features, hazards and obstructions, navigation aids, navigation routes, designated airspace, and airports. Commonly used aeronautical charts are:
1. Sectional Charts – 1:500,000 – Designed for visual navigation of slow or medium speed aircraft. Topographic information on these charts features the portrayal of relief, and a judicious selection of visual check points for VFR flight. Aeronautical information includes visual and radio aids to navigation, airports, controlled airspace, restricted areas, obstructions and related data.
2. VFR Terminal Area Charts – 1:250,000 – Depict Terminal Control Area (TCA) airspace which pro-

vides for the control or segregation of all the aircraft within the TCA. The chart depicts topographic information and aeronautical information which includes visual and radio aids to navigation, airports, controlled airspace, restricted areas, obstructions, and related data.

3. World Aeronautical Charts (WAC) – 1:1,000,000 – Provide a standard series of aeronautical charts covering land areas of the world, at a size and scale convenient for navigation by moderate speed aircraft. Topographic information includes cities and towns, principal roads, railroads, distinctive landmarks, drainage and relief. Aeronautical information includes visual and radio aids to navigation, airports, airways, restricted areas, obstructions and other pertinent data.

4. En Route Low Altitude Charts – Provide aeronautical information for en route instrument navigation (IFR) in the low altitude stratum. Information includes the portrayal of airways, limits of controlled airspace, position identification and frequencies of radio aids, selected airports, minimum en route and minimum obstruction clearance altitudes, airway distances, reporting points, restricted areas and related data. Area charts which are a part of this series furnish terminal data at a larger scale in congested areas.

5. En Route High Altitude Charts – Provide aeronautical information for en route instrument navigation (IFR) in the high altitude stratum. Information includes the portrayal of jet routes, identification and frequencies of radio aids, selected airports, distances, time zones, special use airspace and related information.

6. Area Navigation (RNAV) High Altitude Charts – Provide aeronautical information for en route IFR navigation for high altitude air routes established for aircraft equipped with RNAV systems. Information includes portrayal of RNAV routes, waypoints, track angles, changeover points, distances, selected navigational aids and airports, special use airspace, oceanic routes, and transitional information.

7. Instrument Approach Procedures (IAP) Charts – Portray the aeronautical data which is required to execute an instrument approach to an airport. These charts depict the Procedures, including all related data, and the airport diagram. Each procedure is designated for use with a specific type of electronic navigation system including NDB, TACAN, VOR, ILS, and RNAV. These charts are identified by the primary navigational aid upon which the IAP is predicated.

8. Standard Instrument Departure (SID) Charts – Designed to expedite clearance delivery and to facilitate transition between take-off and en route operations. Each SID procedure is presented as a separate chart and may serve a single airport or more than one airport in a given geographical location.

9. Standard Terminal Arrival Route (STAR) Charts – Designed to expedite air traffic control arrival route procedures and to facilitate transition between en route and instrument approach operations. Each STAR procedure is presented as a separate chart and may serve a single airport or more than one airport in a given geographical location.

10. Airport Taxi Charts – Designed to expedite the efficient and safe flow of ground traffic at an airport. These charts are identified by the official airport name, e.g., Washington National Airport.

*ICAO – AERONAUTICAL CHART* – A representation of a portion of the earth, its culture and relief, specifically designated to meet the requirements of air navigation.

*AFFIRMATIVE* — Yes.

**AIR CARRIER DISTRICT OFFICE / ACDO** — An FAA field office serving an assigned geographical area, staffed with Flight Standards personnel serving the aviation industry and the general public, on matters relating to the certification and operation of scheduled air carriers and other large aircraft operations.

**AIR DEFENSE EMERGENCY** — A military emergency condition declared by a designated authority. This condition exists when an attack upon the continental U.S., Alaska, Canada, or U.S. installations in Greenland by hostile aircraft or missiles is considered probable, is imminent, or is taking place. (Refer to AIM Part 1)

**AIR DEFENSE IDENTIFICATION ZONE / ADIZ** — The area of airspace over land or water, extending upward from the surface, within which the ready identification, the location, and the control of aircraft are required in the interest of national security.
1. Domestic Air Defense Identification Zone – an ADIZ within the United States along an international boundary of the United States.
2. Coastal Air Defense Identification Zone – an ADIZ over the coastal waters of the United States.
3. Distant Early Warning Identification Zone (DEWIZ) – an ADIZ over the coastal waters of the State of Alaska.

ADIZ locations, and operating and flight plan requirements for civil aircraft operations are specified in FAR Part 99. (Refer to AIM Part 1)

**AIR NAVIGATION FACILITY** — Any facility used in, available for use in, or designed for use in, aid of air navigation, including landing areas, lights, any apparatus or equipment for disseminating weather information, for signaling, for radio-directional finding, or for radio or other electrical communication, and any other structure or mechanism having a similar purpose for guiding or controlling flight in the air or the landing and take-off of aircraft. (See Navigation Aid)

**AIR ROUTE SURVEILLANCE RADAR / ARSR** — Air route traffic control center (ARTCC) radar, used primarily to detect and display an aircraft's position while en route between terminal areas. The ARSR enables controllers to provide radar air traffic control service when aircraft are within the ARSR coverage. In some instances, ARSR may enable an ARTCC to provide terminal radar services similar to, but usually more limited than, those provided by a radar approach control.

**AIR ROUTE TRAFFIC CONTROL CENTER / ARTCC / CENTER** — A facility established to provide air traffic control service to aircraft operating on IFR flight plans within controlled airspace and principally during the en route phase of flight. When equipment capabilities and controller workload permit, certain advisory/assistance services may be provided to VFR aircraft. (See NAS Stage A, En Route Air Traffic Control Service) (Refer to AIM Part 1)

**AIR TRAFFIC** — Aircraft operating in the air or on an airport surface, exclusive of loading ramps and parking areas. *ICAO – AIR TRAFFIC –* All aircraft in flight or operating on the manoeuvring area of an aerodrome.

**AIR TRAFFIC CLEARANCE / ATC CLEARANCE / CLEARANCE** — An authorization by air traffic control, for the purpose of preventing collision between known aircraft, for an aircraft to proceed under specified traffic conditions within controlled airspace. (See ATC Instructions)
*ICAO – AIR TRAFFIC CONTROL CLEARANCE –* Authorization for an aircraft to proceed under conditions specified by an air traffic control unit.

**AIR TRAFFIC CONTROL / ATC** — A service operated by appropriate authority to promote the safe, orderly and expeditious flow of air traffic.
*ICAO – AIR TRAFFIC CONTROL SERVICE –* A service provided for the purpose of:
1. Preventing collisions:
   a. Between aircraft, and
   b. On the manoeuvring area between aircraft and obstructions, and
2. Expediting and maintaining an orderly flow of air traffic.

**AIR TRAFFIC CONTROL SERVICE** — (See Air Traffic Control)

**AIR TRAFFIC CONTROL SPECIALIST / CONTROLLER** — A person authorized to provide air traffic control service. This term refers to en route and terminal control personnel. Flight Service personnel are referred to as Flight Service specialists. (See Air Traffic Control Service, Flight Service Station)
*ICAO – CONTROLLER –* A person authorized to provide air traffic control services.

**AIR TRAFFIC CONTROL SYSTEMS COMMAND CENTER / ATCSCC** — An air traffic service facility consisting of four operational units, and located in FAA Headquarters.
1. Central Flow Control Function / CFCF – Responsible for coordination and approval of all major inter-center flow control restrictions on a system basis in order to obtain maximum utilization of the airspace. (See Quota Flow Control, Fuel Advisory Departure)
2. Central Altitude Reservation Function / CARF – Responsible for coordinating, planning and approving special user requirements under the Altitude Reservation (ALTRV) concept. (See Altitude Reservation)
3. Airport Reservation Office / ARO – Responsible for approving IFR flights at designated high density traffic airports (John F. Kennedy, LaGuardia, O'Hare and Washington National) during specified hours. (Refer to FAR Part 93, AIM Part 3)
4. ATC Contingency Command Post – A facility which enables the FAA to manage the ATC system when significant portions of the system's capabilities have been lost or are threatened.

**AIRCRAFT** — A device that is used or intended to be used for flight in the air.
*ICAO – AIRCRAFT –* Any machine that can derive support in the atmosphere from the reactions of the air other than the reactions of the air against the earth's surface.

**AIRCRAFT APPROACH CATEGORY** — A grouping of aircraft based on a speed of 1.3 Vso (at maximum certificated landing weight), or on maximum certificated landing weight. Vso and the maximum certificated landing weight are those values as established for the aircraft by the certificating authority of the country of registry. If an aircraft falls into two categories, it is placed in the higher of the two. The categories are as follows:
1. Category A – Speed less than 91 knots; weight less than 30,001 pounds.
2. Category B – Speed 91 knots or more but less than 121 knots; weight 30,001 pounds or more but less than 60,001 pounds.
3. Category C – Speed 121 knots or more but less than 141 knots; weight 60,001 pounds or more but less than 150,001 pounds.
4. Category D – Speed 141 knots or more but less than 166 knots; weight 150,001 pounds or more.
5. Category E – Speed 166 knots or more; any weight.
(Refer to FAR Parts 1 and 97)

**AIRCRAFT CLASSES** — For the purposes of Wake Turbulence Separation Minima, ATC classifies aircraft as Heavy, Large and Small as follows:
1. Heavy – Aircraft capable of takeoff weights of 300,000 pounds or more whether or not they are operating at this weight during a particular phase of flight.
2. Large – Aircraft of more than 12,500 pounds, maximum certificated takeoff weight, up to 300,000 pounds.
3. Small – Aircraft of 12,500 pounds or less, maximum certificated takeoff weight. (Refer to AIM Part 1)

**AIRMAN'S INFORMATION MANUAL / AIM** — A publication designed primarily as a pilot's operational and instructional manual for use in the National Airspace System of the United States. It consists of the following basic parts which may be purchased separately.
Part 1 – Basic Flight Manual and ATC Procedures.
Part 2 – Airport Directory.
Part 3 and 3A – Operational Data and Notices to Airmen.
Part 4 – Graphic Notices and Supplemental Data.
*ICAO – AERONAUTICAL INFORMATION PUBLICATION –* A publication issued by or with the authority of a state and containing aeronautical information of a lasting character essential to air navigation.

**AIRMET / AIRMAN'S METEOROLOGICAL INFORMATION** — Inflight weather advisories which cover moderate icing, moderate turbulence, sustained winds of 30 knots or more within 2,000 feet of the surface and the initial onset of phenomena producing extensive areas of visiblities below 3 miles or ceilings less than 1,000 feet. It concerns weather phenomena which are of operational interest to all aircraft and potentially hazardous to aircraft having limited capability because of lack of equipment, instrumentation or pilot qualifications. It concerns weather of less severity than SIGMETs. (See SIGMET)

**AIRPORT** — An area of land or water that is used or intended to be used for the landing and takeoff of aircraft, and includes its buildings and facilities, if any.

**AIRPORT ADVISORY AREA** — The area within five statute miles of an airport not served by a control tower, i.e., there is no tower or the tower is not in operation, on which is located a Flight Service Station. (See Airport Advisory Service) (Refer to AIM Part 1)

**AIRPORT ADVISORY SERVICE / AAS** — A service provided by Flight Service Stations at airports not served by a control tower. This service consists of providing information to arriving and departing aircraft concerning wind direction and speed, favored runway, altimeter setting, pertinent known traffic, pertinent known field conditions, airport taxi routes and traffic patterns, and authorized instrument approach procedures. This information is advisory in nature and does not constitute an ATC clearance. (See Airport Advisory Area)

**AIRPORT ELEVATION / FIELD ELEVATION** — The highest point of an airport's usable runways measured in feet from mean sea level. (See Touchdown Zone Elevation)
*ICAO – AERODROME ELEVATION* – The elevation of the highest point of the landing area.

**AIRPORT INFORMATION DESK / AID** — An airport unmanned facility designed for pilot self-service briefing, flight planning, and filing of flight plans. (Refer to AIM Part 1)

**AIRPORT LIGHTING** — Various lighting aids that may be installed on an airport. Types of airport lighting include:
1. Approach Light System / ALS – An airport lighting facility which provides visual guidance to landing aircraft by radiating light beams in a directional pattern by which the pilot aligns the aircraft with the extended centerline of the runway on his final approach for landing.
Condenser-Discharge Sequential Flashing Lights / Sequenced Flashing Lights may be installed in conjunction with the ALS at some airports.
Types of Approach Light Systems are:
   a. ALSF-I – Approach Light System with Sequenced Flashing Lights in ILS CAT-I configuration,
   b. ALSF-II – Approach Light System with Sequenced Flashing Lights in ILS CAT-II configuration,
   c. SSALF – Simplified Short Approach Light System with Sequenced Flashing Lights,
   d. SSALR – Simplified Short Approach Light System with Runway Alignment Indicator Lights,
   e. MALSF – Medium Intensity Approach Light System with Sequenced Flashing Lights,
   f. MALSR – Medium Intensity Approach Light System with Runway Alignment Indicator Lights,
   g. LDIN – Sequenced Flashing Lead-in Lights,
   h. RAIL – Runway Alignment Indicator Lights (Sequenced Flashing Lights which are installed only in combination with other light systems).
2. Runway Lights / Runway Edge Lights – Lights having a prescribed angle of emission used to define the lateral limits of a runway. Runway lights are uniformly spaced at intervals of approximately 200 feet, and the intensity may be controlled or preset.
3. Touchdown Zone Lighting – Two rows of transverse light bars located symmetrically about the runway centerline normally at 100 foot intervals. The basic system extends 3,000 feet along the runway.
4. Runway Centerline Lighting – Flush centerline lights spaced at 50-foot intervals beginning 75 feet from the landing threshold and extending to within 75 feet of the opposite end of the runway.
5. Threshold Lights – Fixed green lights arranged symmetrically left and right of the runway centerline, identifying the runway threshold.
6. Runway End Identifier Lights / REIL – Two synchronized flashing lights, one on each side of the runway threshold, which provide rapid and positive identification of the approach end of a particular runway.
7. Visual Approach Slope Indicator / VASI – An airport lighting facility providing vertical visual approach slope guidance to aircraft during approach to landing by radiating a directional pattern of high intensity red and white focused light beams which indicate to the pilot that he is "on path" if he sees red/white, "above path" if white/white, and "below path" if red/red. Some airports serving large aircraft have three-bar VASIs which provide two visual glide paths to the same runway.
8. Boundary Lights – Lights defining the perimeter of an airport or landing area. (Refer to AIM Part 1)

**AIRPORT ROTATING BEACON / ROTATING BEACON** — A visual NAVAID operated at many airports. At civil airports alternating white and green flashes indicate the location of the airport. The total number of flashes are 12 to 15 per minute. At military airports, the beacons flash alternately white and green, but are differentiated from civil beacons by dualpeaked (two quick) white flashes between the green flashes. Normally, operation of an airport rotating beacon during the hours of daylight means that the reported ground visibility at the airport is less than three miles and/or the reported ceiling is less than 1,000 feet and, therefore, an ATC clearance is required for landing or takeoff. (See Control Zone, Special VFR Operations, Instrument Flight Rules) (Refer to AIM Part 1)
*ICAO – AERODROME BEACON* – Aeronautical beacon used to indicate the location of an aerodrome.

**AIRPORT SURFACE DETECTION EQUIPMENT / ASDE** — Radar equipment specifically designed to detect all principal features on the surface of an airport, including aircraft and vehicular traffic and to present the entire image on a radar indicator console in the control tower. Used to augment visual observation by tower personnel of aircraft and/or vehicular movements on runways and taxiways.

**AIRPORT SURVEILLANCE RADAR / ASR** — Approach control radar used to detect and display an aircraft's position in the terminal area. ASR provides range and azimuth information but does not provide elevation data. Coverage of the ASR can extend up to 60 miles.

**AIRPORT TRAFFIC AREA** — Unless otherwise specifically designated in FAR Part 93, that airspace within a horizontal radius of 5 statute miles from the geographical center of any airport at which a control tower is operating, extending from the surface up to, but not including, an altitude of 3,000 feet above the elevation of the airport. Unless otherwise authorized or required by ATC, no person may operate an aircraft within an airport traffic area except for the purpose of landing at, or taking off from, an airport within that area. ATC authorizations may be given as individual approval of specific operations or may be contained in written agreements between airport users and the tower concerned. (Refer to FAR Parts 1 and 91)

**AIRPORT TRAFFIC CONTROL SERVICE** — A service provided by a control tower for aircraft operating on the movement area and in the vicinity of an airport. (See Movement Area, Tower)
*ICAO – AERODROME CONTROL SERVICE* – Air traffic control service for aerodrome traffic.

**AIRSPEED** — The speed of an aircraft relative to its surrounding air mass. The unqualified term "airspeed" means one of the following:
1. Indicated Airspeed – The speed shown on the aircraft airspeed indicator. This is the speed used in pilot/controller communications under the general term "airspeed." (Refer to FAR Part 1)
2. True Airspeed – The airspeed of an aircraft relative to undisturbed air. Used primarily in flight planning and the en route portion of flight. When used in pilot/controller communications, it is referred to as "true airspeed" and not shortened to "airspeed."

**AIRSTART** — The starting of an aircraft engine while the aircraft is airborne, preceded by engine shutdown during training flights or by actual engine failure.

**AIRWAY / FEDERAL AIRWAY** — A control area or portion thereof established in the form of a corridor, the centerline of which is defined by radio navigational aids. (Refer to FAR Part 71, AIM Part 1)
*ICAO – AIRWAY* – A control area or portion thereof established in the form of corridor equipped with radio navigational aids.

**ALERT AREA** — Airspace which may contain a high volume of pilot training activities or an unusual type of aerial activity. Alert Areas are depicted on aeronautical charts for the information of nonparticipating pilots. All activities within an Alert Area are conducted in accordance with Federal Aviation Regulations and pilots of participating aircraft as well as pilots transiting the area are equally responsible for collision avoidance.

**ALERT NOTICE / ALNOT** — A message sent by a Flight Service Station (FSS) that requests an extensive communication search for overdue, unreported or missing aircraft.

**ALL WEATHER LOW ALTITUDE TRAINING ROUTE** — (See Olive Branch Routes)

**ALPHA-NUMERIC DISPLAY / DATA BLOCK** — Letters and numerals used to show identification, altitude, beacon code and other information concerning a target on a radar display. (See Automated Radar Terminal Systems, NAS Stage A)

**ALTERNATE AIRPORT** — An airport at which an aircraft may land if a landing at the intended airport becomes inadvisable.
*ICAO – ALTERNATE AERODROME* – An aerodrome specified in the flight plan to which a flight may proceed when it becomes inadvisable to land at the aerodrome of intended landing.

**ALTIMETER SETTING** — The barometric pressure reading used to adjust a pressure altimeter for variations in existing atmospheric pressure or to the standard altimeter setting (29.92). (Refer to FAR Part 91, AIM Part 1)

*ALTITUDE* — The height of a level, point or object measured in feet Above Ground Level (AGL) or from Mean Sea Level (MSL). (See Flight Level)
1. MSL Altitude – Altitude, expressed in feet measured from mean sea level.
2. AGL Altitude – Altitude expressed in feet measured above ground level.
3. Indicated Altitude – The altitude as shown by an altimeter. On a pressure or barometric altimeter it is altitude as shown uncorrected for instrument error and uncompensated for variation from standard atmospheric conditions.
*ICAO – ALTITUDE* – The vertical distance of a level, a point, or an object considered as a point, measured from a certain level.

*ALTITUDE READOUT / AUTOMATIC ALTITUDE REPORT* — An aircraft's altitude, transmitted via the Mode C transponder feature, that is visually displayed in 100-foot increments on a radar scope having readout capability. (See Automatic Radar Terminal Systems, NAS Stage A, Alpha Numeric Display) (Refer to AIM Part 1)

**ALTITUDE RESERVATION / ALTRV** — Airspace utilization under prescribed conditions normally employed for the mass movement of aircraft or other special user requirements which cannot otherwise be accomplished. ALTRVs are approved by the appropriate FAA facility. (See Air Traffic Control Systems Command Center)

**ALTITUDE RESTRICTION** — An altitude or altitudes stated in the order flown, which are to be maintained until reaching a specific point or time. Altitude restrictions may be issued by ATC due to traffic, terrain or other airspace considerations.

*ALTITUDE RESTRICTIONS ARE CANCELLED* — Adherence to previously imposed altitude restrictions is no longer required during a climb or descent.

**APPROACH CLEARANCE** — Authorization by ATC for a pilot to conduct an instrument approach. The type of instrument approach for which cleared and other pertinent information is provided in the approach clearance when required. (See Instrument Approach Procedure, Cleared for Approach) (Refer to AIM Part 1, FAR Part 91)

**APPROACH CONTROL / APPROACH CONTROL FACILITY** — A terminal air traffic control facility providing approach control service. (See Approach Control Service, Tower, Terminal Radar Approach Control, Radar Approach Control, Radar Air Traffic Control Facility.)

**APPROACH CONTROL SERVICE** — Air traffic control service provided by an approach control facility for arriving and departing VFR/IFR aircraft and, on occasion, en route aircraft. At some airports not served by an approach control facility, the ARTCC provides limited approach control service. (Refer to AIM Part 1)
*ICAO – APPROACH CONTROL SERVICE* – Air Traffic service for arriving or departing controlled flights.

**APPROACH GATE** — The point on the final approach course which is 1 mile from the final approach fix on the side away from the airport or 5 miles from landing threshold, whichever is farther from the landing threshold. This is an imaginary point used within ATC as a basis for final approach course interception for aircraft being vectored to the final approach course.

**APPROACH LIGHT SYSTEM** — (See Airport Lighting)

**APPROACH SEQUENCE** — The order in which aircraft are positioned while on approach or awaiting approach clearance. (See Landing Sequence)
*ICAO – APPROACH SEQUENCE* – The order in which two or more aircraft are cleared to approach to land at the aerodrome.

*APPROACH SPEED* — The recommended speed contained in aircraft operating manuals used by pilots when making an approach to landing. This speed will vary for different segments of an approach as well as for aircraft weight and configuration.

**APRON / RAMP** — A defined area, on a land airport, intended to accommodate aircraft for purposes of loading or unloading passengers or cargo, refueling, parking or maintenance. With regard to seaplanes, a ramp is used for access to the apron from the water.
*ICAO – APRON* – A defined area, on a land aerodrome, intended to accommodate aircraft for purposes of loading or unloading passengers or cargo, refueling, parking or maintenance.

**ARC** — The track over the ground of an aircraft flying at a constant distance from a navigational aid by reference to distance measuring equipment (DME).

**AREA NAVIGATION / RNAV** — A method of navigation that permits aircraft operations on any desired course within the coverage of station-referenced navigation signals or within the limits of self-contained system capability. (Refer to AIM Parts 1 and 3, FAR Part 71)
   1. Area Navigation Low Route – An area navigation route within the airspace extending upward from 1,200 feet above the surface of the earth to, but not including 18,000 feet MSL.
   2. Area Navigation High Route – An area navigation route within the airspace extending upward from and including 18,000 feet MSL to flight level 450.
   3. Random Area Navigation Routes / Random RNAV Routes – Direct routes, based on area navigation capability, between waypoints defined in terms of degree/distance fixes or offset from published or established routes/airways at specified distance and direction.
   4. RNAV Waypoint / W/P – A predetermined geographical position used for route or instrument approach definition or progress reporting purposes that is defined relative to a VORTAC station position.
*ICAO – AREA NAVIGATION/RNAV* – A method of navigation which permits aircraft operation on any desired flight path within the coverage of station-referenced navigation aids or within the limits of the capability of self-contained aids or a combination of these.

**ARMY AVIATION FLIGHT INFORMATION BULLETIN / USAFIB** — A bulletin that provides air operation data covering Army, National Guard, and Army Reserve aviation activities.

**ARMY RADAR APPROACH CONTROL / ARAC** — An air traffic control facility located at a U.S. Army Airport utilizing surveillance and normally precision approach radar and air/ground communications equipment to provide approach control services to aircraft arriving, departing or transiting the airspace controlled by the facility. Service may be provided to both civil and military airports. Similar to TRACON (FAA), RAPCON (USAF) and RATCF (Navy). (See Approach Control, Approach Control Service, Departure Control)

**ARRESTING SYSTEM** — A safety device consisting of two major components, namely, engaging or catching devices, and energy absorption devices for the purpose of arresting both tail hook and/or non-tail hook equipped aircraft. It is used to prevent aircraft from overrunning runways when the aircraft cannot be stopped after landing or during aborted takeoff. Arresting systems have various names, e.g., arresting gear, hook, device, wire, barrier, cable. (See Abort) (Refer to AIM Parts 1 and 2)

**ARRIVAL TIME** — The time an aircraft touches down on arrival.

**ARTCC** — (See Air Route Traffic Control Center)

**ASR APPROACH** — (See Surveillance Approach)

*ATC ADVISES* — Used to prefix a message of noncontrol information when it is relayed to an aircraft by other than an air traffic controller. (See Advisory)

**ATC ASSIGNED AIRSPACE / ATCAA** — Airspace of defined vertical/lateral limits, assigned by ATC, for the purpose of providing air traffic segregation between the specified activities being conducted within the assigned airspace and other IFR air traffic. (See Military Operations Area, Alert Area)

**ATC CLEARANCE —** (See Air Traffic Clearance)

*ATC CLEARS —* Used to prefix an ATC clearance when it is relayed to an aircraft by other than an air traffic controller.

**ATC INSTRUCTION —** Directives issued by air traffic control for the purpose of requiring a pilot to take specific actions; e.g., "Turn left heading 250", "Go around," "Clear the runway." (Refer to FAR Part 91)

*ATC REQUESTS —* Used to prefix an ATC request when it is relayed to an aircraft by other than an air traffic controller.

**ATCRBS —** (See Radar)

**AUTOMATED RADAR TERMINAL SYSTEMS / ARTS** — The generic term for the ultimate in functional capability afforded by several automation systems. Each differs in functional capabilities and equipment. ARTS plus a suffix Roman Numeral denotes a specific system. A following letter indicates a major modification to that system. In general, an ARTS displays for the terminal controller aircraft identification, flight plan data, other flight associated information, e.g., altitude and speed, and aircraft position symbols in conjunction with his radar presentation. Normal radar co-exists with the alphanumeric display. In addition to enhancing visualization of the air traffic situation, ARTS facilitate intra/interfacility transfer and coordination of flight information. These capabilities are enabled by specially designed computers and subsystems tailored to the radar and communications equipments and operational requirements of each automated facility. Modular design permits adoption of improvements in computer software and electronic technologies as they become available while retaining the characteristics unique to each system:

1. ARTS IA – The functional capabilities and equipment of the New York Common IFR Room Terminal Automation System. It tracks primary as well as secondary targets derived from two radar sources. The aircraft targets are displayed on a radar type console by means of an alphanumeric generator. Aircraft identity is depicted in association with the appropriate aircraft target. When the aircraft is equipped with an encoded altimeter (Mode C), its altitude is also displayed. The system can exchange flight plan information with the ARTCC.

2. ARTS II – A programmable non-tracking, computer aided display subsystem capable of modular expansion. ARTS II systems provide a level of automated air traffic control capability at terminals having low to medium activity. Flight identification and altitude may be associated with the display of secondary radar targets. Also, flight plan information may be exchanged between the terminal and ARTCC.

3. ARTS III – The Beacon Tracking Level (BTL) of the modular programmable automated radar terminal system in use at medium to high activity terminals. ARTS III detects, tracks and predicts secondary radar derived aircraft targets. These are displayed by means of computer generated symbols and alphanumeric characters depicting flight identification, aircraft altitude, ground speed and flight plan data.

Although it does not track primary targets, they are displayed coincident with the secondary radar as well as the symbols and alphanumerics. The system has the capability of communicating with ARTCCs and other ARTS III facilities.

4. ARTS IIIA – The Radar Tracking and Beacon Tracking Level (RT&BTL) of the modular, programmable automated radar terminal system. ARTS IIIA detects, tracks and predicts primary as well as secondary radar derived aircraft targets. An enhancement of the ARTS III, this more sophisticated computer driven system will eventually replace the ARTS IA system and upgrade about half of the existing ARTS III systems. The enhanced system will provide improved tracking, continuous data recording and fail-soft capabilities.

**AUTOMATIC ALTITUDE REPORTING —** That function of a transponder which responds to Mode C interrogations by transmitting the aircraft's altitude in 100-foot increments.

**AUTOMATIC CARRIER LANDING SYSTEM / ACLS** — U.S. Navy final approach equipment consisting of precision tracking radar coupled to a computer data link to provide continuous information to the aircraft, monitoring capability to the pilot and a backup approach system.

**AUTOMATIC DIRECTION FINDER / ADF —** An aircraft radio navigation system which senses and indicates the direction to a L/MF nondirectional radio beacon (NDB) ground transmitter. Direction is indicated to the pilot as a magnetic bearing or as a relative bearing to the longitudinal axis of the aircraft depending on the type of indicator installed in the aircraft. In certain applications, such as military, ADF operations may be based on airborne and ground transmitters in the VHF/UHF frequency spectrum. (See Bearing, Nondirectional Beacon)

**AUTOMATIC TERMINAL INFORMATION SERVICE / ATIS —** The continuous broadcast of recorded noncontrol information in selected terminal areas. Its purpose is to improve controller effectiveness and to relieve frequency congestion by automating the repetitive transmission of essential but routine information, e.g.," Los Angeles Information Alpha. 1300 Greenwich Weather, measured ceiling 2000 overcast, visibility three, haze, smoke, temperature seven one, wind two five zero at five, altimeter two niner niner six. ILS runway two five left approach in use, runway two five right closed, advise you have Alpha." (Refer to AIM Part 1)
*ICAO – AUTOMATIC TERMINAL INFORMATION SERVICE –* The provision of current, routine information to arriving and departing aircraft by means of continuous and repetitive broadcasts throughout the day or a specified portion of the day.

**AUTOROTATION —** A rotorcraft flight condition in which the lifting rotor is driven entirely by action of the air when the rotorcraft is in motion.

1. Autorotative Landing/Touchdown Autorotation – Used by a pilot to indicate that he will be landing without applying power to the rotor.

2. Low Level Autorotation – Commences at an altitude well below the traffic pattern, usually below 100 feet AGL and is used primarily for tactical military training.

3.  180 degree Autorotation – Initiated from a downwind heading and is commenced well inside the normal traffic pattern. "Go around" may not be possible during the latter part of this maneuver.

**AVIATION WEATHER SERVICE** — A service provided by the National Weather Service (NWS) and FAA which collects and disseminates pertinent weather information for pilots, aircraft operators and ATC. Available aviation weather reports and forecasts are displayed at each NWS office and FAA FSS. (See En Route Flight Advisory Service , Transcribed Weather Broadcasts, · Scheduled Weather Broadcasts, Inflight Weather Advisories, Pilots Automatic Telephone Weather Answering Service) (Refer to AIM Parts 1 and 2)

**BASE LEG** — (See Traffic Pattern)

**BEACON** — (See Radar)

**BEARING** — The horizontal direction to or from any point, usually measured clockwise from true north, magnetic north or some other reference point, through 360 degrees. (See Nondirectional Beacon)

**BELOW MINIMUMS** — Weather conditions below the minimums prescribed by regulation for the particular action involved, e.g., landing minimums, takeoff minimums.

**BLAST FENCE** — A barrier that is used to divert or dissipate jet or propeller blast.

**BLIND SPEED** — The rate of departure or closing of a target relative to the radar antenna at which cancellation of the primary radar target by moving target indicator (MTI) circuits in the radar equipment causes a reduction or complete loss of signal.
*ICAO – BLIND VELOCITY* – The radial velocity of a moving target such that the target is not seen on primary radars fitted with certain forms of fixed echo suppression.

**BLIND SPOT / BLIND ZONE** — An area from which radio transmissions and/or radar echoes cannot be received. The term is also used to describe portions of the airport not visible from the control tower.

**BOUNDARY LIGHTS** — (See Airport Lighting)

**BRAKING ACTION (GOOD, MEDIUM OR FAIR, POOR, NIL)** — A report of conditions on the airport movement area providing a pilot with a degree/quality of braking that he might expect. Braking action is reported in terms of good, medium (or fair), poor or nil. (See Runway Condition Reading)

**BROADCAST** — Transmission of information for which an acknowledgement is not expected.
*ICAO – BROADCAST* – A transmission of information relating to air navigation that is not addressed to a specific station or stations.

**CALL-UP** — Initial voice contact between a facility and an aircraft, using the identification of the unit being called and the unit initiating the call. (Refer to AIM Part 1)

**CARDINAL ALTITUDES OR FLIGHT LEVELS** — "Odd" or "Even" thousand-foot altitudes or flight levels; e.g., 5000, 6000, 7000, FL 250, FL 260, FL 270. (See Altitude, Flight Levels)

**CEILING** — The height above the earth's surface of the lowest layer of clouds or obscuring phenomena that is reported as "broken," "overcast," or "obscuration," and not classified as "thin" or "partial".
*ICAO – CEILING* – The height above the ground or water of the base of the lowest layer of cloud below 6000 metres (20,000 feet) covering more than half the sky.

**CELESTIAL NAVIGATION** — The determination of geographical position by reference to celestial bodies. Normally used in aviation as a secondary means of position determination.

*CENTER* — (See Air Route Traffic Control Center)

**CENTER'S AREA** — The specified airspace within which an air route traffic control center (ARTCC) provides air traffic control and advisory service. (See Air Route Traffic Control Center) (Refer to AIM Part 1)

**CHAFF** — Thin, narrow metallic reflectors of various lengths and frequency responses, used to reflect radar energy. These reflectors when dropped from aircraft and allowed to drift downward result in large targets on the radar display.

*CHASE / CHASE AIRCRAFT* — An aircraft flown in proximity to another aircraft normally to observe its performance during training or testing.

**CIRCLE TO LAND MANEUVER / CIRCLING MANEUVER** — A maneuver initiated by the pilot to align the aircraft with a runway for landing when a straight-in landing from an instrument approach is not possible or is not desirable. This maneuver is made only after ATC authorization has been obtained and the pilot has established required visual reference to the airport (See Circle to Runway, Landing Minimums) (Refer to AIM Part 1)

*CIRCLE TO RUNWAY (RUNWAY NUMBER)* — Used by ATC to inform the pilot that he must circle to land because the runway in use is other than the runway aligned with the instrument approach procedure. When the direction of the circling maneuver in relation to the airport/runway is required, the controller will state the direction (eight cardinal compass points) and specify a left or right downwind or base leg as appropriate; e.g., "Cleared VOR Runway 36 approach circle to Runway 22" or "Circle northwest of the airport for a right downwind to Runway 22." (See Circle to Land Maneuver, Landing Minimums) (Refer to AIM Part 1)

**CIRCLING APPROACH** — (See Circle-to-land Maneuver)

**CIRCLING MINIMA** — (See Landing Minimums)

*CLEAR OF TRAFFIC* — Previously issued traffic is no longer a factor.

**CLEAR-AIR TURBULENCE / CAT** — Turbulence encountered in air where no clouds are present. This term is commonly applied to high-level turbulence associated with wind shear. CAT is often encountered in the vicinity of the jet stream. (See Wind Shear, Jet Stream)

*CLEARANCE* — (See Air Traffic Clearance)

**CLEARANCE LIMIT** — The fix, point, or location to which an aircraft is cleared when issued an air traffic clearance.
*ICAO – CLEARANCE LIMIT –* The point of which an aircraft is granted an air traffic control clearance.

*CLEARANCE VOID IF NOT OFF BY (TIME)* — Used by ATC to advise an aircraft that the departure clearance is automatically cancelled if takeoff is not made prior to a specified time. The pilot must obtain a new clearance or cancel his IFR flight plan if not off by the specified time.
*ICAO – CLEARANCE VOID TIME –* A time specified by an air traffic control unit at which a clearance ceases to be valid unless the aircraft concerned has already taken action to comply therewith.

*CLEARED AS FILED* — Means the aircraft is cleared to proceed in accordance with the route of flight filed in the flight plan. This clearance does not include the altitude, SID, or SID Transition. (See Request Full Route Clearance) (Refer to AIM Part 1)

*CLEARED FOR (Type Of) APPROACH* — ATC authorization for an aircraft to execute a specific instrument approach procedure to an airport; e.g., "Cleared for ILS runway 36 approach." (See Instrument Approach Procedure, Approach Clearance) (Refer to AIM Part 1, FAR Part 91)

*CLEARED FOR APPROACH* — ATC authorization for an aircraft to execute any standard or special instrument approach procedure for that airport. Normally, an aircraft will be cleared for a specific instrument approach procedure. (See Instrument Approach Procedure, Cleared for (type of) Approach) (Refer to AIM Part 1, FAR Part 91)

*CLEARED FOR TAKE-OFF* — ATC authorization for an aircraft to depart. It is predicated on known traffic and known physical airport conditions.

*CLEARED FOR THE OPTION* — ATC authorization for an aircraft to make a touch-and-go, low approach, missed approach, stop and go, or full stop landing at the discretion of the pilot. It is normally used in training so that an instructor can evaluate a student's performance under changing situations. (See Option Approach) (Refer to AIM Part 1)

*CLEARED THROUGH* — ATC authorization for an aircraft to make intermediate stops at specified airports without refiling a flight plan while en route to the clearance limit.

*CLEARED TO LAND* — ATC authorization for an aircraft to land. It is predicated on known traffic and known physical airport conditions.

*CLIMB TO VFR* — ATC authorization for an aircraft to climb to VFR conditions within a control zone when the only weather limitation is restricted visibility. The aircraft must remain clear of clouds while climbing to VFR. (See Special VFR) (Refer to AIM Part 1)

**CLIMBOUT** — That portion of flight operation between takeoff and the initial cruising altitude.

**CLOSED RUNWAY** — A runway that is unusable for aircraft operations. Only the airport management/military operations office can close a runway.

**CLOSED TRAFFIC** — Successive operations involving takeoffs and landings or low approaches where the aircraft does not exit the traffic pattern.

**CLUTTER** — In radar operations, clutter refers to the reception and visual display of radar returns caused by precipitation, chaff, terrain, numerous aircraft targets, or other phenomena. Such returns may limit or preclude ATC from providing services based on radar. (See Ground Clutter, Chaff, Precipitation, Target)
*ICAO – Radar Clutter –* The visual indication on a radar display of unwanted signals.

**COASTAL FIX** — A navigation aid or intersection where an aircraft transitions between the domestic route structure and the oceanic route structure.

**CODES / TRANSPONDER CODES** — The number assigned to a particular multiple pulse reply signal transmitted by a transponder. (See Discrete Code)

**COMBINED STATION/TOWER / CS/T** — An air traffic control facility which combines the functions of a flight service station and an airport traffic control tower. (See Tower, Flight Service Station) (Refer to AIM Part 1)

**COMMON ROUTE / COMMON PORTION** — That segment of a North American route between the inland navigation facility and the coastal fix.

**COMPASS LOCATOR** — A low power, low or medium frequency (L/MF) radio beacon installed in conjunction with the outer or middle marker of an instrument landing system (ILS). It can be used for navigation at distances of approximately 15 miles or as authorized in the approach procedure.
1. Outer Compass Locator / LOM – A compass locator installed in conjunction with the outer marker of an instrument landing system. (See Outer Marker)
2. Middle Compass Locator / LMM – A compass locator installed in conjunction with the middle marker of an instrument landing system. (See Middle Marker)
*ICAO – LOCATOR –* An LF/MF NDB used as an aid to final approach.

**COMPASS ROSE** — A circle graduated in degrees, printed on some charts or marked on the ground at an airport. It is used as a reference to either true or magnetic direction.

**COMPOSITE FLIGHT PLAN** — A flight plan which specifies VFR operation for one portion of flight and IFR for another portion. It is used primarily in military operations. (Refer to AIM Part 1)

**COMPULSORY REPORTING POINTS** — Reporting points which must be reported to ATC. They are designated on aeronautical charts by solid triangles or filed in a flight plan as fixes selected to define direct routes. These points are geographical locations which are defined by navigation aids/fixes. Pilots should discontinue position reporting over compulsory reporting points when informed by ATC that their aircraft is in "radar contact."

*CONFLICT ALERT, ADVISE YOU TURN RIGHT/LEFT HEADING (DEGREES) AND/OR CLIMB/DESCEND TO (ALTITUDE) IMMEDIATELY* — (See Radar Safety Advisory)

**CONSOLAN** — A low frequency, long-distance NAVAID used principally for transoceanic navigations.

*CONTACT —*
1. Establish communication with (followed by the name of the facility and, if appropriate, the frequency to be used).
2. A flight condition wherein the pilot ascertains the attitude of his aircraft and navigates by visual reference to the surface. (See Contact Approach, Radar Contact)

**CONTACT APPROACH** — An approach wherein an aircraft on an IFR flight plan, operating clear of clouds with at least 1 mile flight visibility and having received an air traffic control authorization, may deviate from the prescribed instrument approach procedure and proceed to the airport of destination by visual reference to the surface. This approach will only be authorized when requested by the pilot and the reported ground visibility at the destination airport is at least 1 statute mile. (See Visual Approach) (Refer to AIM Part 1)

**CONTERMINOUS U.S.** — The forty-eight adjoining states and the District of Columbia.

**CONTINENTAL CONTROL AREA** — (See Controlled Airspace)

**CONTINENTAL UNITED STATES** — The 49 states located on the continent of North America and the District of Columbia.

**CONTROL AREA** — (See Controlled Airspace)

**CONTROL SECTOR** — An airspace area of defined horizontal and vertical dimensions for which a controller, or group of controllers, has air traffic control responsibility, normally within an air route traffic control center or an approach control facility. Sectors are established based on predominant traffic flows, altitude strata, and controller workload. Pilot-controller communications during operations within a sector are normally maintained on discrete frequencies assigned to the sector. (See Discrete Frequency)

**CONTROL SLASH** — A radar beacon slash representing the actual position of the associated aircraft. Normally, the control slash is the one closest to the interrogating radar beacon site. When ARTCC radar is operating in narrowband (digitized) mode, the control slash is converted to a target symbol.

**CONTROL ZONE** — (See Controlled Airspace)

**CONTROLLED AIRSPACE** — Airspace, designated as a continental control area, control area, control zone, terminal control area, or transition area, within which some or all aircraft may be subject to air traffic control. (Refer to AIM Part 1, FAR Part 71)
*ICAO – CONTROLLED AIRSPACE –* Airspace of defined dimensions within which air traffic control service is provided to controlled flights.
Types of U.S. Controlled Airspace:
1. Continental Control Area – The airspace of the 48 contiguous states, the District of Columbia and Alas-

ka, excluding the Alaska peninsula west of Long. 160° 00' 00"W. at and above 14,500 feet MSL, but does not include:
   a. The airspace less than 1,500 feet above the surface of the earth or,
   b. Prohibited and restricted areas, other than the restricted areas listed in FAR Part 71.
2. Control Area – Airspace designated as Colored Federal Airways, VOR Federal Airways, Terminal Control Areas, Additional Control Areas, and Control Area Extensions, but not including the Continental Control Area. Unless otherwise designated, control areas also include the airspace between a segment of a main VOR airway and its associated alternate segments. The vertical extent of the various categories of airspace contained in control areas are defined in FAR Part 71.
   *ICAO – Control Area –* A controlled airspace extending upward from a specified limit above the earth.
3. Control Zone – Controlled airspace which extends upward from the surface and terminates at the base of the continental control area. Control zones that do not underlie the continental control area have no upper limit. A control zone may include one or more airports and is normally a circular area within a radius of 5 statute miles and any extensions necessary to include instrument approach and departure paths.
   *ICAO – Control Zone –* A controlled airspace extending upwards from the surface of the earth to a specified upper limit.
4. Terminal Control Area / TCA – Controlled airspace extending upward from the surface or higher to specified altitudes, within which all aircraft are subject to operating rules and pilot and equipment requirements specified in FAR Part 91. Graphics of all TCAs, their airspace limitations and associated communications frequencies are shown in AIM Part 4 and DOD FLIP. TCA's are depicted on Sectional, World Aeronautical, En Route Low Altitude and TCA charts. (Refer to FAR Part 91, AIM Parts 1 and 4)
   *ICAO – Terminal Control Area –* A control area normally established at the confluence of ATS routes in the vicinity of one or more major aerodromes.
5. Transition Area – Controlled airspace extending upward from 700 feet or more above the surface of the earth when designated in conjunction with an airport for which an approved instrument approach procedure has been prescribed, or from 1,200 feet or more above the surface of the earth when designated in conjunction with airway route structures or segments. Unless otherwise limited, transition areas terminate at the base of the overlying controlled airspace. Transition areas are designed to contain IFR operations in controlled airspace during portions of the terminal operation and while transiting between the terminal and en route environment.

**CONTROLLER** — (See Air Traffic Control Specialist)

**COORDINATES** — The intersection of lines of reference, usually expressed in degrees/minutes/seconds of latitude and longitude, used to determine position or location.

**COORDINATION FIX** — The fix in relation to which

facilities will handoff, transfer control of an aircraft, or coordinate flight progress data. For terminal facilities it may also serve as a clearance limit for arriving aircraft.

**CORRECTION** — An error has been made in the transmission and the correct version follows.

**COURSE** —
1. The intended direction of flight in the horizontal plane measured in degrees from north.
2. The ILS localizer signal pattern usually specified as front course or back course. (See Bearing, Radial, Instrument Landing Systems)

**CRITICAL ENGINE** — The engine which, upon failure, would most adversely affect the performance or handling qualities of an aircraft.

*CROSS (FIX) AT (ALTITUDE)* — Used by ATC when a specific altitude restriction at a specified fix is required.

*CROSS (FIX) AT OR ABOVE (ALTITUDE)* — Used by ATC when an altitude restriction at a specified fix is required. It does not prohibit the aircraft from crossing the fix at a higher altitude than specified; however, the higher altitude may not be one that will violate a succeeding altitude restriction or altitude assignment. (See Altitude Assignment, Altitude Restriction.) (Refer to AIM Part 1)

*CROSS (FIX) AT OR BELOW (ALTITUDE)* — Used by ATC when a maximum crossing altitude at a specific fix is required. It does not prohibit the aircraft from crossing the fix at a lower altitude; however, it must be at or above the minimum IFR altitude. (See Minimum IFR Altitude, Altitude Restriction) (Refer to FAR Part 91)

**CROSSWIND** —
1. When used concerning the traffic pattern, the word means "crosswind leg." (See Traffic Pattern)
2. When used concerning wind conditions, the word means a wind not parallel to the runway or the path of an aircraft. (See Crosswind Component)

**CROSSWIND COMPONENT** — The wind component measured in knots at 90 degrees to the longitudinal axis of the runway.

*CRUISE* — Used in an ATC clearance to authorize a pilot to conduct flight at any altitude from the minimum IFR altitude up to and including the altitude specified in the clearance. The pilot may level off at any intermediary altitude within this block of airspace. Climb/descent within the block is to be made at the discretion of the pilot. However, once the pilot starts descent and reports leaving an altitude in the block he may not return to that altitude without additional ATC clearance. Further, it is approval for the pilot to proceed to and make an approach at destination airport and can be used in conjunction with:
1. An airport clearance limit at locations with a standard/special instrument approach procedure. The FARs require that if an instrument letdown to an airport is necessary the pilot shall make the letdown in accordance with a standard/special instrument approach procedure for that airport, or
2. An airport clearance limit at locations that are within/below/outside controlled airspace and without a standard/special instrument approach proce-

dure. Such a clearance is NOT AUTHORIZATION for the pilot to descend under IFR conditions below the applicable minimum IFR altitude nor does it imply that ATC is exercising control over aircraft in uncontrolled airspace; however, it provides a means for the aircraft to proceed to destination airport, descend and land in accordance with applicable FARs governing VFR flight operations. Also, this provides search and rescue protection until such time as the IFR flight plan is closed. (See Instrument Approach Procedure)

**CRUISING ALTITUDE / LEVEL** — An altitude or flight level maintained during en route level flight. This is a constant altitude and should not be confused with a cruise clearance. (See Altitude)
*ICAO – CRUISING LEVEL* – A level maintained during a significant portion of a flight.

**DECISION HEIGHT / DH** — With respect to the operation of aircraft means, the height at which a decision must be made, during an ILS or PAR instrument approach, to either continue the approach or to execute a missed approach.
*ICAO – DECISION HEIGHT* – A specified height at which a missed approach must be initiated if the required visual reference to continue the approach to land has not been established.

**DECODER** — The device used to decipher signals received from ATCRBS and SIF transponders to effect their display as select codes. (See Codes, Radar)

**DEFENSE VISUAL FLIGHT RULES / DVFR** — Rules applicable to flights within an ADIZ conducted under the visual flight rules in FAR Part 91. (See Air Defense Identification Zone) (Refer to FAR Part 99)

*DELAY INDEFINITE (REASON IF KNOWN) EXPECT APPROACH/FURTHER CLEARANCE (TIME)* — Used by ATC to inform a pilot when an accurate estimate of the delay time and the reason for the delay cannot immediately be determined; e.g., a disabled aircraft on the runway, terminal or center area saturation, weather below landing minimums. (See Expect Approach Clearance, Expect Further Clearance)

**DEPARTURE CONTROL** — A function of an approach control facility providing air traffic control service for departing IFR and, under certain conditions, VFR aircraft. (See Approach Control) (Refer to AIM Part 1)

**DEPARTURE TIME** — The time an aircraft becomes airborne.

**DEVIATION** —
1. A departure from a current clearance; such as an off course maneuver to avoid weather or turbulence.
2. Where specifically authorized in the FAR's and requested by the pilot ATC may permit pilots to deviate from certain regulations. (Refer to AIM Part 1)

**DF APPROACH PROCEDURE** — Used under emergency conditions where another instrument approach procedure cannot be executed. DF guidance for an instrument approach is given by ATC facilities with DF capability. (See DF Guidance, Direction Finder) (Refer to AIM Parts 1 and 3)

**DF FIX** — The geographical location of an aircraft obtained by one or more direction finders. (See Direction Finder)

**DF GUIDANCE / DF STEER** — Headings provided to aircraft by facilities equipped with direction finding equipment. These headings, if followed, will lead the aircraft to a predetermined point such as the DF station or an airport. DF guidance is given to aircraft in distress or to other aircraft which request the service. Practice DF guidance is provided when workload permits. (See Direction Finder, DF Fix) (Refer to AIM Part 1)

**DIRECT** — Straight line flight between two navigational aids, fixes, points or any combination thereof. When used by pilots in describing off-airway routes, points defining direct route segments become compulsory reporting points unless the aircraft is under radar contact.

**DIRECTION FINDER / DF / UDF / VDF / UVDF** — A radio receiver equipped with a directional sensing antenna used to take bearings on a radio transmitter. Specialized radio direction finders are used in aircraft as air navigation aids, others are ground based primarily to obtain a "fix" on a pilot requesting orientation assistance, or to locate downed aircraft. A location "fix" is established by the intersection of two or more bearing lines plotted on a navigational chart using either two separately located Direction Finders to obtain a fix on an aircraft or by a pilot plotting the bearing indications of his DF on two separately located ground based transmitters both of which can be identified on his chart. UDFs receive signals in the ultra high frequency radio broadcast band. VDFs in the very high frequency band, UVDFs in both bands. ATC provides DF service at those air traffic control towers and Flight Service Stations listed in AIM Part 2, and DOD FLIP IFR En Route Supplement. (See DF Guidance, DF Fix)

**DISCRETE CODE / DISCRETE BEACON CODE** — As used in the Air Traffic Control Radar Beacon System (ATCRBS), any one of the 4096 selectable Mode 3/A aircraft transponder codes except those ending in zero zero; e.g., discrete codes: 0010, 1201, 2317, 7777; non-discrete codes: 0100, 1200, 7700. Non-discrete codes are normally reserved for radar facilities that are not equipped with discrete decoding capability and for other purposes such as emergencies (7700), VFR aircraft (1200), etc. (See Radar) (Refer to AIM Part 1)

**DISCRETE FREQUENCY** — A separate radio frequency for use in direct pilot-controller communications in air traffic control which reduces frequency congestion by controlling the number of aircraft operating on a particular frequency at one time. Discrete frequencies are normally designated for each control sector in en route/terminal ATC facilities. Discrete frequencies are listed in the Airport/Facilities Directory, AIM Part 3, and DOD FLIP IFR En Route Supplement. (See Control Sector)

**DISPLACED THRESHOLD** — A threshold that is located at a point on the runway other than the designated beginning of the runway. (See Threshold) (Refer to AIM Part 1)

**DISTANCE MEASURING EQUIPMENT / DME** — Equipment (airborne and ground) used to measure, in nautical miles, the slant range distance of an aircraft from the DME navigation aid. (See TACAN, VORTAC)

**DME FIX** — A geographical position determined by reference to a navigational aid which provides distance and azimuth information. It is defined by a specific distance in nautical miles and a radial or course (i.e., localizer) in degrees magnetic from that aid. (See Distance Measuring Equipment, Fix)

**DME SEPARATION** — Spacing of aircraft in terms of distances (nautical miles) determined by reference to distance measuring equipment (DME) (See Distance Measuring Equipment).

**DOD FLIP** — Department of Defense Flight Information Publications used for flight planning, en route, and terminal operations. FLIP is produced by the Defense Mapping Agency for world-wide use. United States Government Flight Information Publications (en route charts and instrument approach procedure charts) are incorporated in DOD FLIP for use in the National Airspace System (NAS).

**DOWNWIND LEG** — (See Traffic Pattern)

**DRAG CHUTE** — A parachute device installed on certain aircraft which is deployed on landing roll to assist in deceleration of the aircraft.

**DUTY PRIORITY** — The order of providing various services by air traffic controllers and flight service specialists.
1. Air Traffic Controller Duty Priorities:
   a. Give first priority to separation of aircraft as required in the Controllers Handbook and to the issuance of radar safety advisories.
   b. Give second priority to other services that are required but do not involve separation of aircraft.
   c. Give third priority to additional service to the extent possible.
   d. Priority for the handling of emergencies cannot be prescribed because of the infinite variety of possible situations which may occur. (See Air Traffic Control Specialist, Radar Safety Advisory, Additional Services)
2. Flight Service Specialist Duty Priorities:
   a. Emergency or urgent actions when life or property is in immediate danger.
   b. Actions required by indications of NAVAID malfunctioning.
   c. Services to airborne aircraft.
   d. Weather observations and PIREPs.
   e. Preflight pilot briefings.
   f. Unscheduled broadcasts.
   g. Teletypewriter duties.
   h. Transcribed Weather Broadcasts and Pilot Automatic Telephone Weather Answering Service.
   i. Scheduled broadcasts.
   (See En Route Flight Advisory Service, Airport Advisory Service, Flight Service Station)

**EMERGENCY LOCATOR TRANSMITTER / ELT** — A radio transmitter attached to the aircraft structure which operates from its own power source on 121.5 MHz and 243.0

MHz. It aids in locating downed aircraft by radiating a downward sweeping audio tone, 2–4 times per second. It is designed to function without human action after an accident. (Refer to FAR Part 91, AIM Part 1)

**EN ROUTE AIR TRAFFIC CONTROL SERVICES —** Air traffic control service provided aircraft on an IFR flight plan, generally by centers, when these aircraft are operating between departure and destination terminal areas. When equipment capabilities and controller workload permit, certain advisory/assistance services may be provided to VFR aircraft. (See NAS Stage A, Air Route Traffic Control Center) (Refer to AIM Part 1)

**EN ROUTE CHARTS —** (See Aeronautical Charts)

**EN ROUTE DESCENT —** Descent from the en route cruising altitude which takes place along the route of flight.

**EN ROUTE FLIGHT ADVISORY SERVICE / FLIGHT WATCH —** A service specifically designed to provide, upon pilot request, timely weather information pertinent to his type of flight, intended route of flight and altitude. The FSSs providing this service are listed in AIM Part 3. (See Flight Watch) (Refer to AIM Parts 1 and 3)

**EXECUTE MISSED APPROACH —** Instructions issued to a pilot making an instrument approach which means continue inbound to the missed approach point and execute the missed approach procedure as described on the Instrument Approach Procedure Chart, or as previously assigned by ATC. The pilot may climb immediately to the altitude specified in the missed approach procedure upon making a missed approach. No turns should be initiated prior to reaching the missed approach point. When conducting an ASR or PAR approach, execute the assigned missed approach procedure immediately upon receiving instructions to "execute missed approach." (Refer to AIM Part 1)

**EXPECT (ALTITUDE) AT (TIME) or (FIX) —** Used to inform a pilot when to expect the specified altitude. In the event of a two-way radio communication failure before receiving an en route altitude assignment within the highest route structure filed, the aircraft should begin climb to the expected altitude at the time or fix specified in the clearance.

**EXPECT APPROACH CLEARANCE (TIME) / EAC —** The time at which it is expected that an arriving aircraft will be cleared to commence an approach for landing. It is issued when the aircraft clearance limit is a designated Initial, Intermediate, or Final Approach Fix for the approach in use and the aircraft is to be held. If delay is anticipated, the pilot should be advised of his EAC at least 5 minutes before the aircraft is estimated to reach the clearance limit.

**EXPECT DEPARTURE CLEARANCE (TIME) / EDCT —** Used in Fuel Advisory Departure (FAD) program. The time the operator can expect a gate release. Excluding long distance flight, an EDCT will always be assigned even though it may be the same as the Estimated Time of Departure (ETD). The EDCT is calculated by adding the ground delay factor. (See Fuel Advisory Departure)

**EXPECT FURTHER CLEARANCE (TIME) / EFC —** The time at which it is expected that additional clearance will be issued to an aircraft. It is issued when the aircraft clearance limit is a fix not designated as part of the approach procedure to be executed and the aircraft will be held. If delay is anticipated the pilot should be advised of his EFC at least 5 minutes before the aircraft is estimated to reach the clearance limit.

**EXPECT FURTHER CLEARANCE VIA (AIRWAYS, ROUTES OR FIXES) —** Used to inform a pilot of the routing he can expect if any part of the route beyond a short range clearance limit differs from that filed.

**FAST FILE —** A system whereby a pilot files a flight plan via telephone that is tape recorded and then transcribed for transmission to the appropriate air traffic facility. Locations having a fast file capability are contained in the AIM Part 3. (Refer to AIM Part 1 and 3)

**FEATHERED PROPELLER —** A propeller whose blades have been rotated so that the leading and trailing edges are nearly parallel with the aircraft flight path to stop or minimize drag and engine rotation. Normally used to indicate shutdown of a reciprocating or turboprop engine due to malfunction.

**FEEDER ROUTE —** A route depicted on instrument approach procedure charts to designate routes for aircraft to proceed from the en route structure to the initial approach fix (IAF). (See Instrument Approach Procedure)

**FERRY FLIGHT —** A flight for the purpose of:
1. returning an aircraft to base;
2. delivering an aircraft from one location to another; or
3. moving an aircraft to and from a maintenance base.

Ferry flights, under certain conditions may be conducted under terms of a special flight permit.

**FILED —** Normally used in conjunction with flight plans, meaning a flight plan has been submitted to ATC.

**FINAL —** Commonly used to mean that an aircraft is on the final approach course or is aligned with a landing area. (See Final Approach Course, Final Approach – IFR, Traffic Pattern, Segments of an Instrument Approach Procedure)

**FINAL APPROACH – IFR —** The flight path of an aircraft which is inbound to an airport on a final instrument approach course, beginning at the final approach fix or point and extending to the airport or the point where a circle to land maneuver or a missed approach is executed. (See Segments of an Instrument Approach Procedure, Final Approach Fix, Final Approach Course, Final Approach Point)

**ICAO – FINAL APPROACH -** That part of an instrument approach procedure from the time the aircraft has:
1. completed the last procedure turn or base turn, where one is specified, or
2. crossed a specified fix, or
3. intercepted the last track specified for the procedures; until it has crossed a point in the vicinity of an aerodrome from which:
   a. a landing can be made; or
   b. a missed approach procedure is initiated.

**FINAL APPROACH – VFR —** (See Traffic Pattern)

**FINAL APPROACH COURSE —** A straight line extension of a localizer, a final approach radial/bearing, or a runway centerline, all without regard to distance. (See Final Approach – IFR, Traffic Pattern)

**FINAL APPROACH FIX / FAF** — The designated fix from or over which the final approach (IFR) to an airport is executed. The FAF identifies the beginning of the final approach segment of the instrument approach. (See Final Approach Point, Segments of an Instrument Approach Procedure, Glide Slope Intercept Altitude)

**FINAL APPROACH POINT** — The point, within prescribed limits of an instrument approach procedure, where the aircraft is established on the final approach course and final approach descent may be commenced. A final approach point is applicable only in non-precision approaches where a final approach fix has not been established. In such instances, the point identifies the beginning of the final approach segment of the instrument approach. (See Final Approach Fix, Segments of an Instrument Approach Procedure, Glide Slope Intercept Altitude)

**FINAL APPROACH SEGMENT** — (See Segments of an Instrument Approach Procedure)

**FINAL CONTROLLER** — The controller providing information and final approach guidance during PAR and ASR approaches utilizing radar equipment. (See Radar Approach)

**FIX** — A geographical position determined by visual reference to the surface, by reference to one or more radio NAVAIDs, by celestial plotting, or by another navigational device.

**FIXED-WING SPECIAL IFR** — Aircraft operating in accordance with a waiver and a Letter of Agreement within control zones specified in FAR 93.113. These operations are conducted by IFR qualified pilots in IFR equipped aircraft and by pilots of agricultural and industrial aircraft.

**FLAG / FLAG ALARM** — A warning device incorporated in certain airborne navigation and flight instruments indicating that:
1. instruments are inoperative or otherwise not operating satisfactorily, or
2. signal strength or quality of the received signal falls below acceptable values.

**FLAMEOUT** — Unintended loss of combustion in turbine engines resulting in the loss of engine power.

**FLIGHT INFORMATION REGION / FIR** — An airspace of defined dimensions within which Flight Information Service and Alerting Service are provided.
1. Flight Information Service – A service provided for the purpose of giving advice and information useful for the safe and efficient conduct of flights.
2. Alerting Service – A service provided to notify appropriate organizations regarding aircraft in need of search and rescue aid, and assist such organizations as required.

**FLIGHT INSPECTION / FLIGHT CHECK** — Inflight investigation and evaluation of a navigational aid to determine whether it meets established tolerances. (See Navigational Aid)

*FLIGHT LEVEL* — A level of constant atmospheric pressure related to a reference datum of 29.92 inches of mercury.

Each is stated in three digits that represent hundreds of feet. For example, flight level 250 represents a barometric altimeter indication of 25,000 feet; flight level 255, an indication of 25,500 feet.
*ICAO – FLIGHT LEVELS* – Surfaces of constant atmospheric pressure which are related to a specific pressure datum, 1013.2 mb (29.92 inches) and are separated by specific pressure intervals.

**FLIGHT PATH** — A line, course, or track along which an aircraft is flying or intended to be flown. (See Track, Course)

**FLIGHT PLAN** — Specified information relating to the intended flight of an aircraft, that is filed orally or in writing with an FSS or an ATC facility. (See Fast File, Filed) (Refer to AIM Part 1)

**FLIGHT RECORDER** — A general term applied to any instrument or device that records information about the performance of an aircraft in flight, or about conditions encountered in flight. Flight recorders may make records of airspeed, outside air temperature, vertical acceleration, engine RPM, manifold pressure, and other pertinent variables for a given flight.
*ICAO – FLIGHT RECORDER* -- Any type of recorder installed in the aircraft for the purpose of complementing accident/incident investigation.

**FLIGHT SERVICE STATION / FSS** — Air Traffic Service facilities within the National Airspace System (NAS) which provide preflight pilot briefing and en route communications with VFR flights, assist lost IFR/VFR aircraft, assist aircraft having emergencies, relay ATC clearances, originate, classify, and disseminate Notices to Airmen, broadcast aviation weather and NAS information, receive and close flight plans, monitor radio NAVAIDS, notify search and rescue units of missing VFR aircraft, and operate the national weather teletypewriter systems. In addition, at selected locations, FSSs take weather observations, issue airport advisories, administer airman written examinations, and advise Customs and Immigration of transborder flight. (Refer to AIM Part 1)

**FLIGHT STANDARDS DISTRICT OFFICE / FSDO** — An FAA field office serving an assigned geographical area, staffed with Flight Standards personnel, who serve the aviation industry and the general public on matters relating to the certification and operation of air carrier and general aviation aircraft. Activities include general surveillance of operational safety, certification of airmen and aircraft, accident prevention, investigation, enforcement, etc.

**FLIGHT TEST** — A flight for the purpose of:
1. Investigating the operation/flight characteristics of an aircraft or aircraft component.
2. Evaluating an applicant for a pilot certificate or rating.

**FLIGHT VISIBILITY** — (See Visibility)

**FLIGHT WATCH** — A shortened term for use in air-ground contacts on frequency 122.0 MHz to identify the flight service station providing En Route Flight Advisory Service; e.g., "Oakland Flight Watch." (See En Route Flight Advisory Service)

**FLIP** — (See DOD FLIP)

**FLOW CONTROL** — Measures designed to adjust the flow of traffic into a given airspace, along a given route, or bound for a given aerodrome (airport) so as to ensure the most effective utilization of the airspace. (See Quota Flow Control) (Refer to AIM Part 3)

*FLY HEADING (DEGREES)* — Informs the pilot of the heading he should fly. The pilot may have to turn to, or continue on, a specific compass direction in order to comply with the instructions. The pilot is expected to turn in the shorter direction to the heading, unless otherwise instructed by ATC.

**FORMATION FLIGHT** — More than one aircraft which, by prior arrangement between the pilots, operate as a single aircraft with regard to navigation and position reporting. Separation between aircraft within the formation is the responsibility of the flight leader and the pilots of the other aircraft in the flight. This includes transition periods when aircraft within the formation are maneuvering to attain separation from each other to effect individual control and during join-up and breakaway.

1. A standard formation is one in which a proximity of more than 1 mile laterally or longitudinally and within 100 feet vertically from the flight leader is maintained by each wingman.
2. Nonstandard formations are those operating under any of the following conditions:
   a. When the flight leader has requested and ATC has approved other than standard formation dimensions.
   b. When operating within an authorized altitude reservation (ALTRV) or under the provisions of a Letter of Agreement.
   c. When the operations are conducted in airspace specifically designed for a special activity. (See Altitude Reservation) (Refer to FAR Part 91)

**FSS** — (See Flight Service Station)

**FUEL ADVISORY DEPARTURE / FAD** — Procedures to minimize engine running time for aircraft destined for an airport experiencing prolonged arrival delays. (Refer to AIM Part 1)

**FUEL DUMPING** — Airborne release of usable fuel. This does not include the dropping of fuel tanks. (See Jettisoning of External Stores)

**FUEL SIPHONING / FUEL VENTING** — Unintentional release of fuel caused by overflow, puncture, loose cap, etc.

**GATE HOLD PROCEDURES** — Procedures at selected airports to hold aircraft at the gate or other ground location whenever departure delays exceed or are anticipated to exceed 5 minutes. The sequence for departure will be maintained in accordance with initial call up unless modified by Flow Control restrictions. Pilots should monitor the ground control/clearance delivery frequency for engine startup advisories or new proposed start time if the delay changes. (See Flow Control)

**GENERAL AVIATION** — That portion of civil aviation which encompasses all facets of aviation except air carriers holding a certificate of public convenience and necessity from the Civil Aeronautics Board, and large aircraft commercial operators.
*ICAO – GENERAL AVIATION* – All civil aviation operations other than scheduled air services and nonscheduled air transport operations for remuneration or hire.

**GENERAL AVIATION DISTRICT OFFICE / GADO** — An FAA field office serving a designated geographical area, staffed with Flight Standards personnel, who have responsibility for serving the aviation industry and the general public on all matters relating to the certification and operation of general aviation aircraft.

*GLIDE PATH, (ON/ABOVE/BELOW)* — Used by ATC to inform an aircraft making a PAR approach of its vertical position (elevation) relative to the descent profile. The terms "slightly" and "well" are used to describe the degree of deviation; e.g., "slightly above glidepath." Trend information is also issued with respect to the elevation of the aircraft and may be modified by the terms "rapidly" and "slowly;" e.g., "well above glidepath, coming down rapidly." (See PAR Approach)

**GLIDE SLOPE / GS** — Provides vertical guidance for aircraft during approach and landing. The glide slope consists of the following:
1. Electronic components emitting signals which provide vertical guidance by reference to airborne instruments during instrument approaches such as ILS; or
2. Visual ground aids such as VASI which provides vertical guidance for VFR approach or for the visual portion of an instrument approach and landing.
*ICAO – GLIDE PATH* - A descent profile determined for vertical guidance during a final approach.

**GLIDE SLOPE INTERCEPT ALTITUDE** — The minimum altitude of the intermediate approach segment prescribed for a precision approach which assures required obstacle clearance. It is depicted on instrument approach procedure charts. (See Segments of an Instrument Approach Procedure, Instrument Landing System)

*GO AHEAD* — Proceed with your message. Not to be used for any other purpose.

*GO AROUND* — Instructions for a pilot to abandon his approach to landing. Additional instructions may follow. Unless otherwise advised by ATC, a VFR aircraft or an aircraft conducting visual approach should overfly the runway while climbing to traffic pattern altitude and enter the traffic pattern via the crosswind leg. A pilot on an IFR flight plan making an instrument approach should execute the published missed approach procedure or proceed as instructed by ATC; e.g., "Go Around" (additional instructions, if required). (See Low Approach, Missed Approach)

**GROUND CLUTTER** — A pattern produced on the radar scope by ground returns which may degrade other radar returns in the affected area. The effect of ground clutter is minimized by the use of moving target indicator (MTI) circuits in the radar equipment resulting in a radar presentation which displays only targets which are in motion. (See Clutter)

**GROUND CONTROLLED APPROACH / GCA —** A radar approach system operated from the ground by air traffic control personnel transmitting instructions to the pilot by radio. The approach may be conducted with surveillance radar (ASR) only or with both surveillance and precision approach radar (PAR). Usage of the term "GCA" by pilots is discouraged except when referring to a GCA facility. Pilots should specifically request a "PAR" approach when a precision radar approach is desired or request an "ASR" or "surveillance" approach when a nonprecision radar approach is desired. (See Radar Approach)

**GROUND SPEED —** The speed of an aircraft relative to the surface of the earth.

**GROUND VISIBILITY —** (See Visibility)

**HANDOFF —** Transfer of radar identification of an aircraft from one controller to another, either within the same facility or interfacility. Actual transfer of control responsibility may occur at the time of the handoff, or at a specified time, point or altitude.

*HAVE NUMBERS —* Used by pilots to inform ATC that they have received runway and wind information only.

**HEAVY (AIRCRAFT) —** (See Aircraft Classes)

**HEIGHT ABOVE AIRPORT / HAA —** The height of the Minimum Descent Altitude above the published airport elevation. This is published in conjunction with circling minimums. (See Minimum Descent Altitude)

**HEIGHT ABOVE LANDING / HAL —** The height above a designated helicopter landing area used for helicopter instrument approach procedures. (Refer to FAR Part 97)

**HEIGHT ABOVE TOUCHDOWN / HAT —** The height of the Decision Height or Minimum Descent Altitude above the highest runway elevation in the touchdown zone (first 3,000 feet of the runway). HAT is published on instrument approach charts in conjunction with all straight-in minimums. (See Decision Height, Minimum Descent Altitude)

**HELICOPTER / COPTER —** Rotorcraft that, for its horizontal motion, depends principally on its engine-driven rotors.
*ICAO – HELICOPTER –* A heavier-than-air aircraft supported in flight by the reactions of the air on one or more power-driven rotors on substantially vertical axes.

**HELIPAD —** That part of the landing and takeoff area designed for helicopters.

**HELIPORT —** An area of land, water, or structure used or intended to be used for the landing and takeoff of helicopters.

**HERTZ / Hz —** The standard radio equivalent of frequency in cycles per second of an electromagnetic wave. Kilohertz (kHz) is a frequency of one thousand cycles per second. Megahertz (MHz) is a frequency of one million cycles per second.

**HIGH FREQUENCY / HF —** The frequency band between 3 and 30 MHz. (See High Frequency Communications)

**HIGH FREQUENCY COMMUNICATIONS / HF COMMUNICATIONS —** High radio frequencies (HF) between 3 and 30 MHz used for air-to-ground voice communication in overseas operations.

**HIGH SPEED TAXIWAY / EXIT / TURNOFF —** A long radius taxiway designed and provided with lighting or marking to define the path of aircraft, traveling at high speed (up to 60 knots), from the runway center to a point on the center of a taxiway. Also referred to as long radius exit or turn-off taxiway. The high speed taxiway is designed to expedite aircraft turning off the runway after landing, thus reducing runway occupancy time.

**HOLD / HOLDING PROCEDURE —** A predetermined maneuver which keeps aircraft within a specified airspace while awaiting further clearance from air traffic control. Also used during ground operations to keep aircraft within a specified area or at a specified point while awaiting further clearance from air traffic control. (See Holding Fix) (Refer to AIM Part 1)

**HOLDING FIX —** A specified fix identifiable to a pilot by NAVAIDS or visual reference to the ground used as a reference point in establishing and maintaining the position of an aircraft while holding. (See Fix, Hold, Visual Holding) (Refer to AIM Part 1)
*ICAO – HOLDING POINT -* A specified location, identified by visual or other means, in the vicinity of which the position of an aircraft in flight is maintained in accordance with air traffic control clearances.

**HOMING —** Flight toward a NAVAID, without correcting for wind, by adjusting the aircraft heading to maintain a relative bearing of zero degrees. (See Bearing)
*ICAO – HOMING –* The procedure of using the direction-finding equipment of one radio station with the emission of another radio station, where at least one of the stations is mobile, and whereby the mobile station proceeds continuously towards the other station.

*HOW DO YOU HEAR ME? —* A question relating to the quality of the transmission or to determine how well the transmission is being received.

*I SAY AGAIN —* The message will be repeated.

**IDENT —** A request for a pilot to activate the aircraft transponder identification feature. This will help the controller to confirm an aircraft identity or to identify an aircraft. (Refer to AIM Part 1)

**IDENT FEATURE —** The special feature in the Air Traffic Control Radar Beacon System (ATCRBS) equipment and the "I/P" feature in certain Selective Identification Feature (SIF) equipment. It is used to immediately distinguish one displayed beacon target from other beacon targets. (See IDENT)

*IF FEASIBLE, REDUCE SPEED TO (SPEED) —* (See Speed Adjustment)

**IF NO TRANSMISSION RECEIVED FOR (TIME)** — Used by ATC in radar approaches to prefix procedures which should be followed by the pilot in event of lost communications. (See Lost Communications)

**IFR AIRCRAFT / IFR FLIGHT** — An aircraft conducting flight in accordance with instrument flight rules.

**IFR CONDITIONS** — Weather conditions below the minimum for flight under visual flight rules. (See Instrument Meteorological Conditions)

**IFR OVER THE TOP** — The operation of an aircraft over the top on an IFR flight plan when cleared by air traffic control to maintain "VFR conditions" or "VFR conditions on top." (See VFR on Top)

**IFR TAKEOFF MINIMUMS AND DEPARTURE PROCEDURES** — FAR, Part 91, prescribes standard takeoff rules for certain civil users. At some airports, obstructions or other factors require the establishment of nonstandard takeoff minimums, departure procedures, or both, to assist pilots in avoiding obstacles during climb to the minimum en route altitude. Those airports are listed in NOS/DOD Instrument Approach Charts (IAPs) under a section entitled "IFR Takeoff Minimums and Departure Procedures." The NOS/DOD IAP chart legend illustrates the symbol used to alert the pilot to nonstandard takeoff minimums and departure procedures. When departing IFR from such airports, or from any airports where there are no departure procedures, SIDs, or ATC facilities available, pilots should advise ATC of any departure limitations. Controllers may query a pilot to determine acceptable departure directions, turns, or headings after takeoff. Pilots should be familiar with the departure procedures and must assure that their aircraft can meet or exceed any specified climb gradients.

**ILS CATEGORIES —**

1. ILS Category I – An ILS approach procedure which provides for approach to a height above touchdown of not less than 200 feet and with runway visual range of not less than 1800 feet.

2. ILS Category II – An ILS approach procedure which provides for approach to a height above touchdown of not less than 100 feet and with runway visual range of not less than 1200 feet.

3. ILS Category III.
   a. IIIA – An ILS approach procedure which provides for approach without a decision height minimum and with runway visual range of not less than 700 feet.
   b. IIIB – An ILS approach procedure which provides for approach without a decision height minimum and with runway visual range of not less than 150 feet.
   c. IIIC – An ILS approach procedure which provides for approach without a decision height minimum and without runway visual range minimum.

**IMMEDIATELY** — Used by ATC when such action is required to avoid an imminent situation.

**INCREASE SPEED TO (SPEED)** — (See Speed Adjustment)

**INFORMATION REQUEST / INREQ** — A request originated by an FSS for information concerning an overdue VFR aircraft.

**INITIAL APPROACH FIX / IAF** — The fix(s) depicted on instrument approach procedure charts that identifies the beginning of the initial approach segment(s). (See Fix, Segments of an Instrument Approach Procedure)

**INITIAL APPROACH SEGMENT** — (See Segments of an Instrument Approach Procedure)

**INNER MARKER / IM / INNER MARKER BEACON** — A marker beacon used with an ILS (CAT II) precision approach located between the middle marker and the end of the ILS runway, transmitting a radiation pattern keyed at six dots per second and indicating to the pilot, both aurally and visually, that he is at the designated decision height (DH), normally 100 feet above the touchdown zone elevation, on the ILS CAT II approach. It also marks progress during a CAT III approach. (See Instrument Landing System) (Refer to AIM Part 1)

**INSTRUMENT APPROACH PROCEDURE / IAP / INSTRUMENT APPROACH** — A series of predetermined maneuvers for the orderly transfer of an aircraft under instrument flight conditions from the beginning of the initial approach to a landing, or to a point from which a landing may be made visually. It is prescribed and approved for a specific airport by competent authority. (See Segments of an Instrument Approach Procedure) (Refer to FAR Part 91, AIM Part 1)

1. U. S. civil standard instrument approach procedures are approved by the FAA as prescribed under FAR, Part 97, and are available for public use.
2. U.S. military standard instrument approach procedures are approved and published by the Department of Defense.
3. Special instrument approach procedures are approved by the FAA for individual operators, but are not published in FAR, Part 97, for public use.

*ICAO – INSTRUMENT APPROACH PROCEDURE* – A series of predetermined maneuvers for the orderly transfer of an aircraft under instrument flight conditions from the beginning of the initial approach to a landing, or to a point from which a landing may be made visually.

**INSTRUMENT FLIGHT RULES / IFR** — Rules governing the procedures for conducting instrument flight. Also a term used by pilots and controllers to indicate type of flight plan. (See Visual Flight Rules, Instrument Meteorological Conditions, Visual Meteorological Conditions) (Refer to AIM Part 1)

*ICAO – INSTRUMENT FLIGHT RULES* -- A set of rules governing the conduct of flight under instrument meteorological conditions.

**INSTRUMENT LANDING SYSTEM / ILS** — A precision instrument approach system consisting of the following electronic components and visual aids:

1. Localizer (See Localizer)
2. Glide Slope (See Glide Slope)
3. Outer Marker (See Outer Marker)
4. Middle Marker (See Middle Marker)
5. Approach Lights (See Airport Lighting)

(Refer to FAR Part 91, AIM Part 1)

**INSTRUMENT METEOROLOGICAL CONDITIONS / IMC** — Meteorological conditions expressed in terms of visibility, distance from cloud, and ceiling less than the minima specified for visual meteorological conditions. (See Visual Meteorological Conditions, Instrument Flight Rules, Visual Flight Rules)

**INSTRUMENT RUNWAY** — A runway equipped with electronic and visual navigation aids for which a precision or nonprecision approach procedure having straight-in landing minimums has been approved.

*ICAO – INSTRUMENT RUNWAY* – A runway intended for the operation of aircraft using nonvisual aids and comprising:

1. Instrument Approach Runway – An instrument runway served by a nonvisual aid providing at least directional guidance adequate for a straight-in approach.
2. Precision Approach Runway, Category I – An instrument runway served by ILS or GCA approach aids and visual aids intended for operations down to 60 metres (200 feet) decision height and down to an RVR of the order of 800 metres (2600 feet.)
3. Precision Approach Runway, Category II – An instrument runway served by ILS and visual aids intended for operations down to 30 metres (100 feet) decision height and down to an RVR of the order of 400 metres (1200 feet.)
4. Precision Approach Runway, Category III – An instrument runway served by ILS (no decision height being applicable) and:
   a. By visual aids intended for operations down to an RVR of the order of 200 metres (700 feet);
   b. By visual aids intended for operations down to an RVR of the order of 50 metres (150 feet);
   c. Intended for operations without reliance on external visual reference.

**INTERMEDIATE APPROACH SEGMENT** — (See Segments of an Instrument Approach Procedure)

**INTERMEDIATE FIX / IF** — The fix that identifies the beginning of the intermediate approach segment of an instrument approach procedure. The fix is not normally identified on the instrument approach chart as an intermediate fix (IF). (See Segments of an Instrument Approach Procedure)

**INTERNATIONAL AIRPORT** — Relating to international flights, it means:

1. An airport of entry which has been designated by the Secretary of Treasury or Commissioner of Customs as an international airport for customs service,
2. A landing rights airport at which specific permission to land must be obtained from customs authorities in advance of contemplated use,
3. Airports designated under the Convention on International Civil Aviation as an airport for use by international commercial air transport and/or international general aviation. (Refer to AIM Part 2)

*ICAO – INTERNATIONAL AIRPORT* – Any airport designated by the Contracting State in whose territory it is situated as an airport of entry and departure for international air traffic, where the formalities incident to customs, immigration, public health, animal and plant quarantine and similar procedures are carried out.

**INTERNATIONAL CIVIL AVIATION ORGANIZA-TION / ICAO** — A specialized agency of the United Nations whose objective is to develop the principles and techniques of international air navigation and to foster planning and development of international civil air transport.

**INTERROGATOR** — The ground-based surveillance radar beacon transmitter-receiver which normally scans in synchronism with a primary radar, transmitting discrete radio signals which repetitiously requests all transponders, on the mode being used, to reply. The replies received are mixed with the primary radar returns and displayed on the same plan position indicator (radar scope). Also applied to the airborne element of the TACAN/DME system. (See Transponder) (Refer to AIM Part 1)

**INTERSECTING RUNWAYS** — Two or more runways which cross or meet within their lengths. (See Intersection)

**INTERSECTION** —
1. A point defined by any combination of courses, radials or bearings of two or more navigational aids.
2. Used to describe the point where two runways cross, a taxiway and a runway cross or, two taxiways cross.

**INTERSECTION DEPARTURE / INTERSECTION TAKEOFF** — A takeoff or proposed takeoff on a runway from an intersection. (See Intersection)

**JAMMING** — Electronic or mechanical interference which may disrupt the display of aircraft on radar or the transmission/reception of radio communications/navigation.

**JET BLAST** — Jet engine exhaust (thrust stream turbulence). (See Wake Turbulence)

**JET ROUTE** — A route designed to serve aircraft operations from 18,000 feet MSL up to and including flight level 450. The routes are referred to as "J" routes with numbering to identify the designated route; e.g., J 105. (See Route) (Refer to FAR Part 71)

**JET STREAM** — A migrating stream of high-speed winds present at high altitudes.

**JETTISONING OF EXTERNAL STORES** — Airborne release of external stores; e.g., tiptanks, ordnances. (See Fuel Dumping) (Refer to FAR Part 91)

**JOINT USE RESTRICTED AREA** — (See Restricted Area)

**JUMBO (JET)** — DC-10, B-747, L-1011, C-5A and future similar or larger aircraft. (See Aircraft Classes)

**KNOWN TRAFFIC** — With respect to ATC clearances, means aircraft whose altitude, position and intentions are known to ATC.

**LANDING AREA** — Any locality either of land or water, including airports and intermediate landing fields, which is used, or intended to be used, for the landing and takeoff of aircraft, whether or not facilities are provided for the shelter, servicing, or repair of aircraft, or for receiving or discharging passengers or cargo.

*ICAO – LANDING AREA* – That part of the movement area intended for the landing and takeoff of aircraft.

**LANDING DIRECTION INDICATOR** — A device which visually indicates the direction in which landings and takeoffs should be made. (See Tetrahedron) (Refer to AIM Part 1)

**LANDING MINIMUMS / IFR LANDING MINIMUMS** — The minimum visibility prescribed for landing a civil aircraft while using an instrument approach procedure. The minimum applies with other limitations set forth in FAR Part 91, with respect to the Minimum Descent Altitude (MDA) or Decision Height (DH) prescribed in the instrument approach procedures as follows:

1. Straight-in landing minimums – A statement of MDA and visibility, or DH and visibility, required for straight-in landing on a specified runway; or
2. Circling minimums – A statement of MDA and visibility required for the circle-to-land maneuver.

Descent below the established MDA or DH is not authorized during an approach unless the aircraft is in a position from which a normal approach to the runway of intended landing can be made, and adequate visual reference to required visual cues is maintained. (See Straight-in Landing, Circle-to-Land Maneuver, Decision Height, Minimum Descent Altitude, Visibility, Instrument Approach Procedure) (Refer to FAR Part 91)

**LANDING ROLL** — The distance from the point of touchdown to the point where the aircraft can be brought to a stop, or exit the runway.

**LANDING SEQUENCE** — The order in which aircraft are positioned for landing. (See Approach Sequence)

**LAST ASSIGNED ALTITUDE** — The last altitude/flight level assigned by ATC and acknowledged by the pilot. (See Maintain) (Refer to FAR Part 91)

**LATERAL SEPARATION** — The lateral spacing of aircraft at the same altitude by requiring operation on different routes or in different geographical locations. (See Separation)

**LIGHT GUN** — A handheld directional light signaling device which emits a brilliant narrow beam of white, green, or red light as selected by the tower controller. The color and type of light transmitted can be used to approve or disapprove anticipated pilot actions where radio communication is not available. The light gun is used for controlling traffic operating in the vicinity of the airport and on the airport movement area. (Refer to AIM Part 1)

**LIGHTED AIRPORT** — An airport where runway and obstruction lighting is available. (See Airport Lighting) (Refer to AIM Parts 1 and 2)

**LIMITED REMOTE COMMUNICATIONS OUTLET / LRCO** — An unmanned satellite air/ground communications facility which may be associated with a VOR. These outlets effectively extend the service range of the FSS and provide greater communications reliability. LRCOs are depicted on En Route Charts. (See Remote Communications Outlet)

**LOCAL TRAFFIC** — Aircraft operating in the traffic pattern or within sight of the tower, or aircraft known to be departing or arriving from flight in local practice areas, or, aircraft executing practice instrument approaches at the airport. (See Traffic Pattern)

**LOCALIZER** — The component of an ILS which provides course guidance to the runway. (See Instrument Landing System) (Refer to AIM Part 1)
*ICAO – LOCALIZER COURSE (ILS)* - The locus of points, in any given horizontal plane, at which the DDM (difference in depth of modulation) is zero.

**LOCALIZER TYPE DIRECTIONAL AID / LDA** — A NAVAID, used for nonprecision instrument approaches with utility and accuracy comparable to a localizer but which is not a part of a complete ILS and is not aligned with the runway. (Refer to AIM Part 1)

**LOCALIZER USABLE DISTANCE** — The maximum distance from the localizer transmitter at a specified altitude, as verified by flight inspection, at which reliable course information is continuously received. (Refer to AIM Part 1)

**LONGITUDINAL SEPARATION** — The longitudinal spacing of aircraft at the same altitude by a minimum distance expressed in units of time or miles. (See Separation) (Refer to AIM Part 1)

**LORAN / LONG RANGE NAVIGATION** — An electronic navigational system by which hyperbolic lines of position are determined by measuring the difference in the time of reception of synchronized pulse signals from two fixed transmitters. Loran A operates in the 1750 – 1950 kHz frequency band. Loran C and D operate in the 100 – 110 kHz frequency band. (Refer to AIM Part 1)

**LOST COMMUNICATIONS / TWO-WAY RADIO COMMUNICATIONS FAILURE** — Loss of the ability to communicate by radio. Aircraft are sometimes referred to as NORDO (No Radio). Standard pilot procedures are specified in FAR Part 91. Radar controllers issue procedures for pilots to follow in the event of lost communications during a radar approach, when weather reports indicate that an aircraft will likely encounter IFR weather conditions during the approach. (Refer to FAR Part 91, AIM Part 1)

**LOW ALTITUDE AIRWAY STRUCTURE / FEDERAL AIRWAYS** — The network of airways serving aircraft operations up to but not including 18,000 feet MSL. (See Airway) (Refer to AIM Part 1)

*LOW ALTITUDE ALERT, ADVISE YOU CLIMB IMMEDIATELY TO (ALTITUDE)* — (See Radar Safety Advisory)

**LOW APPROACH** — An approach over an airport or runway following an instrument approach or a VFR approach including the go-around maneuver where the pilot intentionally does not make contact with the runway. (Refer to AIM Part 1)

**LOW FREQUENCY / LF** — The frequency band between 30 and 300 kHz. (Refer to AIM Part 1)

**MACH NUMBER** — The ratio of true airspeed to the speed of sound, e.g., MACH .82, MACH 1.6. (See Airspeed)

*MAINTAIN* —
1. Concerning altitude/flight level, the term means to remain at the altitude/flight level specified. The phrase "climb and" or "descend and" normally precede "maintain" and the altitude assignment; e.g., "descend and maintain 5000." If a SID procedure is assigned in the initial or subsequent clearance, the altitude restrictions in the SID, if any, will apply unless otherwise advised by ATC.
2. Concerning other ATC instructions, the term is used in its literal sense; e.g., maintain VFR.

*MAKE SHORT APPROACH* — Used by ATC to inform a pilot to alter his traffic pattern so as to make a short final approach. (See Traffic Pattern)

**MARKER BEACON** — An electronic navigation facility transmitting a 75 MHz vertical fan or boneshaped radiation pattern. Marker beacons are identified by their modulation frequency and keying code and when received by compatible airborne equipment, indicate to the pilot, both aurally and visually, that he is passing over the facility. (See Outer Marker, Middle Marker, Inner Marker) (Refer to AIM Part 1)

**MAXIMUM AUTHORIZED ALTITUDE / MAA** — A published altitude representing the maximum usable altitude or flight level for an airspace structure or route segment. It is the highest altitude on a Federal airway, Jet route, area navigation low or high route, or other direct route for which an MEA is designated in FAR Part 95, at which adequate reception of navigation and signals is assured.

*MAYDAY* — The international radiotelephony distress signal. When repeated three times, it indicates imminent and grave danger and that immediate assistance is requested. (See PAN) (Refer to AIM Part 1)

**MICROWAVE LANDING SYSTEM / MLS** — An instrument landing system operating in the microwave spectrum which provides lateral and vertical guidance to aircraft having compatible avionics equipment. (See Instrument Landing System) (Refer to AIM Part 1)

**MID-FIELD RVR** — (See Visibility)

**MIDDLE COMPASS LOCATOR** — (See Compass Locator)

**MIDDLE MARKER / MM** — A marker beacon that defines a point along the glide slope of an ILS normally located at or near the point of decision height (ILS Category I). It is keyed to transmit alternate dots and dashes, two per second, on a 1300 HZ tone which is received aurally and visually by compatible airborne equipment. (See Marker Beacon, Instrument Landing System) (Refer to AIM Part 1)

**MILITARY AUTHORITY ASSUMES RESPONSIBILITY FOR SEPARATION OF AIRCRAFT / MARSA** — A condition whereby the military services involved assume responsibility for separation between participating military aircraft in the ATC system. It is used only for required IFR operations which are specified in Letters of Agreement or other appropriate FAA or military documents.

**MILITARY OPERATIONS AREA / MOA** — An airspace area established for the purpose of segregating certain military training activities from airspace containing IFR aircraft. Nonparticipating IFR traffic may be cleared through an active MOA if IFR separation can be provided by ATC. VFR pilots should exercise caution while flying within an active MOA. These areas are depicted on sectional, VFR terminal and low altitude en route charts. (Refer to AIM Part 1)

**MINIMUM CROSSING ALTITUDE / MCA** — The lowest altitude at certain fixes at which an aircraft must cross when preceding in the direction of a higher minimum en route IFR altitude (MEA). (See Minimum En Route IFR Altitude)

**MINIMUM DESCENT ALTITUDE / MDA** — The lowest altitude, expressed in feet above mean sea level, to which descent is authorized on final approach or during circle-to-land maneuvering in execution of a standard instrument approach procedure where no electronic glide slope is provided. (See Nonprecision Approach Procedure)

**MINIMUM EN ROUTE IFR ALTITUDE / MEA** — The lowest published altitude between radio fixes which assures acceptable navigational signal coverage and meets obstacle clearance requirements between those fixes. The MEA prescribed for a Federal airway or segment thereof, area navigation low or high route, or other direct route, applies to the entire width of the airway, segment or route between the radio fixes defining the airway, segment or route. (Refer to FAR Parts 91 and 95, AIM Part 1)

**MINIMUM FUEL** — Indicates that an aircraft's fuel supply has reached a state where, upon reaching the destination, it can accept little or no delay. This is not an emergency situation but merely indicates an emergency situation is possible should any undue delay occur.

**MINIMUM HOLDING ALTITUDE / MHA** — The lowest altitude prescribed for a holding pattern which assures navigational signal coverage, communications, and meets obstacle clearance requirements.

**MINIMUM IFR ALTITUDES** — Minimum altitudes for IFR operations as prescribed in FAR Part 91. These altitudes are published on aeronautical charts and prescribed in FAR Part 95, for airways and routes, and FAR Part 97, for standard instrument approach procedures. If no applicable minimum altitude is prescribed in FAR Parts 95 or 97, the following minimum IFR altitude applies:
1. In designated mountainous areas, 2000 feet above the highest obstacle within a horizontal distance of 5 statute miles from the course to be flown; or
2. Other than mountainous areas, 1000 feet above the highest obstacle within a horizontal distance of 5 statute miles from the course to be flown; or
3. As otherwise authorized by the Administrator or assigned by ATC. (See Minimum En Route IFR Altitude, Minimum Obstruction Clearance Altitude, Minimum Crossing Altitude, Minimum Safe Altitude, Minimum Vectoring Altitude) (Refer to FAR Part 91)

**MINIMUM OBSTRUCTION CLEARANCE ALTITUDE / MOCA** — The lowest published altitude in effect between radio fixes on VOR airways, off-airway routes, or route segments which meets obstacle clearance requirements for the entire route segment and which assures acceptable navigational signal coverage only within 22 nautical miles of a VOR. (Refer to FAR Part 91 and 95)

**MINIMUM RECEPTION ALTITUDE / MRA** — The lowest altitude at which an intersection can be determined. (Refer to FAR Part 95)

**MINIMUM SAFE ALTITUDE / MSA** —

1. The minimum altitudes specified in FAR Part 91, for various aircraft operations.
2. Altitudes depicted on instrument approach charts and identified as minimum sector altitudes or emergency safe altitudes which provide a minimim of 1000 feet obstacle clearance within a specified distance from the navigation facility upon which an instrument approach procedure is predicated. These altitudes are for EMERGENCY USE ONLY and do not necessarily guarantee NAVAID reception. Minimum sector altitudes are established for all procedures (except localizers without an NDB) within a 25 nautical mile radius of the navigation facility. Emergency safe altitudes are established for some military procedures within a 100 nautical mile radius of the navigation facility.

*ICAO – MINIMUM SECTOR ALTITUDE* - The lowest altitude which may be used under emergency conditions which will provide a minimum clearance of 300 meters (1000 feet) above all obstacles located in an area contained within a sector of a circle of 25 nautical miles radius centred on a radio aid to navigation.

**MINIMUM VECTORING ALTITUDE / MVA** — The lowest MSL altitude at which an IFR aircraft will be vectored by a radar controller, except as otherwise authorized for radar approaches, departures and missed approaches. The altitude meets IFR obstacle clearance criteria. It may be lower than the published MEA along an airway or J-route segment. It may be utilized for radar vectoring only upon the controllers' determination that an adequate radar return is being received from the aircraft being controlled. Charts depicting minimum vectoring altitudes are normally available only to the controllers and not to pilots. (Refer to AIM Part 1)

**MINIMUMS / MINIMA** — Weather condition requirements established for a particular operation or type of operation; e.g., IFR takeoff or landing, alternate airport for IFR flight plans, VFR flight. (See Landing Minimums, IFR Takeoff Minimums, VFR Conditions, IFR Conditions) (Refer to FAR Part 91, AIM Part 1)

*MISSED APPROACH—*

1. A maneuver conducted by a pilot when an instrument approach cannot be completed to a landing. The route of flight and altitude are shown on instrument approach procedure charts. A pilot executing a missed approach prior to the Missed Approach Point (MAP) must continue along the final approach to the MAP. The pilot may climb immediately to the altitude specified in the missed approach procedure.
2. A term used by the pilot to inform ATC that he is executing the missed approach.

3. At locations where ATC radar service is provided the pilot should conform to radar vectors, when provided by ATC, in lieu of the published missed approach procedure. (See Missed Approach Point) (Refer to AIM Part 1)

*ICAO – MISSED APPROACH PROCEDURE* - The procedure to be followed if, after an instrument approach, a landing is not effected, and occuring normally:

1. When the aircraft has descended to the decision height and has not established visual contact, or
2. When directed by air traffic control to pull up or to go around again.

**MISSED APPROACH POINT / MAP** — A point prescribed in each instrument approach procedure at which a missed approach procedure shall be executed if the required visual reference does not exist. (See Missed Approach, Segments of an Instrument Approach Procedure)

**MISSED APPROACH SEGMENT** — (See Segments of an Instrument Approach Procedure)

**MODE** — The letter or number assigned to a specific pulse spacing of radio signals transmitted or received by ground interrogator or airborne transponder components of the Air Traffic Control Radar Beacon System (ATCRBS). Mode A (military Mode 3) and Mode C (altitude reporting) are used in air traffic control. (See Transponder, Interrogator, Radar) (Refer to AIM Part 1)

*ICAO – MODE (SSR MODE)* - The letter or number assigned to a specific pulse spacing of the interrogation signals transmitted by an interrogator. There are 4 modes, A, B, C and D corresponding to four different interrogation pulse spacings.

**MOVEMENT AREA** — The runways, taxiways, and other areas of an airport which are utilized for taxiing, takeoff, and landing of aircraft, exclusive of loading ramp and parking areas. At those airports with a tower, specific approval for entry onto the movement area must be obtained from ATC.

*ICAO – MOVEMENT AREA* - That part of an aerodrome intended for the surface movement of aircraft, including the manoeuvring area and aprons.

**MOVING TARGET INDICATOR / MTI** — An electronic device which will permit radar scope presentation only from targets which are in motion. A partial remedy for ground clutter.

**NAS STAGE A** — The en route ATC system's radar, computers and computer programs, controller plan view displays (PVDs/Radar Scopes), input/output devices, and the related communications equipment which are integrated to form the heart of the automated IFR air traffic control system. This equipment performs Flight Data Processing (FDP) and Radar Data Processing (RDP). It interfaces with automated terminal systems and is used in the control of en route IFR aircraft. (Refer to AIM Part 1)

**NAS STAGE A CONFLICT ALERT FUNCTION** — An aid to the radar controller in detecting impending situations where loss of radar separation minimums for IFR aircraft may occur. Using inputs from automatic tracking, the Conflict Alert Function provides the ARTCC controller with an alert on the radar display when aircraft flying above 18,000 feet MSL are positioned so that violations of separation standards could occur in a short time (approximately 2 minutes).

**NATIONAL AIRSPACE SYSTEM / NAS** — The common network of U.S. airspace; air navigation facilities, equipment and services, airports or landing areas; aeronautical charts, information and services; rules, regulations and procedures, technical information, and manpower and material. Included are system components shared jointly with the military.

**NATIONAL FLIGHT DATA CENTER / NFDC** — A facility in Washington D.C., established by FAA to operate a central aeronautical information service for the collection, validation, and dissemination of aeronautical data in support of the activities of government, industry, and the aviation community. The information is published in the National Flight Data Digest. (See National Flight Data Digest)

**NATIONAL FLIGHT DATA DIGEST / NFDD** — A daily (except weekends and federal holidays) publication of flight information appropriate to aeronautical charts, aeronautical publications, Notices to Airmen or other media serving the purpose of providing operational flight data essential to safe and efficient aircraft operations.

**NATIONAL SEARCH AND RESCUE PLAN** — An interagency agreement which provides for the effective utilization of all available facilities in all types of search and rescue missions.

**NAVAID CLASSES** — VOR, VORTAC, and TACAN aids are classed according to their operational use. The three classes of NAVAIDS are:

T – Terminal
L – Low altitude
H – High altitude

The normal service range for T, L, and H class aids is found in AIM Part 1. Certain operational requirements make it necessary to use some of these aids at greater service ranges than specified. Extended range is made possible through flight inspection determinations. Some aids also have lesser service range due to location, terrain, frequency protection, etc. Restrictions to service range are listed in AIM Part 3.

**NAVIGABLE AIRSPACE** — Airspace at and above the minimum flight altitudes prescribed in the FARs including airspace needed for safe take-off and landing. (Refer to FAR Part 91)

**NAVIGATIONAL AID / NAVAID** — Any visual or electronic device airborne or on the surface which provides point to point guidance information or position data to aircraft in flight. (See Air Navigation Facility)

**NDB** — (See Nondirectional beacon)

*NEGATIVE* — "No" or "Permission not granted" or "That is not correct."

*NEGATIVE CONTACT* —
Used by pilots to inform ATC that:
1. Previously issued traffic is not in sight. It may be followed by the pilot's request for the controller to provide assistance in avoiding the traffic.
2. They were unable to contact ATC on a particular frequency.

**NIGHT** — The time between the end of evening civil twilight and the beginning of morning civil twilight, as published in the American Air Almanac, converted to local time.

*ICAO – NIGHT* - The hours between the end of evening civil twilight and the beginning of morning civil twilight or such other period between sunset and sunrise as may be specified by the appropriate authority.

*NO GYRO APPROACH/VECTOR* — A radar approach/vector provided in case of a malfunctioning gyrocompass or directional gyro. Instead of providing the pilot with headings to be flown, the controller observes the radar track and issues control instructions "turn right/left" or "stop turn" as appropriate. (Refer to AIM Part 1)

**NONDIRECTIONAL BEACON / RADIO BEACON / NDB** — An L/MF or UHF radio beacon transmitting nondirectional signals whereby the pilot of an aircraft equipped with direction finding equipment can determine his bearing to or from the radio beacon and "home" on or track to or from the station. When the radio beacon is installed in conjunction with the Instrument Landing System marker, it is normally called a Compass Locator. (See Compass Locator, Automatic Direction Finder)

**NONPRECISION APPROACH PROCEDURE / NONPRECISION APPROACH** — A standard instrument approach procedure in which no electronic glide slope is provided; e.g., VOR, TACAN, NDB, LOC, ASR, LDA, or SDF approaches.

**NONRADAR** — Precedes other terms and generally means without the use of radar, such as:

1. Nonradar Route – A flight path or route over which the pilot is performing his own navigation. The pilot may be receiving radar separation, radar monitoring or other ATC services while on a nonradar route. (See Radar Route)
2. Nonradar Approach – Used to describe instrument approaches for which course guidance on final approach is not provided by ground based precision or surveillance radar. Radar vectors to the final approach course may or may not be provided by ATC. Examples of nonradar approaches are VOR, ADF, TACAN, and ILS, approaches. (See Final Approach – IFR, Final Approach Course, Radar Approach, **Instrument Approach Procedure**)
3. Nonradar Separation – The spacing of aircraft in accordance with established minima without the use of radar; e.g., vertical, lateral, or longitudinal separation. (See Radar Separation)
   *ICAO – Non-Radar Separation* - The separation used when aircraft position information is derived from sources other than radar.
4. Nonradar Arrival – An arriving aircraft that is not being vectored to the final approach course for an instrument approach or towards the traffic pattern for a visual approach. The aircraft may or may not be in a radar environment and may or may not be receiving radar separation, radar monitoring or other services provided by ATC. (See Radar Arrival, Radar Environment)

**NONRADAR APPROACH CONTROL** — An ATC facility providing approach control service without the use of radar. (See Approach Control, Approach Control Service)

*NORDO* — (See Lost Communications)

**NORTH AMERICAN ROUTE** — A numerically coded route preplanned over existing airway and route systems to and from specific coastal fixes serving the North Atlantic. North American Routes consist of the following:

1. Common Route/Portion – That segment of a North American route between the inland navigation facility and the coastal fix.
2. Non-Common Route/Portion – That segment of a North American route between the inland navigation facility and a designated North American terminal.
3. Inland Navigation Facility – A navigation aid on a North American route at which the common route and/or the non-common route begins or ends.
4. Coastal Fix – A navigation aid or intersection where an aircraft transitions between the domestic route structure and the oceanic route structure.

**NOTICE TO AIRMEN / NOTAM** — A notice containing information (not known sufficiently in advance to publicize by other means) concerning the establishment, condition or change in any component (facility, service, or procedure of, or hazard in the National Airspace System) the timely knowledge of which is essential to personnel concerned with flight operations.

1. NOTAM(D) – A NOTAM given (in addition to local dissemination) distant dissemination via teletypewriter beyond the area of responsibility of the Flight Service Station. These NOTAMS will be stored and repeated hourly until cancelled.
2. NOTAM(L) – A NOTAM given local dissemination by voice, (Teletypewriter where applicable), and a wide variety of means such as: Teleautograph, teleprinter, facsimile reproduction, hot line, telecopier, telegraph, and telephone to satisfy local user requirements.
3. FDC NOTAM – A notice to airmen, regulatory in nature, transmitted by NFDC and given all-circuit dissemination.

*ICAO – NOTAM* - A notice, containing information concerning the establishment, condition or change in any aeronautical facility, service, procedure or hazard, the timely knowledge of which is essential to personnel concerned with flight operations.

*NUMEROUS TARGETS VICINITY (LOCATION)* — A traffic advisory issued by ATC to advise pilots that targets on the radar scope are too numerous to issue individually. (See Traffic Advisories)

**OBSTACLE** — An existing object, object of natural growth, or terrain at a fixed geographical location, or which may be expected at a fixed location within a prescribed area, with reference to which vertical clearance is or must be provided during flight operation.

**OBSTRUCTION** — An object which penetrates an imaginary surface described in FAR Part 77. (Refer to FAR Part 77)

**OBSTRUCTION LIGHT** — A light, or one of a group of lights, usually red or white, frequently mounted on a surface structure or natural terrain to warn pilots of the presence of an obstruction.

**OFF-ROUTE VECTOR** — A vector by ATC which takes an aircraft off a previously assigned route. Altitudes assigned by ATC during such vectors provide required obstacle clearance.

**OFFSET PARALLEL RUNWAYS** — Staggered runways having centerlines which are parallel.

**OLIVE BRANCH ROUTES / OB ROUTES / ALL WEATHER LOW ALTITUDE ROUTES / AWLARs** — Training routes used by USAF and USN jet aircraft in both VFR and IFR weather conditions from the surface to the published altitude. Routes, their altitudes and times of operation are shown in AIM Part 4 and the DOD FLIP. Graphic information and weather requirements necessary for IFR and VFR training operations are also shown in AIM Part 4. The current operational status of a particular route may be obtained by calling an FSS near the route. (Refer to AIM Part 4)

*ON COURSE* —
1. Used to indicate that an aircraft is established on the route centerline.
2. Used by ATC to advise a pilot making a radar approach that his aircraft is lined up on the final approach course. (See On-Course Indication)

**ON-COURSE INDICATION** — An indication on an instrument which provides the pilot a visual means of determining that the aircraft is located on the centerline of a given navigational track; or an indication on a radar scope that an aircraft is on a given track.

**OPTION APPROACH** — An approach requested and conducted by a pilot which will result in either a touch-and-go, missed approach, low approach, stop-and-go or full stop landing. (See Cleared for the option) (Refer to AIM Part 1)

**ORGANIZED TRACK SYSTEM** — A moveable system of oceanic tracks that traverses the North Atlantic between Europe and North America the physical position of which is determined twice daily taking the best advantage of the winds aloft.

*OUT* — The conversation is ended and no response is expected.

**OUTER COMPASS LOCATOR** — (See Compass Locator)

**OUTER FIX** — A general term used within ATC to describe fixes in the terminal area, other than the final approach fix. Aircraft are normally cleared to these fixes by an Air Route Traffic Control Center or an Approach Control Facility. Aircraft are normally cleared from these fixes to the final approach fix or final approach course.

**OUTER MARKER / OM** — A marker beacon at or near the glide slope intercept altitude of an ILS approach. It is keyed to transmit two dashes per second on a 400 Hz tone which is received aurally and visually by compatible airborne equipment. The OM is normally located four to seven miles from the runway threshold on the extended centerline of the runway. (See Marker Beacon, Instrument Landing System) (Refer to AIM Part 1)

*OVER* — My transmission is ended; I expect a response.

**OVERHEAD APPROACH / 360 OVERHEAD** — A series of predetermined maneuvers prescribed for VFR arrival of military aircraft (often in formation) for entry into the VFR traffic pattern and to proceed to a landing. The pattern usually specifies the following:
1. The radio contact required of the pilot.
2. The speed to be maintained.
3. An initial approach 3 to 5 miles in length.
4. An elliptical pattern consisting of two 180 degree turns.
5. A break point at which the first 180 degree turn is started.
6. The direction of turns.
7. Altitude (at least 500 feet above the conventional pattern).
8. A "Roll-out" on final approach not less than 1/4 mile from the landing threshold and not less than 300 feet above the ground.

*PAN* — The international radio-telephony urgency signal. When repeated three times indicates uncertainty or alert, followed by nature of urgency. (See MAYDAY) (Refer to AIM Part 1)

**PAR APPROACH** — A precision instrument approach wherein the air traffic controller issues guidance instructions, for pilot compliance, based on the aircraft's position in relation to the final approach course (azimuth), the glide slope (elevation), and the distance (range) from the touchdown point on the runway as displayed on the controller's radar scope. (See Precision Approach Radar, Glide Path) (Refer to AIM Part 1)

**PARALLEL OFFSET ROUTE** — A parallel track to the left or right of the designated or established airway/route. Normally associated with Area Navigation (RNAV) operations. (See Area Navigation)

**PARALLEL RUNWAYS** — Two or more runways at the same airport whose centerlines are parallel. In addition to runway number, parallel runways are designated as L (left) and R (right) or, if three parallel runways exist, L (left), C (center), and R (right).

**PERMANENT ECHO** — Radar signals reflected from fixed objects on the earth's surface; e.g., buildings, towers, terrain. Permanent echoes are distinguished from "ground clutter" by being definable locations rather than large areas. Under certain conditions they may be used to check radar alignment.

**PILOT BRIEFING / PRE-FLIGHT PILOT BRIEFING** — A service provided by the FSS to assist pilots in flight planning. Briefing items may include weather information, NOTAMS, military activities, flow control information and other items as requested. (Refer to AIM Part 1)

**PILOT IN COMMAND** — The pilot responsible for the operation and safety of an aircraft during flight time. (Refer to FAR Part 91)

**PILOT SELF-BRIEFING TERMINAL / PSBT** — A device that provides users direct access to a computer for preflight briefing information and for filing flight plans.

**PILOT WEATHER REPORT / PIREP** — A report of meteorological phenomena encountered by aircraft in flight. (Refer to AIM Part 1)

*PILOT'S DISCRETION* — When used in conjunction with altitude assignments, means that ATC has offered the pilot the option of starting climb or descent whenever he wishes and conducting the climb or descent at any rate he wishes. He may temporarily level off at any intermediary altitude. However, once he has vacated an altitude he may not return to that altitude.

**PILOTS AUTOMATIC TELEPHONE WEATHER ANSWERING SERVICE / PATWAS** — A continuous telephone recording containing current and forecast weather information for pilots. (See Flight Service Station) (Refer to AIM Parts 1 and 2)

**POSITION REPORT / PROGRESS REPORT** — A report over a known location as transmitted by an aircraft to ATC. (Refer to AIM Part 1)

**POSITION SYMBOL** — A computer generated indication shown on a radar display to indicate the mode of tracking.

**POSITIVE CONTROL** — The separation of all air traffic, within designated airspace, by air traffic control. (See Positive Control Area)

**POSITIVE CONTROL AREA / PCA** — Airspace designated in FAR Part 71 wherein aircraft are required to be operated under Instrument Flight Rules (IFR). Vertical extent of PCA is from 18,000 feet to and including flight level 600 throughout most of the conterminous United States and from flight level 240 to and including flight level 600 in designated portions of Alaska. (Refer to FAR Parts 71 and 91, AIM Parts 1 and 4, DOD FLIP)

**PRECIPITATION** — Any or all forms of water particles (rain, sleet, hail, or snow), that fall from the atmosphere and reach the surface.

**PRECISION APPROACH PROCEDURE / PRECISION APPROACH** — A standard instrument approach procedure in which an electronic glide slope is provided; e.g., ILS and PAR. (See Instrument Landing System, Precision Approach Radar)

**PRECISION APPROACH RADAR / PAR** — Radar equipment in some ATC facilities serving military airports, which is used to detect and display the azimuth, range, and elevation of an aircraft on the final approach course to a runway. It is used by air traffic controllers to provide the pilot with a precision approach, or to monitor certain nonradar approaches. (See PAR Approach)

*ICAO - PRECISION APPROACH RADAR / PAR* - Primary radar equipment used to determine the position of an aircraft during final approach, in terms of lateral and vertical deviations relative to a nominal approach path, and in range relative to touchdown.

**PREFERENTIAL ROUTES** — Preferential routes (PDRs, PARs, and PDARs) are adapted in ARTCC computers to accomplish inter/intra-facility controller coordination and to assure that flight data is posted at the proper control positions. Locations having a need for these specific inbound and outbound routes normally publish such routes in local facility bulletins and their use by pilots minimizes flight plan route amendments. When the workload or traffic situation permits, controllers normally provide radar vectors or assign requested routes to minimize circuitous routing. Preferential routes are usually confined to one ARTCC's area and are referred to by the following names or acronyms:

1. Preferential Departure Route / PDR – A specific departure route from an airport or terminal area to an en route point where there is no further need for flow control. It may be included in a Standard Instrument Departure (SID) or a Preferred IFR Route.
2. Preferential Arrival Route / PAR – A specific arrival route from an appropriate en route point to an airport or terminal area. It may be included in a Standard Terminal Arrival Route (STAR) or a Preferred IFR Route. The abbreviation "PAR" is used primarily within the ARTCC and should not be confused with the abbreviation for Precision Approach Radar.
3. Preferential Departure and Arrival Route / PDAR – A route between two terminals which are within or immediately adjacent to one ARTCC's area. PDARs are not synonomous with Preferred IFR Routes but may be listed as such as they do accomplish essentially the same purpose. (See Preferred IFR Routes, NAS Stage A)

**PREFERRED IFR ROUTES** — Routes established between busier airports to increase system efficiency and capacity. They normally extend through one or more ARTCC areas and are designed to achieve balanced traffic flows among high density terminals. IFR clearances are issued on the basis of these routes except when severe weather avoidance procedures or other factors dictate otherwise. Preferred IFR Routes are listed in the AIM Part 3. If a flight is planned to or from an area having such routes but the departure or arrival point is not listed in the AIM Part 3, pilots may use that part of a Preferred IFR Route which is appropriate for the departure or arrival point that is listed. Preferred IFR Routes are correlated with SIDs and STARs and may be defined by airways, jet routes, direct routes between NAVAIDS. Waypoints, NAVAID radials/DME, or any combinations thereof. (See Standard Instrument Departure, Standard Terminal Arrival Route, Preferential Routes, Center Area) (Refer to AIM Parts 3 and 3A)

**PREVAILING VISIBILITY** — (See Visibility)

**PROCEDURE TURN / PT** — The maneuver prescribed when it is necessary to reverse direction to establish an aircraft on the intermediate approach segment or final approach course. The outbound course, direction of turn, distance within which the turn must be completed, and minimum altitude are specified in the procedure. However, the point at which the turn may be commenced, and the type and rate of turn, are left to the discretion of the pilot.
*ICAO – PROCEDURE TURN* - A manoeuvre in which a turn is made away from a designated track followed by a turn

in the opposite direction, both turns being executed so as to permit the aircraft to intercept and proceed along the reciprocal of the designated track.

**PROCEDURE TURN INBOUND** — That point of a procedure turn maneuver where course reversal has been completed and an aircraft is established inbound on the intermediate approach segment or final approach course. A report of "procedure turn inbound" is normally used by ATC as a position report for separation purposes. (See Final Approach Course, Procedure Turn, Segments of an Instrument Approach Procedure)

**PROHIBITED AREA** — Designated airspace within which the flight of aircraft is prohibited. (Refer to En Route Charts, AIM Part 1)
*ICAO – PROHIBITED AREA* - An airspace of defined dimensions, above the land areas or territorial waters of a state, within which the flight of aircraft is prohibited.

**PROPOSED BOUNDARY CROSSING TIME / PBCT** — Each center has a PBCT parameter for each internal airport. Proposed internal flight plans are transmitted to the adjacent center if the flight time along the proposed route from the departure airport to the center boundary is less than or equal to the value of PBCT or if airport adaptation specifies transmission regardless of PBCT.

**PUBLISHED** — Depicted on aeronautical charts or described in other aeronautical publications and approved for use in the National Airspace System.

**PUBLISHED ROUTE** — a route for which an IFR altitude has been established, and published; e.g., Federal Airways, Jet Routes, Area Navigation Routes, Specified Direct Routes. (See Published)

**QUICK LOOK** — A feature of NAS Stage A and ARTS which provides the controller the capability to display full data blocks of tracked aircraft from other control positions.

**QUOTA FLOW CONTROL / QFLOW** — A flow control procedure by which the Central Flow Control Facility (CFCF) restricts traffic to the ARTC Center area having an impacted airport thereby avoiding sector/area saturation. (See Air Traffic Control Systems Command Center) (Refer to AIM Part 3)

**RADAR / RADIO DETECTION AND RANGING** — A device which, by measuring the time interval between transmission and reception of radio pulses and correlating the angular orientation of the radiated antenna beam or beams in azimuth and/or elevation, provides information on range, azimuth and/or elevation of objects in the path of the transmitted pulses.

1. Primary Radar – A radar system in which a minute portion of a radio pulse transmitted from a site is reflected by an object and then received back at that site, for processing and display at an air traffic control facility.
2. Secondary Radar / Radar Beacon / ATCRBS – A radar system in which the object to be detected is fitted with cooperative equipment in the form of a

radio receiver/transmitter (transponder). Radar pulses transmitted from the searching transmitter/receiver (interrogator) site are received in the cooperative equipment and used to trigger a distinctive transmission from the transponder. This reply transmission rather than a reflected signal, is then received back at the transmitter/receiver site for processing and display at an air traffic control facility. (See Transponder, Interrogator) (Refer to AIM Part 1)

*ICAO – RADAR* - A radio detection device which provides information on range, azimuth and/or elevation of objects.

1. *Primary Radar* – A radar system which uses reflected radio signals.
2. *Secondary Radar* – A radar system wherein a radio signal transmitted from a radar station initiates the transmission of a radio signal from another station.

**RADAR ADVISORY** — The provision of advice and information based on radar observations. (See Advisory Service)

**RADAR AIR TRAFFIC CONTROL FACILITY / RATCF** — An air traffic control facility, located at a U.S. Navy or Marine Corps Air Station, utilizing surveillance and, normally, precision approach radar and air/ground communications equipment to provide approach control services to aircraft arriving, departing or transiting the airspace controlled by the facility. The facility may be operated by the FAA, the USN, or USMC and service may be provided to both civil and military airports. Similar to TRACON (FAA), RAPCON (USAF), RATCF (Navy), and ARAC (Army) (See Approach Control, Approach Control Service, Departure Control)

**RADAR APPROACH** — An instrument approach procedure which utilizes Precision Approach Radar (PAR) or Airport Surveillance Radar (ASR). (See PAR Approach, Surveillance Approach, Airport Surveillance Radar, Precision Approach Radar, Instrument Approach Procedure) (Refer to AIM Part 1)

*ICAO – RADAR APPROACH* - An approach, executed by an aircraft, under the direction of a radar controller.

**RADAR APPROACH CONTROL / RAPCON** — An air traffic control facility, located at a U.S. Air Force Base, utilizing surveillance and, normally, precision approach radar and air/ground communications equipment to provide approac control services to aircraft arriving, departing or transiting the airspace controlled by the facility. The facility may be operated by the FAA or the USAF and service may be provided to both civil and military airports. Similar to TRACON (FAA), RATCF (Navy) and ARAC (Army.) (See Approach Control, Approach Control Service, Departure Control)

**RADAR ARRIVAL** — An arriving aircraft which is being vectored to the final approach course for an instrument approach, or toward the traffic pattern for a visual approach. (See Radar Approach, Visual Approach)

**RADAR BEACON** — (See Radar)

*RADAR CONTACT* —

1. Used by ATC to inform an aircraft that it is identified on the radar display and radar service may be provided until radar identification is lost or radar service is terminated. When a pilot is informed of "radar contact" he automatically discontinues reporting over compulsory reporting points.
2. The term an air traffic controller uses to inform the transferring controller that the target being transferred is identified on his radar display. (See Radar Service, Radar Contact Lost, Radar Service Terminated) (Refer to AIM Part 1)

*ICAO – RADAR CONTACT* - The situation which exists when the radar blip of a particular aircraft is seen and identified on a radar display.

**RADAR CONTACT LOST** — Used by ATC to inform a pilot that radar identification of his aircraft has been lost. The loss may be attributed to several things including the aircraft merging with weather or ground clutter, the aircraft flying below radar line of sight, the aircraft entering an area of poor radar return or a failure of the aircraft transponder or ground radar equipment. (See Clutter, Radar Contact)

**RADAR ENVIRONMENT** — An area in which radar service may be provided. (See Radar Contact, Radar Service, Additional Services, Traffic Advisories)

**RADAR FLIGHT FOLLOWING** — The observation of the progress of radar identified aircraft, whose primary navigation is being provided by the pilot, wherein the controller retains and correlates the aircraft identity with the appropriate target or target symbol displayed on the radar scope. (See Radar Contact, Radar Service) (Refer to AIM Part 1)

**RADAR IDENTIFICATION** — The process of ascertaining that an observed radar target is the radar return from a particular aircraft. (See Radar Contact, Radar Service)

*ICAO – RADAR IDENTIFICATION* – The process of correlating a particular radar blip with a specific aircraft.

**RADAR SERVICE** — A term which encompasses one or more of the following services based on the use of radar which can be provided by a controller to a pilot of a radar identified aircraft.

1. Radar Separation. – Radar spacing of aircraft in accordance with established minima.
2. Radar Navigational Guidance. – Vectoring aircraft to provide course guidance.
3. Radar Monitoring. – The radar flight following of aircraft, whose primary navigation is being performed by the pilot, to observe and note deviations from its authorized flight path, airway, or route. When being applied specifically to radar monitoring of instrument approaches, i.e., with precision approach radar (PAR) or radar monitoring of simultaneous ILS approaches, it includes advice and instructions whenever an aircraft nears or exceeds the prescribed PAR safety limit or simultaneous ILS no transgression zone. (See Additional Services, Traffic Advisories, Duty Priority)

*ICAO – RADAR SERVICE* - Term used to indicate a service provided directly by means of radar.

*ICAO – RADAR SEPARATION* - The separation used when aircraft position information is derived from radar sources.

*ICAO – RADAR MONITORING* - The use of radar for the purpose of providing aircraft with information and advice relative to significant deviations from nominal flight path.

**RADAR SERVICE TERMINATED** — Used by ATC to inform a pilot that he will no longer be provided any of the services that could be received while under radar contact. Radar service is automatically terminated and the pilot is not advised in the following cases:

1. When the aircraft cancels its IFR flight plan.
2. At the completion of a radar approach.
3. When an arriving aircraft receiving Stage I, II, or III service is advised to contact the tower.
4. When an aircraft conducting a visual approach is advised to contact the tower.
5. When an aircraft vectored to a final approach course for an instrument approach has landed or the tower has the aircraft in sight, whichever occurs first.

**RADAR SURVEILLANCE** — The radar observation of a given geographical area for the purpose of performing some radar function.

**RADAR TRAFFIC ADVISORIES** — (See Traffic Advisories)

**RADAR TRAFFIC INFORMATION SERVICE** — (See Traffic Advisories)

**RADIAL** — A magnetic bearing extending from a VOR/VORTAC/TACAN navigation facility.

**RADAR IDENTIFIED AIRCRAFT** — An aircraft, the position of which has been correlated with an observed target or symbol on the radar display. (See Radar Contract, Radar Contact Lost)

**RADAR MONITORING** — (See Radar Service)

**RADAR NAVIGATIONAL GUIDANCE** — (See Radar Service)

**RADAR POINT OUT / POINT OUT** — Used between controllers to indicate radar handoff action where the initiating controller plans to retain communications with an aircraft penetrating the other controller's airspace and additional coordination is required.

**RADAR ROUTE** — A flight path or route over which an aircraft is vectored. Navigational guidance and altitude assignments are provided by ATC. (See Flight Path, Route, Radar Vector)

**RADAR SAFETY ADVISORY** — A radar advisory issued by ATC to radar identified aircraft under their control if an aircraft is observed to be at an altitude which, in the controller's judgment, places the aircraft in unsafe proximity to terrain, obstructions or other aircraft. The controller may discontinue the issuance of further advisories if the pilot advises he is taking action to correct the situation or has the other aircraft in sight.

1. Terrain/Obstruction Advisory – A Radar advisory issued by ATC to aircraft under their control if an aircraft is observed at an altitude which, in the controller's judgment, places the aircraft in unsafe proximity to terrain/obstructions; e.g., "Low Altitude Alert, advise you climb immediately to three thousand."
2. Aircraft Conflict Advisory – A radar advisory issued by ATC to aircraft under their control if an aircraft not under their control is observed at an altitude which, in the controller's judgment, places both aircraft in unsafe proximity to each other. With the alert, ATC will offer the pilot an alternate course of action when feasible, e.g., – "Conflict Alert, advise you turn right heading zero niner zero" or "climb to eight thousand immediately."

The issuance of a radar safety advisory is contingent upon the capability of the controller to observe an unsafe condition. The course of action provided will be predicated on other traffic under ATC control. Once the advisory is issued, it is solely the pilot's prerogative to determine what course of action, if any, he will take. (See Radar Advisory) (Refer to AIM Part 1)

**RADAR SEPARATION** — (See Radar Service)

**RADIO** —
1. A device used for communication.
2. Used to refer to a Flight Service Station, e.g., "Seattle Radio" is used to call Seattle FSS.

**RADIO ALTIMETER / RADAR ALTIMETER** — Aircraft equipment which makes use of the reflection of radio waves from the ground to determine the height of the aircraft above the surface.

**RADIO BEACON** — (See Nondirectional Beacon)

**RADIO MAGNETIC INDICATOR / RMI** — An aircraft navigational instrument coupled with a gyro compass or similar compass that indicates the direction to a selected NAVAID and indicates bearing with respect to the heading of the aircraft.

**RAMP** — (See Apron)

**READ BACK** — Repeat my message back to me.

**RECEIVING CONTROLLER/FACILITY** — A controller/facility receiving control of an aircraft from another controller/facility.

**REDUCE SPEED TO (SPEED)** — (See Speed Adjustment)

**RELEASE TIME** — A departure time restriction issued to a pilot by ATC when necessary to separate a departing aircraft from other traffic.

*ICAO – RELEASE TIME* – Time prior to which an aircraft should be given further clearance or prior to which it should not proceed in case of radio failure.

**REMOTE COMMUNICATIONS AIR/GROUND FACILITY / RCAG** — An unmanned VHF/UHF transmitter/receiver facility which is used to expand ARTCC air/ground communications coverage and to facilitate direct contact between pilots and controllers. RCAG facilities are sometimes not equipped with emergency frequencies 121.5 MHz and 243.0 MHz. (Refer to AIM Part 1)

**REMOTE COMMUNICATIONS OUTLET / RCO** — An unmanned air/ground communications station remotely controlled, providing UHF and VHF transmit and receive capability to extend the service range of the FSS.

*REPORT* — Used to instruct pilots to advise ATC of specified information, e.g., "Report passing Hamilton VOR."

**REPORTING POINT** — A geographical location in relation to which the position of an aircraft is reported. (See Compulsory Reporting Point) (Refer to AIM Part 1)
*ICAO – REPORTING POINT* - A specified geographical location in relation to which the position of an aircraft can be reported.

*REQUEST FULL ROUTE CLEARANCE / FRC* — Used by pilots to request that the entire route of flight be read verbatim in an ATC clearance. Such request should be made to preclude receiving an ATC clearance based on the originally filed flight plan when a filed IFR flight plan has been revised by the pilot, company, or operations prior to departure.

**RESCUE COORDINATION CENTER / RCC** — A search and rescue (SAR) facility equipped and manned to coordinate and control SAR operations in an area designated by the SAR plan. The U.S. Coast Guard and the U.S. Air Force have responsibility for the operation of RCCs.
*ICAO – RESCUE CO-ORDINATION CENTRE* - A unit responsible for promoting efficient organization of search and rescue service and for co-ordinating the conduct of search and rescue region.

**RESTRICTED AREA** — Airspace designated under FAR Part 73 within which the flight of aircraft, while not wholly prohibited, is subject to restriction. Most restricted areas are designated joint use and IFR/VFR operations in the areas may be authorized by the controlling ATC facility when it is not being utilized by the using agency. Restricted areas are depicted on en route charts. Where joint use is authorized the name of the ATC controlling facility is also shown. (Refer to FAR Part 73, AIM Part 1)
*ICAO – RESTRICTED AREA* - Airspace of defined dimensions, above the land areas or territorial waters of a State, within which the flight of aircraft is restricted in accordance with certain specified conditions.

*RESUME NORMAL NAVIGATION* — Used by ATC to advise a pilot to resume his own navigational responsibility. It is issued after completion of a radar vector or when radar contact is lost while the aircraft is being radar vectored. (See Radar Contact Lost, Radar Service Terminated)

**RNAV** — (See Area Navigation)

**RNAV APPROACH** — An instrument approach procedure which relies on aircraft area navigation equipment for navigational guidance. (See Instrument Approach Procedure, Area Navigation)

*ROGER* — I have received all of your last transmission. It should not be used to answer a question requiring a yes or no answer. (See Affirmative, Negative)

**ROLLOUT RVR** — (See Visibility)

**ROUTE** — A defined path, consisting of one or more courses in a horizontal plane, which aircraft traverse over the surface of the earth. (See Jet Route, Airway, Published Route, Unpublished Route)

**ROUTE SEGMENT** — As used in Air Traffic Control, a part of a route that can be defined by two navigational fixes, two NAVAIDs, or a fix and a NAVAID. (See Route, Fix)
*ICAO – ROUTE SEGMENT* - A portion of a route to be flown, as defined by two consecutive significant points specified in a flight plan.

**RUNWAY** — A defined rectangular area, on a land airport prepared for the landing and takeoff run of aircraft along its length. Runways are normally numbered in relation to their magnetic direction rounded off to the nearest 10 degrees. e.g., Runway 25, Runway 01. (See Parallel Runways)
*ICAO – RUNWAY* - A defined rectangular area, on a land aerodrome prepared for the landing and takeoff run of aircraft along its length.

**RUNWAY CENTERLINE LIGHTING** — (See Airport Lighting)

**RUNWAY CONDITION READING / RCR** — Numerical decelerometer readings relayed by air traffic controllers at USAF and certain civil bases for use by the pilot in determining runway braking action. These readings are routinely relayed only to USAF and Air National Guard Aircraft. (See Braking Action)

**RUNWAY END IDENTIFIER LIGHTS** — (See Airport Lighting)

**RUNWAY GRADIENT** — The average slope, measured in percent, between two ends or points on a runway. Runway gradient is depicted on Government aerodrome sketches when total runway gradient exceeds 0.3%.

**RUNWAY IN USE / ACTIVE RUNWAY / DUTY RUNWAY** — Any runway or runways currently being used for takeoff or landing. When multiple runways are used, they are all considered active runways.

**RUNWAY LIGHTS** — (See Airport Lighting)

**RUNWAY MARKINGS** —
1.  Basic marking – Markings on runways used for operations under visual flight rules consisting of centerline marking and runway direction numbers and, if required, letters.
2.  Instrument marking – Markings on runways served by nonvisual navigation aids and intended for landings under instrument weather conditions, consisting of basic marking plus threshold marking.
3.  All-weather (precision instrument) marking – Markings on runways served by nonvisual precision approach aids and on runways having special operational requirements, consisting of instrument markings plus landing zone marking and side strips. (Refer to AIM Part 1)

**RUNWAY VISIBILITY VALUE** — (See Visibility)

**RUNWAY VISUAL RANGE** — (See Visibility)

**RVAV** — (See Area Navigation)

*SAY AGAIN* — Used to request a repeat of the last transmission. Usually specifies transmission or portion thereof not understood or received, e.g., "Say again all after ABRAM VOR."

*SAY ALTITUDE* — Used by ATC to ascertain an aircraft's specific altitude/flight level. When the aircraft is climbing or descending, the pilot should state the indicated altitude rounded to the nearest 100 feet.

*SAY HEADING* — Used by ATC to request an aircraft heading. The pilot should state the actual heading of the aircraft.

**SEARCH AND RESCUE / SAR** — A service which seeks missing aircraft and assists those found to be in need of assistance. It is a cooperative effort using the facilities and services of available Federal, state and local agencies. The U.S. Coast Guard is responsible for coordination of search and rescue for the Maritime Region and the U.S. Air Force is responsible for search and rescue for the Inland Region. Information pertinent to search and rescue should be passed through any air traffic facility or be transmitted directly to the Rescue Coordination Center by telephone. (See Flight Service Station, Rescue Coordination Center) (Refer to AIM Part 1)

**SEARCH AND RESCUE FACILITY** — A facility responsible for maintaining and operating a search and rescue (SAR) service to render aid to persons and property in distress. It is any SAR unit, station, NET or other operational activity which can be usefully employed during an SAR Mission, e.g., a Civil Air Patrol Wing or a Coast Guard Station. (See Search and Rescue)

**SEE AND AVOID** — A visual procedure wherein pilots of aircraft flying in visual meteorological conditions (VMC), regardless of type of flight plan, are charged with the responsibility to observe the presence of other aircraft and to maneuver their aircraft as required to avoid the other aircraft. Right-of-way rules are contained in FAR, Part 91. (See Instrument Flight Rules, Visual Flight Rules, Visual Meteorological Conditions, Instrument Meteorological Conditions)

**SEGMENTED CIRCLE** — A system of visual indicators designed to provide traffic pattern information at airports without operating control towers. (Refer to AIM Part 1)

**SEGMENTS OF AN INSTRUMENT APPROACH PROCEDURE** — An instrument approach procedure may have as many as four separate segments depending on how the approach procedure is structured.
1. Initial Approach – The segment between the initial approach fix and the intermediate fix or the point where the aircraft is established on the intermediate course or final approach course.
2. Intermediate Approach – The segment between the intermediate fix or point and the final approach fix.
3. Final Approach – The segment between the final approach fix or point and the runway, airport or missed approach point.
4. Missed Approach – The segment between the missed

approach point, or point of arrival at decision height, and the missed approach fix at the prescribed altitude. (Refer to FAR Part 97)

*ICAO*
1. *Initial Approach* – That part of an instrument approach procedure consisting of the first approach to the first navigational facility associated with the procedure, or to a predetermined fix.
2. *Intermediate Approach* – That part of an instrument approach procedure from the first arrival at the first navigational facility or predetermined fix, to the beginning of the final approach.
3. *Final Approach* – That part of an instrument approach procedure from the time the aircraft has:
   a. Completed the last procedure turn or base turn where one is specified, or
   b. crossed a specified fix, or
   c. intercepted the last track specified for the procedures; until it has crossed a point in the vicinity of an aerodrome from which:
      (1) a landing can be made; or
      (2) a missed approach procedure is initiated.
4. *Missed Approach Procedure* – The procedure to be followed if, after an instrument approach, a landing is not effected and occurring normally:
   a. When the aircraft has descended to the decision height and has not established visual contact, or
   b. when directed by air traffic control to pull up or to go around again.

**SEPARATION** — In air traffic control, the spacing of aircraft to achieve their safe and orderly movement in flight and while landing and taking off. (See Separation Minima)
*ICAO – SEPARATION* – Spacing between aircraft, levels or tracks.

**SEPARATION MINIMA** — The minimum longitudinal, lateral, or vertical distances by which aircraft are spaced through the application of air traffic control procedures. (See Separation)

**SEVERE WEATHER AVOIDANCE PLAN / SWAP** — A plan to reroute traffic to avoid severe weather in the New York ARTCC area to provide the least disruption to the ATC system when large portions of airspace are unusable due to severe weather. (Refer to AIM Part 3)

**SHORT RANGE CLEARANCE** — A clearance issued to a departing IFR flight which authorizes IFR flight to a specific fix short of the destination while air traffic control facilities are coordinating and obtaining the complete clearance.

**SHORT TAKEOFF AND LANDING AIRCRAFT / STOL AIRCRAFT** — An aircraft which, at some weight within its approved operating weight, is capable of operating from a STOL runway in compliance with the applicable STOL characteristics, airworthiness, operations, noise, and pollution standards. (See Vertical Takeoff and Landing Aircraft)

**SIDESTEP MANEUVER** — A visual maneuver accomplished by a pilot at the completion of an instrument approach to permit a straight-in landing on a parallel runway not more than 1200 feet to either side of the runway to which the instrument approach was conducted. (Refer to AIM Part 1)

**SIGMET / SIGNIFICANT METEOROLOGICAL INFORMATION** — A weather advisory issued concerning weather significant to the safety of all aircraft. SIGMET advisories cover tornadoes, lines of thunderstorms, embedded thunderstorms, large hail, severe and extreme turbulence, severe icing, and widespread dust or sandstorms that reduce visibility to less than 3 miles. (See AIRMET) (Refer to AIM Part 1)

*ICAO – SIGMET INFORMATION* - Information prepared by a meteorological watch office regarding the occurrence or expected occurrence of one or more of the following phenomena:

    1.  At subsonic cruising levels:
      Active thunderstorm area
      Tropical revolving storm
      Severe line squall
      Heavy hail
      Severe turbulence
      Severe icing
      Marked mountain waves
      Widespread sandstorm/duststorm.
    2.  At transonic levels and supersonic cruising levels:
      Moderate or severe turbulence
      Cumulonimbus clouds
      Hail.

**SIMPLIFIED DIRECTIONAL FACILITY / SDF** — A NAVAID used for nonprecision instrument approaches. The final approach course is similar to that of an ILS localizer except that the SDF course may be offset from the runway, generally not more than 3 degrees, and the course may be wider than the localizer, resulting in a lower degree of accuracy. (Refer to AIM Part 1)

**SIMULATED FLAMEOUT / SFO** — A practice approach by a jet aircraft (normally military) at idle thrust to a runway. The approach may start at a relatively high altitude over a runway (high key) and may continue on a relatively high and wide downwind leg with a high rate of descent and a continuous turn to final. It terminates in a landing or low approach. The purpose of this approach is to simulate a flameout. (See Flameout)

**SIMULTANEOUS ILS APPROACHES** — An approach system permitting simultaneous ILS approaches to airports having parallel runways separated by at least 4,300 feet between centerlines. Integral parts of a total system are ILS, radar, communications, ATC procedures, and appropriate airborne equipment. (See Parallel Runways) (Refer to AIM Part 1)

**SINGLE DIRECTION ROUTES** — Preferred IFR routes which are sometimes depicted on high altitude en route charts and which are normally flown in one direction only. (See Preferred IFR Route) (Refer to AIM Part 3)

**SINGLE FREQUENCY APPROACH / SFA** — A service provided under a Letter of Agreement to single piloted military aircraft which permits use of a single UHF frequency during approach for landing. Pilots will not normally be required to change frequency from the beginning of the approach to touchdown except that pilots conducting an en route descent are required to change frequency when control is transferred from the air route traffic control center to the terminal facility. The abbreviation "SFA" in the DOD FLIP IFR Supplement under "Communications" indicates this service is available at an aerodrome.

**SINGLE FREQUENCY OUTLETS / SFO and SIMULTANEOUS SINGLE FREQUENCY OUTLETS / SSFO** Frequency Outlets commissioned at locations in Alaska not served by air traffic control facilities and remotely controlled by adjacent FSSs. They are subject to undetected and prolonged outages.

**SINGLE-PILOTED AIRCRAFT** —
A military aircraft possessing one set of flight controls, tandem cockpits or two sets of flight controls but operated by one pilot is considered single-piloted by ATC when determining the appropriate air traffic service to be applied. (See Single Frequency Approach)

**SLASH** — A radar beacon reply displayed as an elongated target.

***SPEAK SLOWER*** — Used in verbal communications as a request to reduce speech rate.

**SPECIAL IFR** — (See Fixed-Wing Special IFR)

**SPECIAL INSTRUMENT APPROACH PROCEDURE** — (See Instrument Approach Procedure)

**SPECIAL VFR CONDITIONS** — Weather conditions in a control zone which are less than basic VFR and in which some aircraft are permitted flight under Visual Flight Rules. (See Special VFR Operations) (Refer to FAR Part 91)

**SPECIAL VFR OPERATIONS** — Aircraft operating in accordance with clearances within control zones in weather conditions less than the basic VFR weather minima. Such operations must be requested by the pilot and approved by ATC. (See Special VFR Conditions)

*ICAO – SPECIAL VFR FLIGHT* – A controlled VFR flight authorized by air traffic control to operate within a control zone under meteorological conditions below the visual meteorological conditions.

**SPEED** — (See Airspeed, Groundspeed)

**SPEED ADJUSTMENT** — An ATC procedure used to request pilots to adjust aircraft speed to a specific value for the purpose of providing desired spacing. Speed adjustments are always expressed as indicated airspeed and pilots are expected to maintain a speed of plus or minus 10 knots of the specified speed.

Examples of Speed Adjustments are:

    1.  "Increase speed to (speed)" or "Increase speed (number of) knots" – Used by ATC to request a pilot to increase the indicated airspeed of the aircraft.
    2.  "Reduce speed to (speed)" or "Reduce speed (number of) knots" – Used by ATC to request a pilot to reduce the indicated airspeed of the aircraft.
    3.  "If feasible, reduce speed to (speed)" or "If feasible,

reduce speed (number of) knots" – Used by ATC to request a pilot to reduce the indicated airspeed of the aircraft below specified speeds. (Refer to AIM Part 1, FAR Part 91)

**SPEED BRAKES / DIVE BRAKES —** Moveable aerodynamic devices on aircraft that reduce airspeed during descent and landing.

*SQUAWK (Mode, Code, Function) —* Activate specific modes/codes/functions on the aircraft transponder, e.g., "Squawk Three/Alpha, Two one zero five, Low." (See Transponder)

**STAGE I/II/III SERVICE —** (See Terminal Radar Program)

*STAND BY —* Means the controller or pilot must pause for a few seconds, usually to attend to other duties of a higher priority. Also means to "wait" as in "stand by for clearance." If a delay is lengthy, the caller should re-establish contact.

**STANDARD INSTRUMENT APPROACH PROCEDURE —** (See Instrument Approach Procedure)

**STANDARD INSTRUMENT DEPARTURE / SID —** A preplanned instrument flight rule (IFR) air traffic control departure procedure printed for pilot use in graphic and/or textual form. SIDs provide transition from the terminal to the appropriate en route structure. (See IFR Takeoff Minima and Departure Procedures) (Refer to AIM Part 1)

**STANDARD RATE TURN —** A turn of three degrees per second.

**STANDARD TERMINAL ARRIVAL ROUTE / STAR —** A preplanned instrument flight rule (IFR) air traffic control arrival route published for pilot use in graphic and/or textual form. STARs provide transition from the en route structure to a fix or point from which an approach can be made. (See Preferential Routes) (Refer to AIM Part 1)

**STATIONARY RESERVATIONS —** Altitude reservations which encompass activities in a fixed area. Stationary reservations may include activities such as special test of weapons systems or equipment, certain U.S. Navy carrier, fleet, and anti-submarine operations, rocket, missile and drone operations, and certain aerial refueling or similar operations.

*STOP ALTITUDE SQUAWK —* Used by ATC to inform an aircraft to turn-off the automatic altitude reporting feature of its transponder. It is issued when the verbally reported altitude varies 300 feet or more from the automatic altitude report. (See Altitude Readout, Transponder)

**STOP AND GO —** A procedure wherein an aircraft will land, make a complete stop on the runway, and then commence a takeoff from that point. (See Low Approach, Option Approach)

*STOP SQUAWK (Mode or Code) —* Used by ATC to tell the pilot to turn specified functions of the aircraft transponder off. (See Stop Altitude Squawk, Transponder)

*STOP STREAM/BURST/BUZZER —* Used by ATC to request a pilot to suspend electronic countermeasure activity. (See Jamming)

**STOP-OVER FLIGHT PLAN —** A flight plan which includes two or more separate en route flight segments with a stopover at one or more intermediate airports.

**STRAIGHT-IN APPROACH – IFR —** An instrument approach wherein final approach is begun without first having executed a procedure turn. Not necessarily completed with a straight-in landing or made to straight-in landing minimums. (See Straight-in Landing, Landing Minimums, Straight-in Approach-VFR)

**STRAIGHT-IN APPROACH – VFR —** Entry into the traffic pattern by interception of the extended runway centerline (final approach course) without executing any other portion of the traffic pattern. (See Traffic Pattern)

**STRAIGHT-IN LANDING —** A landing made on a runway aligned within 30° of the final approach course following completion of an instrument approach. (See Straight-in Approach-IFR)

**STRAIGHT-IN LANDING MINIMUMS / STRAIGHT-IN MINIMUMS —** (See Landing Minimums)

**SUBSTITUTE ROUTE —** A route assigned to pilots when any part of an airway or route is unusable because of NAVAID status. These routes consist of:
1. Substitute routes which are shown on U.S. Government Charts.
2. Routes defined by ATC as specific NAVAID radials or courses.
3. Routes defined by ATC as direct to or between NAVAIDs.

**SUNSET AND SUNRISE —** The mean solar times of sunset and sunrise as published in the Nautical Almanac, converted to local standard time for the locality concerned. Within Alaska, the end of evening civil twilight and the beginning of morning civil twilight, as defined for each locality.

**SURVEILLANCE APPROACH —** An instrument approach wherein the air traffic controller issues instructions, for pilot compliance, based on aircraft position in relation to the final approach course (azimuth), and the distance (range) from the end of the runway as displayed on the controller's radar scope. The controller will provide recommended altitudes on final approach if requested by the pilot. (See PAR Approach) (Refer to AIM Part 1)

**T-VOR / TERMINAL-VERY HIGH FREQUENCY OMNIDIRECTIONAL RANGE STATION —** A very high frequency terminal omnirange station located on or near an airport and used as an approach aid. (See VOR, Navigational Aid)

**TACAN / TACTICAL AIR NAVIGATION —** An ultrahigh frequency electronic rho-theta air navigation aid which provides suitably equipped aircraft a continuous indication of bearing and distance to the TACAN station. (See VORTAC) (Refer to AIM Part 1)

**TACAN ONLY AIRCRAFT —** An aircraft, normally military, possessing TACAN with DME but no VOR navigational system capability. Clearances must specify TACAN or VORTAC fixes and approaches.

**TARGET** — The indication shown on a radar display resulting from a primary radar return or a radar beacon reply. (See Target Symbol, Radar)

*ICAO – TARGET –* In radar:
1. Generally, any discrete object which reflects or retransmits energy back to the radar equipment;
2. specifically, an object of radar search or surveillance.

**TARGET SYMBOL** — A computer generated indication shown on a radar display resulting from a primary radar return or a radar beacon reply.

**TAXI PATTERNS** — Patterns established to illustrate the desired flow of ground traffic for the different runways or airport areas available for use.

**TERMINAL AREA** — A general term used to describe airspace in which approach control service or airport traffic control service is provided.

**TERMINAL AREA FACILITY** — A facility providing air traffic control service for arriving and departing IFR, VFR, Special VFR, Special IFR aircraft and on occasion, en route aircraft. (See Approach Control, Tower)

**TERMINAL CONTROL AREA** — (See Controlled Airspace)

**TERMINAL RADAR APPROACH CONTROL / TRACON** — An FAA air traffic control facility using radar and air/ground communications to provide approach control services to aircraft arriving, departing or transiting the airspace controlled by the facility. Service may be provided to both civil and military airports. A TRACON is similar to a RAPCON (USAF) RATCF (Navy) and ARAC (Army). (See Approach Control, Approach Control Service, Departure Control).

**TERMINAL RADAR PROGRAM** — A national program instituted to extend the terminal radar services provided IFR aircraft to VFR aircraft. Pilot participation in the program is urged but is not mandatory. The progressive stages of the program are referred to as Stage I, Stage II and Stage III. The stage service provided at a particular location is contained in AIM, Part 3.

1. Stage I / Radar Advisory Service for VFR Aircraft – Provides traffic information and limited vectoring to VFR aircraft on a workload permitting basis.
2. Stage II / Radar Advisory and Sequencing for VFR Aircraft – Provides, in addition to Stage I service, vectoring and sequencing on a full-time basis to arriving VFR aircraft. The purpose is to adjust the flow of arriving IFR and VFR aircraft into the traffic pattern in a safe and orderly manner and to provide traffic advisory to departing VFR aircraft.
3. Stage III / Radar Sequencing and Separation Service for VFR Aircraft – Provides, in addition to Stage II services, separation between all participating aircraft. The purpose is to provide separation between all participating VFR aircraft and all IFR aircraft operating within the airspace defined as a Terminal Radar Service Area (TRSA), or Terminal Control Area (TCA). (See Terminal Radar Service Area, Controlled Airspace) (Refer to AIM Parts 1, 3 and 4)

**TERMINAL RADAR SERVICE AREA / TRSA** — Airspace surrounding designated airports wherein ATC provides radar vectoring, sequencing and separation on a full-time basis for all IFR and participating VFR aircraft. Service provided in a TRSA is called Stage III Service. AIM Part 1 contains an explanation of TRSA. Graphics depicting TRSA layouts and communications frequencies are shown in AIM, Parts 3 and 4. Pilot participation is urged but is not mandatory. (See Terminal Radar Program) (Refer to AIM Parts 1, 3 and 4)

**TETRAHEDRON** — A device located on the airport which is used as a landing direction indicator. The small end of a tetrahedron points in the direction of landing and/or wind direction. (See Segmented Circle) (Refer to AIM Part 1)

***THAT IS CORRECT*** — The understanding you have is right.

**THRESHOLD** — The beginning of that portion of the runway usable for landing. (See Airport Lighting, Displayed Threshold)

**THRESHOLD CROSSING HEIGHT / TCH** — The height of the glide slope above the runway threshold. (See Threshold, Glide Slope)

**THRESHOLD LIGHTS** — (See Airport Lighting)

**TIME GROUP** — Four digits representing the hour and minutes from the 24-hour clock. Time group without time zone indicators are understood to be GMT (Greenwich Mean Time); e.g., "0205." A time zone designator is used to indicate local time, e.g., "0205M." The end and beginning of the day are shown by "2400" and "0000," respectively.

**TORCHING** — The burning of fuel at the end of an exhaust pipe or stack of a reciprocating aircraft engine, the result of an excessive richness in the fuel air mixture.

**TOUCH AND GO / TOUCH AND GO LANDING** — An operation by an aircraft that lands and departs on a runway without stopping or exiting the runway.

**TOUCHDOWN** —
1. The point at which an aircraft first makes contact with the landing surface.
2. Concerning a precision radar approach (PAR), it is the point where the glide path intercepts the landing surface.

*ICAO – TOUCHDOWN* - The point where the nominal glide path intercepts the runway.

**TOUCHDOWN RVR** — (See Visibility)

**TOUCHDOWN ZONE** — The first 3,000 feet of the runway beginning at the threshold. The area is used for determination of Touchdown Zone Elevation in the development of straight-in landing minimums for instrument approaches.

**TOUCHDOWN ZONE ELEVATION / TDZE** — The highest elevation in the first 3,000 feet of the landing surface.

TDZE is indicated on the instrument approach procedure chart when straight-in landing minimums are authorized. (See Touchdown Zone)

**TOUCHDOWN ZONE LIGHTING —** (See Airport Lighting)

**TOWER / AIRPORT TRAFFIC CONTROL TOWER —** A terminal facility which through the use of air/ground communications, visual signaling, and other devices, provides ATC services to airborne aircraft operating in the vicinity of an airport and to aircraft operating on the movement area. (See Airport Traffic Area, Airport Traffic Control Service, Approach Control Service, VFR Tower, Movement Area) *ICAO – AERODROME CONTROL TOWER* - A unit established to provide air traffc control service to aerodrome traffic.

**TOWER EN ROUTE CONTROL SERVICE / TOWER TO TOWER —** The control of IFR en route traffic within delegated airspace between two or more adjacent approach control facilities. This service is designed to expedite traffic and reduce control and pilot communication requirements.

**TPX 42 —** A numeric beacon decoder equipment/system. It is designed to be added to terminal radar systems for beacon decoding. It provides rapid target identification, reinforcement of the primary radar target and altitude information from Mode C. (See Automatic Radar Terminal Systems, Transponder)

**TRACK —** The actual flight path of an aircraft over the surface of the earth. (See Course, Route, Flight Path) *ICAO – TRACK* – The projection on the earth's surface of the path of an aircraft, the direction of which path at any point is usually expressed in degrees from north (true, magnetic or grid).

**TRAFFIC ADVISORIES —** Advisories issued to alert a pilot to other known or observed IFR/VFR air traffic which may be in such proximity to his aircraft's position or intended route of flight to warrant his attention. Such advisories may be based on:

1.  visual observation from a control tower,
2.  observation of radar identified and nonidentified aircraft targets on an ARTCC/ Approach Control radar scope, or,
3.  verbal reports from pilots or other facilities.

Controllers use the word "traffic" followed by additional information, if known, to provide such advisories, e.g., "Traffic, 2 o'clock, one zero miles, southbound, fast moving, altitude readout seven thousand five hundred."
Traffic advisory service will be provided to the extent possible depending on higher priority duties of the controller or other limitations, e.g., radar limitations, volume of traffic, frequency congestion or controller workload. Radar/nonradar traffic advisories do not relieve the pilot of his responsibility for continual vigilance to see and avoid other aircraft. IFR and VFR aircraft are cautioned that there are many times when the controller is not able to give traffic advisories concerning all traffic in the aircraft's proximity; in other words, when a pilot requests or is receiving traffic advisories, he should not assume that all traffic will be issued.

*TRAFFIC IN SIGHT —* Used by pilots to inform a controller that previously issued traffic is in sight. (See Negative Contact, Traffic Advisories)

**TRAFFIC INFORMATION —** (See Traffic Advisories)

**TRAFFIC PATTERN —** The traffic flow that is prescribed for aircraft landing at, taxiing on, or taking off from an airport. The components of a typical traffic pattern are upwind leg, crosswind leg, downwind leg, base leg and final approach.

1.  Upwind Leg – A flight path parallel to the landing runway in the direction of landing.
2.  Crosswind Leg – A flight path at right angles to the landing runway off its takeoff end.
3.  Downwind Leg – A flight path parallel to the landing runway in the direction opposite to landing. The downwind leg normally extends between the crosswind leg and the base leg.
4.  Base Leg – A flight path at right angles to the landing runway off its approach end. The base leg normally extends from the downwind leg to the intersection of the extended runway centerline.
5.  Final Approach – A flight path in the direction of landing along the extended runway centerline. The final approach normally extends from the base leg to the runway. An aircraft making a straight-in approach VFR is also considered to be on final approach.

(See Taxi Patterns, Straight-In Approach-VFR) (Refer to AIM Part 1, FAR Part 91) *ICAO – AERODROME TRAFFIC CIRCUIT* - The specified path to be flown by aircraft operating in the vicinity of an aerodrome.

**TRANSCRIBED WEATHER BROADCAST / TWEB —** A continuous recording of meteorological and aeronautical information that is broadcast on L/MF and VOR facilities for pilots. (Refer to AIM Parts 1 and 2)

**TRANSFER OF CONTROL —** That action whereby the responsibility for the separation of an aircraft is transferred from one controller to another. *ICAO – TRANSFER OF CONTROL* - Transfer of responsibility for providing air traffic control service.

**TRANSFERRING CONTROLLER/FACILITY —** A controller/facility transferring control of an aircraft to another controller/facility. *ICAO – TRANSFERRING UNIT/CONTROLLER* - Air Traffic Control Unit/Air Traffic Controller in the process of transferring the responsibility for providing air traffic control service to an aircraft to the next air traffic control unit/air traffic controller along the route of flight.

**TRANSITION —**
1.  The general term that describes the change from one phase of flight or flight condition to another, e.g., transition from en route flight to the approach or transition from instrument flight to visual flight.
2.  A published route (SID Transition) used to connect the basic SID to one of several en route airways/jet routes; or, a published route (STAR Transition) used

to connect one of several en route airways/jet routes to the basic STAR. (Refer to SID/STAR Charts)

**TRANSITION AREA** — (See Controlled Airspace)

**TRANSMISSOMETER** — An apparatus used to determine visibility by measuring the transmission of light through the atmosphere. It is the measurement source for determining runway visual range (RVR) and runway visibility value (RVV). (See Visibility)

*TRANSMITTING IN THE BLIND / BLIND TRANS-MISSION* — A transmission from one station to other stations in circumstances where two-way communication cannot be established, but where it is believed that the called stations may be able to receive the transmission.

**TRANSPONDER** — The airborne radar beacon receiver/transmitter portion of the Air Traffic Control Radar Beacon System (ATCRBS) which automatically receives radio signals from interrogators on the ground, and selectively replies with a specific reply pulse or pulse group only to those interrogations being received on the mode to which it is set to respond. (See Interrogator) (Refer to AIM Part 1)
*ICAO – TRANSPONDER* – A receiver/transmitter which will generate a reply signal upon proper interrogation; the interrogation and reply being on different frequencies.

**TWO WAY RADIO COMMUNICATIONS FAILURE** — (See Lost Communications)

**ULTRAHIGH FREQUENCY / UHF** — The frequency band between 300 and 3,000 MHz. The bank of radio frequencies used for military air/ground voice communications. In some instances this may go as low as 225 MHz and still be referred to as UHF.

*UNABLE* — Indicates inability to comply with a specific instruction, request, or clearance.

**UNCONTROLLED AIRSPACE** — That portion of the airspace that has not been designated as continental control area, control area, control zone, terminal control area, or transition area. (See Controlled Airspace)

*UNDER THE HOOD* — Indicates that the pilot is using a hood to restrict visibility outside the cockpit while simulating instrument flight. An appropriately rated pilot is required in the other control seat while this operation is being conducted. (Refer to FAR Part 91)

**UNICOM** — A non-government air/ground radio communication facility which may provide airport advisory service at certain airports. Locations and frequencies of UNICOMs are shown on aeronautical charts and publications. (Refer to AIM Parts 1 and 2)

*UNPUBLISHED ROUTE* — A route for which no minimum altitude is published or charted for pilot use. It may include a direct route between NAVAIDS, a radial, a radar vector, or a final approach course beyond the segments of an instrument approach procedure. (See Route, Published Route)

**UPWIND LEG** — (See Traffic Pattern)

**VECTOR** — A heading issued to an aircraft to provide navigational guidance by radar.
*ICAO – RADAR VECTORING* - Provision of navigational guidance to aircraft in the form of specific headings, based on the use of radar.

*VERIFY* — Request confirmation of information; e.g., "verify assigned altitude."

*VERIFY SPECIFIC DIRECTION OF TAKEOFF (OR TURNS AFTER TAKEOFF)* — Used by ATC to ascertain an aircraft's direction of takeoff and /or direction of turn after takeoff. It is normally used for IFR departures from an airport not having a control tower. When direct communication with the pilot is not possible, the request and information may be relayed through an FSS, dispatcher, or by other means. (See IFR Takeoff Minimums and Departure Procedures)

**VERTICAL SEPARATION** — Separation established by assignment of different altitudes or flight levels. (See Separation)
*ICAO – VERTICAL SEPARATION* – Separation between aircraft expressed in units of vertical distance.

**VERTICAL TAKEOFF AND LANDING AIRCRAFT / VTOL AIRCRAFT** — Aircraft capable of vertical climbs and/or descents and of using very short runways or small areas for takeoff and landings. These aircraft include, but are not limited to, helicopters. (See Short Takeoff and Landing Aircraft)

**VERY HIGH FREQUENCY / VHF** — The frequency band between 30 and 300 MHz. Portions of this band, 108 to 118 MHz, are used for certain NAVAIDS; 118 to 136 MHz are used for civil air/ground voice communications. Other frequencies in this band are used for purposes not related to air traffic control.

**VERY LOW FREQUENCY / VLF** — The frequency band between 3 and 30 kHz.

**VFR AIRCRAFT / VFR FLIGHT** — An aircraft conducting flight in accordance with visual flight rules. (See Visual Flight Rules)

**VFR LOW ALTITUDE TRAINING ROUTES** — Routes flown by the military services at or below 1,500 feet above the surface at speeds in excess of 250 knots airspeed. These routes are flown only when weather conditions are equal to or better than 3,000 feet ceiling and visibility is greater than 5 miles. These routes are selected to avoid congested airspace. Their descriptions are available at flight service stations and at many airports. (See Olive Branch Routes) (Refer to AIM Part 1)

**VFR ON TOP / VFR CONDITIONS ON TOP** — An authorized IFR altitude issued upon pilot's request in which the aircraft is flown at any appropriate VFR altitude which is at least 1000 feet above any cloud, smoke, haze, or fog layer. (Refer to FAR Part 91)

**VFR OVER THE TOP** — The operation of an aircraft over-the-top under VFR when it is not being operated on an IFR flight plan. (See VFR On Top)

## VFR TOWER / NON-APPROACH CONTROL TOWER
— An airport traffic control tower that does not provide approach control service. (See Tower, Approach Control Service)

**VIDEO MAP** — An electronically displayed map on the radar display that may depict data such as: airports, heliports, runway centerline extensions, hospital emergency landing areas, NAVAIDS and fixes, reporting points, airway/route centerlines, boundaries, handoff points, special use tracks, obstructions, prominent geographic features, map alignment indicators, range accuracy marks, minimum vectoring altitudes.

**VISIBILITY** — The ability, as determined by atmospheric conditions and expressed in units of distance, to see and identify prominent unlighted objects by day and prominent lighted objects by night. Visibility is reported as statute miles, hundreds of feet or meters. (Refer to FAR Part 91, AIM Part 1)

1.  Flight Visibility – The average forward horizontal distance, from the cockpit of an aircraft in flight, at which prominent unlighted objects may be seen and identified by day and prominent lighted objects may be seen and identified by night.
    *ICAO – Flight Visibility – The visibility forward from the cockpit of an aircraft in flight.*
2.  Ground Visibility – Prevailing horizontal visibility near the earth's surface as reported by the United States National Weather Service or an accredited observer.
    *ICAO – Ground Visibility – The visibility at an aerodrome, as reported by an accredited observer.*
3.  Prevailing Visibility – The greatest horizontal visibility equaled or exceeded throughout at least half the horizon circle which need not necessarily be continuous.
4.  Runway Visibility Value / RVV – The visibility determined for a particular runway by a transmissometer. A meter provides a continuous indication of the visibility (reported in miles or fractions of miles) for the runway. RVV is used in lieu of prevailing visibility in determining minimums for a particular runway.
5.  Runway Visual Range / RVR – An instrumentally derived value, based on standard calibrations, that represents the horizontal distance a pilot will see down the runway from the approach end; it is based on the sighting of either high intensity runway lights or on the visual contrast of other targets whichever yields the greater visual range. RVR, in contrast to prevailing or runway visibility, is based on what a pilot in a moving aircraft should see looking down the runway. RVR is horizontal visual range, not slant visual range. It is based on the measurement of a transmissometer made near the touchdown point of the instrument runway and is reported in hundreds of feet.
    a.  Touchdown RVR – The RVR visibility readout values obtained from RVR equipment serving the runway touchdown zone.
    b.  Mid-Field RVR – The RVR readout values obtained from RVR equipment located midfield of the runway.

c.  Rollout RVR – The RVR readout values obtained from RVR equipment located nearest the rollout end of the runway.
*ICAO – Runway Visual Range – The maximum distance in the direction of takeoff or landing at which the runway or the specified lights or markers delineating it can be seen from a position above a specified point on its centreline at a height corresponding to the average eye-level of pilots at touchdown.*

**VISUAL APPROACH** — An approach wherein an aircraft on an IFR flight plan, operating in VFR conditions under the control of an air traffic control facility and having an air traffic control authorization, may deviate from the prescribed instrument approach procedure and proceed to the airport of destination under VFR conditions.
*ICAO – VISUAL APPROACH – An approach by an IFR flight when either part or all of an instrument approach procedure is not completed and the approach is executed in visual reference to terrain.*

**VISUAL APPROACH SLOPE INDICATOR** — (See Airport Lighting)

**VISUAL FLIGHT RULES / VFR** — Rules that govern the procedures for conducting flight under visual conditions. The term "VFR" is also used in the United States to indicate weather conditions that are equal to or greater than minimum VFR requirements. In addition, it is used by pilots and controllers to indicate type of flight plan. (See Instrument Flight Rules, Instrument Meteorological Conditions, Visual Meteorological Conditions) (Refer to FAR Part 91, AIM Part 1)

**VISUAL HOLDING** — The holding of aircraft at selected, prominent, geographical fixes which can be easily recognized from the air. (See Hold, Holding Fixes)

**VISUAL METEOROLOGICAL CONDITIONS / VMC** — Meteorological conditions expressed in terms of visibility, distance from cloud, and ceiling equal to or better than specified minima. (See Instrument Meteorological Conditions, Visual Flight Rules, Instrument Flight Rules)

**VISUAL SEPARATION** — A means employed by ATC to separate IFR aircraft in terminal areas. There are two ways to effect this separation:
1.  The tower controller sees the aircraft involved and issues instructions, as necessary, to ensure that the aircraft avoid each other.
2.  A pilot sees the other aircraft involved and upon instructions from the controller provides his own separation by maneuvering his aircraft as necessary to avoid it. This may involve following another aircraft or keeping it in sight until it is no longer a factor. (See See and Avoid) (Refer to FAR Part 91)

**VOR / VERY HIGH FREQUENCY OMNIDIRECTIONAL RANGE STATION** — A ground-based electronic navigation aid transmitting very high frequency navigation signals, 360 degrees in azimuth, oriented from magnetic north. Used as the basis for navigation in the national airspace system. The VOR periodically identifies itself by morse code and may have an additional voice identification feature. Voice features may be used by ATC or FSS for transmitting instructions/information to pilots. (See Navigational Aid) (Refer to AIM Part 1)

**VORTAC / VHF OMNIDIRECTIONAL RANGE/TACTICAL AIR NAVIGATION** A navigation aid providing VOR azimuth, TACAN azimuth, and TACAN distance measuring equipment (DME) at one site. (See VOR, Distance Measuring Equipment, TACAN, Navigational Aid) (Refer to AIM Parts 1 and 3)

**VORTICES / WING TIP VORTICES —** Circular patterns of air created by the movement of an airfoil through the air when generating lift. As an airfoil moves through the atmosphere in sustained flight, an area of high pressure is created beneath it and an area of low pressure is created above it. The air flowing from the high pressure area to the low pressure area around and about the tips of the airfoil tends to roll up into two rapidly rotating vortices, cylindrical in shape. These vortices are the most predominant parts of aircraft wake turbulence and their rotational force is dependent upon the wing loading, gross weight, and speed of the generating aircraft. The vortices from medium to heavy aircraft can be of extremely high velocity and hazardous to smaller aircraft. (See Wake Turbulence, Aircraft Classes) (Refer to AIM Part 1)

**VOT / VOR TEST SIGNAL —** A ground facility which emits a test signal to check VOR receiver accuracy. The system is limited to ground use only. (Refer to FAR Part 91, AIM Parts 1 and 3)

**WAKE TURBULENCE —** Phenomena resulting from the passage of an aircraft through the atmosphere. The term includes vortices, thrust stream turbulence, jet blast, jet wash, propeller wash and rotor wash, both on the ground and in the air. (See Jet Blast, Aircraft Classes) (Refer to AIM Part 1)

**WARNING AREA —** Specified international airspace within which there may exist activities constituting a potential danger to aircraft. Warning areas are depicted on aeronautical charts.

**WAYPOINT —** (See Area Navigation)

**WEATHER ADVISORY / INFLIGHT WEATHER ADVISORY —** (See SIGMET, AIRMET)

**WEATHER ADVISORY / WS / WA —** In aviation weather forecast practice, an expression of hazardous weather conditions not predicted in the area forecast, as they affect the operation of air traffic and as prepared by the NWS.

*WILCO —* I have received your message, understand it, and will comply with it.

**WIND SHEAR —** A change in wind speed and/or wind direction in a short distance, resulting in a tearing or shearing effect. It can exist in a horizontal or vertical direction and occasionally in both.

*WORDS TWICE —*
1. As a request: "Communication is difficult. Please say every phrase twice."
2. As information: "Since communications are difficult, every phrase in this message will be spoken twice."

# ABBREVIATIONS

Note: "s" may be added for plural, or as appropriate.

## A

AAS..... Airport Advisory Service
A/C..... Approach Control
acft..... aircraft
ADCUS.. Advise Customs
ADF..... Automatic Direction Finder
AGL..... above ground level
AID..... Airport Information Desk
AIM..... Airman's Information Manual
ALS..... Approach light system
apch..... approach
apchg... approaching
aprx..... approximate
arpt..... airport
ARSR.... Air Route Surveillance Radar
ARTCC... Air Route Traffic Control Center
ASDE.... airport surface detection equipment
ASR..... Arpt Surveillance Radar
ATC..... air traffic control
ATCT.... air traffic control tower
ATIS..... Automatic Terminal Information Service
avbl..... available
awy..... airway

## B

BC..... back course
bcn..... beacon
bcst..... broadcast
bldg..... building
brg..... bearing
btn..... between

## C

CFR..... crash fire rescue
CIFFR... common IFR room
clsd..... closed
cmsnd... commissioned
cntr..... center
cntrln.... centerline
Comlo... Compass locator
const.... construction
crcg..... circling
CS/T.... combined station/tower
ctc..... contact
CTLZ.... Control Zone

## D

dalgt.... daylight
dcmsnd.. decommissioned
degs..... degrees
DF...... direction finder

DME..... UHF standard (TACAN compatible) distance measuring equipment
dsplcd... displaced
durg..... during
DVFR.... Defense Visual Flight Rule

## E

E....... east
elev..... elevation
emerg... emergency
equip.... equipment

## F

FL....... Flight Level
FM...... fan marker
freq..... frequency
FSS..... Flight Service Station

## G

GS...... glide slope
GWT.... gross weight

## H

HIRL.... High intensity Runway Lights
hwy..... highway

## I

ident.... identification
IFR...... Instrument Flight Rules
IFSS..... International Flight Service Station
ILS...... instrument landing system
info..... information
intl...... international
ISMLS... Interim Standard Microwave Landing System

## J

J-bar.... jet runway barrier

## K

kHz..... kilohertz

## L

lat...... latitude
lctd..... located
LDA..... Localizer type directional aid
lgts..... lights
lgtd..... lighted
LMM..... compass locator at middle marker ILS
lndg.... landing
loc...... localizer

LOM.... compass locator at outer marker ILS
long..... longitude
LRCO.... Limited Remote Communications Outlet

## M

MAA.... maximum authorized altitude
mag..... magnetic
maint.... maintain, maintenance
MALS... Medium Intensity Approach Light System
MALSR.. Medium Intensity Simplified Short Approach Light System with Rail
max..... maximum
MCA.... minimum crossing altitude
MEA.... minimum enroute IFR altitude
MHz.... megahertz
min..... minimum or minute
MIRL.... Medium Intensity Runway Edge Lights
MLS.... Microwave Landing System
MM..... middle marker ILS
MOCA... minimum obstruction clearance altitude
MRA.... minimum reception altitude
MSL..... mean sea level
muni.... municipal

## N

N....... north
NA...... not authorized
natl..... national
navaid.. navigational aid
NDB.... Non-directional Radio Beacon
ngt..... night
NM..... nautical mile(s)
Nr....... number

## O

obstn.... obstruction
OM..... outer marker ILS
oper.... operate
opn..... operation
OTS.... Out of Service
ovrn..... overrun

## P

PAR..... Precision Apch Radar

permly... permanently

## Q

quad.... quadrant

## R

RAIL.... Runway Alignment Indicator Lights
RAPCON. radar approach control (USAF)
RATCF.. radar air traffic control facility (USN)
RCAG... Remote Center air/ground
RCLS.... Runway Centerline Lights System
RCO.... Remote Communications Outlet
rcv..... receive
rcvg.... receiving
rcvr.... receiver
REIL.... Runway End Identifier Lights
req..... request
rqrd.... required
rgt...... right
RRP.... Runway Reference Point
ruf...... rough
RTS..... returned to service
RVR..... runway visual range
RVRC.... Runway Visual Range Center
RVRT.... Runway Visual Range Touchdown
RVRR.... Runway Visual Range Rollout
RVV.... runway visibility values
RWY.... Runway

## S

S........ south
SDF.... Simplified Directional Facility
sfc...... surface
SID..... Standard Instrument Departure
SM...... statute mile(s)
SR....... sunrise
SS....... sunset
STAR.... Standard Terminal Arrival Route
STOL.... Short take-off & landing rnwy
svc..... service

## T

T....... true (after a bearing)

## TACAN

TACAN.. UHF navigational facility—omnidirectional course and distance information
TCA..... Terminal Control Area
TCH..... Threshold Crossing Height
tfc...... traffic
thr...... threshold
tkof..... take-off
tmprly... temporarily
tmpry... temporary
TPA..... Traffic Pattern Altitude
TRACON. Terminal Radar approach control
TRSA.... Terminal Radar Service Area
tsmt..... transmit
tsmtg.... transmitting
tsmtr.... transmitter
TV....... television
TWEB.... transcribed weather bcst
twr..... tower
twy..... taxiway

## U

UHF..... Ultra high frequency
unavbl... unavailable
unctld... uncontrolled
unlgtd... unlighted

## V

VASI.... Visual Approach Slope Indicator
VFR..... visual flight rules
VGS..... Visual Guidance System
VHF..... Very high frequency
VOR.... VHF Omni-Directional Radio Range
VORTAC. Combined VOR and TACAN System
VOT..... a VOR Receiver testing facility
vsby..... visibility

## W

W....... west
WS...... Weather Service
wt....... weight

## Z

Z....... Greenwich mean time

# DEAD RECKONING COMPUTERS

## Section 1. THE E-6B TYPE COMPUTER THE CIRCULAR SLIDE RULE

### GENERAL

*Use.* In air navigation, a circular slide rule is used for solving problems in multiplication and division, distance conversions, corrections of altitude and airspeed for temperature and pressure variations, rate-time distance problems, fuel consumption, radius of action formula, and off-course corrections. The slide rule face of the E-6B computer is standard with all major types of navigation computers.

*Ratio and Proportion Scales.* The outer and inner scales on the slide rule face of the E-6B computer are used to express the ratio and proportion of numerical values of any denomination.

*(1) Ratio.* The ratio of any two numbers is the quotient when the first number is divided by the second number. This ratio may be expressed as a fraction. The first number will be the numerator and the second the denominator. For example, the ratio of 6 to 8 may be expressed as $\frac{6}{8}$. In solving ratio problems on the E-6B computer, consider a number on the outer scale of the computer as a numerator and a number on the inner scale as a denominator.

*(2) Proportion.* A proportion is the expression of equality of two or more ratios. For example the ratio of 6 to

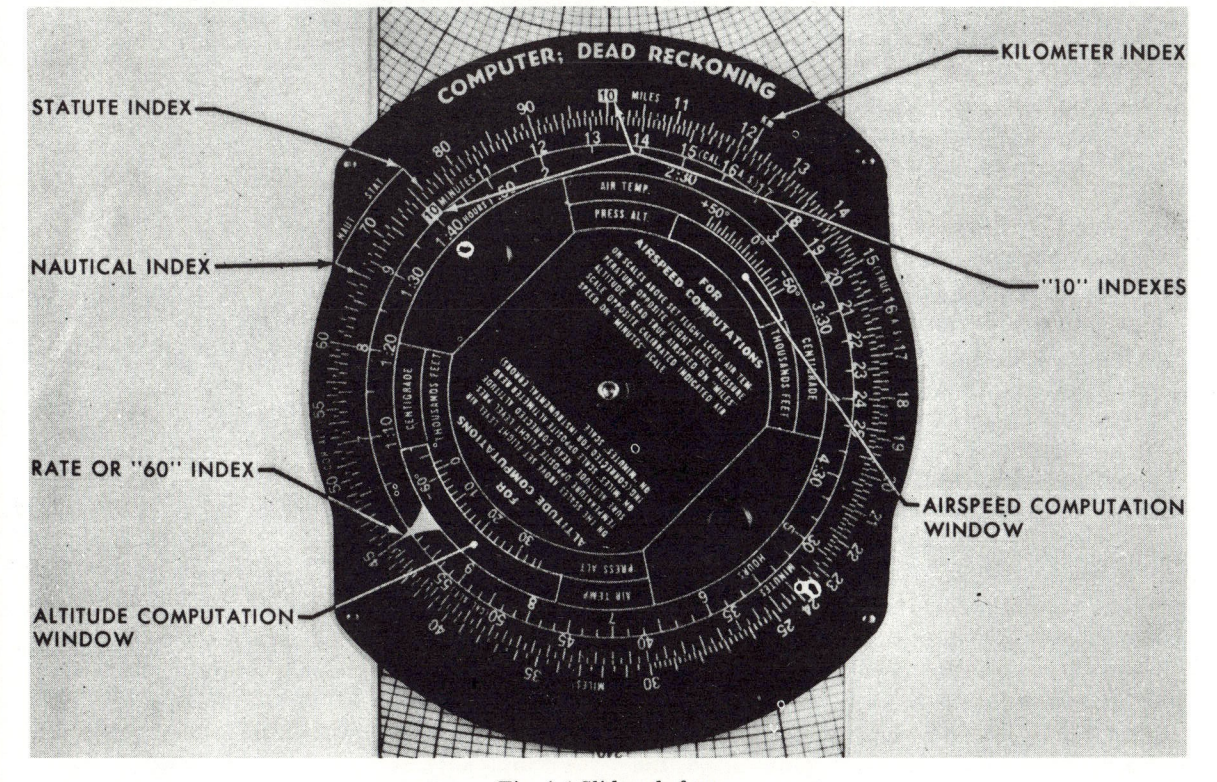

*Fig. A-1 Slide rule face.*

8 is equal to the ratio of 3 to 4 ($\frac{6}{8}$ equals $\frac{3}{4}$) When any ratio is set on the computer, any other equal ratios may be read. If 6 is set over 8, 12 is over 16, 18 over 24, etc. This gives a proportion as follows:

$$\frac{6}{8} = \frac{12}{16} = \frac{18}{24} \text{ etc.}$$

(3) Applying ratios and proportions to practical problems. To what number does 45 have the same ratio as 6 to 8? By setting 6 over 8, the unkown number is found under 45. In this case, the un-known number is 60. It can then be said that 45 is to 60 as 6 is to 8 or $\frac{45}{60} = \frac{6}{8}$. For example, if 6 men earn 8 dollars, at the same rate 9 men will earn 12 dollars; if an aircraft climbs 6,000 feet in 8 minutes, at the same rate it will climb 9,000 feet in 12 minutes; if 6 gallons of fuel are consumed in 8 minutes, 9 gallons will be consumed in 12 minutes.

## DESCRIPTION

*Scale.* The numbers on any computer scales, as on any slide rule, represent multiples of 10 of the values shown. For example, the number 24 on the outer scale may represent 0.24, 2.4, 24, 240, or 2400. The numbers on the inner scale for 24 have like values. On the inner scale, minutes may be converted to hours by reference to the adjacent hour scale. For example, adjacent to 24, in this case meaning 240 minutes, is found 4 hours.

*Scale Divisions.* Relative values should be kept in mind when reading the computer. For example, the numbers 21 and 22 on either scale are separated by 5 dividing lines, spaced 2 units apart. The second division past 21, would be read as 21.4, 2140, etc. Spacing of these divisions should be studied, as the breakdown of dividing lines may be into units of 1, 2, 5, or 10.

*Indexes.* The computer has four indexes on the outer scale. Three of the indexes are used in establishing proportional relationship be-tween statute miles, nautical miles, and kilo-meters; the fourth is an index for multipli-cation and division. These indexes are called the nautical index marked "Naut" at 66, the statute index marked "Stat" at 76, and the kilometer index marked "Km" at 122. A rate arrow on the inner scale is also known as the 60 index. The 10's on both scales are indexes for multiplication and division.

*Airspeed and Altitude Computation Windows.* There are two windows for use in making temperature and pressure corrections for airspeed and altitude. One is marked "FOR AIRSPEED COMPUTATIONS"; the other, "FOR ALTITUDE COMPUTATIONS."

## DISTANCE CONVERSION

Problem. How many statute miles equal 90 nautical miles? How many kilometers equal 90 nautical miles?

Solution. Using an E-6B computer, refer to A-2 and solve as follows:

(1) Set 90 on inner scale to "Naut" index.

(2) Read 104 under "Stat" index (104 statute miles).

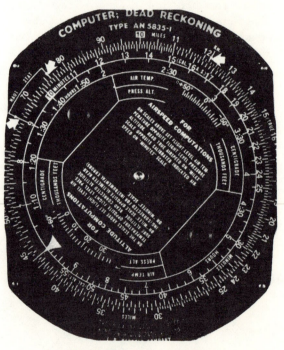

Fig. A-2 Distance conversion.

(3) Read 166 under "Km" index (166 kilo-meters).

## DETERMINING GROUNDSPEED

Ground speed equals distance divided by time.

Problem. What is the groundspeed if it takes 35 minutes to fly 80 nautical miles?

Solution: Using an E-6B computer, refer to A-3 and solve as follows:

(1) Set 35 (inner scale) opposite 80 (outer scale).

(2) Over 60 index read groundspeed (137 knots).

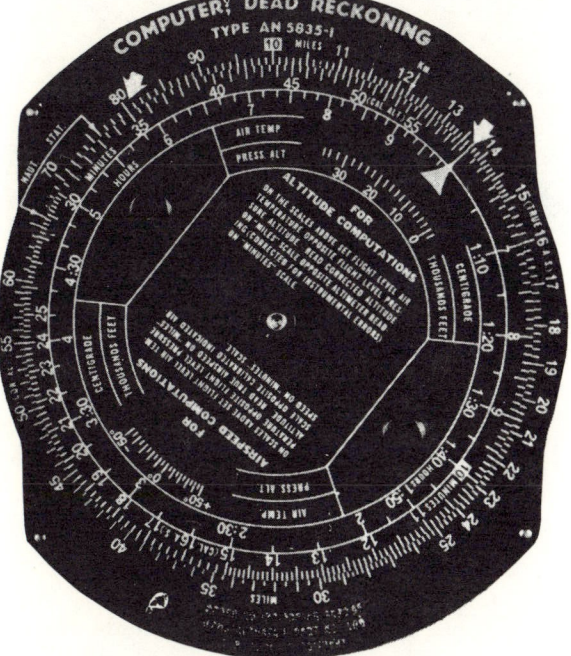

*Fig. A-3 Determining groundspeed.*

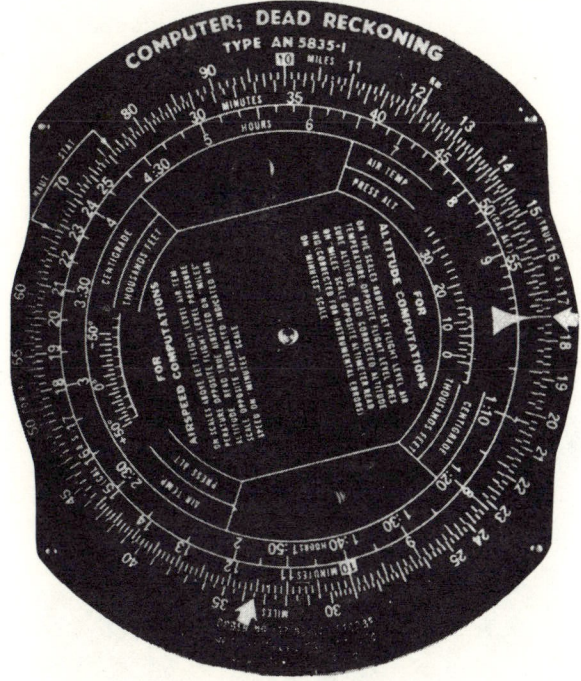

*Fig. A-4 Determining time.*

### DETERMINING TIME REQUIRED

Time equals distance divided by groundspeed.

Problem. How much time is required to fly 335 nautical miles at a groundspeed of 174 knots?

Solution. Using an E-6B computer, refer to A-4 and solve as follows:

(1) Set rate or 60 index on 174 (outer scale)

(2) Under 335 (outer scale) read 115 minutes (inner scale) 1 + 55 (Hours scale).

### DETERMINING DISTANCE

Distance equals groundspeed multiplied by time.

Problem: How far does an aircraft travel in 2 hours, 15 minutes at a groundspeed of 133 knots?

Solution. Using an E-6B computer, refer to A-5 and solve as follows:

(1) Set 60 index at 133 (outer scale).

(2) Over 135 (inner scale) or 2 hours, 15 minutes (hours scale), read 300 nautical miles (outer scale).

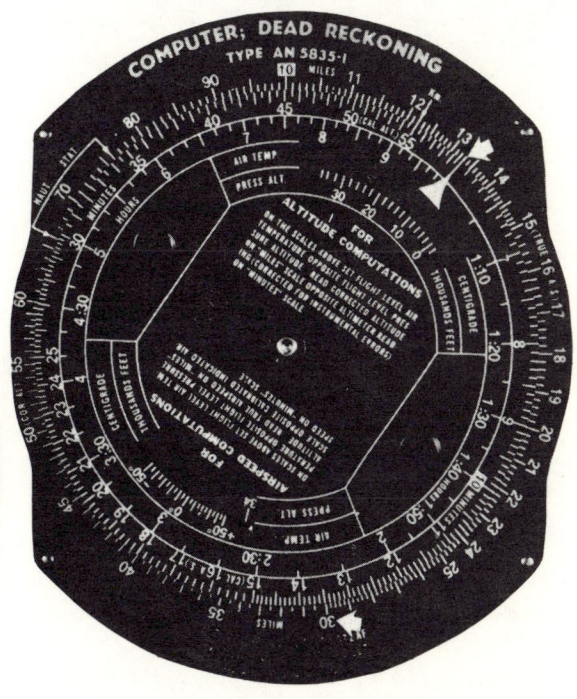

*Fig. A-5 Determining distance.*

### DETERMINING RATE OF FUEL CONSUMPTION

Rate of fuel consumption equals gallons of fuel consumed divided by time.

Problem. What is the rate of fuel consumption if 30 gallons of fuel are consumed in 111 minutes (1 hour and 51 minutes)?

Solution. Using an E-6B computer, refer to A-6 and solve as follows:

(1) Set 111 (inner scale) under 30 on outer scale (in this case outer scale is used to represent gallons).

(2) Opposite 60 index, read 16.2 gallons per hour (gph).

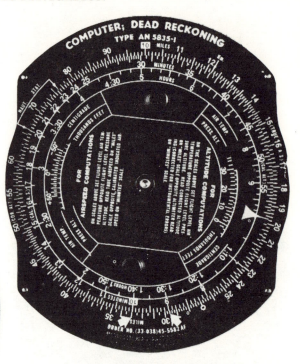

Fig. A-7 Altitude gained in a given time at a given rate of climb.

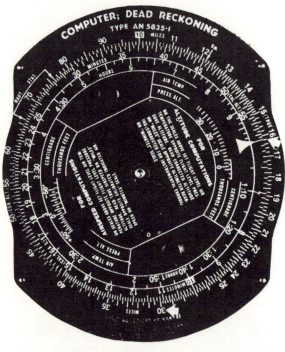

Fig. A-6-Determining rate of fuel consumption.

## DETERMINING GALLONS USED IN A GIVEN TIME

Place 60 index under rate (gph) and read gallons used over time.

## ALTITUDE IN A GIVEN TIME AT A GIVEN RATE OF CLIMB

Outer scale is used to represent rate of climb and height climbed. The 10 index is used instead of 60 index.

Problem. How many feet of altitude are gained in 9 minutes if an aircraft climbs at a rate of 300 feet per minute?

Solution. Using an E-6B computer refer to A-7 and solve as follows:

(1) Set 10 index under 330 (outer scale).

(2) Over 9 minutes (inner scale) read 2,970 feet climbed.

## AVERAGE RATE OF CLIMB

Rate of climb equals height climbed divided by time in minutes.

Problem. What is the average rate of climb

in feet per minute if an aircraft climbs from 4,000 feet to 5,220 feet altitude in 4 minutes?

Solution. Using an E-6B computer refer to A-8 and solve as follows:

(1) 5,200 - 4,000 = 1,220 feet (number of feet climbed).

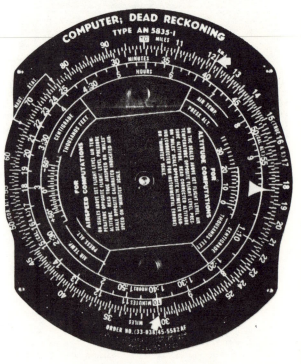

Fig. A-8 Average rate of climb.

(2) Set 4 minutes under 1,200 feet (outer scale).

(3) Over 10 index, read 305 from (outer scale).

## TIME-DISTANCE

Time-distance problems are worked on the inner (minutes) scale and the outer (miles) scale.

Problem. If 50 minutes are required to travel 120 nautical miles, how many minutes are required to travel 86 nautical miles at the same rate?

Solution. Using an E-6B computer, refer to A-9 and solve as follows:

(1) Set 50 (inner scale) under 120 (outer scale).

(2) Under 86 (outer scale), read 36 (inner scale) minutes required.

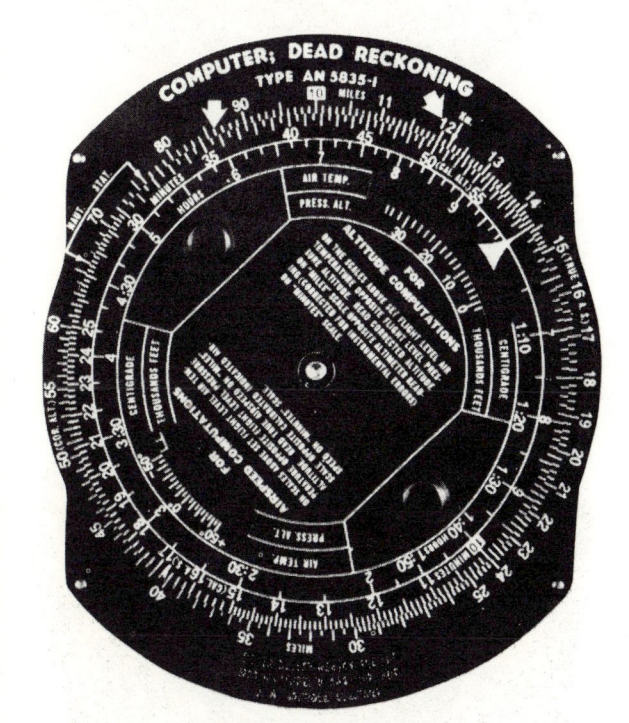

*Fig. A-9 Time and distance.*

## USE OF THE 36 INDEX

The number 36 on the inner scale is used in solving rate-time-distance problems in instrument flight when time must be calculated in seconds and minutes, instead of minutes and hours. An example of such a problem is to determine the time required to fly from the outer marker to the middle marker or from the middle marker to the point of touchdown during an instrument approach.

*Formula.* Problems where seconds must be used as a unit of time may be solved by the formula

$$\frac{GS}{36} = \frac{Distance}{Seconds}$$

in which GS is the groundspeed; 36 represents the number of seconds in one hour (3,600); distance is the number of miles or decimal parts of miles to be flown; and seconds is the time required to fly that distance.

Problems involving less than 60 seconds.

(1) Problem. What is the time required to from the middle marker to the point of touchdown if the groundspeed is 100 knots and the distance between these points is 0.5 nautical miles?

(2) Solution. Set 36 (inner scale) under groundspeed (100 on the outer scale). Under 5 (0.5) on the outer scale, read 18 seconds on the inner scale.

Problems involving more than 60 seconds.

(1) Problem. What is the time required fly from the outer marker to the middle marker if the groundspeed is 95 knots and the distance between the two points if 5 nautical miles?

(2) Solution. Set 36 (inner scale) under the groundspeed (95 knots on the outer scale). Under 5 on the outer scale, read 190 seconds, or 3 minutes, 10 seconds.
NOTE: When using the minutes scale as a second scale, the hour scale becomes a minute scale.

## FUEL CONSUMPTION

Use same scales as used with the time-distance problems discussed and solve the following fuel consumption problem:

Problem. Forty gallons of fuel have been consumed in 135 minutes (2 hours + 15 minutes) flying time. How much longer can the aircraft continue flying if 25 gallons of available fuel (usable fuel not including reserve) remain and the rate of consumption remains unchanged?

Solution. Using an E-6B computer, refer to A-10 and solve as follows:

(1) Set 135 (inner scale) under 40 (outer scale).

(2) Under 25 (outer scale), read 84½ (inner scale) minutes fuel remaining.

*Fig. A-10 Fuel consumption.*

to converge? In what direction is the correction applied?

*Fig. A-11 Off-course correction to parallel.*

## OFF-COURSE CORRECTION

An aircraft headed one degree off-course will be one mile off-course for each 60 miles flown. This is the rule of 60. Inversely, for each miles an aircraft is off-cours after each 60 miles of flight, one degree of correction will be required to parallel the intended course. Applied to other distances (multiples of 60), such as 1½ miles off-course in 90 miles, 2 miles off-course in 120 miles, or 2½ miles off-course 150 miles, a correction of one degree will be required to parallel the intended course. To converge at destination, an extra correction must be made based on the same rule of 60.

*Formula.* The degrees correction required to converge at destination is determined by adding the results of the following formulas:

$$\text{Correction to parallel course} = \frac{\text{miles off-course}}{\text{miles flown}} = \frac{\text{degrees correction}}{60}.$$

Additional correction to converge.

$$\frac{\text{miles off-course}}{\text{miles to fly}} = \frac{\text{degrees correction}}{60}$$

Problem. An aircraft is 10 nautical miles to the left of course when 150 nautical miles from departure point A. How many degrees correction are required to parallel course? If 80 nautical miles remain to destination B, how many additional degrees are required

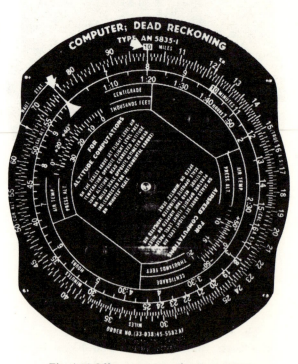

*Fig. A-12 Off-course correction to coverage.*

Solution. Using an E-6B computer, refer to A-11 and A-12, and solve as follows:

(1) Set 150 (Inner scale) under 10 (outer scale) (A-11).

344

(2) Over the 60 index, read 4° (correction required to parallel).

(3) Set 30 (inner scale) under 10 (outer scale) (A-12).

(4) Over 60 index, read $7\frac{1}{2}°$ to converge.

(5) $4° + 7\frac{1}{2}° = 11.5°$, total correction to converge at destination. Since aircraft is off-course to the left, correction will be made to the right, or added to the original heading. For example, if the original heading was 090°, the new heading is $101\frac{1}{2}°$ or 102° to the nearest degree.

RADIUS OF ACTION (FIXED BASE)

As discussed, radius of action to the same base refers to the maximum distance an aircraft can be flown on a given course and still be able to return to the starting point within a given time. The amount of available fuel (not including reserve fuel) is usually the factor determining time.

Problem. The groundspeed on the outbound leg of the flight is 160 knots; on the return leg, 130 knots. Available fuel permits $4\frac{1}{2}$ hours (270 minutes) total time for the flight. How many minutes will be available for the outbound leg of the flight? How many minutes will be required for the return leg of the flight? What is the radius of action?

Solution. The sum of the groundspeed out

($GS_1$) and the groundspeed on the return leg ($GS_2$) is to the total time in minutes (T), as the groundspeed on the return leg ($GS_2$) is to the time in minutes on the outbound leg ($t_1$).

Minutes on the outbound leg of the flight can be calculated by the formula

$$\frac{GS_1 + GS_2}{T} = \frac{GS_2}{t_1}.$$

The formula for calculating time required for the return leg of the flight is

$$\frac{GS_1 + GS_2}{T} = \frac{GS_1}{t_2},$$

in which $t_2$ is the time required for the return leg of the flight.

These formulas can be calculated on the E-6B computer as ratio and proportion problems and appear on the E-6B computer as they appear in mathematical form. To solve radius of action fixed base problems with the E-6B computer, use the problem given in a above, referring to figures A-13 and A-14 and proceed as follows:

(1) Find the sum of the groundspeeds (160 + 130 = 290)

(2) Set the total time (T=$4\frac{1}{2}$ hours or 270 minutes) under the sum of the groundspeeds (290) (A-13).

(3) Under 130 ($GS_2$), read the time on the outbound leg, 2 hours + 1 minute or 121 minutes (A-13).

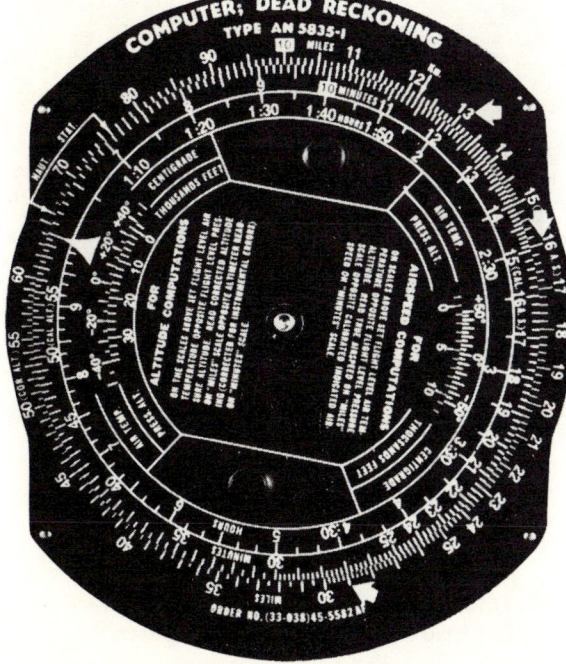

Fig. A-13 Reading time for outbound and inbound leg.

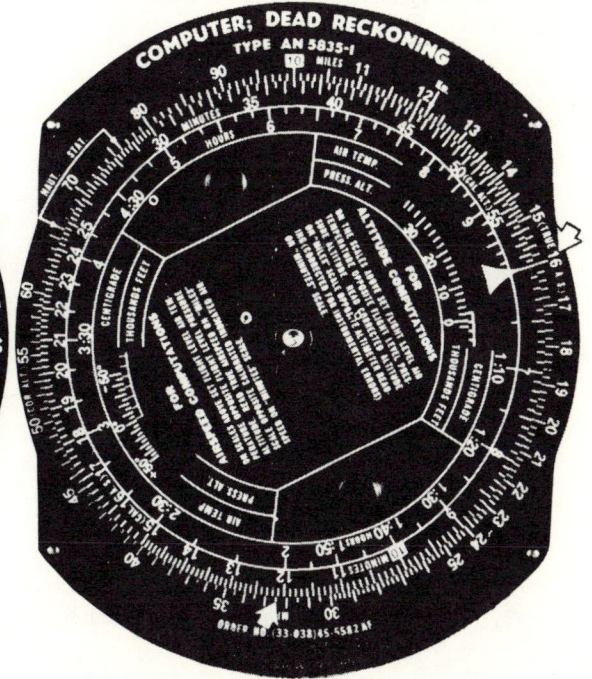

Fig. A-14 Finding radius of action.

(4) Without changing the setting of the computer, under 160 (GS$_1$), read the time required for the return leg, 2 hours + 29 minutes or 149 minutes (A-13).

(5) These two amounts of time should be equivalent to the total amount of time of the flight.

(6) Place the 60 index under 160 (GS$_1$) and over 121 minutes (time on the outbound leg), read the radius of action, 324 nautical miles (A-14).

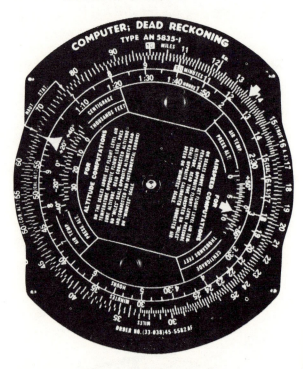

*Fig. A-15 Airspeed computation.*

AIRSPEED COMPUTATIONS

The window marked FOR AIRSPEED COMPUTATIONS provides a means for computing true airspeed when indicating airspeed, temperature, and altitude are known or vica versa. To change from one to the other, it is necessary to correct for altitude and temperature differences existing from those that are standard at sea level. Free air temperature is read from a free air thermometer and the pressure altitude is found by setting the altimeter at 29.92″ Hg and reading the altimeter directly.

Problem. The indicated airspeed is 125 knots, free air temperature is —15° centigrade, and the pressure altitude is 8,000 feet. What is the true airspeed?

Solution. Using an E-6B computer, refer to

A-15 and solve as follows:

(1) Set 8,000 against —15°C. in the airspeed computation window.

(2) Over 125 knots (inner scale), read true airspeed 137 knots (outer scale).

ALTITUDE COMPUTATIONS

The window marked FOR ALTITUDE COMPUTATIONS provides a means for computing corrected altitude by applying any variations from standard temperature to indicated (or calibrated) altitude.

Problem. The pressure altitude is 9,000 feet, indicated altitude is 9,100 feet, and the free air temperature is —15°C. What is the corrected altitude?

Solution. Using an E-6B computer, refer to A-16 and solve as follows:

(1) Set 9,000 against —15°C. in the altitude computation window.

(2) Above 9,100 feet (calibrated) indicated altitude (inner scale), read corrected altitude 8,700 on the outer scale (corrected altitude).

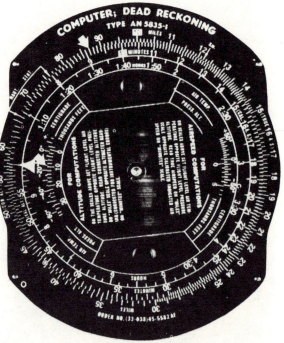

*Fig. A-16 Altitude computation.*

## Section II. GRID SIDE OF E-6B COMPUTER

PLOTTING DISC AND CORRECTION SCALES

The grid side of the E-6B computer (A - 17) enables the pilot to quickly and conveniently solve triangle of velocity problems during flight. It consists of a transparent, rotatable plotting disc mounted in a metal or plastic frame on the reverse side of the circular slide rule. A compass rose is located on the periphery of the plotting disc. The correction scale, on the top frame of the circular grid, is graduated in degrees right and left of the true index. This scale is used for calculating drift or drift correction and is labeled drift right and drift left. The scale may also be used for correcting for variation east or west. A small reference circle or grommet is located at the center of the plotting disc.

### SLIDING GRID

A reversible sliding grid (A - 17) inserted between the circular slide rule and the plotting disc is used with the plotting disc for computing triangles of velocities. The slide is imprinted with converging lines space 2° apart between concentric circles marked 30 to 100 and 1° above the 100 concentric circle. The concentric circles are used for calculations of speed and are spaced 2 units (usually knots or miles per hour) apart. Direction of the centerline coincides with the true index. The common center of the concentric circles

and the point at which all converging lines meet is at a point below the lower end of the slide. On one side of the sliding grid the speed circles are numbered from 30 to 300; on the reverse side from 230 to 400. This reverse side also has a rectangular grid numbered from 0 to 90.

### Section III. TRIANGLE OF VELOCITIES (CENTERLINE AS GROUND VECTOR)

### GENERAL

In solving a triangle of velocities on the computer, part of the triangle is plotted on the transparent surface of the circular disc. Lines which are printed on the slide are used for the other two sides of the triangle. Actually, there is not room on the computer for the whole triangle, for the center of the concentric speed circles (A - 18) is one vertex of the triangle. There are many methods applicable for computing any one problem but the following method for each type of problem is standard for use by the pilot. This section includes problems where the centerline is used as ground vector and the wind vector is plotted above the grommet.

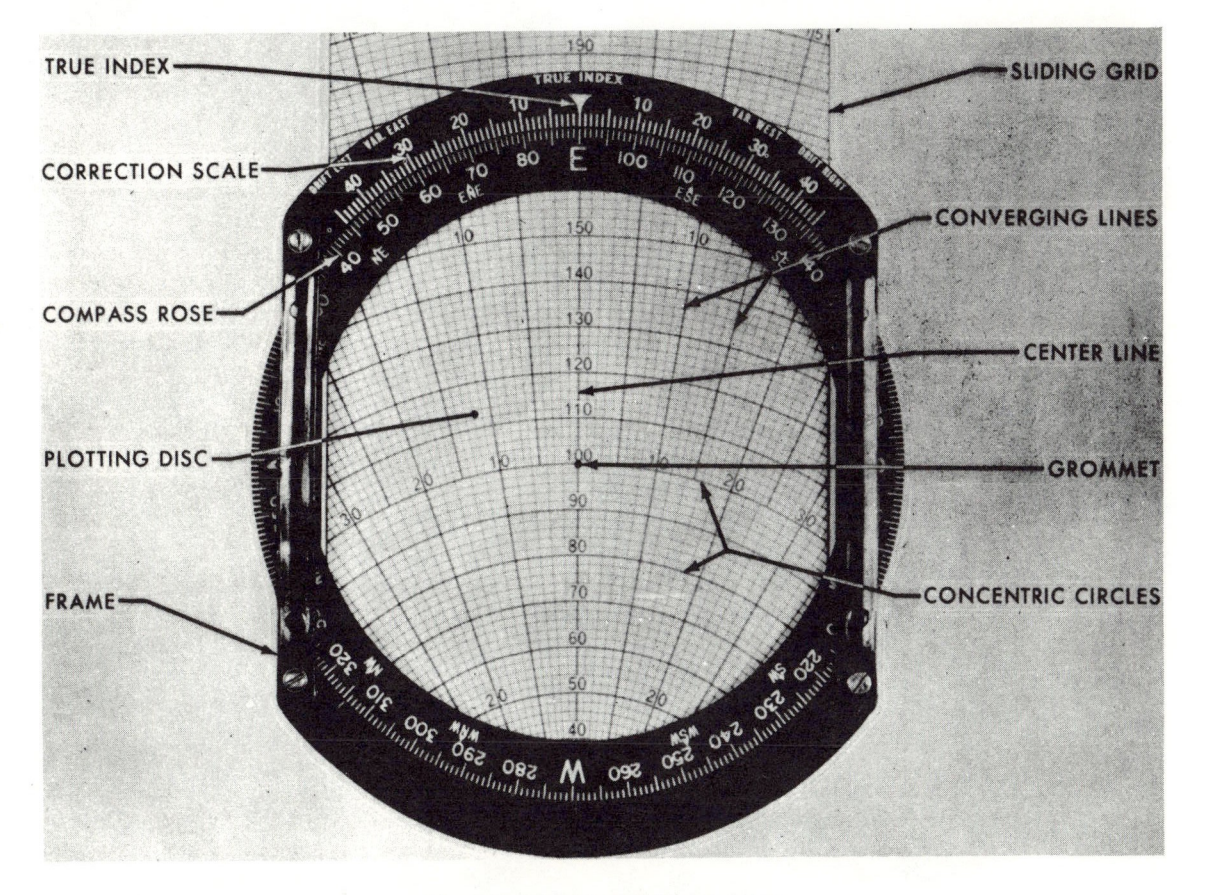

*Fig. A-17 Grid side of E-6B computer.*

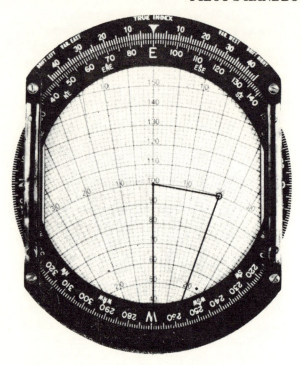

*Fig. A-18 Triangle of velocity plotted on the E-6B computer.*

**HEADING AND GROUNDSPEED COMPUTATION**

When plotting heading and groundspeed with course, true airspeed and wind velocity known, the following method is used:

Problem. The wind is from 160°/30 knots, the true airspeed 120 knots, true course 090°.

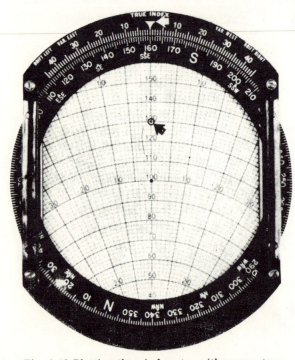

*Fig. A-19 Plotting the wind vector with course, true airspeed, and wind velocity known.*

What is the heading and groundspeed?

Solution. Using an E-6B computer, refer to A-19 and A-20 and solve as follows:

(1) Set 160° (direction from which the wind is blowing) to the true index (A-19).

(2) Plot the wind vector above the grommet 30 units (wind speed) and place a dot within a circle at this point.

(3) Set 090° (true course) at the true index (A-20).

(4) Adjust sliding grid so that the true airspeed (120 knots) is at the wind dot.

(5) Note that the dot is at the 14° converging line to the right of centerline.

(6) Under the 14° correction scale (labeled "Drift" scale) to the right of center at the top of the computer, read the true heading (104°).

(7) Under the grommet, read the groundspeed (106 knots).

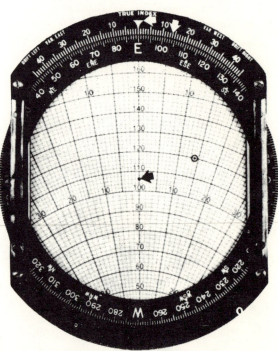

*Fig. A-20 Reading heading, wind correction, and groundspeed.*

**HEADING AND TRUE AIRSPEED COMPUTATION**

This problem was solved graphically in earlier problem. Here, it is solved by the E-6B computer.

Problem. Wind is from 090°/20 knots, true course 120°, groundspeed 90 knots. What is the true heading and true airspeed?

Solution. Using an E-6B computer, refer to A-21 and solve as follows:

(1) Set 090 (wind direction) under the true

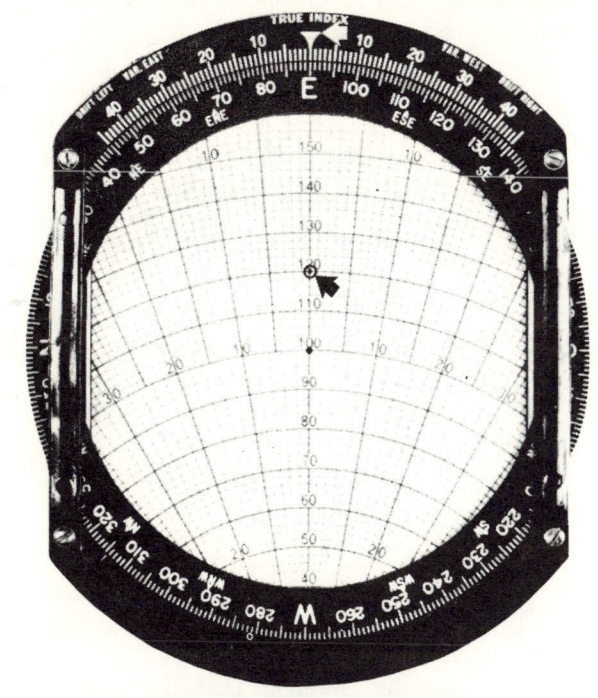

*Fig. A-21 Plotting wind vector.*

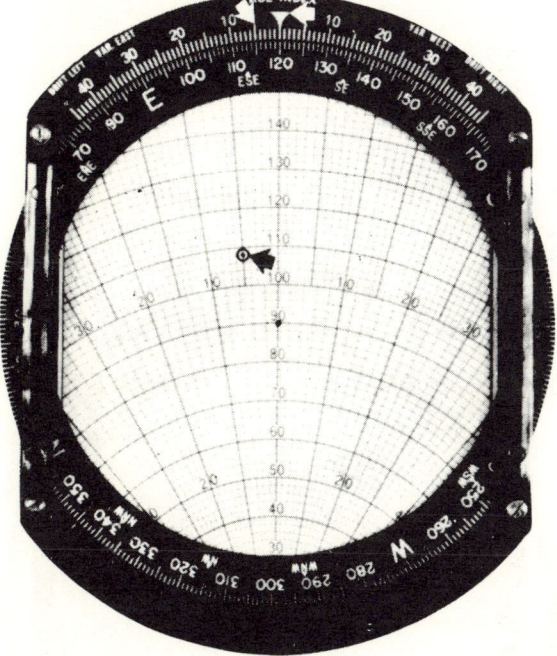

*Fig. A-22 Reading true heading, drift correction, and true airspeed.*

index and plot wind vector 20 units above the grommet using dot whithin circle (A - 21).

(2) Set true course (120°) to the true index. (A - 22)

(3) Move sliding grid so that groundspeed (90 knots) concentric circle is at the grommet.

(4) The wind dot is now on the converging line 5° to the left of true index, read the true heading (115°).

(5) Under the wind dot, read the true airspeed (108 knots).

WIND VELOCITY COMPUTATION

This is identical to the problem solved graphically earlier. Here, it is solved on the E-6B computer.

Problem. True heading 130°, true airspeed 100 knots, track 140°, groundspeed 90 knots. What is the wind velocity?

Solution. Using an E-6B computer, refer to A-23 and A-24 and solve as follows:

(1) Set track (1409) at true index and grommet over the groundspeed (90 knots).

(2) Since the true heading is 10° less than the track, find where the 10° converging line to the left of centerline crosses the 100 knots (true airspeed) line and place a dot within a circle at this point (A - 23).

(3) Turn circular grid until the dot is directly above the grommet (A - 24).

(4) Under the true index, read direction from which the wind is blowing (075°). The distance in units between the dot and the grommet indicate the speed of the wind (20 knots).

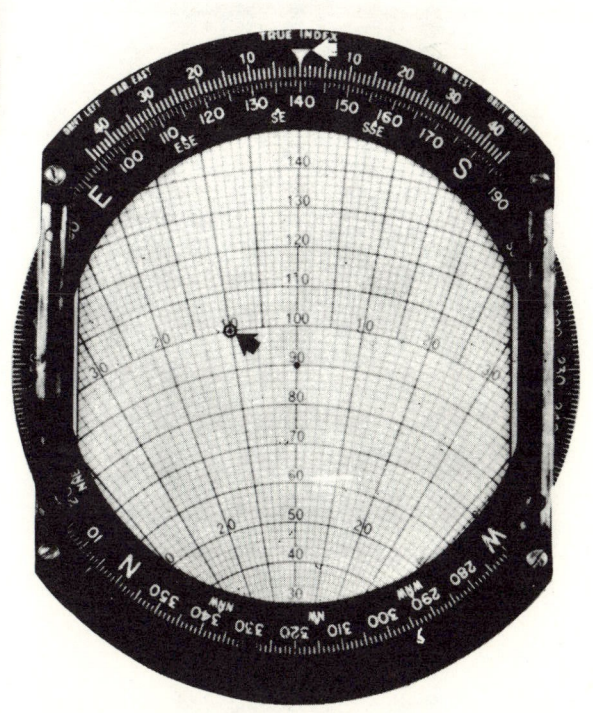

*Fig. A-23 Preparing computer.*

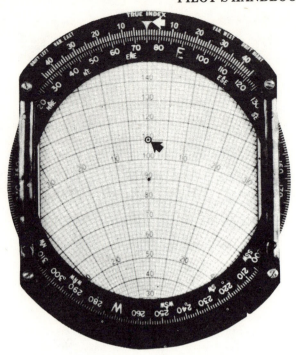

*Fig. A-24 Reading wind velocity.*

RADIUS OF ACTION (FIXED BASE)
COMPUTATION

This problem was solved graphically earlier.
Here, the same problem is solved on the E-6B
computer.

Problem. The true course out is 090°, true
airspeed 100 knots, wind velocity 030°/20
knots, total flight time 2 hours. What is the

true heading and groundspeed out? The true
and groundspeed back?

Solution. Using an E-6B computer, refer to
A-25 and A-26 and solve as follows:

(1) Set (not shown) the wind direction
(030°) to the true index and plot the
wind vector 20 units above the grommet using a dot within a circle.

NOTE: This wind dot is also used to plot
return leg.

(2) Set true course outbound (090°) under
the true index.

(3) Move sliding grid so that wind dot is at
the true airspeed (100 knots).

(4) The wind dot is now on the converging
line 10° to the left of centerline
(A-25). Under the 10° left correction
scale, read the true heading on the
outbound leg (080°).

(5) Under the grommet, read the groundspeed on the outbound leg (89 knots).

(6) Under the true index, place the reciprocal of the outbound course (090° +
180°) 270° (A-26).

(7) Move sliding grid so that 100 knots is
under the wind dot.

(8) The wind dot is now on the converging
line 10° to the right of centerline
(A-26). Under the 10° right correction

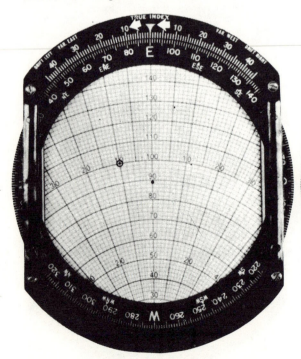

*Fig. A-25 Solving true heading and groundspeed
for outbound leg.*

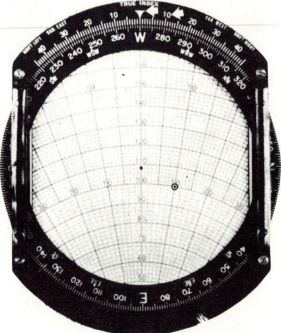

*Fig. A-26 Solving true heading and groundspeed
for return leg.*

scale read true heading for return flight (280°).

(9) Under the grommet, read groundspeed for the return leg (108 knots).

## Section IV. TRIANGLE OF VELOCITIES (CENTERLINE AIR VECTOR)

### GENERAL

The previous problems have been basically course-to-heading problems. In those problems, the centerline was always used as the ground vector and the wind vector was plotted above the grommet. For the preceding type of problems this system is preferable because it deletes the necessity of juggling the computer. There are certain problems, however, that would require juggling if they were plotted in such manner. The following are such problems and the centerline is used as an air vector and the wind is plotted below the grommet.

### FINDING TRACK AND GROUNDSPEED

This is the identical problem solved graphically earlier in the appendix. Here, it is solved on the E-6B computer.

Problem: Wind 300°/20 knots, true heading 045°, true airspeed 100 knots. What is the track and groundspeed?

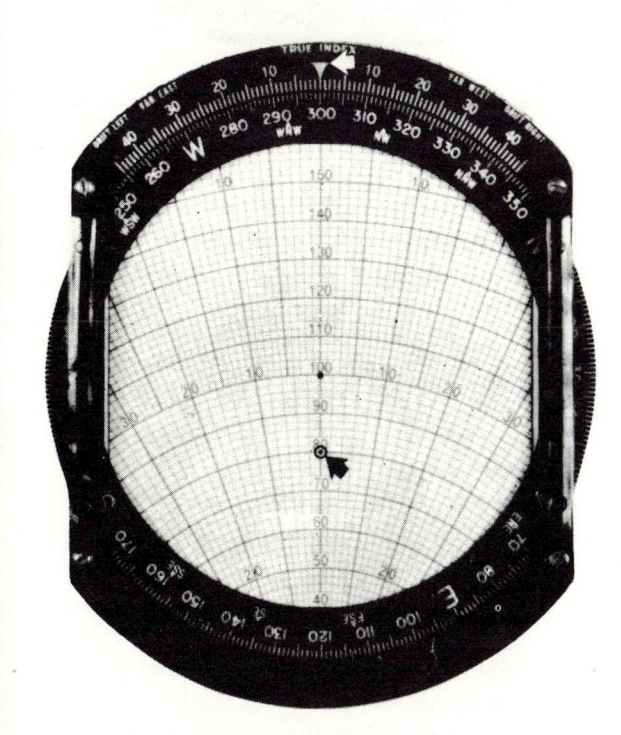

*Fig. A-27 Plotting the wind below the grommet.*

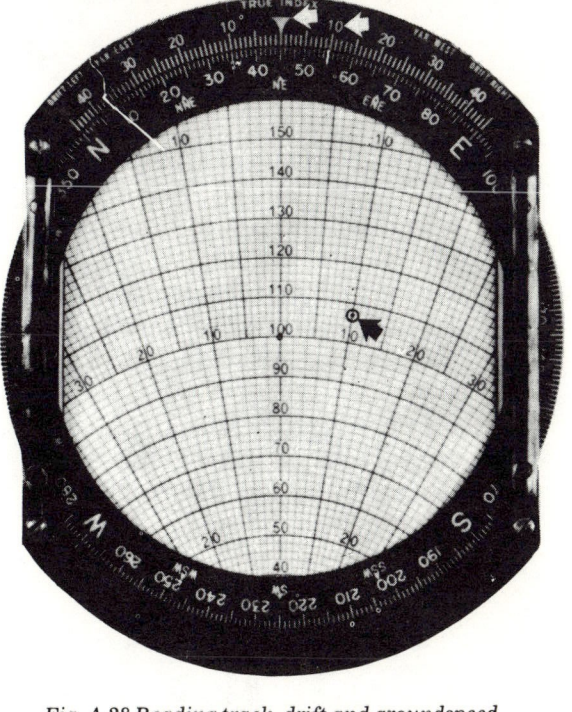

*Fig. A-28 Reading track, drift and groundspeed.*

Solution. Using an E-6B computer, refer to A-27 and A-28 and solve as follows:

(1) Set wind direction (300°) to the true index.

(2) Plot wind vector 20 units below the grommet, using dot inside of circle (A-27).

(3) Set true heading (045°) to the true index (A-28).

(4) Move the sliding grid so that the grommet is at the true airspeed (100 knots).

> NOTE: In the preceding problems, the dot was placed on the true airspeed. In all of the following problems, the grommet is used for the true airspeed, and the dot indicates groundspeed.

(5) The dot is now on the 10° converging line to the right of the centerline. Under 10° on the correction scale to the right of the true index, read the track 055°.

(6) Under the wind dot, read the groundspeed (107 knots).

### DOUBLE DRIFT WIND

Problem. True airspeed of aircraft on all headings 100 knots. On a true heading of 120°, 4° of right drift is observed; on a true heading of 220°, there is 5° left drift. What is the wind velocity?

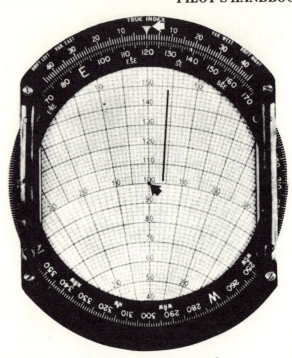

Fig. A-29 Drift on first heading.

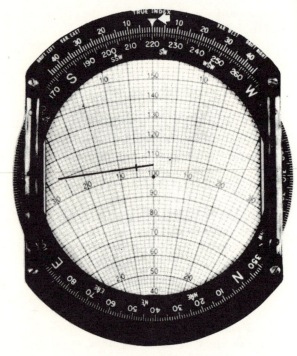

Fig. A-30 Drift on second heading.

(3) Set second true heading (220°) under the true index and draw a short line 5° to the left of centerline crossing the line previously drawn (A - 30).
(4) Turn the cross until it is directly below the grommet (A - 31).
(5) Under the true index, read the wind direction, 347° (A - 31).
(6) By counting the units between the cross and the grommet, determine the windspeed, 11 knots (A - 31).

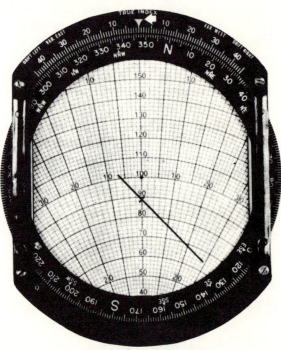

Fig. A-31 Reading wind direction and speed.

**OFF-COURSE CORRECTION**

Problem. Airport B is 260 nautical miles from airport A. After some time, while maintaining a true heading of 060°, the pilot determines that he is 6 nautical miles to the right of his intended course and 110 nautical miles from point of departure. What is the correct heading to reach destination?

Solution. Using an E-6B computer, refer to A - 32, A - 33 and A - 34, and solve as follows:
(1) Place the rectangular grid under the plotting disc.
(2) Set the true heading (060°) under the true index.
(3) Using the grommet as a starting point, draw a line 6 nautical miles to the right of the centerline (each division has a value of 2 nautical miles) (A - 32)

Solution. Using an E-6B computer, refer to (A - 29, A - 30, and A - 31) and solve as follows:
(1) Set first true heading (120°) under the true index and true airspeed (100 knots) under the grommet (A - 29).
(2) Draw line along the converging line 4° to the right of centerline (A - 29).

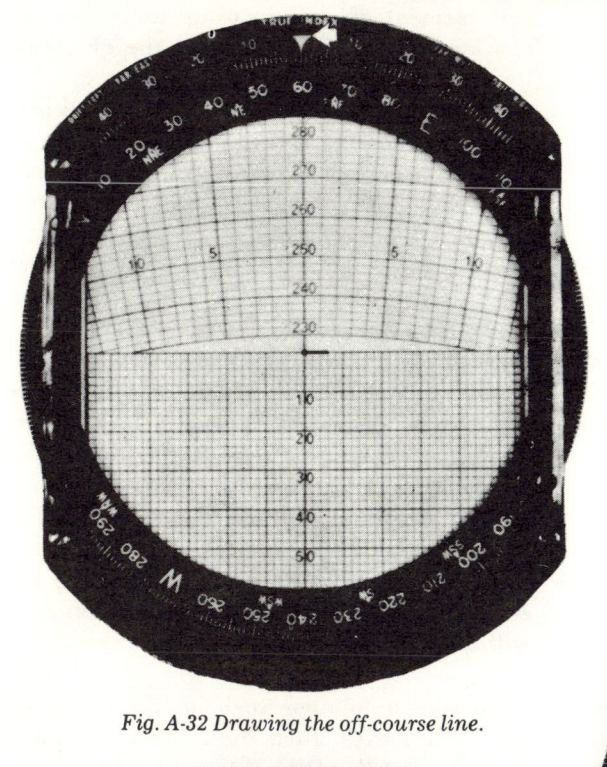

*Fig. A-32 Drawing the off-course line.*

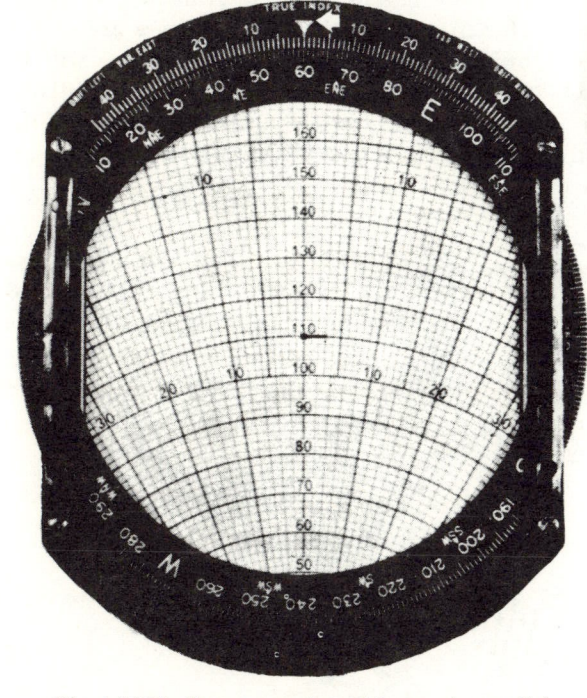

*Fig. A-33 Finding the number of degrees to parallel.*

(4) Reverse the sliding grid and set the distance flown (110 nautical miles) under the grommet (A - 33).

(5) From the converging line at the end of the horizontal line, determine 3° to the right of course (A-33).

(6) Set the distance to go 150 nautical miles (260 nautical miles (total dis-

tance) minus 110 nautical miles flown) under the grommet (A - 34).

(7) From the converging line at the end of the horizontal line, determine 2° necessary for convergence on destination (A - 34).

(8) Since the aircraft is to the right of course, correction is made to the left.

(9) Adding the 3° to parallel to the 2° to converge gives 5° of correction to converge at destination.

(10) Since correction is to the left, correction must be substracted from the original true heading. The new true heading is 055° (060° — 5° = 055°).

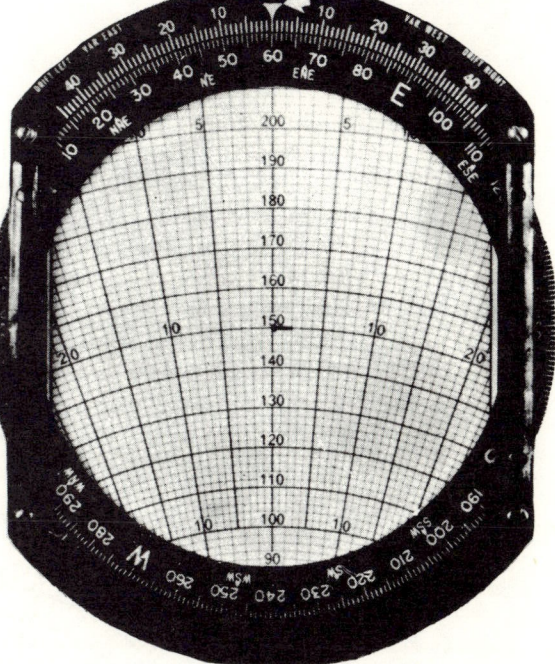

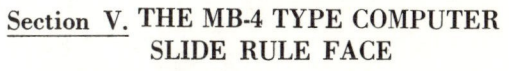

*Fig. A-34 Finding the number of degrees further necessary to converge.*

## Section V. THE MB-4 TYPE COMPUTER SLIDE RULE FACE

Whether you will be flying a high-speed aircraft or a cub, you will be continually solving navigation and engineering problems. How far? How fast? What time? What direction? How much fuel? Each of these calculations will take only a fraction of a minute, but together they will occupy almost all of your time while you are in the air.

You will find the solution to these problems quite easy once you have mastered the operations of the Dead Reckoning Computers. A Dead Reckoning Computer is simply a combination of two devices, one a specially designed instrument for the solution of the wind triangle, and the other, a circular slide rule for the solution of mathematical problems.

The slide rule is a standard device for the mechanical solution of various mathematical problems. It consists of several scales, fixed and movable, which may be moved into one desired relationship to obtain the result of the operation on the known quantities. Slide rules operate on the basis of logarithms, but no knowledge of logarithms is necessary for the successful use of the instrument. Slide rules are either straight or circular; the one on the dead reckoning computer is of the circular type.

In the case of the dead reckoning computer, the slide rule consists of two circular scales, one a stationary scale (called the MILES scale), the other a rotating scale (called the MINUTES scale). Each scale has a reference mark, the 10 called the "index." The rotating scale also has a black arrowhead at the sixty minute or one hour point; this is known as the "Speed Index." Graduations of both scales are identical. The graduations are numbered from 10 to 100, and the unit intervals decrease in size as the numbers increase in size. Not all unit intervals are numbered, and the

first element of skill in using the computer is a knowledge of how to read the numbers.

The unit intervals which are numbered present no difficulty. The whole problem lies in giving the correct values to the many small lines which come between numbered intervals. There are no numbers given between 25 and 30, for example, but it is obvious that the larger intermediate divisions are 26, 27, 28, and 29. Between 25 and (unnumbered) 26 there are five smaller divisions, each of which would be .2 of the larger units 10 and 11 one finds that the interval is divided into ten smaller divisions; each of these would be .1 of the larger unit, and the longer mark at the center would indicate the .5 value. The fractional value of each mark between numbered intervals must thus be found by the student. A little practice in reading numbers will quickly give him the required skill.

To find a number such as 278 first find the 27 (for 270), then the 28 (for 280), and then the fourth of the intermediate divisions (for 278), since the divisions here are of two units each. Obviously, unit divisions will be found only for numbers made up of relatively few integers; for a number such as 27812, the point already chosen for 278 would be used. This same point would also be used for 27.8,

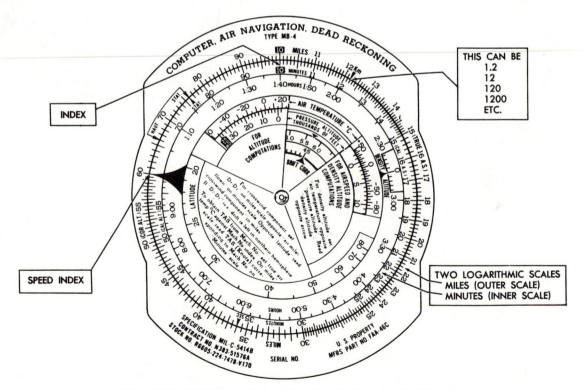

*Fig. A-35 The MB-4 DR computer is fundamentally a circular slide rule.*

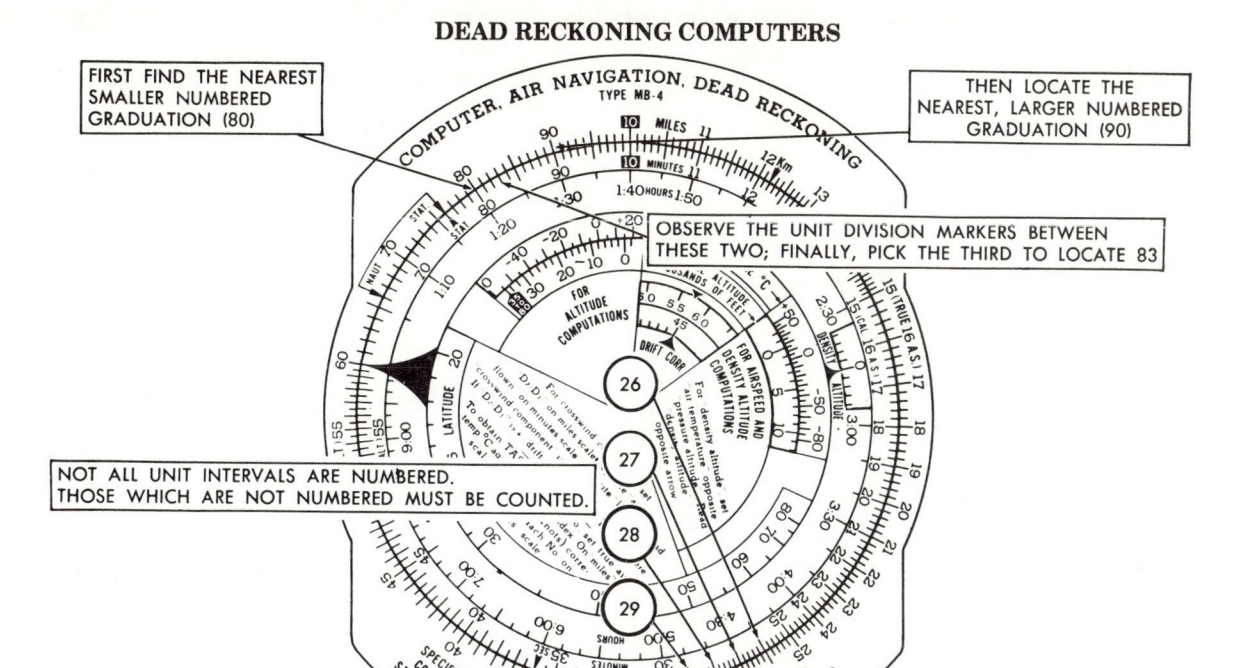

FIRST FIND THE NEAREST SMALLER NUMBERED GRADUATION (80)

THEN LOCATE THE NEAREST, LARGER NUMBERED GRADUATION (90)

OBSERVE THE UNIT DIVISION MARKERS BETWEEN THESE TWO; FINALLY, PICK THE THIRD TO LOCATE 83

NOT ALL UNIT INTERVALS ARE NUMBERED. THOSE WHICH ARE NOT NUMBERED MUST BE COUNTED.

*Fig. A-36*

2.78, etc.; there are thus several numbers for each point on the scale. With a little practice, the student will be able to make the correct interpretation of each reading. The matter of the use and place of the decimal point will be largely one of common sense.

The slide rule face of the MB-4 computer also contains special scales. One is marked FOR AIRSPEED AND DENSITY ALTITUDE COMPUTATIONS for applying temperature to pressure altitude to find true airspeed; one is marked FOR ALTITUDE COMPU- TATIONS for applying temperature to pressure altitude to find true altitude; and, one is marked DRIFT CORR to solve for the correction needed to parallel course and to correct into destination when the aircraft is not making good its preflight true course.

## Section VI. PROBLEMS ON THE SLIDE RULE FACE

### SIMPLE PROPORTION

The slide rule face of the dead reckoning computer is so constructed that any relation- ship between two numbers, one on the station- ary scale and one on the movable scale, will hold true for all other numbers on the two scales. If the two 10's are placed opposite each

other, all other numbers will be identical around the whole circle. If 20 on the inner scale is placed opposite on the outer scale, all numbers on the inner scale will be double those on the outer scale. If 12 on the outer scale is placed opposite 10 on the outer scale, all numbers will be in a three-to-four relation- ship. This feature of the slide rule permits you to supply the fourth terms of any mathemati- cal problem in proportion. The simplest and most useful application of this feature of the computer is in the solution of time-speed- distance problems.

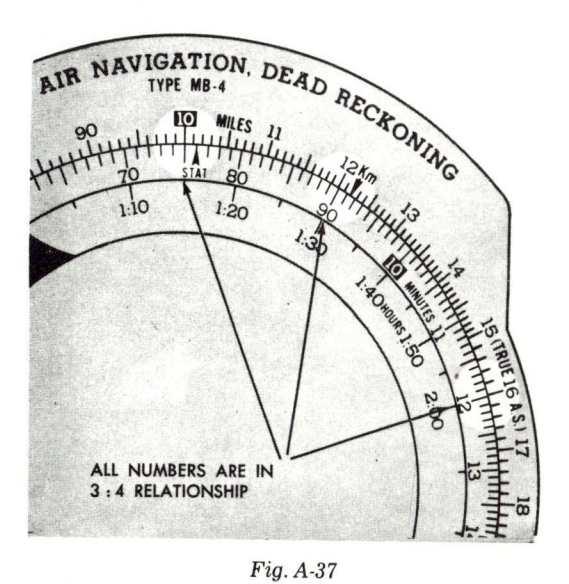

ALL NUMBERS ARE IN 3 : 4 RELATIONSHIP

*Fig. A-37*

TIME, SPEED, AND DISTANCE (nautical miles)

A plane has traveled 24 nm in 8 minutes. How many minutes will be required to travel 150 nm? The answer is the missing number in a simple proportion:

$$\frac{24}{8} = \frac{150}{?}.$$

In the latter form, the proportion may be set on the computer exactly as it appears, with 24 on the outer MILES scale opposite 8 on the inner MINUTES scale; one then reads the number opposite 150 nm, and finds it to be 50 minutes. This is the basis of all-time-speed-distance computations on the computer. In most of these problems the arrowhead "speed index" is used for setting or reading the rate. In addition to the minutes scale, an inner HOURS scale is used for reading times in excess of one hour.

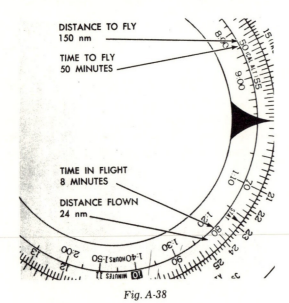

Fig. A-38

NOTE: In these problems the unit for measuring distance and speed is immaterial. If statute miles (mi) are used for distance, the speed will be expressed in miles per hour (mph); if nautical miles (nm) are used, the speed will be expressed in knots (k).

To solve for distance when speed and time are given:

  Given: Speed, 174 knots
      Time, 40 minutes
  Find: Distance

(1) Set arrow (60 minutes) on the inner scale opposite the speed (174 k) on the outer scale.

(2) Opposite the time (40 min.) on the inner scale read the distance in miles (116 nm) on the outer scale.

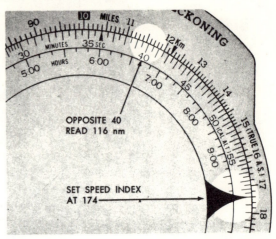

Fig. A-39

To solve for time when speed and distance are known:

  Given: Speed, 113 knots
      Distance, 180 nm
  Find: Time

(1) Set arrow opposite the speed (113 k) on the outer scale.

(2) Opposite the distance (180 nm) on the outer scale read the time on the inner scale (1 : 36, or one hour and thirty-six minutes).

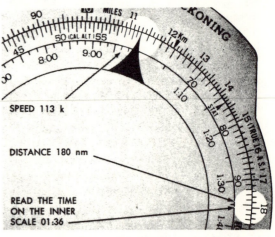

Fig. A-40

To solve for speed when time and distance are known:

  Given: Distance, 266 nm
      Time, 1 : 25
  Find: Speed

(1) Set the known time (1 : 25) on the inner scale opposite the known distance (266 nm) on the outer scale. Do not use the arrow to set the inner scale.

(2) Read the speed (188 k) opposite the arrow.

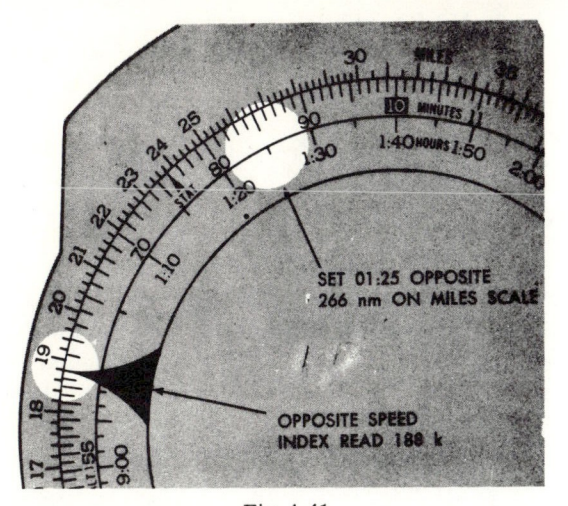

*Fig. A-41*

## FUEL CONSUMPTION

Problems in fuel consumption are solved in exactly the same manner as problems in speed, except that the rate is expressed as gallons per hour (gph) instead of miles per hour. Gallons are therefore treated as miles, and read on the outer scale. The arrow is used as a rate index, in the usual manner.

To find hours of flying time when rate of fuel consumption and amount of fuel are known:

Given:  Rate of fuel consumption, 85 gph.
Remaining fuel, 150 gals.
Find:  Remaining flight time

(1) Set the arrow to the rate of consumption (85 gph) on the miles scale.

(2) Under the fuel available (150 gals.) on the miles scale read the time of flight on the hours scale (1 : 46).

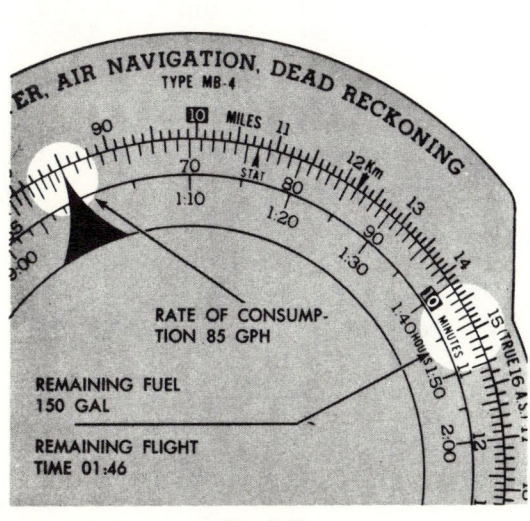

*Fig. A-42*

To find the fuel consumed per hour (rate of consumption) when the total fuel consumed and the flying time are known:

Given:  Fuel consumed, 135 gals.
Time in flight, 1 : 28
Find:  Rate of fuel consumption

(1) Under the total fuel consumed (135 gals.) on the outer scale set the time (1 : 28) on the inner scale. Do not use the arrow to set the inner scale.

(2) Read the rate of consumption (92 gph) on the outer scale opposite the arrow.

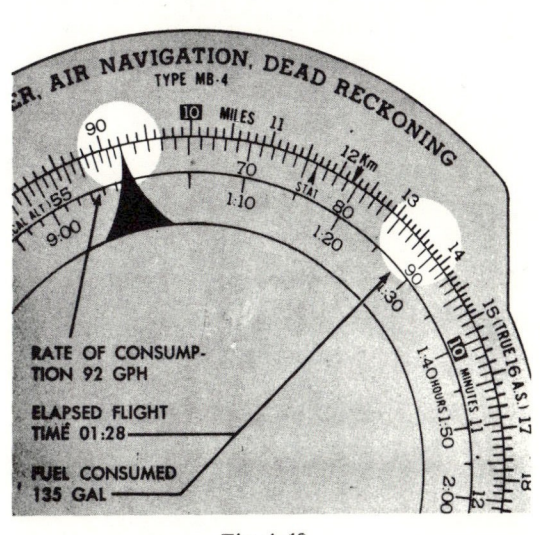

*Fig. A-43*

To find the amount of gas required when rate of fuel consumption and hours of flying time are known:

Given:  Rate of consumption, 85 gph
Time 1 : 30
Find:  Total fuel required.

(1) Set the arrow at the rate of consumption (85 gph) on the outer scale.

(2) Over the time (1 : 30) on the inner hour scale read the total fuel required on the outer scale (127 gals.).

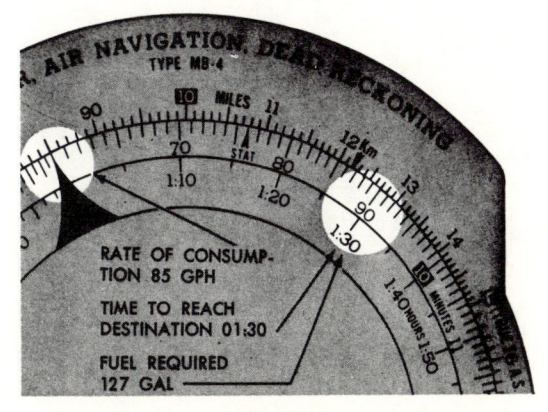

*Fig. A-44*

SECOND INDEX

Pilots often find it necessary to know how long it will take them to fly a very short distance for radio range problems, and they will usually want to know in seconds. Navigators also need to be able to figure how much before an ETA to a turning point they will have to start a precision procedure turn. The SEC (seconds) index on the DR computer gives a quick conversion. A small arrow has been placed on the minutes or inner scale at 36. This represents 3600 seconds or the equivalent of one hour. By placing this index opposite a known groundspeed, the time to fly a given distance may then be read on the minutes scale under the distance to fly. The time will be in seconds.

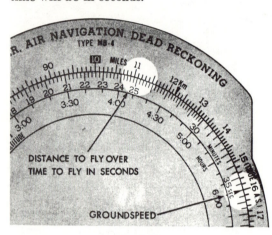

*Fig. A-45*

Given:   Groundspeed 180 knots.
         Distance to fly 2.7 nautical
         miles

Find:    Time in seconds

(1) Set SEC index under 180 on the miles scale. The SEC index is located at 36 on the minutes scale.

(2) On the minutes scale directly under the distance to fly (2.7 nm), read the time to fly in seconds (54 seconds).

OFF-COURSE CORRECTION

One form of a simple proportion has been set up for quick usage in low level navigation where correction must be immediate or in a case where the pilot is unable to determine his exact wind for precise corrections. This is the DRIFT CORR window.

The proportion is simply:

$$\frac{\text{Naut. miles off-course}}{\text{distance flown}} = \frac{\text{correction to parallel}}{57.3}$$

and

$$\frac{\text{Naut. miles off-course}}{\text{distance to fly}} = \frac{\text{correction to converge}}{57.3}$$

The window on the MB-4 was derived from prior usage of the old rule of thumb that one mile off course in sixty requires a one degree correction. The angular corrections given in the window are derived from the actual value of 47.3 (one radian).

The illustrations show the procedure in using the DRIFT CORR window with a sample problem.

Given:   Course distance traveled, 400
         nm
         Distance off course, 30 nm

Find:    Correction angle to parallel
         desired course and correction

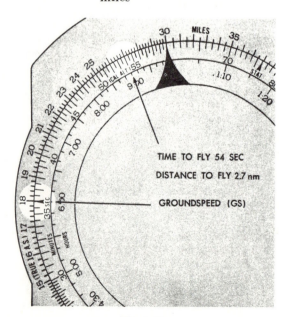

*Fig. A-46*

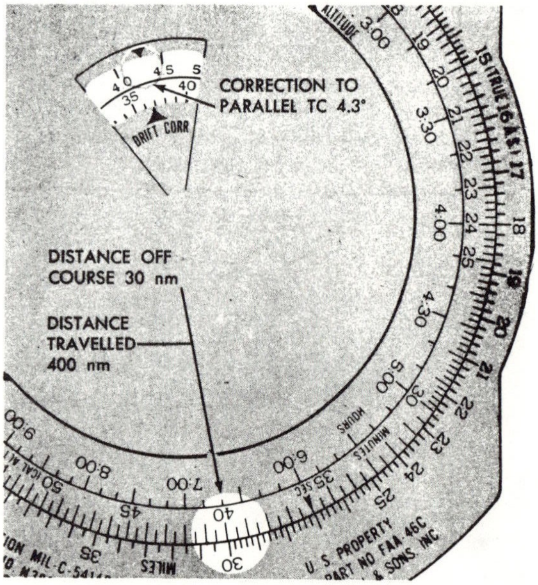

*Fig. A-47*

angle to intercept desired course in 150 additional nautical miles.

(1) Set the miles off course (30 nm on the MILES scale) over distance traveled (400 nm on the MINUTES scale). Read the angle to parallel in the drift correction window. The answer is approximately 4.3°.

(2) To find the angle to intercept, place miles off course (30 nm on the MILES scale) over course miles to interception point (150 nm on the MINUTES scale). Read the additional angle to intercept in the drift correction window. The answer is approximately 11.3°.

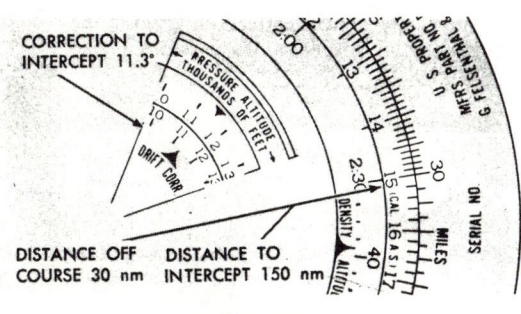

*Fig. A-48*

(3) To find the total correction angle, add this to the correction parallel. The total correction is 15.6° (4.3 + 11.3).

> NOTE: Notice that it is still necessary to determine the decimal point. If the miles flown or miles to fly is more than 10 times the miles off course, use the smaller angle. For example: 9 miles off course in 300 miles, correct 1.7° to parallel; for 90 miles off course in 300 miles, correct 17° to parallel.

### RADIUS OF ACTION FORMULAS

The solution for radius of action—the distance which a plane can fly out to return to the same base in a given total time or consuming a given amount of available fuel—consists of two parts:

(1) Wind triangle solutions, which provide certain needed data, and

(2) The use of these data in standard formulas.

The solution of the wind triangles will be treated later. As for the formulas, the ones most generally used are the following:

$$t = \frac{GS_2 + T}{GS_1 + GS_2}.$$

Radius of action = t + GS$_1$

t  - time to turn

T  - total available fuel hours (after

allowance for reserve)

GS$_1$ - groundspeed out

GS$_2$ - groundspeed back

The formula for time to turn may be converted into a proportion easily set up on the computer merely by dividing both sides of the above equation by T, thus:

$$\frac{t}{T} = \frac{GS_2}{GS_1 + GS_2}.$$

The solution as set up on the computer looks exactly like the equation.

To solve for time to turn:

Given:  GS$_1$, 136 knots

GS$_2$, 94 knots

T, 3 hours (after allowance for reserve)

Find:  Time to turn

Radius of action

(1) Locate GS$_2$ (94 k) on the outer miles scale.

(2) Opposite GS$_2$ on the outer scale set the sum of GS$_1$ and GS$_2$ (230 k) on the inner scale.

(3) Without moving the inner scale, find T (3 : 00) on the inner scale.

(4) Opposite T on the inner scale read time to turn (in minutes) on the outer scale (73.5 minutes or 1 : 13.5).

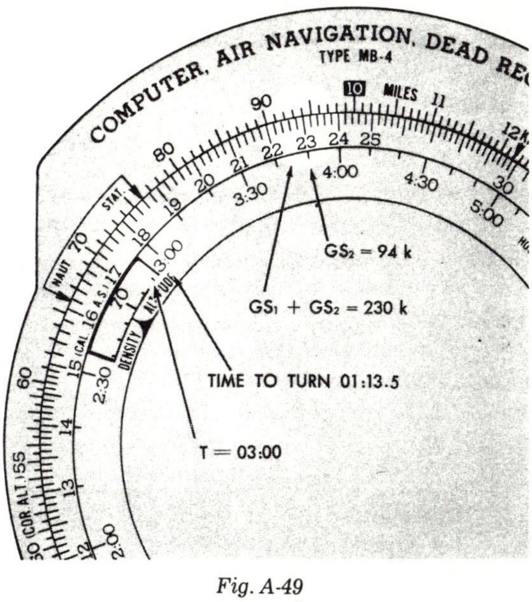

*Fig. A-49*

> NOTE: In this solution only one setting of the computer is made. The arrow is not used. The answer as read on the outer scale is always in minutes.

The radius of action is found in the usual way for time-speed-distance solutions, by setting the arrow at $GS_1$ (136 k) and reading the distance (168 nm) opposite the time (1 : 14) on the minutes scale.

A similar formula may be used to complete solutions of problems on radius of action to an alternate base. In this formula, $S_1$ is the rate of departure and $S_2$ is the rate of closure.

$$\frac{t}{T} = \frac{S_2}{S_1 + S_2}$$

## CONVERSION OF KNOTS TO NAUTICAL MILES PER MINUTE

Another useful computation for the pilot is the conversion of knots to nautical miles per minute. This he usually does mentally. But he may use the computer as a check on his mental calculations, or when mental division would be difficult. On the computer, the problem is again one of simple proportion:

$$\frac{knots}{60} = \frac{naut.\ miles/min.}{1}$$

To convert knots to nautical miles/minute:

Given:  Speed, 180 knots

Find:  Speed in nautical miles/min.

(1) Set the arrow opposite knots on the miles scale (180 k).

(2) Over 10 on the minutes scale read nautical miles/min. on the miles scale. Since the 10 is used for 1, the answer on the miles scale must be divided by ten.

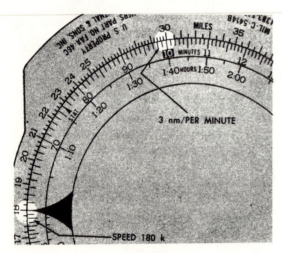

*Fig. A-50*

## STATUTE MILES, NAUTICAL MILES, AND KILOMETERS

The navigator's work is usually done in terms of nautical miles and knots. In certain European theaters, distances are expressed in

kilometers. The dead reckoning computer slide rule face provides a device for converting both nautical miles and kilometers into the statute miles with which the pilot is familiar. This is done by the use of the NAUT., STAT., and KM. markings on the

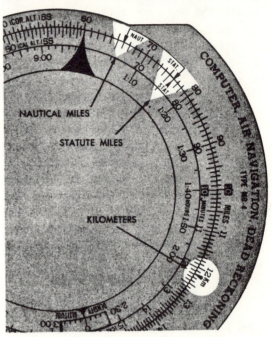

*Fig. A-51*

outer scale.

The arithmetical factor for converting nautical miles to statute miles is 1.15; for converting statute miles to nautical miles, .87. It will be found that 76 (the STAT. mark) divided by 66 (the NAUT. mark) gives 1.15, and that 66 divided by 76 gives .87. In the same way, the conversion factor for statute miles to kilometers is .62, and 76 (the STAT. mark) divided by 122 (the KM. mark) gives the same conversion factor. The conversion on the computer is thus the familiar simple proportion:

$$\frac{66}{Nautical\ miles} = \frac{76}{Statute\ miles}$$

To convert statute miles to nautical miles:

Given:  136 miles.

Find:  Number of nautical miles

Number of kilometers

(1) Place statute miles (136 mi.) on the inner scale opposite the small STAT. arrow on the outer scale.

(2) Opposite the small NAUT. arrow on the outer scale read the number of nautical miles (118 nm) on the inner scale. Opposite

the Km. arrow read the number of kilometers (218 km.).

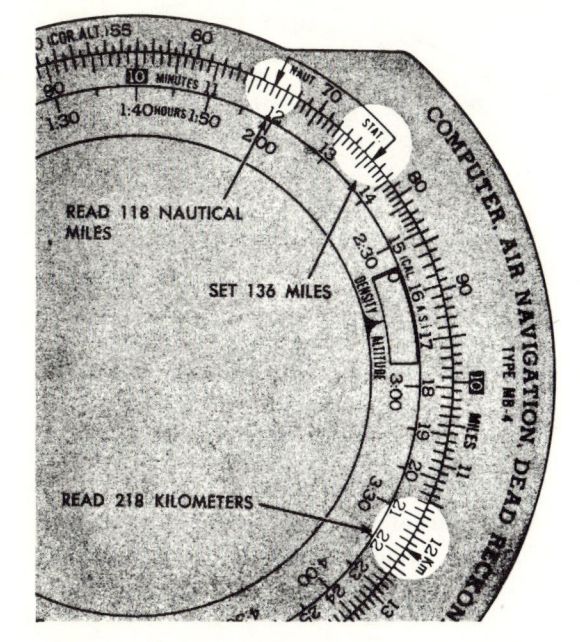

*Fig. A-52*

Conversions of any of the other units are made in the same way. The known figure is placed under the arrow for the unit in which that figure was expressed and the unknown quantity is read under the appropriate arrow.

CORRECTION OF AIRSPEED READINGS

The airspeed indicator is constructed to operate correctly under standard atmospheric conditions, at a pressure of 29.92 inches of mercury and a temperature of 15°C. In flight, these conditions will rarely be found. The pilot will therefore have to correct the readings of his airspeed indicator for pressure and temperature in order to discover his true airspeed. The slide rule face of the dead reckoning computer has a special scale which will enable him to make the necessary corrections; it is labeled FOR AIRSPEED AND DENSITY ALTITUDE COMPUTATIONS.

The most accurate use of this scale requires the setting of pressure altitude (the difference between 29.92" Hg and the actual barometric pressure at flight level, in terms of feet) opposite free air temperature. But in most cases the pilot will not know pressure altitude, and to find it will complicate the correction. Therefore, in following outline of procedures, indicated altitude (or calibrated altitude) will be used instead of pressure altitude. This introduces a slight error, but one which will not be significant for the uses to which the

pilot puts his airspeed readings under most circumstances.

True airspeed is obtained in one setting. Using the scale marked "FOR AIRSPEED AND DENSITY ALTITUDE COMPUTATIONS," take the following steps:

Given: Free air temperature, —10°C.
Indicated altitude, 20,000 ft.
Calibrated airspeed, 200 knots.
Find: True airspeed

(1) Locate the proper free air temperature (—10°C) on the small scale marked "AIR TEMPERATURE °C."

(2) By rotating the disk, set this temperature opposite the proper indicated altitude (20,000') in the window marked PRESSURE ALTITUDE THOUSANDS OF FEET. Each of the marks here is or one thousand feet.

(3) Opposite indicated (or calibrated) airspeed on the minutes scale (200 k) read the true airspeed on the miles scale (282 k).

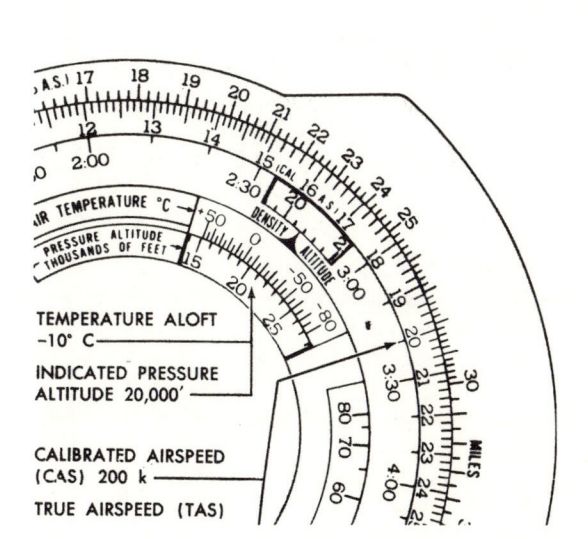

*Fig. A-53*

(This solution disregards the Pitot-static installation. Compressibility, and Heat of Compression errors. These errors are fully explained elsewhere in the appendix; they are insignificant at lower altitudes and airspeeds.)

Two mistakes are frequently made: One is using the wrong "window" and the other is forgetting that the temperature scale is reversed. The plus temperatures are to the left.

At times, especially in case of controlled

groundspeed, the true airspeed will be known and the pilot will want to know at what indicated airspeed to fly in order to make good the given true airspeed. This solution on the dead reckoning computer is the same as the preceding one, except that after free air temperature and indicated altitude have been set in the proper relationship, the indicated airspeed is read on the minutes scale opposite the known true airspeed on the miles scale.

Given:  Free air temperature, 0°C.
        Indicated altitude, 5,000 ft.
        True airspeed, 200 knots
Find:   Indicated airspeed (188 k)

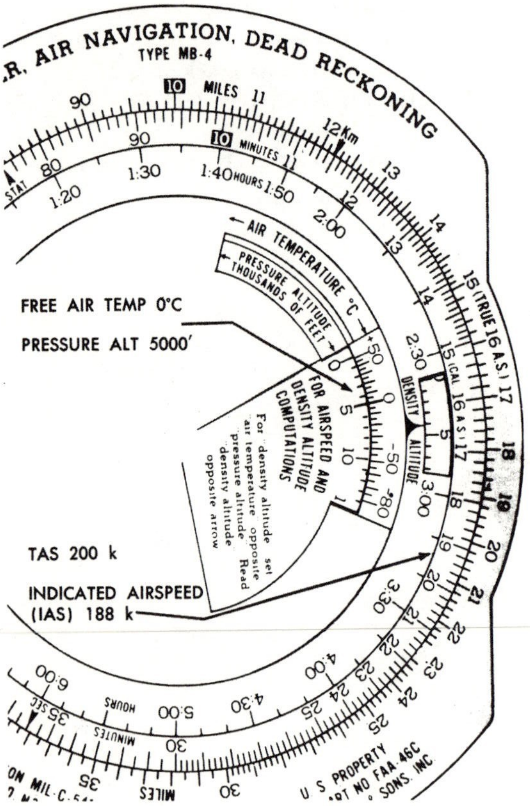

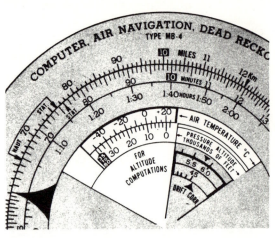

Fig. A-55

pilot without introducing any appreciable error.

True altitude is obtained in one setting. Using the scale marked FOR ALTITUDE COMPUTATIONS, take the following steps:

Given:  Indicated altitude, 20,000 feet
        Free air temperature, —10°C.
Find:   True altitude

(1) Locate the indicated altitude (20,000′) on the small scale marked PRESSURE ALTITUDE. Each of the marks here is for two thousand feet, and the "10" therefore means 10,000 feet.

(2) Rotate the disk until this indicated altitude is opposite the proper free air temperature (—10°C) in the window marked AIR TEMPERATURE °C.

(3) Opposite indicated (or calibrated) altitude (20,000′) on the minutes scale read the true altitude (21,200′) on the miles scale.

Fig. A-54

CORRECTION OF ALTIMETER READINGS

The altimeter is constructed to operate in terms of an arbitrary scale, the standard pressure lapse rate. Hence it must not only be reset whenever weather changes in order to operate correctly under local conditions, but its readings while in the air must be corrected for pressure and temperature. The dead reckoning computer has a small scale, labeled FOR ALTITUDE COMPUTATIONS, on which these corrections can be made. As in the case of airspeed corrections, the computer calls for pressure altitude; but, once again, indicated altitude may be used by the

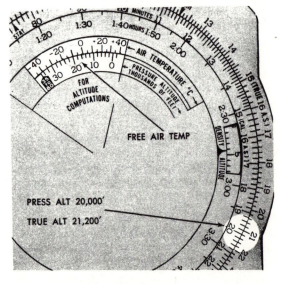

Fig. A-56

## MULTIPLICATION

Some of the problems already discussed really involved simple multiplication and division. Any other problems using these arithmetical operations may also be solved. Two things should be noted about these solutions on the slide rule face:

(1) The rule may have a slight error and hence the answers obtained may be incorrect. However, the slide rule result is sufficiently accurate for practical navigational work.

(2) Multiplication and division on the computer make use of the 10 index marks; they do not use the arrow or "speed index."

Furthermore, the value to be assigned to the answer must be checked by common-sense mental calculations. If you are multiplying 20 by 15, the "30" in the answer will be 300; but if you are multiplying 20 by 1.5, the "30" in the answer will actually be 30. This possibility of different values makes it necessary to think twice before putting down any answer for multiplication or division.

To multiply:

    Given: Divide 14 by 70
    Find: Quotient

(1) Place the 10 index mark on the minutes scale under one of the numbers to be multiplied (14) on the miles scale.

(2) Using the minutes scale, find the other number to be multiplied (70). Opposite this number, on the miles scale, read the product (980).

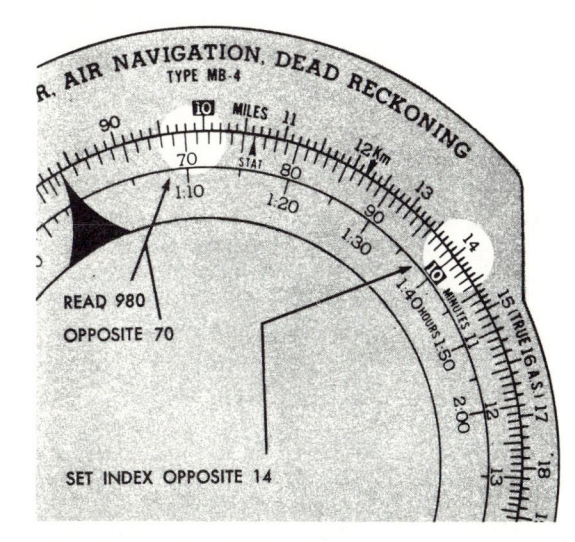

*Fig. A-57*

NOTE: Of the two numbers to be multiplied, one is on the miles scale and the other on the minutes scale.

## DIVISION

Division is the inverse of multiplication. To divide:

    Given: Divide 180 by 15
    Find: Quotient

(1) Under the dividend on the miles scale (180) set the divisor on the minutes scale (15).

(2) Opposite the 10 index mark on the minutes scale read the quotient on the miles scale (12).

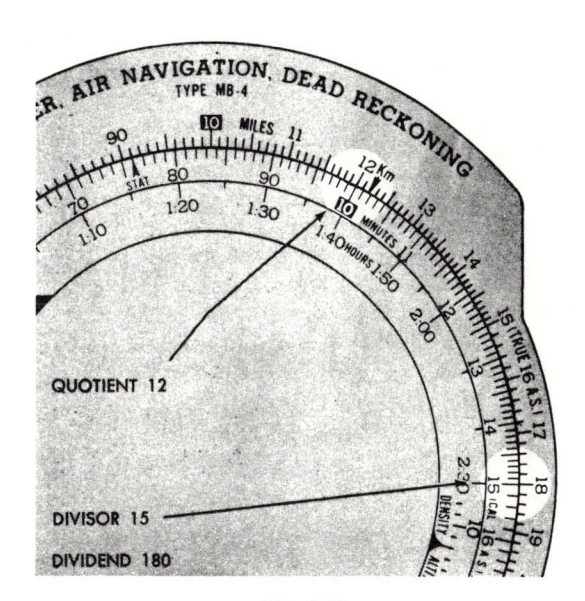

*Fig. A-58*

As set up on the computer, the problem looks just like the problem in division stated as a fraction; the quotient is read on the miles scale opposite the 10 index mark on the minutes scale.

## COMBINED MULTIPLICATION AND DIVISION

Many calculations involve both multi-

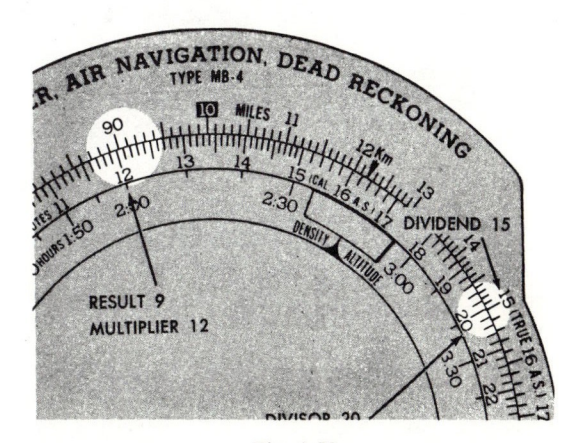

*Fig. A-59*

plication and division; for example,

$$\frac{15 \quad 12}{20}$$

Problems of this type may be solved with one setting of the computer.

(1) Divide 15 by 20 by setting 15 on the miles scale over 20 on the minutes scales.

(2) Multiply the answer by 12; over the 12 on the minutes scale read the final result, 9, on the miles scale. The multiplication is done in the usual way.

### Section VII.  THE WIND FACE

All problems involving wind depend upon the solution of the wind triangle. In its simple form, this triangle is made up of three simple vectors, the elements of which are always the same:

(1) A wind vector, made up of the direction and speed of the wind.

(2) A "ground" vector, representing the movement of the plane with respect to the ground, and made up of the true course (or track) and the groundspeed.

(3) An "air" vector, representing the movement of the plane with respect to the air mass, and made up of the true heading and the true airspeed.

length is measured by some standard scale, the same for all three vectors. The triangle can be determined in one of three ways:

(1) By two sides and an included angle.
(2) By two angles and an included side.
(3) By three sides.

NOTE: The quantities representing direction are all true, i.e., measured from true north; in the same way, airspeed is "true," i.e., corrected for temperature and altitude. If any four of these quantities are known, the other two may be found very easily, either by a graphical or by a computer solution. This solution is always in terms of one hour.

The wind face of the dead reckoning computer is a simple arrangement for setting up wind vector triangles of the standard type. It consists of a movable disk graduated in degrees from 0° to 360°, making a complete azimuth scale, which can be used to set desired angular directions measured from true north. There is also a secondary scale giving the sixteen compass points in which wind directions are stated; these may be used for setting wind when wind is stated in terms of these points rather than in degrees. The center of this disk is of transparent material so that lines may be drawn on it; at the very center

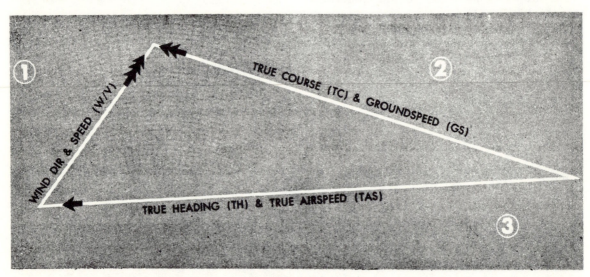

*Fig. A-60*

There are, in all, six quantities involved: wind direction, wind speed, true course (track), groundspeed, true heading, and true airspeed.

Each of the vectors in the triangle has a direction and length. The direction is measured as an angle from true north. The

is a dot or grommet. The second part is a movable plastic slide on which are printed two sets of lines. One set is a series of concentric arcs, known as "speed circles." These are at equal distances, each representing two units of distance, and numbered from 80 to 400 on one side of the slide and from 250 to 800 on the other side. Since these are

relationship and reading the unknown ones from the completed triangle.

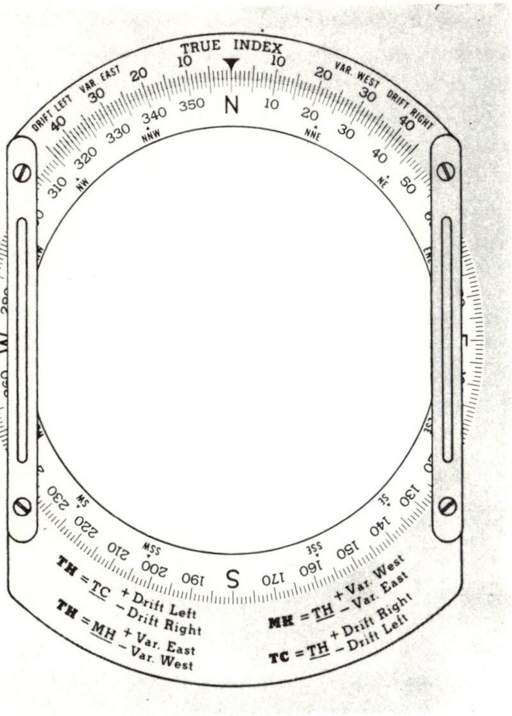

*Fig. A-61*

| PRESS. ALT FEET | F CORRECTION FACTORS FOR TAS | | | | | | | |
|---|---|---|---|---|---|---|---|---|
| | CALIBRATED AIRSPEED KNOTS | | | | | | | |
| | 200 | 250 | 300 | 350 | 400 | 450 | 500 | 550 |
| 10,000 | 1.0 | 1.0 | .99 | .99 | .98 | .98 | .97 | .97 |
| 20,000 | .99 | .98 | .97 | .97 | .96 | .95 | .94 | .93 |
| 30,000 | .97 | .96 | .95 | .94 | .92 | .91 | .90 | .89 |
| 40,000 | .96 | .94 | .92 | .90 | .88 | .87 | .87 | .86 |
| 50,000 | .93 | .90 | .87 | .86 | .84 | .84 | .84 | .84 |

**DIRECTIONS**

USE CALIBRATED AIRSPEED AND PRESS. ALT. TO OBTAIN F FACTOR. MULTIPLY F FACTOR BY TAS OBTAINED WITH COMPUTER TO OBTAIN TAS CORRECTED FOR COMPRESSIBILITY.

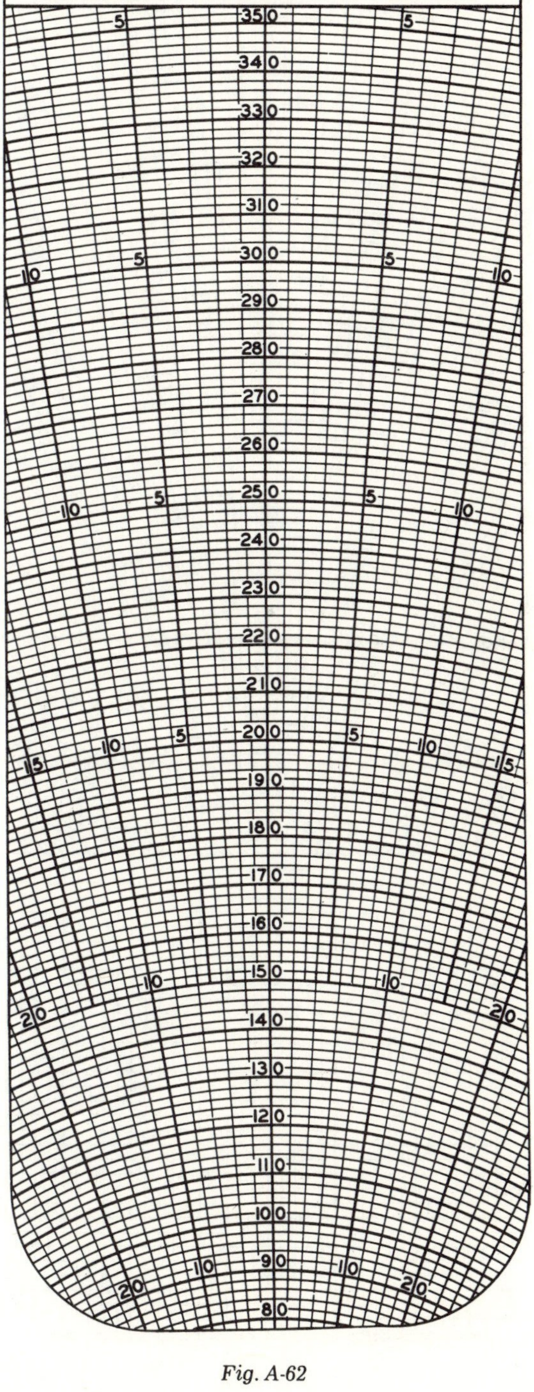

"speed circles," they are used for all measurements of speeds for the wind triangle: wind speed, groundspeed, true airspeed. The second set of lines is a series of converging straight lines, which do not actually meet on the slide itself but which converge toward a point outside the slide. These are called "track lines," since they show the number of degrees by which track (or true course) differs from true headings as read on the center line and under the "True Index" marker. On the high speed side of the wind slide, the lines are five degrees apart for speeds up to 150 knots, two degrees apart between 150 and 250 knots, and one degree apart for all speeds above 250 knots. (The high speed side also has a rectangular grid, using the same scale as the speed circles, which is employed in certain special problems.) On the low speed side, the track, or drift lines, are two degrees apart up to 150 knots and one degree apart above 150 knots.

These two parts of the wind face make it possible to set up speeds, angular directions from true north, and angular difference between true course and true heading - all the elements necessary for the solution of wind vector triangles. The handling of the computer is a matter of placing the known quantities of the computer in the proper

*Fig. A-62*

In solving the wind triangle graphically, only one basic triangle is correct since the vectors have both direction and quantity.

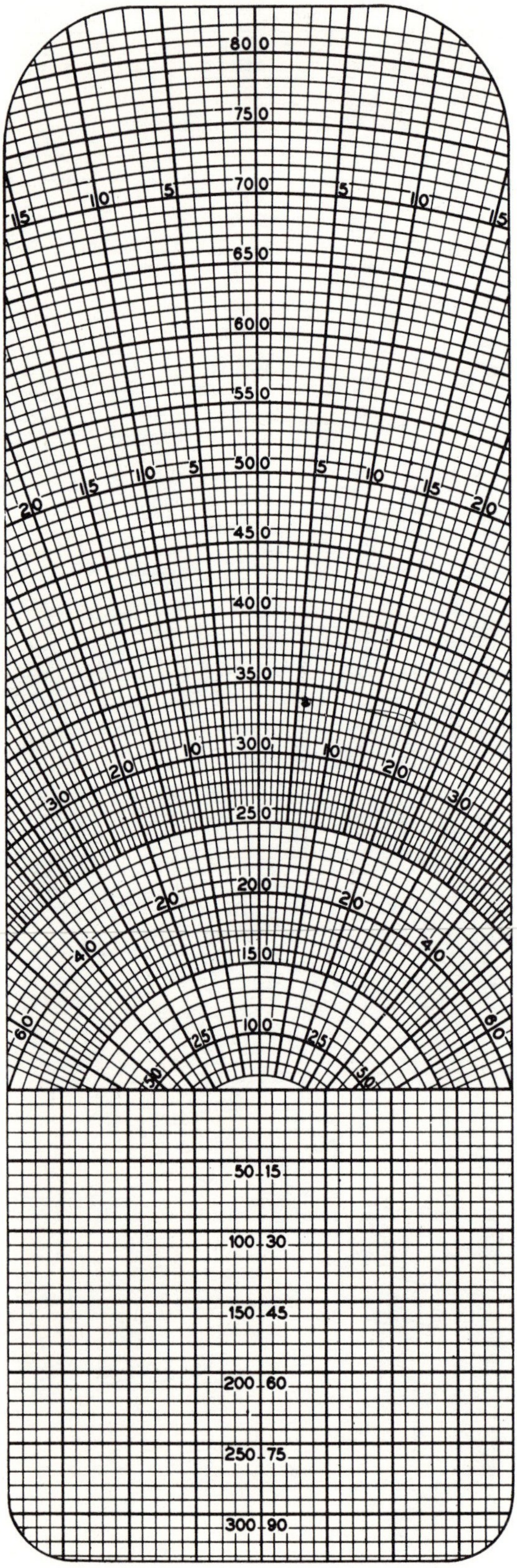

The three sides of the triangle are known as the wind vector (wind direction and velocity-W/V), the air vector (true heading and true airspeed - TH and TAS), and the ground vector (true course or track and groundspeed - TC or Tr and GS).

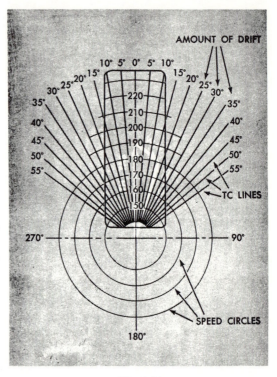

Fig. A-64

Normally there will be three primary wind triangle problems with a fourth that may be encountered. The first is to solve for TC and GS. This solution is basic to pure dead reckoning when no fixing aids are available. The wind vector is plotted and the head of the air vector attached to the tail of the wind vector. Departure point A is established and TC and GS can be measured between points A and C.

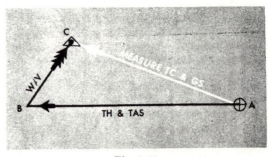

Fig. A-65

The next solution is for TH and GS. Time problem is most common to pre-flighting a mission. To solve it graphically, the wind

Fig. A-63

vector is plotted first, the TC plotted on a reciprocal bearing from the head of the wind vector, and the TAS swung in an arc across the TC vector. Point A is again established.

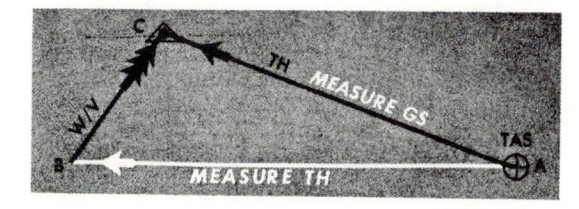

*Fig. A-66*

The third problem may be encountered if the mission requires a certain GS. Again the wind vector is plotted first. The ground vector is established with its head to the head of the wind vector. The components of the air vector, TH and TAS, are then measured between points A and B.

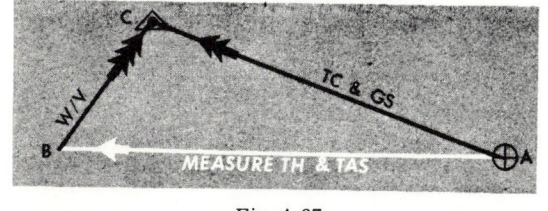

*Fig. A-67*

The last problem, and probably most important, is to solve for a wind. By establishing a departure point, A, and plotting the air and ground vectors from it, points B and C, are located. Wind is measured from B to C since the aircraft will always be blown from heading to track.

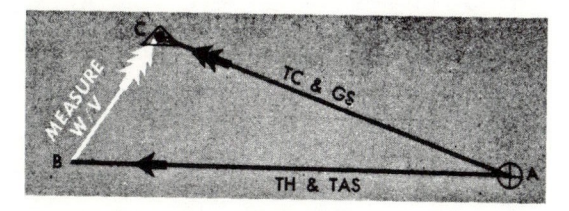

*Fig. A-68*

It is possible to use different methods than the following ones to be shown on the computer. These alternative methods are explained in the appendix earlier. However, it should be remembered they are still solutions to the basic wind triangle. Only the methods of setting up the problem on the computer is different.

## Section VIII. PROBLEMS ON THE WIND FACE

### SIMPLE WIND VECTOR PROBLEMS

Solution for Track and Groundspeed. - The triangle most easily set up on the DR computer is the one in which track and groundspeed are unknown. Practically, this solution is useful to the pilot in the air when he knows the wind, knows his true heading (the compass reading corrected back for variation and deviation) and his true airspeed (the airspeed indicator reading corrected for temperature and pressure), but has no means of visual reference to determine his groundspeed or his track. With these data, the problem would be solved by the following steps:

Given:  Wind direction, 320°
        Wind speed, 30 knots
        True heading, 175°
Find:   Track
        Groundspeed

(1) Set wind direction (320°) under the TRUE INDEX.

(2) Draw wind vector down from center of disk. The length the vector (3 k) is read between speed circles along the center line. Any two speed lines may be used, but it is usually easiest to set one of the heavy lines under the grommet and read down.

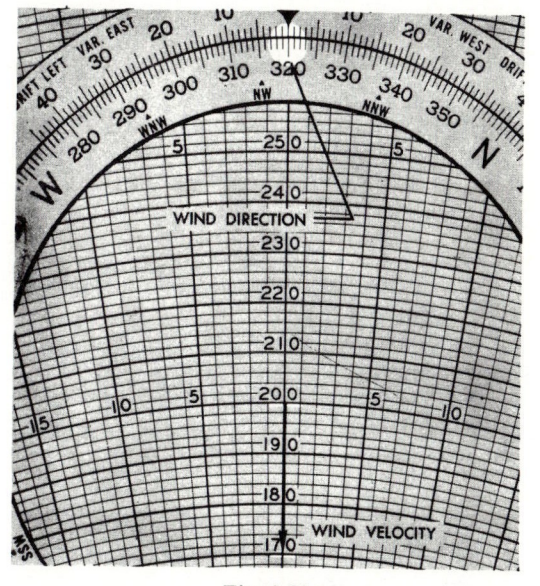

*Fig. A-69*

(3) Rotate the compass rose until the true heading (175°) is under the TRUE INDEX.

(4) Move the slide up or down until the true airspeed (180 k) is read under the center of the disk.

CAUTION

Step (4) is the one most frequently over-looked!

(5) Read drift angle (5° left) on the radial line at the head of the wind vector. The number of degrees between the center line and the head of the vector (wind arrow) is drift. When the vector is left, drift is left; when the vector is right, drift is right.

(6) Apply drift correction to find track. Remember that left drift gives a plus correction and right drift a minus correction and that drift correction is always applied to track or true course. In this case our drift is left, our correction is plus, and therefore a track of 170° + 5° =175°. When TH and DC are known, you should think of it as "To what number do I apply my DC to give me a certain TH."

(7) Read groundspeed (206 k) on the speed circle at the head of the wind vector.

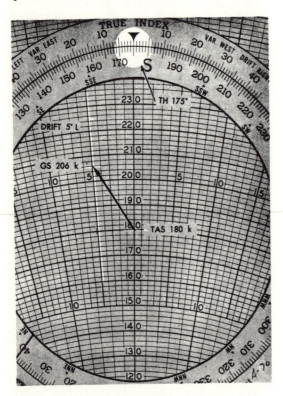

*Fig. A-70*

Solution for True Heading and Ground-speed. - On the ground, in your preflight calculations, the most important problem for you is the discovery of your true heading and groundspeed. This is done by vector solution on the basis of available information about the wind. On the computer, this so-lution is not as simple as the one for track and groundspeed, for the reason that true

heading—which is always set on the center line—is not yet known. Some other quantity must be set tentatively under the true index, and adjustments made until true heading ultimately appears. There are various ways of doing this. One of these, the one most consistent with the other solutions suggested in this manual, is outlined below. The others are described earlier on "Alternate Solutions."

Given: Wind direction, 240°
Wind speed, 30 knots
True course, 195°
True airspeed, 180 knots
Find: True heading
Groundspeed

(1) Set wind direction (240°) under the TRUE INDEX.

(2) Draw wind vector down from center of disk to the proper length for 30 knots.

(3) Move the slide until true airspeed (180 k) is read under the center of the disk.

(4) Rotate the compass rose until the true course (195°) is under the TRUE IN-DEX.

NOTE: At this point we have a problem on the computer which is incorrect for wind tri-angle solution—the true course and the true airspeed are on the same line. This cannot stand. We must make adjustments on the computer until the proper relationships are established. No readings should be taken until these adjustments are made!

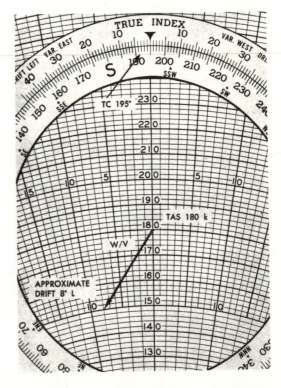

*Fig. A-71*

| | Tentative TH | Resulting Drift | Resulting TC | Remarks | Necessary Further Readjustment |
|---|---|---|---|---|---|
| 1st Try | 195° | 8° L | 187° | TC is 8° too far to the left | Try a TH 8° farther to the right |
| 2nd Try | 203° | 7° L | 196° | TC is 1° too far to the right | Try a TH 1° farther to the left |
| 3rd Try | 202° | 7° L | 195° | TC agrees with desired TC | None |

**By common sense.** - To make good a TC of 195°, you have to find a TH which when placed under the index will give you a TC of 195°. Then try different TH's until you get the one you want. The most logical TH to try first is the TC itself. In other words, you will see what would happen if you headed down the course without making any allowance for wind. You would drift about 8° to the left. Make other tries as shown in the table. above.

The adjustments must be made by common sense or by the numbers

**By the numbers.** - Place a pencil mark on the TC (195°) on the compass rose. The wind arrow shows 8° left drift. We move the compass rose until the TC mark is under 8° DRIFT LEFT. You keep on moving the rose until the number of degrees DRIFT LEFT or DRIFT RIGHT above the TC mark agrees with the number of degrees DRIFT LEFT or DRIFT RIGHT under the wind arrow.

(5) Read the true heading (202°) under the TRUE INDEX.

(6) Read the groundspeed (158 k) on the speed circle at the head of the wind vector.

NOTE: In preparing for a triangular cross-country flight you would have to draw the wind vector on the computer only once (unless the wind changed); the solutions for all three true headings would be made as outlined above, with the same wind vector and without moving the slide.

**Solution for True Heading and True Airspeed.** - At times you may not only have to make it good in a given time. Your groundspeed will be determined for you in advance. The problem of preflight calculations then becomes one of finding true heading and true airspeed. This solution differs only very slightly from the one for true heading and groundspeed. The problems and difficulties are the same.

Given:    Wind direction, 90°
          Wind speed, 20 knots
          True course, 240°
          Groundspeed, 140 knots
Find:     True heading
          True airspeed

(1) Set wind direction (090°) under the TRUE INDEX.

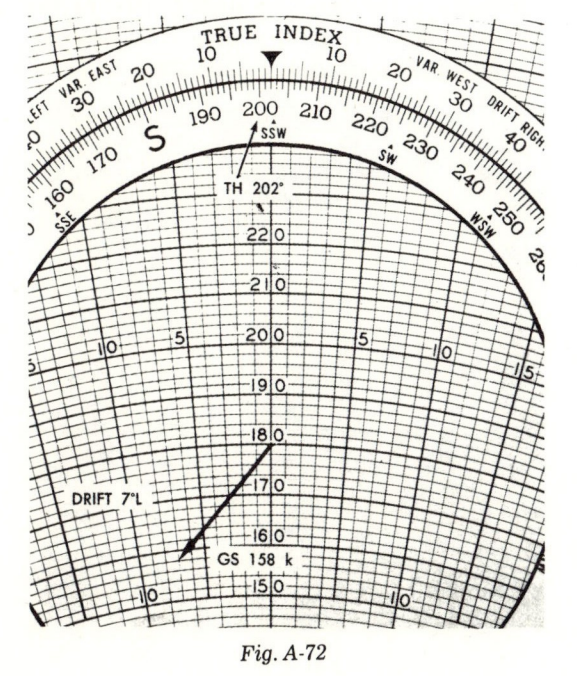

*Fig. A-72*                    *Fig. A-73*

(2) Draw wind vector down from center of disk to the proper length for 20 knots.

(3) Rotate the compass rose until the true course (240°) is under the TRUE INDEX.

(4) Move the slide until the head of the wind vector (not the grommet) is on the proper speed circle (140 k).

(5) Read the approximate drift (4° right) at the head of the wind vector.

(6) Rotate the compass rose in the direction of the position of the wind vector (here, to the right) the number of degrees of approximate drift (4°). When this is done, move the slide again until the head of the wind vector is on the 140 knots speed circle.

(7) Check the drift. It is now about 4½°. A further adjustment makes the correction and the drift reading about 5°.

(8) Read the true heading (235°) under the TRUE INDEX.

(9) Read the true airspeed (123 k) on the centerline under the grommet.

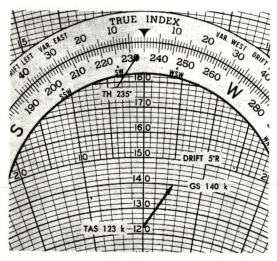

*Fig. A-74*

Solution for Wind Direction and Wind Speed. - All the solution previously discussed supposed that you knew the wind and could use it for preflight or flight calculations. But in many cases—perhaps in most cases—you will not know the wind direction and speed. You must find them from information obtained in flight. If you are able to determine your ground position, you will know your track and groundspeed. With true heading (compass reading corrected for variation and deviation) and true airspeed (airspeed reading corrected for temperature and pressure), you will have the four elements needed for a wind vector solution.

Given: True heading, 118°
     True airspeed, 160 knots
     Track, 122°
     Groundspeed, 154 knots
Find: Wind direction
     Wind speed

(1) Set the true heading (118°) under the TRUE INDEX.

(2) Move the slide until the true airspeed (160 k) is under the center of the disk.

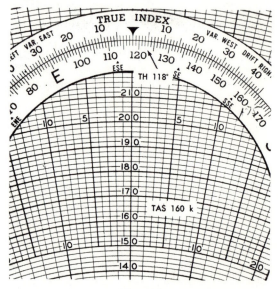

*Fig. A-75*

(3) Compare the figures given for true track. The difference between true heading and track is the drift. When track is greater than true heading, drift is right. When track is less than true heading, drift is left. In this problem, track is greater than true heading, and the drift is 4° right.

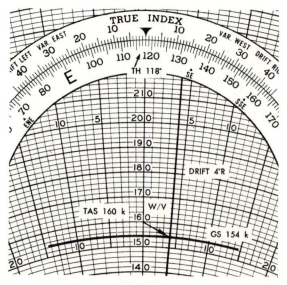

*Fig. A-76*

(4) Locate the track line 4° to the right of center.

(5) Along this line, locate the speed circle for 154 knots. Place a mark at the intersection of the track line and the groundspeed circle.

(6) Connect the marked point with the center of the disk. This line is the wind vector.

(7) Rotate the compass rose until the wind vector points down along the centerline. Read wind direction (058°) at the TRUE INDEX and wind speed (13 k) as the length, of the wind vector.

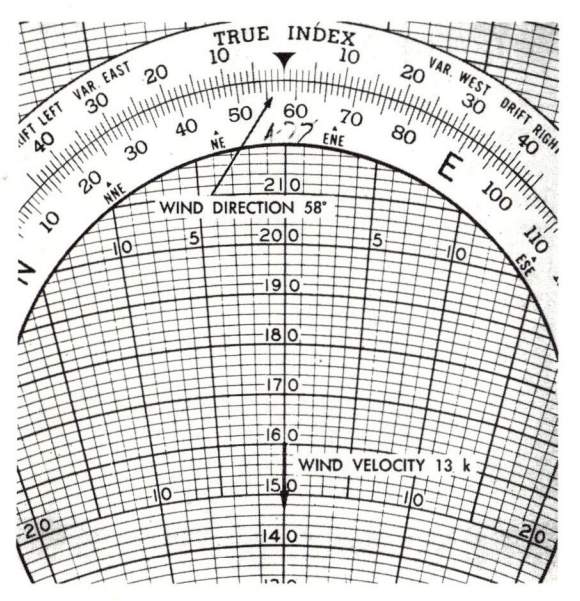

*Fig. A-77*

Radius of Action to Same Base. - For the solution of the radius of action formulas for time to turn and radius of action, most of the data in the formulas must be supplied by wind triangle solutions. To fly the problem, true headings out and back must be known. To find the time to turn, groundspeeds out and back must be known. Two triangles are therefore involved, with solution for true heading and groundspeed. The wind will be the same for both, and the true course back will be the reciprocal of the true course out. Graphically, the two triangles may be combined into a single diagram, thus:

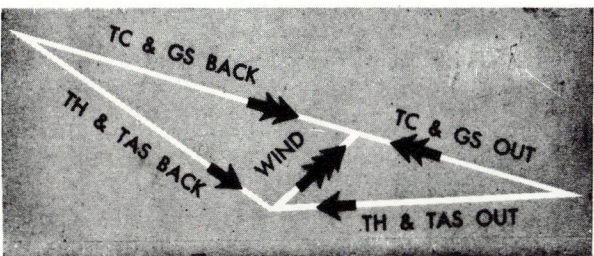

*Fig. A-78*

Similarly on the computer the wind vector need only be drawn once, and then the two triangles may be solved in the following manner:

Given:   Wind direction, 190°
             Wind speed, 20 knots
             True course out, 120°
             True airspeed, 140 knots
             Available fuel hours, 2 : 15

Find:     Groundspeed out
             True heading out
             Groundspeed back
             True heading back
             t (time to turn)
             R/A (radius of action)

(1) Set the wind direction (190°) under the TRUE INDEX

(2) Draw wind vector down from center of the disk to the proper length for 20 knots.

(3) Move the slide until the true airspeed (140 k) is read under the center of the disk.

(4) Set the true course out (120°) under the TRUE INDEX. Read the approximate drift, rotate the compass rose in the direction of the drift, and make the usual adjustments in solving for true heading.

(5) Read the true heading out (128°) under the TRUE INDEX and the groundspeed out (132 k) at the head of the wind vector.

(6) Set the true course back (300°, reciprocal of the true course out) under the TRUE INDEX. Read the approximate drift,

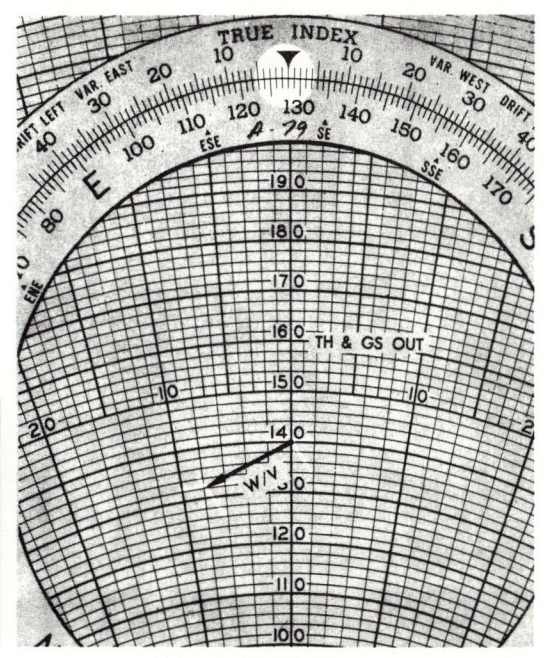

*Fig. A-79*

rotate the compass rose in the direction of the drift, and make the usual adjustments in solving for true heading.

(7) Read the true heading back (292°) under the TRUE INDEX and the ground-speed back (146 k) at the head of the wind vector.

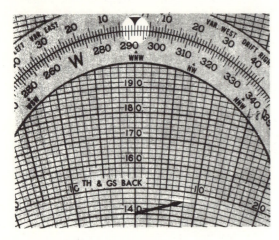

*Fig. A-80*

(8) Turning to the slide rule face of the computer, set up the formula in the way outlined on "Radius of Action Formulas," and get the time to turn (71 minutes, or 1 hour, 11 minutes).

(9) Use the slide rule for finding radius of action (156 nm).

**Wind and Groundspeed from Double or Multiple Drift.** - When wind information is unknown and a ground fix is unavailable to provide the data necessary for a regular vector solution, very accurate wind speed and direction may be determined by the double drift method. This method can be used only with a drift meter. It consists of reading drift on the original heading and on two headings—specially flown—at 045°, to the original heading. The drifts are plotted on the dead reckoning computer, and groundspeed as well as wind information is discovered. With the information obtained from it, the solution is made in the following manner:

NOTE: Before beginning the solution, note that the value and the sign read on the drift meter are drift correction rather than drift. Remember that plus drift correction is left drift, minus drift correction is right drift. To keep this straight, it might be well to mark a plus on the left side of scale above the compass rose and a minus on the right side.

Given: On-course true heading, 060°

Drift correction on course, +3°

Drift correction on right leg, +8°

Drift correction on left leg, -4°

True airspeed, 125 knots

Find: Wind direction

Wind speed

Groundspeed on course

(1) Move the slide until the true air-speed (125 k) is read under the center of the disk.

(2) Set the true heading on the right leg (060° + 045° = 105°) under the TRUE INDEX.

(3) Draw a straight line along the +8° drift correction line (the 8° left drift line). (In order to set the true heading on the right leg it is not necessary to add 45° to the original heading. Simply set the on-course true heading under the 45° mark to the left of the index on the upper scale.)

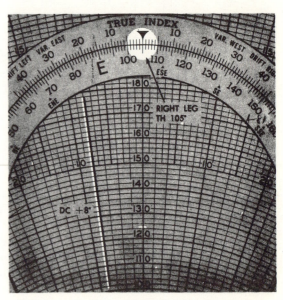

*Fig. A-81*

(4) For the left leg, set the on-course true heading under the 45° mark on the right side of the upper scale.

(5) Draw a straight line along the —4° drift correction line (the 4° right drift line).

(6) Set the on-course true heading (060°) under the TRUE INDEX.

(7) Draw a straight line along the +3° drift correction line (the 3° left drift line).

(8) If the drift readings are all correct,

all track lines will intersect at a point. Draw the wind vector from the center of the disk to this point. If small errors in reading drift have been made, the intersections of the track lines will form a small triangle. In most cases, the small triangle rather than the single point will result. Draw the wind vector to the center of this triangle. In the event that the three lines form a rather large triangle, the likelihood is that the error in reading drift has been considerable. In such cases the solution should be discarded and the problem worked over.

(9) With the on-course heading still set under the TRUE INDEX, read the on-course groundspeed at the head of the wind vector (145 k).

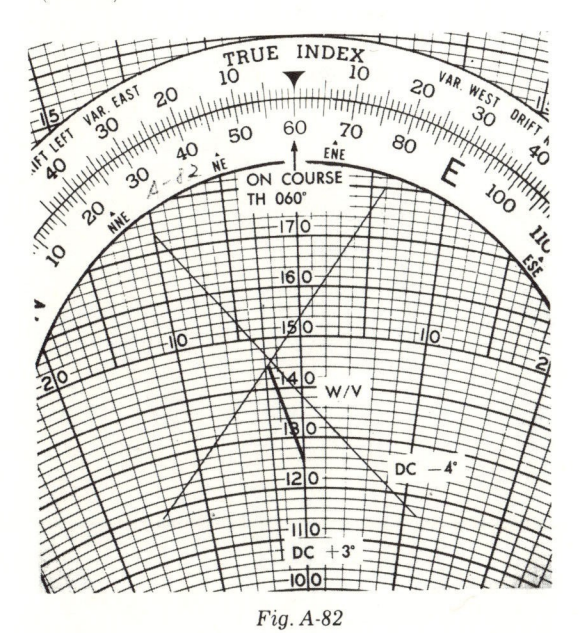

*Fig. A-82*

(10) Rotate the compass rose to place the wind vector down along the centerline. Read the wind direction (224½°) at the TRUE INDEX and the wind speed (21 k) as the length of the wind vector.

**Wind and Groundspeed from Drift on Two Headings.** - Without flying the special pattern necessary for the double drift problem, you may determine wind by reading drift on two headings. These may be two headings flown to make good two legs of your mission. The results are somewhat less accurate than those of the double drift solution. Procedure is essentially the same.

Given: First true heading, 064°
Drift correction, +6°
Second true heading, 150°

Drift correction, +4°
True airspeed, 120 knots
Find: Wind direction
Wind speed
Groundspeed on either heading
Groundspeed on any heading.

(1) Move the slide until the true airspeed (120 k) is read under the center of the disk.

(2) Rotate the compass rose to set the first true heading (64°) under the TRUE INDEX.

(3) Since drift correction is +6°, the track line is 6° left of the centerline. Draw in this track line on the plotting disk.

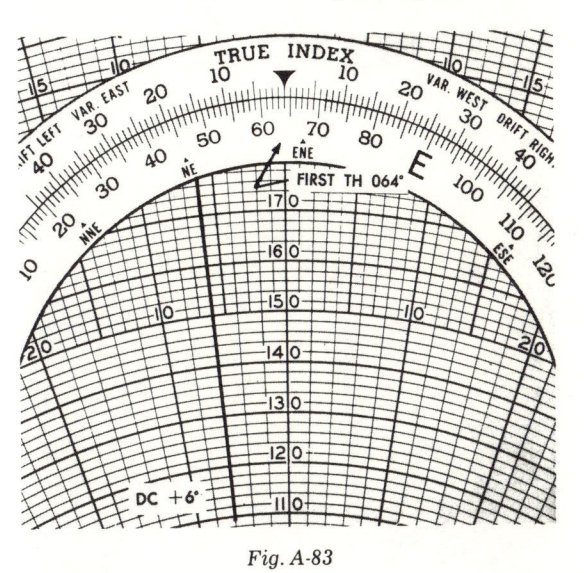

*Fig. A-83*

(4) Rotate the compass rose to set the second true heading (150°) under the TRUE INDEX.

(5) Since the drift correction is +4°, locate the track line 4° left of centerline and

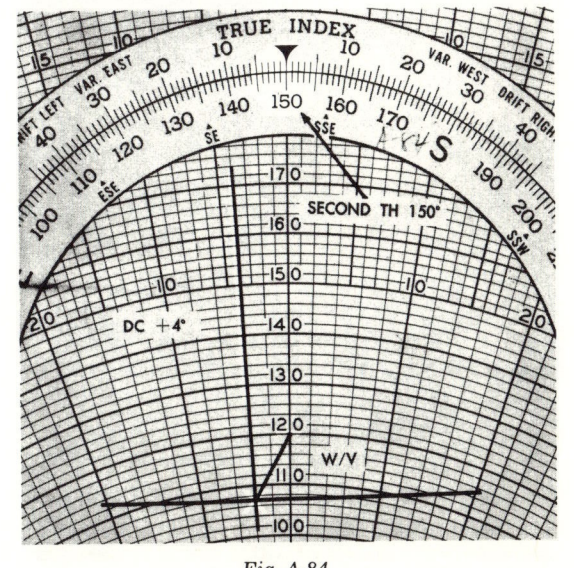

*Fig. A-84*

mark the intersection of the two track lines.

(6) Draw the wind vector from the center of the disk to the intersection of the two track lines.

(7) Rotate the compass rose to bring the wind vector down along the centerline. Read the wind direction (180°) under the TRUE INDEX and the wind speed (15 k) as the length of the wind vector.

(8) To find the groundspeed on the first true heading, set true heading under the TRUE INDEX and read groundspeed (127 k) at the end of the wind vector.

(9) To find groundspeed on any other heading set that heading under the TRUE INDEX and read groundspeed at the end of the wind vector. The groundspeed on the second heading is 107 knots.

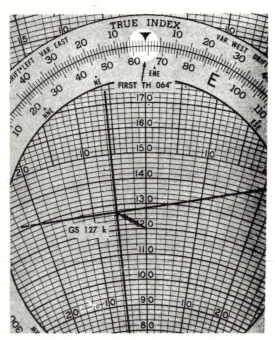

Fig. A-85

**Wind from Groundspeed Measurements.** Usually, if the groundspeed can be determined, so can the heading and track, which would be sufficient to plot the wind at once. However, there are occasions, such as at night or when flying a radio range, when only an approximate heading is maintained. Fairly accurate groundspeed measurements are obtainable by timing between lighted cities, marker beacons, intersections of radio ranges, etc. In such cases, intersecting groundspeed arcs may give a useful plot of the wind. But it must be remembered that such information can be of little value unless the airspeed indicator is accurately calibrated and

the air density corrections are properly applied to its readings.

Given:  True airspeed, 180 knots
         Groundspeed, 160 knots on true heading of 120°
         Groundspeed, 170 knots on true heading of 230°
         Groundspeed, 210 knots on true heading of 335°

Find:  Wind direction
       Wind speed

(1) Move the slide until the true airspeed (180 k) is read under the center of the disk.

(2) Set the first true heading (120°) under the TRUE INDEX.

(3) Trace an arc on the plotting disk along the 160 knots speed circle.

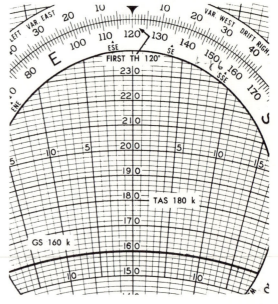

Fig. A-86

(4) Set the second true heading (230°) under the TRUE INDEX and trace an arc on the plotting disk along the 170 knots speed circle.

(5) Set the third true heading (335°) under the TRUE INDEX and trace an arc on the plotting disk along the 210 knots speed circle.

(6) Draw the wind vector from the center of the disk to the point of intersection of the three arcs.

(7) Rotate the disk until the wind vector lies along the centerline, down from the grommet.

(8) Read the wind direction (165°) under the TRUE INDEX and the wind speed (30 k) as the length of the wind vector.

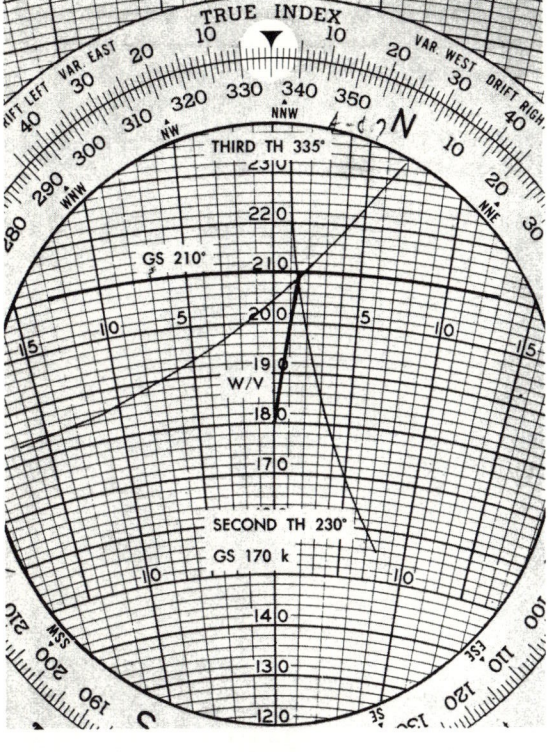

*Fig. A-87*

### PROBLEMS ON THE RECTANGULAIR GRID

One side of the movable slide contains, in addition to the speed circles and track lines which continue those on the other side, a rectangular grid. For all its divisions this rectangular grid uses the same scale as the interval distance between speed circles. The arrangement of the grid makes it possible to draw lines parallel to the centerline or perpendicular to the centerline, to scale off nautical miles (or other units) at right angles to the centerline, and to scale off nautical miles along any line parallel to the centerline. It thus becomes possible to work problems involving parallel lines and problems introducing measurements other than those of groundspeed, true airspeed, and wind speed.

Off-course Corrections. - Since a plane one mile off course after sixty miles of flight is also approximately one degree off course (the "one-in-sixty" rule of thumb), it is possible to work off-course problems graphically through the use of a miles scale and drift lines. (The arithmetical solution of the same problem is treated above in connection with the slide rule face of the computer.) The rectangular grid furnishes the necessary mechanism. By measuring the drift to right

or left of the centerline and setting miles flown, drift in degrees can be discovered:

Given:  Distance flown, 16 nm  
           Distance off course, 12 nm right  
           Distance to destination, 200 nm  
Find:  Correction necessary to parallel the true course  
           Total correction necessary to converge on destination

(1) Reverse the slide to put the rectangular grid under the disk.

The compass rose may be in any position but it must not be moved during the solution of the problem.

(2) From the center point draw a line at right angles to the centerline. The length of the line equals the distance off course.

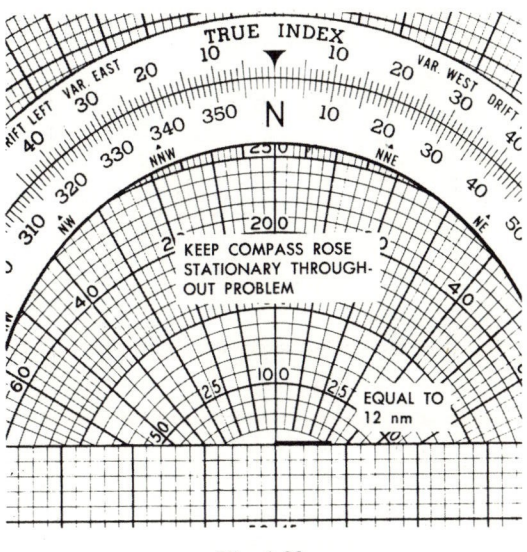

*Fig. A-88*

(3) Reverse the slide and set the distance flown (160 nm) under the center of the disk.

(4) Read the number of degrees of correction required to parallel the course from the drift scale (4½°). In this problem, since drift is to the right, the heading will have to be altered to the left.

(5) To find the additional correction to converge on destination move the slide until 200 is read under the center of the disk. Then read the additional correction to converge (3½°).

(6) Add the two corrections and find the total correction angle (8°)

Correcting the Reported Wind. - A pilotage fix, furnishing information on track and groundspeed, may be used as the basis for finding the wind with a standard wind vector solution. The same information may also be used for correcting the reported wind, using the rectangular grid. In this procedure a true heading is flown which should, on the basis of the reported wind, make good the true course. A pilotage fix is used to discover distance and direction off course in a given time of flight, and these data are translated into an additional wind component.

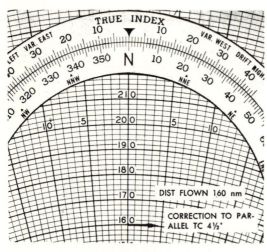

Fig. A-89

Given: Reported wind, 30 knots from 125°

Flying time, 30 minutes

Pilotage fix, 6 nm north of on-course DR position

Find: Corrected wind

(1) Place the rectangular grid of the slide under the transparent disk.

(2) Rotate the compass rose until the wind direction (125°) is under the TRUE INDEX.

(3) Draw the wind vector down from the center of the disk the proper length for 30 knots.

(4) Figure the additional wind component. If the plane is 6 nm off course in 30 minutes, the wind component is 12 knots. If the plane is to the north of the course, the wind is blowing from the south. Therefore the additional wind effect has been 12 nm from the south.

(5) Rotate the compass rose until "S" appears under the TRUE INDEX.

NOTE: For the other approximate wind directions the sixteen compass points on the inner scale of the compass rose may be used.

(6) From the end of the original wind vector, plot the wind correction vector vertically downward 12 knots to scale.

(7) Connect the head of this correction vector with the center of the disk. This gives the corrected wind vector.

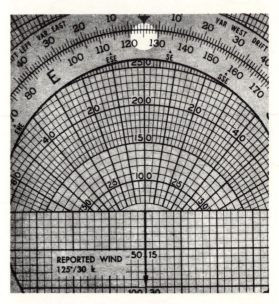

Fig. A-90

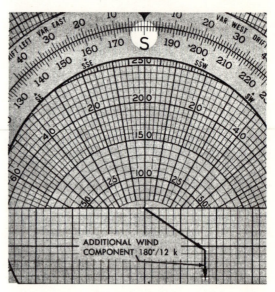

Fig. A-91

(3) Rotate the compass rose until the corrected wind vector lies along the centerline, downward from the center of the disk. Read the wind direction (140°) under the TRUE INDEX and the wind speed (38 k) as the length of the vector.

Air Plot. — Air plot is a technique in which true heading and true airspeed (rather than track and groundspeed) are plotted on a chart to record the movements of the plane through the air mass (rather than over the ground). If the wind is known, a wind vector for the actual time of flight may be plotted from the last air position, and a dead reckoning ground position obtained. If the wind is unknown, it may be discovered by connecting an air position with an actual pilotage fix and resolving this line into a one-hour wind vector. On the dead reckoning computer, solutions may be found for both types of problems but it is not possible to complete these problems without use of the chart.

to the distance flown (here, 5 spaces), along the vertical grid line.

(3) Set the TRUE INDEX at the second true heading, 355°. From the end of the first line draw another down the proper distance, again along the vertical grid or parallel to it.

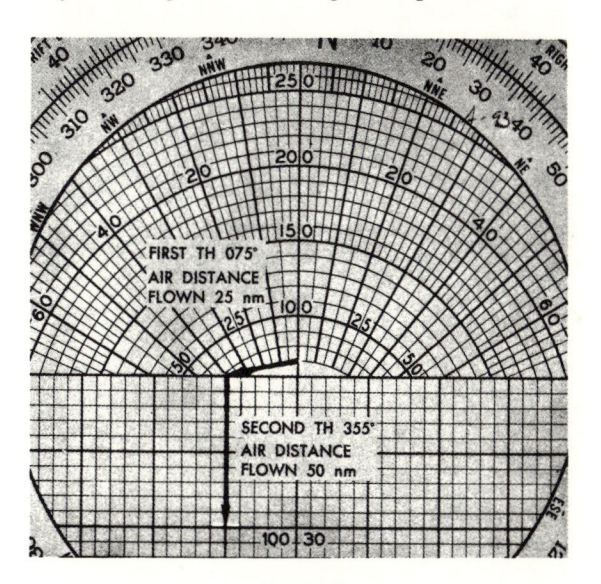

*Fig. A-93*

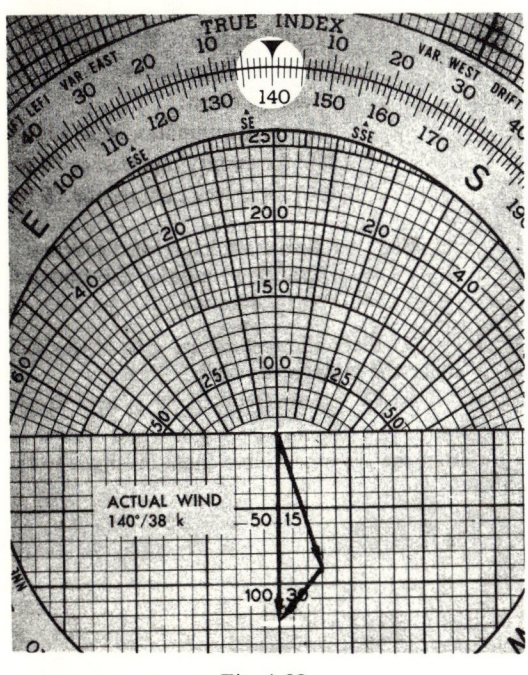

*Fig. A-92*

(4) Set the TRUE INDEX at the third true heading, 105°. From the end of the second line drawn, draw down the proper distance, again along the vertical grid or parallel to it. This completes the air plot. If more headings had been flown, the same procedure would be followed until all had been plotted.

(5) Set the TRUE INDEX at the wind direction (080°). From the end of the third

Given:   True airspeed, 150 knots
           Wind direction, 080°
           Wind speed, 25 knots
           Headings flown:
           075° for ten minutes, or 25 nm
           355° for ten minutes, or 50 nm
           105° for thirty minutes, or 75 nm

Find:    Resultant track from departure
           Distance from point of departure
           Present position

(1) Place the rectangular grid of the slide under the transparent disk.

(2) Set the TRUE INDEX at 075°. Select a scale (e.g., five miles per space) and draw down from the grommet a line equal in length

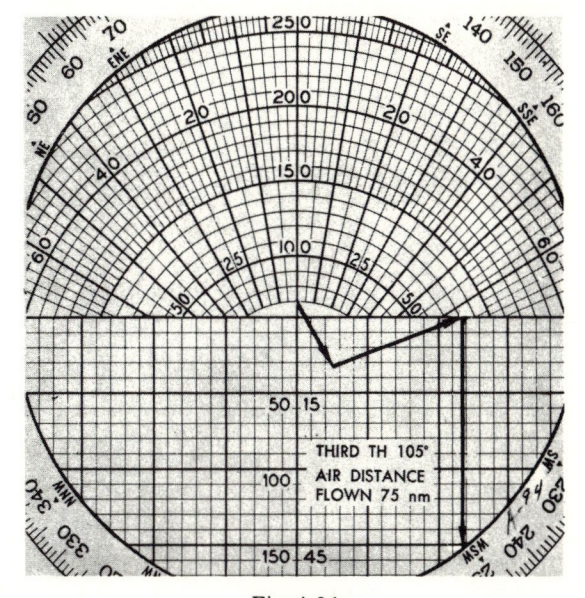

*Fig. A-94*

line drawn, a wind vector up equal in length to the wind speed for the total time of the trip (in this case, one hour). However, two cautions are to be noted in connection with this wind vector. It is not a onehour vector unless the actual time of flight is one hour. (If the flight had taken a total of one and one-half hours and the wind speed was 25 knots, the wind vector would be plotted to a scale distance of 37 nm.) The scale used for plotting the wind vector must be the same as that used for plotting the other lines.

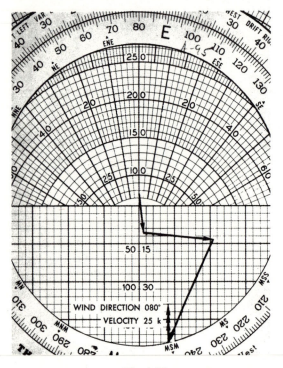

*Fig. A-95*

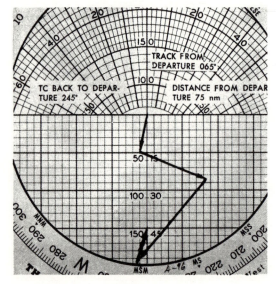

*Fig. A-96*

(6) Rotate the compass rose until the head of the wind vector is on the center vertical line of the grid below the grommet. The resultant departure track now appears under the TRUE INDEX (065°), and the distance from departure will be the distance between the end of the wind line and the grommet (75 nm), again using the same scale.

NOTE: In some cases it may be impossible to plot distance flown on the computer. This would be true if the aircraft were flying at very high speeds or the period of airplot was over a long period. The navigator and pilot may find it much more convenient to plot time on the rectangular grid instead of distance. For instance, in the preceding problem, ten, twenty, and thirty minutes may be plotted in place of 25, 50, and 75 nm with the length of each vector thus proportionately shorter. The resultant vector is then in time and is easily converted into air distance by using true airspeed on the slide rule side of the computer.

(7) This completes the computer part of the solution. To find the present position, plot the resultant track and miles flown from departure or from the point where the airplot was started. The major use of airplot will be in a situation where weather conditions dictate frequent changes of heading. The navigator or pilot can easily follow these heading changes until the weather has been penetrated and thus have a fairly accurate position from which to resume a more normal or straight line flight.

(8) If the wind is not known, the compass rose would be rotated until the head of the last air vector (plotted in the preceding step 4) was on the central grid down from the grommet. This would give a no-wind position, in terms of air miles and average true heading flown. Now, if an actual position were found by a map reading fix, a line connecting the last air position and the actual ground position for the same time, would give wind direction and effect for the time of flight. A simple proportion would resolve this effect into a wind speed for one hour.

(9) It is also possible that reciprocal heading back to the starting point of airplot may be needed. In that case the head of the last air vector may be set on the center line directly above the grommet and the reciprocal heading read under the true index.

## Section IX. TEMPERATURE, ALTITUDE, AND AIRSPEED

You will soon become familiar with the various factors used in determining altitudes and airspeeds. Air navigation cannot be satisfactorily performed without a working knowledge of them and precise navigation in the high-performance aircraft of today demands the utmost accuracy in all computations.

Some of the factors (compressibility, heat of compression, and equivalent airspeed) are not significant to flight at lower altitudes and airspeeds. However, you will need to know them and relate their significance to high performance flight. This chapter is devoted to defining and analyzing the factors affecting altitude and airspeed.

### TEMPERATURE

Indicated Air Temperature (IAT). - Indicated air temperature is read from the free air temperature gage. The most common is the bimetallic thermometer with a dial reading from —60°C to +50°C.

Basic Air Temperature (BAT). - Basic air temperature is IAT corrected for scale error. To calibrate the gage, a series of readings are taken and compared with a master thermometer (known to be accurate). A scale correction card is prepared to show the proper correction at any given temperature. Whenever a temperature reading is taken, this card should be consulted and the proper correction applied to give you basic air temperature.

True Air temperature (TAT). - True air temperature is BAT corrected for heat of compression error. Since the gage must measure the temperature of the air through which the aircraft is passing, it must be installed so the sensitive element is in the airstream. The rapid movement of air around the sensitive element produces added heat by compression and friction. This causes the gage to indicate a higher, than actual, air temperature. Heat of compression error depends on the true airspeed of the aircraft. As true airspeed increases, compression and friction increase, and the error becomes greater. The formula for computing the error is:

Correction in degrees centigrade =
-.0001055 x (TAS in knots)$^2$ or
-.00008    x (TAS in mph)$^2$.

In using your computer to solve for heat of compression error, use a constant of 950 when using knots and 1250 when using miles per hour. Place the constant on the TIME scale below the value of TAS on the miles scale. Locate TAS on the TIME scale and read heat of compression error above it on the MILES scale. Since you do not know your true airspeed when computing heat of compression error, you will have to use BAT, pressure altitude, and equivalent airspeed to get a trial true airspeed for use in the solution for true air temperature; then re-compute.

### ALTITUDE AND THE ALTIMETER

The altimeter is an aneroid barometer which measures atmospheric pressure and converts that pressure to an altitude indication above an adjustable reference plane. This conversion is based on a fixed set of values known as the U.S. Standard atmosphere. This standard approximates the average pressure and temperature observations of 45° North Latitude in the United States.

Whatever its value may be, the pressure set into the Kollsman window is the pressure level from which the altimeter is measuring. Unless instructed otherwise, pilots are required to keep their altimeters set to the latest available altimeter setting (surface pressure converted to sea level pressure) for the area in which they are flying. This gives them an accurate reading near the surface and in close proximity to the station. It cannot correct for the distortion at higher altitudes due to nonstandard temperature and pressure conditions of the upper air; since all pilots flying in a given area will use the same altimeter setting, each altimeter will be equaly affected by temperature and pressure variations and separation will be maintained between aircraft.

Indicated Altitude. - Indicated altitude is the altitude read on the altimeter with the current altimeter setting in the Kollsman window. It is indicated altitude above sea level.

NOTE: Your altimeter scale may or may not be correctly oriented to standard pressure conditions. Altimeters are checked periodically in pressure chambers and a scale correction card prepared. In addition to scale error, static pressure-port installation error affects your altimeter. This error is created by the change of alignment of the static port to the relative wind. This change is due to changes of altitude, airspeed, or location of the port in a disturbed pressure area. The installation error differs with each type of aircraft; and may usually be found in the pilot's handbook for the particular aircraft. You should apply corrections for both of these errors to all altitude readings.

Pressure Altitude (PA). - Pressure altitude is the pressure difference between flight level and the standard datum plane. This pressure is converted to feet in accordance with U.S. Standard temperature and pressure lapse rate When 29.92' Hg is set in the Kollsman window, your altimeter indicates pressure altitude.

Density Altitude (DA). - Density altitude is pressure altitude corrected for nonstandard temperature at flight level. Since aircraft performance is quite sensitive to changes in air density, it is often desirable to compute density altitude. In the standard atmosphere, each altitude has a given density. Aircraft performance is expressed in terms of these standard day pressure altitudes. For maximum efficiency under nonstandard conditions, it is necessary to select the pressure altitude and temperature combination that will produce the desired density altitude. For a given pressure altitude, an increase of temperature increases density altitude. The E-10 computer has adequate instructions for computing density altitude.

True Altitude. - True altitude is the actual height above sea level. A fair estimate of true altitude can be made by using your computer to correct for nonstandard flight level temperature, and nonstandard pressure at the surface. The prerequisites for this solution are: pressure altitude, indicated altitude, and true air temperature. The solution is as follows:

(1) In the window marked FOR ALTITUDE COMPUTATIONS place the true air temperature opposite pressure altitude.

(2) Opposite the indicated altitude on the MINUTES scale, read true altitude on the MILES scale.

It must be remembered that this is a mathematical solution that assumes standard pressure and temperature lapse rates exist between the aircraft and the surface. Frequently pressure and temperature variations will exist which cause this computer solution to be dangerously inaccurate. For an example, take a dense polar air mass with higher standard pressures reported at the surface, and lower than standard pressures aloft. With such a system centered over mountainous terrain, a dangerous situation would exist for the pilot who elects to fly the minimum altitude. When the pilot wishes to know if 11,000 feet indicated will give him the required 2,000-foot terrain clearance over an 8,000-foot mountain peak, he should determine this in advance at the weather station before take-off. A computer solution to this problem will merely create double error while attempting to correct for one.

Remember, your altimeter indicates altitude according to the pressure it senses. In other words, it only measures the weight of the air above it. Low temperature and/or decreased pressure lowers a particular pressure level. Therefore, an aircraft flying at that pressure will be lower than indicated. Hence the saying, "If it's cold or low, look out below."

AIRSPEEDS
Indicated Airspeed (IAS). - Indicated airspeed is the uncorrected reading taken from the face of the indicator. It can be read in miles per hour or in knots depending upon the scale of the dial.

Basic Airspeed (BAS). - Basic airspeed is the indicated airspeed corrected for instrument error. Each airspeed indicator has its own characteristics which make it different from every other airspeed indicator. These differences are caused by slightly different hair-spring tensions, flexibility of the diaphragm, scale markings on the dial, etc. The accuracy of the instrument is also affected by the length and curvature of the pressure line from the pitot head. These errors are lumped together under the title of instrument error. The total instrument error can be determined by various calibration procedures and an airspeed correction card prepared. This is used to find basic airspeed.

**Calibrated Airspeed (CAS).** - Calibrated airpeed is basic airspeed corrected for pitot-static error and/or attitude of the aircraft.

All pitot-static systems have some degree of error built into them because of the impossibility of picking static and impact air sources which will provide accurate pressures for all flight conditions and attitudes. Total impact pressure picked up by a properly installed pitot is generally quite accurate. In most attitudes, however, the static source is not in a region of true static pressure. When the static source is taken from the pitot head, changes of attitude cause the flow of air to strike the static inlet ports at different angles. Consequently the pressure within the static system changes. Different types and locations of the static source will cause different errors. It is immaterial whether the static source is located in the pitot-static head or at some ,flush mounting on the aircraft, a change of attitude will alter the airflow by the inlet ports, thereby affecting the pressure within the static system. Furthermore, the error for a given aircraft attitude will vary with the square of the airspeed. The correction for pitot-static error and/or attitude, of the aircraft will depend upon the primary factors affecting attitude changes in normal flight.

**Equivalent Airspeed (EAS).** - Equivalent airspeed is calibrated airspeed corrected for compressibility. Compressibility error has come into prominence with the advent of highspeed and high-altitude flight. As the air density decreases and airspeed increases, the impact pressure causes the molecules of air to become compressed within pitot chamber and pilot lines. As the air becomes compressed it causes the dynamic pressure to be greater than it should be, making the airspeed indicator read too high.

You can compensate for the compressibility error in various ways. Using the EAS, pressure altitude, and true air temperature you can solve for true airspeed on the slide rule side of your computer.

Another, and more common method is to use the correction table printed on the slide of your computer. To use this table, solve for true airspeed uncorrected for compressibility using CAS, PA, and true air temperature. Multiply the result by the F factor, obtained from the table, to get true airspeed corrected for compressibility.

**True airspeed (TAS).** - True airspeed is equivalent airspeed corrected for density altitude (pressure and temperature). It is the actual speed of the aircraft through the air mass. To find true airspeed it is necessary to work a computer solution. The prerequisites are EAS, pressure altitude, and true air temperature.

## TRUE AIRSPEED ON THE MB-4 COMPUTER

TRUE AIRSPEED COMPUTATION

The following practice exercises have instructions and sample problems for the solution of True Airspeed. Previous sections adequately cover the other uses of this computer.

True Airspeed Computation.

Given: Pressure altitude, 25,000 ft.

Basic air temperature, —30°C
Calibrated airspeed, 240 knots

Find: True air temperature
True airspeed

(1) Compute a trial TAS by using BAT, PA, and CAS. This trial TAS (363 k) will be used to compute temperature rise (heat of compression error).

(2) Use the formula

$$\frac{\text{Temperature TAS}}{950} = \frac{\text{Temperature Rise}}{\text{TAS}}$$

by placing 950 on the minute scale under trial TAS of 363 on the miles scale. Read the temperature rise of 14°C on the MILES scale opposite 363 on the MINUTES scale. True air temperature would be approximately -44°C (BAT - Temperature Rise). Remember that TAT is always colder than indicated on basic air temperature.

(3) Using TAT, PA, and CAS, compute

381

TAS uncorrected for compressibility. Your answer should be 351 knots.

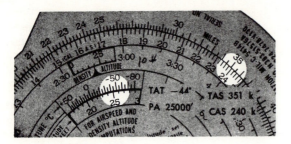

*Fig. A-97*

(4) Determine the F factor on the slide of your computer. Interpolating by inspection, you find it to be .97.

| PRESS. ALT FEET | F CORRECTION FACTORS FOR TAS | | | | | | | |
|---|---|---|---|---|---|---|---|---|
| | CALIBRATED AIRSPEED KNOTS | | | | | | | |
| | 200 | 250 | 300 | 350 | 400 | 450 | 500 | 550 |
| 10,000 | 1.0 | 1.0 | .99 | .99 | .98 | .98 | .97 | .97 |
| 20,000 | .99 | .98 | .97 | .97 | .96 | .95 | .94 | .93 |
| 30,000 | .97 | .96 | .95 | .94 | .92 | .91 | .90 | .89 |
| 40,000 | .96 | .94 | .92 | .90 | .88 | .87 | .87 | .86 |
| 50,000 | .93 | .90 | .87 | .86 | .84 | .84 | .84 | .84 |

**DIRECTIONS**

USE CALIBRATED AIRSPEED AND PRESS ALT TO OBTAIN F FACTOR. MULTIPLY F FACTOR BY **TAS** OBTAINED WITH COMPUTER TO OBTAIN **TAS** CORRECTED FOR COMPRESSIBILITY.

*Fig. A-98*

(5) Multiply TAS uncorrected for compressibility by the F factor. TAS is therefore 342 knots.

NOTE: For accuracy at higher airspeeds, it is necessary to re-compute the last four steps using the result of step number 5 in computing true air temperature.

## TRUE AIRSPEED USING MACH INDEX AND TRUE FREE AIR TEMPERATURE

The Mach number index appears to the right of 80 on the pressure altitude scale in the pressure altitude window.

Given: True free air temperature, -50°C.
Mach, 0.80.

Find: True airspeed

(1) Set true air temperature (-50°C) opposite the MACH NUMBER INDEX.

(2) Opposite the Mach number on the MINUTE scale, read (467 k) true airspeed on the MILES scale. The compressibility factor is automatically compensated in the Mach indicator.

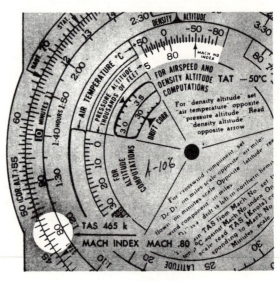

*Fig. A-99*

# SUPPLEMENTARY REFERENCE MATERIAL

## AIRMAN'S INFORMATION MANUAL (AIM)

The Airman's Information Manual has been designed primarily as a pilot's operational and information manual for use in the National Airspace System of the United States (unless otherwise indicated). It is divided into five basic parts, each of which may be purchased separately. Frequency of issuance, area of coverage, annual subscription costs and highlights of the contents of each part follow.

### Part 1—Basic Flight Information and ATC Procedures

*Issued:* Semiannually (Jan. and July). *Coverage:* Entire U.S. unless otherwise indicated.

This part contains the basic fundamentals required to fly in the U.S. National Airspace System. Among other data it also contains adverse factors affecting Safety of Flight; Health and Medical Facts of interest to pilots; ATC information affecting rules, regulations and procedures; a Pilot/Controller Glossary; Air Defense Identification Zones (ADIZ); Designated Mountainous Areas; and Emergency Procedures.

*Annual Subscription:* $5.25 for U.S., plus $1.30 for other foreign mailing.

### Part 2—Airport Directory

*Issued:* Semiannually (Mar. and Sept). *Coverage:* Conterminous U.S., Puerto Rico and Virgin Islands (Note: similar information for Alaska and Hawaii appears in Alaska Supplement and Pacific Chart Supplement, respectively—See Special Notice Section, Part 3 for availability.)

Part 2 contains a Directory of all airports, seaplane bases, and heliports available for civil use. It includes all their services, *except communications*, in codified form. (Those airports with communications are also listed in Part 3.) A list of new and permanently closed airports which updates Part 2 is contained in Part 3. Also included in Part 2 are U.S. Entry and Departure Procedures, including Airports of Entry and Landing Rights Airports; and a listing of Flight Service Station and National Weather Service Telephone Numbers.

*Annual Subscription:* $7.00 for U.S., Canada and Mexico, plus $1.75 for other foreign mailing.

### Part 3—Operational Data and Special Notices

*Issued:* Part 3, every 56 days. *Coverage:* Conterminous U.S., Puerto Rico and Virgin Islands (Note: Similar information for Alaska and Hawaii appears in Alaska Supplement and Pacific Chart Supplement, respectively—(For sale by National Ocean Survey, Distribution Division, C44, Riverdale, Md. 20840).

Part 3 contains an Airport-Facility Directory of all major airports with control towers and/or instrument landing systems; a tabulation of Air Navigation Radio Aids; Special, General, and Area Notices; a tabulation of New and Permanently Closed Airports (which updates Part 2); North Atlantic Routes; Locations of VOR Receiver Check Points (both ground and airborne); Restrictions to Enroute Navigation Aids; Preferred Routes; Area Navigation Routes, and supplemental data to Part 4.

*Annual Subscription:* $30.50 for U.S., plus $7.65 for other foreign mailing.

### Part 3A—Notices to Airmen

*Issued:* Every 14 days. *Coverage:* Same as Part 3.

Part 3A contains current Notices to Airmen considered essential to the safety of flight as well as supplemental data to Parts 3 and 4.

*Annual Subscription:* $20.55 for U.S., plus $5.15 for other foreign mailing.

### Part 4—Graphic Notices and Supplemental Data

*Issued:* Quarterly (Jan., April, July, Oct.). *Coverage:* Conterminous U.S., Puerto Rico and Virgin Islands (Note: similar information for Alaska and Hawaii appears in Alaska Supplement and Pacific Chart Supplement, respectively—(For sale by National Ocean Survey, Distribution Division, C44, Riverdale, Md. 20840).

Part 4 contains a list of abbreviations used in the AIM; a tabulation of Parachute Jump Areas; Special Notice—Area Graphics; Terminal Area Graphics; Olive Branch Routes and other data not requiring frequent change.

*Annual Subscription:* $14.40 for U.S., plus $3.60 for other foreign mailing.

### Where to Purchase AIM

The five basic parts described above are available from the Superintendent of Documents, Government Printing Office, Washington, D.C. 20402. Orders should be accompanied by check or money order made payable to the Superintendent of Documents.

### Errors, Omissions, or Changes

Errors, omissions, or suggested changes should be forwarded to the Federal Aviation Administration, National Flight Data Center, AAT–430, Washington, D.C. 20591.

# STATUS AND AVAILABILITY OF FEDERAL AVIATION REGULATIONS (FARs), ADVISORY CIRCULARS (ACs) AND VFR/IFR EXAM-O-GRAMS

## GENERAL

This section contains information on how to obtain the current status and availability of FARs, ACs and Exam-O-Grams. The information relating to the FARs and ACs is extracted from AC 00–2, *Advisory Circulars and Status of Federal Aviation Regulations.* AC 00–2 is issued every four months and can be obtained free-of-charge from:

U.S. Department of Transportation
Publications Section TAD 443.1
Washington, D.C.  20590

Many of the FARs and ACs listed in AC 00–2 are cross-referenced in the AIM, Part 1. These regulatory and non-regulatory references cover a wide range of subjects and are a source of detailed information that is of value to airmen.

## THE FEDERAL AVIATION REGULATIONS

The FAA publishes the Federal Aviation Regulations to make readily available to the aviation community the regulatory requirements placed upon them. These Regulations are sold as individual Parts by the Superintendent of Documents.

The more frequently amended Parts are sold on subscription service (that is, subscribers will receive Changes automatically as issued), while the less active Parts are sold on a single-sale basis. Changes to single-sale Parts will be sold separately as issued. Information concerning these Changes will be furnished by FAA through its "Status of the Federal Aviation Regulations, AC 00–4."

## ADVISORY CIRCULARS

The FAA issues advisory circulars to inform the aviation public in a systematic way of nonregulatory material of interest. Unless incorporated into a regulation by reference, the contents of an advisory circular are not binding on the public. Advisory circulars are issued in a numbered-subject system corresponding to the subject areas of the Federal Aviation Regulations (14 CFR Ch. I)

The AC 00–2 checklist contains advisory circulars that are for sale as well as those distributed free of charge by the Federal Aviation Administration.

## EXAM-O-GRAMS

Exam-O-Grams are prepared on subjects which prove particularly troublesome to applicants in written tests. They provide information on items which are operationally important but commonly misunderstood.

VFR Exam-O-Grams—single copy (Free).
IFR Exam-O-Grams—single copy (Free).

Single copies of Exam-O-Grams may be obtained free of charge from:

FAA Aeronautical Center
Flight Standards Technical Division
Operations Branch
P.O. Box 25082
Oklahoma City, Oklahoma  73125